T0172843

Advances in
Applied Human Modeling
and Simulation

Advances in Human Factors and Ergonomics Series

Series Editors

Gavriel Salvendy
Professor Emeritus
School of Industrial Engineering
Purdue University

Chair Professor & Head
Dept. of Industrial Engineering
Tsinghua Univ., P.R. China

Waldemar Karwowski
Professor & Chair
Industrial Engineering and
Management Systems
University of Central Florida
Orlando, Florida, U.S.A.

3rd International Conference on Applied Human Factors and Ergonomics (AHFE) 2010

Advances in Applied Digital Human Modeling
Vincent G. Duffy

Advances in Cognitive Ergonomics
David Kaber and Guy Boy

Advances in Cross-Cultural Decision Making
Dylan D. Schmorrow and Denise M. Nicholson

Advances in Ergonomics Modeling and Usability Evaluation
Halimahtun Khalid, Alan Hedge, and Tareq Z. Ahram

Advances in Human Factors and Ergonomics in Healthcare
Vincent G. Duffy

Advances in Human Factors, Ergonomics, and Safety in Manufacturing and Service Industries
Waldemar Karwowski and Gavriel Salvendy

Advances in Occupational, Social, and Organizational Ergonomics
Peter Vink and Jussi Kantola

Advances in Understanding Human Performance: Neuroergonomics, Human Factors Design, and Special Populations
Tadeusz Marek, Waldemar Karwowski, and Valerie Rice

4th International Conference on Applied Human Factors and Ergonomics (AHFE) 2012

Advances in Affective and Pleasurable Design
Yong Gu Ji

Advances in Applied Human Modeling and Simulation
Vincent G. Duffy

Advances in Cognitive Engineering and Neuroergonomics
Kay M. Stanney and Kelly S. Hale

Advances in Design for Cross-Cultural Activities Part I
Dylan D. Schmorrow and Denise M. Nicholson

Advances in Design for Cross-Cultural Activities Part II
Denise M. Nicholson and Dylan D. Schmorrow

Advances in Ergonomics in Manufacturing
Stefan Trzcielinski and Waldemar Karwowski

Advances in Human Aspects of Aviation
Steven J. Landry

Advances in Human Aspects of Healthcare
Vincent G. Duffy

Advances in Human Aspects of Road and Rail Transportation
Neville A. Stanton

Advances in Human Factors and Ergonomics, 2012-14 Volume Set:
Proceedings of the 4th AHFE Conference 21-25 July 2012
Gavriel Salvendy and Waldemar Karwowski

Advances in the Human Side of Service Engineering
James C. Spohrer and Louis E. Freund

Advances in Physical Ergonomics and Safety
Tareq Z. Ahram and Waldemar Karwowski

Advances in Social and Organizational Factors
Peter Vink

Advances in Usability Evaluation Part I
Marcelo M. Soares and Francisco Rebelo

Advances in Usability Evaluation Part II
Francisco Rebelo and Marcelo M. Soares

Advances in
Applied Human Modeling
and Simulation

Edited by
Vincent G. Duffy

CRC Press
Taylor & Francis Group
Boca Raton London New York

CRC Press is an imprint of the
Taylor & Francis Group, an **informa** business

CRC Press
Taylor & Francis Group
6000 Broken Sound Parkway NW, Suite 300
Boca Raton, FL 33487-2742

First issued in paperback 2019

© 2012 by Taylor & Francis Group, LLC
CRC Press is an imprint of Taylor & Francis Group, an Informa business

No claim to original U.S. Government works

ISBN-13: 978-1-4398-7031-0 (hbk)
ISBN-13: 978-0-367-38112-7 (pbk)

This book contains information obtained from authentic and highly regarded sources. Reasonable efforts have been made to publish reliable data and information, but the author and publisher cannot assume responsibility for the validity of all materials or the consequences of their use. The authors and publishers have attempted to trace the copyright holders of all material reproduced in this publication and apologize to copyright holders if permission to publish in this form has not been obtained. If any copyright material has not been acknowledged please write and let us know so we may rectify in any future reprint.

Except as permitted under U.S. Copyright Law, no part of this book may be reprinted, reproduced, transmitted, or utilized in any form by any electronic, mechanical, or other means, now known or hereafter invented, including photocopying, microfilming, and recording, or in any information storage or retrieval system, without written permission from the publishers.

For permission to photocopy or use material electronically from this work, please access www.copyright.com (http://www.copyright.com/) or contact the Copyright Clearance Center, Inc. (CCC), 222 Rosewood Drive, Danvers, MA 01923, 978-750-8400. CCC is a not-for-profit organization that provides licenses and registration for a variety of users. For organizations that have been granted a photocopy license by the CCC, a separate system of payment has been arranged.

Trademark Notice: Product or corporate names may be trademarks or registered trademarks, and are used only for identification and explanation without intent to infringe.

Visit the Taylor & Francis Web site at
http://www.taylorandfrancis.com

and the CRC Press Web site at
http://www.crcpress.com

Table of Contents

Section I: Human Model Fidelity and Sensitivity

1   Human modeling and simulation with high biofidelity   3
        Z. Cheng, S. Mosher, J. Parakkat and K. Robinette, USA

2   Sensitivity analysis of achieving a reach task considering joint angle   13
        and link length variability
        A. Cloutier, J. Gragg and J. Yang, USA

3   Probabilistic and simulation-based methods for study of slips, trips,   23
        and falls - State of the art
        J. Gragg, J. Yang and D. Liang, USA

4   Helmet risk sssessment for top and side impact in construction   33
        sectors
        J. Long, Z. Lei, J. Yang and D. Liang, USA

5   Stochastic optimization applications in robotics and human modeling   43
        - A literature survey
        Q. Zou, J. Yang and D. Liang, USA

Section II: Problem Solving Applications

6   Service management system based on computational customer   57
        models using large-scale log data of chain stores
        H. Koshiba, T. Takenaka and Y. Motomura, Japan

7   Development of the database on daily living activities for the   65
        realistic biomechanical simulation
        H. Kakara and Y. Nishida, Japan, S. Yoon, Korea, Y. Miyazaki,
        H. Mizoguchi and T. Yamanaka, Japan

8   Social system design using models of social and human mind   75
        network - CVCA, WCA and Bayesian network modeling
        T. Maeno, T. Yasui and S. Shirasaka, Japan and O. Bosch,
        Australia

9   Promotion of social participation among the elderly based on   85
        happiness structure analysis using extended ICF
        M. Inoue, K. Kitamura and Y. Nishida, Japan

10   Barcode scanning data acquisition systems in temporary housing area   95
     to support victim community
         T. Nishimura, H. Koshiba, Y. Motomura and K. Ohba, Japan

Section III: Information Processing and Intelligent Agents

11   Analysis of the effect of an information processing task on goal-   103
     directed arm movements
         T. Alexander, Germany

12   Cognitive and affective modeling in intelligent virtual humans for   113
     training and tutoring applications
         R. Sottilare and J. Hart, USA

13   Intelligent agents and serious games for the development of   123
     contextual sensitivity
         A. Tremori, Italy, C. Baisini and T. Enkvist, Sweden,
         A. Bruzzone, Italy, and J. Nyce, USA

14   Outdoor augmented reality for adaptive mental model support   134
         J. Neuhöfer and T. Alexander, Germany

**Section IV: Human Surface Scan, Data Processing and Shape
Modeling**

15   Topology free automated landmark detection   147
         C. Chuang, J. Femiani, A. Razdan and B. Corner, USA

16   Geometric analysis of 3D torso scans for personal protection   159
     applications
         J. Hudson, G. Zehner and B. Corner, USA

17   Shape description of the human body based on discrete cosine   169
     transformation
         P Li, B. Corner and S. Paquette, USA

Section V: Student Models in Adaptive Modern Instructional Settings

18   Personalized refresher training based on a model of competency   181
     acquisition and decay
         W. Johnson and A. Sagae, USA

19   Modeling student arguments in research reports   191
         C. Lynch and K. Ashley, USA

20   Modeling student behaviors in an open-ended environment for   202

learning environment
G. Biswas, J. Kinnebrew and J. Segedy, USA

21   Detailed modeling of student knowledge in a simulation context   212
A. Munro, D. Surmon, A. Koenig, M. Iseli, J. Lee and
W. Bewley, USA

22   Framework for instructional technology   222
P. Durlach and R. Spain, USA

Section VI: Advanced in Modeling for User Centered Design

23   Assessment of manikin motions in IMMA   235
E. Bertilsson, A. Keyvani, D. Högberg, and L. Hanson, Sweden

24   The use of volumetric projection in digital human modelling   245
software for the identification of Category N3 vehicle blind spots
S. Summerskill, R. Marshall and S. Cook, UK

25   The use of DHM based volumetric view assessments in the   255
evaluation of car A-Pillar obscuration
R. Marshall, S. Summerskill and S. Cook, UK

26   Using ergonomic criteria to adaptively define test manikins for   265
design problems
P. Mårdberg, J. Carlson, R. Bohlin, L. Hanson and D. Högberg,
Sweden

27   Automatic fitting of a 3D character to a computer manikin   275
S. Gustafsson, S. Tafuri, N. Delfs, R. Bohlin and J. Carlson,
Sweden

28   Comparison of algorithms for automatic creation of virtual manikin   285
motions
N. Delfs, R. Bohlin, P. Mårdberg, S. Gustafsson and J. Carlson,
Sweden

29   An algorithm for shoe-last bottom flattening   295
X. Ma, Y. Zhang, M. Zhang and A. Luximon, Hong Kong

**Section VII: Validation for Human Interaction in Various Consumer,
Ground Transport and Space Vehicle Applications**

30   Occupant calibration and validation methods   307
J. Pellettiere and D. Moorcroft, USA

31   Model for predicting the performance of planetary suit hip bearing   **317**
     designs
         M. Cowley, S. Margerum, L. Harvill and S. Rajulu, USA

32   Integration of strength models with optimization-based posture   327
     prediction
         T. Marler, L. Frey-Law, A. Mathai, K. Spurrier, K. Avin and
         E. Cole, USA

33   Effects of changes in waist of last on the high-heeled in-shoe plantar   337
     pressure distribution
         J. Zhou, China/Czech Republic, W. Zhang, China, P. Hlaváček,
         Czech Republic, B. Xu, L. Yang and W. Chen, China

34   Fingertips deformation under static forces: Analysis and experiments   348
         E. Peña-Pitarch, N. Ticó-Falguera, A. Vinyes-Casasayas and
         D. Martinez-Carmona, Spain

35   The arch analysis with 3D foot model under different weight-loading   357
         Y.-C. Lin, Taiwan

36   Multi-scale human modeling for injury prevention   364
         S. Sultan and T. Marler, USA

37   Comparative ergonomic evaluation of spacesuit and space vehicle   374
     design
         S. England, E. Benson, M. Cowley, L. Harvill, C. Blackledge,
         E. Perez and S. Rajulu, USA

38   Evaluation of the enlarging method with haptic using multi-touch   384
     interface
         K. Makabe, M. Sakurai and S. Yamamoto, Japan

39   Reconstruction of skin surface models for individual subjects   392
         Y. Endo, N. Miyata, M. Tada, M. Kouchi and M. Mochimaru,
         Japan

40   Effects of high heel shape on gait dynamics and comfort perception   401
         M. Kouchi and M. Mochimaru, Japan

Section VIII: Cognitive and Social Aspects: Modeling, Monitoring, Decision and Response

41   Affective LED lighting color schemes based on product type   407
         J. Park and J. Rhee, South Korea

42   Human activity and social simulation   416
     Y. Haradji, G. Poizat and F. Sempé, France

43   Effects of the order of contents in the voice guidance when operating   426
     machine
     S. Tahara, M. Sakurai, S. Fukano, M. Sakata and S. Yamamoto,
     Japan

44   Framing the socio-cultural context to aid course of action generation   436
     for counterinsurgency
     M. Farry, B. Stark, S. Mahoney, E. Carlson and D. Koelle, USA

45   The *vmStrat* domain specific language   447
     J. Ozik, N. Collier, M. North, W. Rivera, E. Palomaa and
     D. Sallach, USA

46   Decision support system for generating optimal sized and shaped   460
     tool handles
     G. Harih, B. Dolšak and J. Kaljun, Slovenia

47   Ergonomic work analysis applied to chemical laboratories on an oil   471
     and gas research center
     C. Guimarães, G. Cid, A. Paranhos, F. Pastura, G. Franca,
     V. Santos, M. Zamberlan, P. Streit, J. Oliveira and G. Correa,
     Brazil

48   Evaluation of the map for the evacuation in disaster using PDA   478
     based on the model of the evacuation behavior
     Y. Asami, M. Sakurai, D. Kobayashi, K. Ishihara, N. Nojima and
     S. Yamamoto, Japan

Section IX: New Methods and Modeling in Future Applications

49   Diagrammatic User Interfaces   489
     R. Nakatsu, USA

50   A study on effect of coloration balance on operation efficiency and   496
     accuracy, based on visual attractiveness of color
     A. Yokomizo, T. Ikegami and S. Fukuzumi, Japan

51   The integration of ethnography and movement analysis in disabled   507
     workplace development
     G. Andreoni, F. Costa, C. Frigo, E. Gruppioni, E. Pavan,
     M. Romero, B. Saldutto, L. Scapini and G. Verni, Italy

52 Multi-source community pulse dashboard   517
     T. Darr, S. Ramachandran, P. Benjamin, A. Winchell and
     B. Gopal, USA

53 Development of a 3-D kinematic model for analysis of ergonomic   527
     risk for rotator cuff injury in aircraft mechanics
     E. Irwin and K. Streilein, USA

54 Analysis of a procedural system for automatic scenario generation   536
     G. Martin, C. Hughes and J. Moshell, USA

55 Motion synthesizer platform for moving manikins   545
     A. Keyvani, H. Johansson and M. Ericsson, Sweden

56 Human engineering modeling and performance: Capturing humans   555
     and opportunities
     K. Stelges and B. Lawrence, UA

Index of Authors   565

Section I

Human Model Fidelity and Sensitivity

Human Modeling and Simulation with High Biofidelity

Zhiqing Cheng and Stephen Mosher
Infoscitex Corporation
4027 Colonel Glenn Highway, Dayton, OH 45431, USA

Julia Parakkat and Kathleen Robinette
711 Human Performance Wing, Air Force Research Laboratory
2800 Q Street, Wright-Patterson AFB, OH 45433, USA

ABSTRACT

The human modeling and simulation (M&S) technology currently used in most virtual reality systems lacks sufficient biofidelity and thus is not able to describe and demonstrate the nuances of human activities and human signatures. In this paper, the problem is addressed from the perspectives of major factors affecting biofidelity, including static and dynamic shape modeling and motion capture and analysis. An investigation on replicating and creating human activities with high biofidelity is described with examples demonstrated.

Keywords: Human modeling, biofidelity, human shape, human motion, human activity, virtual reality

1. INTRODUCTION

Human modeling and simulation (M&S) plays an important role in simulation-based training and virtual reality (VR). However, the human M&S technology currently used in most simulation-based training tools and VR systems lacks sufficient realism. Biofidelic human M&S provides true-to-life, virtual representations of humans based upon human anthropometry and biomechanics. Biofidelity is critical for virtually describing and demonstrating the nuances of human activities and human signatures. Biofidelity has high impact on VRs and serious games used for human-centered training, such as human threat recognition training and dismount detection training.

Human M&S describes and represents various kinds of human features or signatures. Since human body shape and motion comprise the human signatures that can be observed from a distance and are inherent to human activities, they often become the main focus of human M&S. In this paper, major issues of M&S of human body shape, motion, and activity will be addressed, with an emphasis on the biofidelity of shape and motion. From the perspective of the motion status of a subject to be modeled, human shape modeling can be classified as either static or dynamic. Static shape modeling creates a model to describe human shape at a particular pose. Dynamic shape modeling deals with shape variations due to pose changes or while a subject is moving. Motion capture (mocap) technologies can be marker-based or vision-based. The challenges for motion analysis involve inverse kinematics (IK) and motion mapping and creation. Using M&S, human activities can be either replicated based on shape and motion data collected from the same subject or created using the shape and motion data from different subjects.

2. STATIC SHAPE MODELING

With advances in surface digitization technology, a three dimensional (3D) whole body surface can be scanned in a few seconds. Whole body 3D surface scan provides a very detailed capture of the body shape. Based on body scan data, human shape modeling with high biofidelity becomes possible. The major issues involved in static shape modeling include surface registration, shape parameterization and characterization, and shape reconstruction.

2.1 Surface Registration

Surface registration or point-to-point correspondence among the scan data of different subjects is essential to many problems of human shape modeling, such as shape parameterization and characterization (Allen et al 2003, Azouz et al 2005), and pose modeling and animation (Allen et al 2002, Anguelov et al 2005a) where multiple subjects or multiple poses are involved. Establishing point-to-point correspondence among different scan data sets or models is usually called non-rigid registration. Given a set of markers between two meshes, non-rigid registration brings the meshes into close alignment while simultaneously aligning the markers. Allen et al (2002, 2003) solved the correspondence problem between subjects by deforming a template model which is a hole-free, artist-generated mesh to fit individual scans. The resulting individually fitted scans all have the same number of triangles and point-to-point correspondences. Anguelov et al (2005b) developed an unsupervised algorithm for registering 3D surface scans that is called correlated correspondence (CC). The algorithm registers two meshes with significant deformations by optimizing a joint probabilistic model over all point-to-point correspondences between them. Azouz et al (2004) used a volumetric representation of human 3D surface to establish the correspondences between the scan data of different subjects. By converting their polygonal mesh descriptions to a volumetric representation, the 3D scans of different subjects are aligned inside a volume of fixed dimensions, which is sampled to a set of voxels.

2.2 Shape Parameterization and Characterization

For human shape modeling, it is desirable to have a set of parameters to describe or characterize human shape and its variation among different subjects. Human body shape can be parameterized in three different levels.

- Using surface elements. After surface registration of scan data among all subjects, the same set of vertices or other surface elements can be used to describe different body shapes (3D surfaces) (Allen et al 2003, Angulov et al 2005a). While this method of characterization usually incurs a large number of parameters, a body shape can be directly generated from these parameters.
- Using principal component coefficients. After principal compoenet analysis (PCA), human body shape space is characterized by principal components. Each shape can be projected onto the eigenspace formed by principal components. Within this space, a human shape can be parameterized by its projection coefficients (Allen et al 2003, Azouz et al 2005). If the full eigenspace is used, perfect reconstruction can be achieved.
- Using anthropometric features. The relationship between eigenvectors and human anthropometric features (e.g., height and weight) can be established through regression analysis (Allen et al 2003, Azouz et al 2004 & 2005), and then a body shape can be parameterized by these features. This type of parameterization is not an exact mapping between a human body shape and its anthropometric features. Perfect reconstruction of a body shape usually cannot be achieved given a limited number of features.

2.3 Shape Reconstruction

Given a number of scan data sets of different subjects, a novel human shape can be created that will have resemblance to the samples but is not the exact copy of any existing one. This can be realized in three ways.

- Interpolation or morphing. One shape can be gradually morphed to another by interpolating between their vertices or other graphic entities (Allen et al 2003). In order to create a faithful intermediate shape between two individuals, it is critical that all features are well-aligned; otherwise, features will cross-fade instead of move.
- Reconstruction from eigenspace. After PCA analysis, the features of sample shapes are characterized by eigenvectors or eigenpersons which form an eigenspace. Any new shape model can be generated from this space by combining a number of eigenpersons with appropriate weighting factors (Azouz et al 2005).

- Feature-based synthesis. Once the relationship between human anthropometric features and eigenvectors is established, a new shape model can be constructed from the eigenspace with desired features by editing multiple correlated attributes, such as height and weight (Allen et al 2003) or fat percentage and hip-to-waist ratio (Seo et al 2003).

3. DYNAMIC SHAPE MODELING

Two major issues of dynamic shape modeling are pose modeling and dynamic shape capture and reconstruction.

3.1 Pose Modeling

During pose changing or body movement, muscles, bones, and other anatomical structures continuously shift and change the body shape. For pose modeling, scanning the subject in every pose is impractical; instead, body shapes can be scanned in a set of key poses, and then those body shapes corresponding to intermediate poses are determined by smoothly interpolating among these poses. The issues involved in pose modeling include pose definition, skeleton model derivation, surface deformation (skinning), and pose mapping.

- Pose definition. The human body can assume various poses. In order to have a common basis for pose modeling, a distinct, unique description of different poses is required. One approach is to use joint angle changes as the measures to characterize human pose changing and gross motion. As such, the body shape variations caused by pose changing and motion will consist of both rigid and non-rigid deformation. Rigid deformation is associated with the orientation and position of segments that connect joints. Non-rigid deformation is related to the changes in shape of soft tissues associated with segments in motion, which, however, excludes local deformation caused by muscle action alone. One method for measuring and defining joint angles is using a skeleton model.

- Skeleton model. Allen et al (2002) constructed a kinematic skeleton model to identify the pose of a scan data set using markers captured during range scanning. Anguelov et al (2004) developed an algorithm that automatically recovers from 3D range data a decomposition of the object into approximately rigid parts, the location of the parts in the different poses, and the articulated object skeleton linking the parts. Robertson and Trucco (2006) developed an evolutionary approach to estimate upper-body posture from multi-view markerless sequences. Sundaresan and Chellappa (2007) proposed a general approach that uses Laplacian eigenmaps and a graphical model of the human body to segment 3D voxel data of a body into articulated chains. Hasler et al (2010) developed a method for estimating a rigid skeleton, including skinning weights, skeleton connectivity, and joint positions from a sparse set of example poses.

- Surface deformation modeling. Body surface deformation modeling is also referred to as skinning in animation. Two main approaches for modeling body deformations are anatomical modeling and example-based modeling. The anatomical modeling is based on an accurate representation of the major bones, muscles, and other interior structures of the body (Aubel and Thalmann 2001). The finite element method is the primary modeling technique used for anatomical modeling. In the example-based approach, a model of some body part in several different poses with the same underlying mesh structure can be generated by an artist. These poses are correlated to various degrees of freedom, such as joint angles. An animator can then supply values for the degrees of freedom of a new pose and the body shape (surface) for that new pose is interpolated appropriately. Lewis et al (2000) and Sloan et al (2001) developed similar techniques for applying example-based approaches to meshes. Instead of using artist-generated models, recent work on the example-based modeling uses range-scan data. Allen et al (2002) presented an example-based method for calculating skeleton-driven body deformations based on range scans of a human in a variety of poses. Anguelov et al (2005a) developed a method that incorporates both articulated (rigid) and non-rigid deformations. A pose deformation model was constructed from training scan data that derives the non-rigid surface deformation as a function of skeleton pose. The method (model) is referred to as the SCAPE (Shape Completion and Animation for People). Hasler et al (2009) developed a unified model that describes pose and body shape where the muscle deformation is accurately represented as a function of subject pose and physique.

- Pose mapping. For pose modeling, it is impossible to acquire the body surface deformation for each person at each pose. Instead, surface deformation can be transferred from one person to another for a given pose. Anguelov et al (2005a) addressed this issue by integrating a pose model with a shape model reconstructed from eigenspace. As such, a mesh can be generated for any body shape in their PCA space in any pose. By training their model using 550 full body 3D laser scans taken of 114 subjects and using a rotation invariant encoding of the acquired exemplars, Hasler et al (2009) developed a model that simultaneously encodes pose and body shape thus enabling pose mapping from one subject to another.

3.2 Dynamic Shape Capture and Reconstruction

- Dynamic shape capture. During dynamic activities, the surface of the human body moves in many subtle but visually significant ways: bending, bulging, jiggling, and stretching. Park and Hodgins (2006) developed a technique for capturing and animating those motions using a commercial motion capture system with approximately 350 markers. Supplemented with a detailed, actor specific surface model, the motion of the skin was then computed by segmenting the markers into the motion of a set of rigid parts and a residual deformation.   Sand et al (2003) developed a method (a needle model) for the acquisition of deformable human geometry from silhouettes. New technologies are emerging that can capture body shape and motion simultaneously at a fairly high frame rate (Nguyen and Wang 2010, Izadi et al 2011).

- Shape reconstruction from two dimensional (2D) imagery. Seo et al (2006) presented a data-driven, parameterized, deformable shape model for reconstructing human body models from one or more 2D photos. Guan et al (2009) developed a method for estimating human body shape from a single photograph or painting. One recent work was done by Balan et al (2007) where the human body shape is represented by the SCAPE (Anguelov et al 2005a) and the parameters of the model are directly estimated from image data. Hasler et al (2009) developed a method to estimate the detailed 3D body shape of a person even if heavy or loose clothing is worn. Within a space of human shapes learned from a large database of registered body scans, the method fits a template model (a 3D scan model of a person wearing clothes) to the silhouettes of video images using iterated closest point (ICP) registration and Laplacian mesh deformation.

4. MOTION CAPTURE AND ANALYSIS

4.1. Marker-Based Motion Capture

As a traditional technique, marker-based motion capture technology has been developed to an advanced level that provides accurate and consistent measurements of body motion. The markers used in motion capture can be aligned with those used during body scanning thus providing some correspondence between body shape and skeleton motion. Various software tools are available for the analysis of motion capture data. The major limitations of marker-based motion capture technology include (a) it can only be used in a laboratory environment; (b) it has a limited coverage space; and (c) it requires subject cooperation. Several new technologies are emerging that use accelerometers (Tautges et al 2010) and mini-cameras (Shiratori et al 2011) mounted on the body, enabling open-field motion capture.

4.2. Markerless Motion Capture

As an active research area in computer vision for decades, markerless or vision-based human motion analysis has the potential to provide an inexpensive, unobtrusive solution for the estimation of body poses and motions. Extensive research efforts have been performed in this domain (Moeslund and Granum 2001, Moeslund et al 2006). Agarwal and Triggs (2006) developed a learning-based method for recovering 3D human body pose from single images and monocular image sequences by direct nonlinear regression against shape descriptor vectors extracted automatically from image silhouettes. A recent development is capturing motion and dynamic body shape simultaneously from video imagery. Using SCAPE, Balan et al (2007) developed a method for estimating the model parameters directly from image data. Their results showed that such a rich generative model as SCAPE enables the automatic recovery of detailed human shape and pose from images. Hasler et al (2009) presented an approach for markerless motion capture of articulated objects which are recorded with multiple unsynchronized moving cameras.

4.3. Inverse kinematics

Inverse kinematics, the process of computing the pose of a human body from a set of constraints, is widely used in computer animation. However, the problem is often underdetermined. While many poses are possible, some poses are more likely than others. In general, the likelihood of poses depends on the body shape and style of the individual person. Grochow et al (2004) developed an inverse kinematics system based on a learned model of human poses that can produce the most likely pose satisfying the prescribed constraints in real time. A common task of IK is to derive joint angles from markers, for which OpenSim (https://simtk.org/home/opensim), an open source software package can be used.

4.4. Motion Mapping and Creation

Motion mapping and motion generation are two issues related to IK but have independent significance. It is desirable to map the motion from one subject to another, because it is not feasible to do motion capture for every subject and for every motion or activity. By assuming that different subjects will take the same key poses in an action or motion, joint angles can be mapped from one subject to another. While this way of motion mapping may be fairly natural and realistic, it may not be able to provide sufficiently high biofidelity, because the difference between human bodies and interaction between human body and boundaries are ignored. Wei et al (2011) showed that statistical motion priors can be combined seamlessly with physical constraints for human motion modeling and generation. The key idea of the approach is to learn a nonlinear probabilistic force field function from prerecorded motion data with Gaussian processes and combine it with physical constraints in a probabilistic framework. Some tools were developed for motion creation based on biomechanics and physics, such as DANCE (http://www.arishapiro.com/).

5. ACTIVITY REPLICATION AND CREATION

5.1. Replication

Activity replication is using 3D modeling to replicate a human activity recorded from a human subject in a laboratory. Technologies that are capable of capturing human motion and 3D dynamic shapes of a subject during motion are not yet ready for practical use. Data that can be readily used for 3D activity replication are not currently available. Alternatively, a motion capture system can be used to capture markers on the body during motion and a 3D body scanner can be used to capture the body shape in a pose. Based on the body scan data and motion capture data, animation techniques can be used to build a digital model to replicate a human activity in 3D space.

In this paper, open source software was used for activity replication. MeshLab (http://meshlab.sourceforge.net/) was used to process 3D scan data (smoothing and approximation). OpenSim was used to derive skeleton models and the associated

10

joint angles (BVH files) from motion capture data (TRC files), and Blender (http://www.blender.org/) was used to create an animation model that integrated body shape and motion. Human subject testing for data collection on human activities was conducted in the 3D Human Signatures Laboratory (HSL) at the Air Force Research Laboratory (AFRL). The data collected included scans and mocap data. Figure 1 shows the models created for four activities (jogging, limping, shooting, and walking) at a particular frame.

Figure 1. Replication of four activities: limping, jogging, shooting, and walking.

5.2. Creation

Virtual human activity creation is common nowadays, as numerous commercial software products are available for use. However, a major challenge is the biofidelity of the virtual activities created, which is crucial when a VR is human centered. Activity creation involves motion creation and dynamic shape creation. While some methods have been developed for motion creation, many issues remain. Creating a dynamic shape for any pose or activity is still a challenging task. Alternatively, in the following example, by matching body shape data with mocap data, two activities (diving-rolling and running-ducking) were created using body scan data and mocap data collected from different subjects. The mocap data for the two activities were derived from the Carnegie Mellon University (CMU) mocap database (http://mocap.cs.cmu.edu/). Using the lengths of major segments as the search criteria, the body shape data were derived from CAESAR data-base. Then, 3D animation models were created using Blender which fuses the shape and motion information together and deforms the body shape in accordance with body motion, as shown in Figure 2.

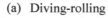

    (a) Diving-rolling   (b) Running-ducking
Figure 2. Activity creation using body scan and mocap data from different subjects.

6. CONCLUSIONS

Biofidelity is a critical factor when human M&S is used in a virtual reality or a human-centered training system. In order to attain high biofidelity, a concerted effort for accurate human data collection, motion analysis, and shape modeling must be undertaken. Human activities can be replicated in 3D space with fairly high biofidelity. However, it is not feasible to collect data for every subject and for every activity. Therefore, it is necessary to develop technologies for creating activities. Activity creation relies on dynamic shape modeling and motion creation, for which further investigations are needed to overcome remaining technical obstacles.

REFERENCES

Agarwal, A., Triggs, B. (2006). Recovering 3D Human Pose from Monocular Images. IEEE Transactions on Pattern Analysis and Machine Intelligence, Vol. 28, No.1.

Allen, B., Curless, B., Popovic, Z. (2002). Articulated Body Deformation from Range Scan Data. In: ACM SIGGRAPH 2002, 21-26, San Antonio, TX, USA.

Allen, B., Curless, B., Popvic, Z. (2003). The space of human body shapes: reconstruction and parameterization from range scans. In: ACM SIGGRAPH 2003, 27-31, San Diego, CA, USA.

Anguelov, D., Koller, D., Pang, H., Srinivasan, P., Thrun, S. (2004). Recovering Articulated Object Models from 3D Range Data. In: Proceedings of the 20th conference on Uncertainty in artificial intelligence, pp 18 – 26.

Anguelov, D., Srinivasan, P., Koller, D., Thrun, S., Rodgers, J., Davis, J. (2005a). SCAPE: Shape Completion and Animation of People. ACM Transactions on Graphics (SIGGRAPH) 24(3).

Anguelov, D., Srinivasan, P., Pang, H., Koller, D., Thrun, S., Davis, J. (2005b). The correlated correspondence algorithm for unsupervised registration of nonrigid surfaces. Advances in Neural Information Processing Systems 17, 33-40.

Aubel, A., Thalmann, D. (2001). Interactive modeling of the human musculature. In: Proc. of Computer Animation.

Azouz, Z., Rioux, M., Shu, C., Lepage, R. (2004). Analysis of Human Shape Variation using Volumetric Techniques. In: Proc. of 17th Annual Conference on Computer Animation and Social Agents Geneva, Switzerland.

Azouz, Z., Rioux, M., Shu, C., Lepage, R. (2005). Characterizing Human Shape Variation Using 3-D Anthropometric Data. International Journal of Computer Graphics, volume 22, number 5, pp. 302-314.

Balan, A., Sigal, L., Black, M., Davis, J., Haussecker, H. (2007). Detailed Human Shape and Pose from Images. In: IEEE Conf. on Comp. Vision and Pattern Recognition (CVPR).

Grochow, K., Martin, S., Hertzmann, A., Popovic, Z. (2004). Style-Based Inverse Kinematics. In: Proc. of SIGGRAPH'04.

Guan, P., Weiss, A., Balan, A., Black, M. (2009). Estimating Human Shape and Pose from a Single Image. In: Proceedings of ICCV., 1381-1388.

Hasler, N., Stoll, C., Sunkel, M., Rosenhahn, B., Seidel, H. (2009a). A Statistical Model of Human Pose and Body Shape. Eurographics, Vol. 28 No. 2.

Hasler, N., Stoll, C., Rosenhahn, B., Thormahlen, T., Seidel, H. (2009b). Estimating Body Shape of Dressed Humans. Computers & Graphics Vol 33, No 3, pp211-216.

Hasler, N., Rosenhahn, B., Thormahlen, T., Wand, M., Gall, J., Seidel, H. (2009c). Markerless Motion Capture with Unsynchronized Moving Cameras. IEEE Conference on Computer Vision Pattern Recognition, CVPR.

Hasler, N., Thormahlen, T., Rosenhahn, B., Seidel, H. (2010). Learning Skeletons for Shape and Pose. In: Proceedings of the 2010 ACM SIGGRAPH symposium on Interactive 3D Graphics and Games, ISBN: 978-1-60558-939-8, New York.

Izadi, S., Newcombe, R., Kim, D., Hilliges, O., Molyneaux, D., Hodges, S., Kohli, P., shotton, J., Davison, A., Fitzgibbon, A. (2011). KinectFusion: Real-Time Dynamic 3D Surface Reconstruction and Interaction. A Technical talk on Siggraph 2011, Vancouver, Canada.

Lewis, J., Cordner, M., Fong, N. (2000). Pose space deformations: A unified approach to shape interpolation and skeleton-driven deformation. In: Proceedings of ACM SIGGRAPH 2000, pp. 165–172.

Moeslund, T., Granum, E. (2001). A Survey of Computer Vision-Based Human Motion Capture. Computer Vision and Image Understanding, Vol. 81, No. 3, pp. 231-268.

Moeslund, T., Hilton, A., Krüger, V. (2006). A survey of advances in vision-based human motion capture and analysis. Computer Vision and Image Understanding, Vol. 104, No. 2-3, pp. 90-126.

Nguyen, D., Wang, Z. (2010). High Speed 3D Shape and Motion Capturing System. A Poster of Siggraph 2010, Los Angles, USA.

Park, S., Hodgins, J. (2006). Capturing and Animating Skin Deformation in Human Motion. ACM Transaction on Graphics (SIGGRAPH 2006), 25(3), pp 881-889.

Robertson, C., Trucco, E. (2006). Human body posture via hierarchical evolutionary optimization. In: BMVC06.

Sand, P., McMilla, L., Popovic, J. (2003). Continuous Capture of Skin Deformation. In: ACM Transactions on Graphics 22 (3), 578-586.

Seo, H., Cordier, F., Thalmann, N. (2003). Synthesizing Animatable Body Models with Parameterized Shape Modifications. In: Eurographics/SIGGRAPH Symposium on Computer Animation.

Seo, H., Yeo, Y., Wohn, K. (2006). 3D Body Reconstruction from Photos Based on Range Scan. Lecture Notes in Computer Science, 2006, No. 3942, pp 849-860, Springer-Verlag.

Shiratori, T., Parky, H., Sigal, L., Sheikhy, Y., Hodginsy, J. (2011). Motion Capture from Body-Mounted Cameras. ACM Trans. Graph. 30, 4, Article 31.

Sloan, P., Rose, C., Cohen, M. (2001). Shape by example. In: Proceedings of 2001 Symposium on Interactive 3D Graphics.

Sundaresan, A., Chellappa, R. (2007). Model driven segmentation of articulating humans in Laplacian Eigenspace. IEEE Transactions on Patter Analysis and Machine Intelligence.

Tautges, J., Zinke, A., Krüger, B., Baumann, J., Weber, A., Helten, T., Müller, M., Seidel, H., Eberhardt, B. (2010). Motion Reconstruction Using Sparse Accelerometer Data, ACM TOG 30(3), article 18.

Wei, X., Min J., Chai, J. (2011). Physically Valid Statistical Models for Human Motion Generation. ACM Trans. Graph. Vol 30, No 3, Article 19.

CHAPTER 2

Sensitivity Analysis of Achieving a Reach Task Considering Joint Angle and Link Length Variability

Aimee Cloutier, Jared Gragg, Jingzhou (James) Yang

Human-Centric Design Research Laboratory
Department of Mechanical Engineering
Texas Tech University, Lubbock, TX 79409, USA
E-mail: james.yang@ttu.edu

ABSTRACT

Due to variability in human anthropometry, no two people will perform a task in exactly the same way—different people will adapt different postures to perform the same reach task, and even for the same person and reach task, postures will vary with time. It is therefore important to consider uncertainty in methods of posture prediction. Currently, techniques in digital human modeling applications primarily employ deterministic methods which do not properly account for variability. An alternative to deterministic methods is probabilistic/reliability design. This study presents a sensitivity approach to gain insights into how uncertainty affects reach tasks. Sensitivity levels are found to determine the importance of each joint to the final posture. A digital human upper body model with 30 degrees of freedom (DOFs) is introduced to demonstrate the sensitivity approach for reach tasks using both arms. Thirty-six different reach tasks (eighteen for male and eighteen for female) are used to compare the sensitivities due to joint angle and link length uncertainty. The results show that the importance of each joint and link length is dependent on the nature of the reach task; sensitivities for joint angles and link lengths are different for each reach task.

Keywords: probability; reach task; uncertainty; joint angle variability; link length variability

1   INTRODUCTION

Variability in human anthropometry plays a large role in creating distinctions among the ways different people perform tasks. Each person's physical characteristics are unique, and given a physical task, no two people will perform it in exactly the same way. Even a single person's performance on the same task will vary over time. Because of these differences in height, weight, stature, and many other anthropometric factors, there is inherent variability associated with problems in digital human modeling. Accounting for this variability in human anthropometry can help lead to the design of products which ultimately accommodate the largest number of people.

Currently, methods in the literature for digital human modeling are primarily deterministic. Deterministic methods do not properly account for variability; instead, they try to limit uncertainty through optimization or simply ignore it. Because the accuracy of deterministic methods depends on the accuracy of their input parameters, any error in the input values will propagate through to the output.

Probabilistic design, an alternative to deterministic methods, accounts for uncertainty by modeling each variable input parameter according to a probability distribution. The corresponding results are probabilities of failure quantifying how the combined variability of all parameters will affect the output. However, it is also useful to identify how much the uncertainty of individual parameters will affect the output's performance. For this purpose, sensitivity analysis, a component of probabilistic analysis, can be used. Sensitivity analysis provides a quantitative result showing how much each input parameter's uncertainty affects the probability of failure. Knowing the sensitivities of each value of the input can help improve the design of products by accounting for variability in those aspects with high sensitivities.

Although there has been little progress towards applying probabilistic methods to applications in digital human modeling, there is a significant amount of research in orthopedics and biomechanics employing probabilistic and sensitivity techniques (Laz and Browne, 2009; Zhang et al., 2010). Probabilistic methods have been favored extensively in the areas of prosthesis and implants, often employing sensitivity analysis in conjunction with finite element modeling (Dar et al., 2002; Browne et al., 1999). Numerous sources of uncertainty have been considered, including patient geometry, material properties, component alignment, loading conditions, joint mechanics, and clinical outcomes (Easley et al., 2007; Fitzpatrick et al., 2010). The application of principal component analysis (PCA) combined with probabilistic techniques and finite element (FE) analysis has also proven to be useful for developing patient-specific models for implants (Rajamani et al., 2007; Barratt et al., 2008). Probabilistic and sensitivity techniques have also been used in knee replacement implants to improve implant alignment and loading on the joints (Pal et al., 2009) and to identify the most significant factors affecting the knee's use during gait (Nicolella et al., 2006). In several studies, probabilistic methods have been employed to analyze the behavior of both cemented (Dopico-Gonzalez et al., 2009a) and uncemented hip prosthesis (Dopico-Gonzalez et al., 2009b; Galbusera et al., 2010). Recently, probabilistic methods have even been used to improve the design of implants in the spine (Rohlmann et al., 2010).

The importance of accuracy in anatomical landmark placement and joint location has been recognized by several researchers. To improve accuracy in joint descriptions, probabilistic methods have been applied to determine the variability in anatomical landmark location for the shoulder (Langenderfer et al., 2009) and knee (Morton et al., 2007) kinematic descriptions. Probabilistic techniques have also been used to analyze the variability of joint parameters in the knee (Holden and Stanhope, 1998) and during analyses of gait (Rao et al., 2006; Reinbolt et al., 2007). Several probabilistic assessments have been conducted to better determine the behavior of finger and thumb kinematics (Valero-Cuervas et al., 2003; Santos and Valero-Cuervas, 2006; Li and Zhang, 2010). Incorporating uncertainty into analyses of bone shape, mechanics, and fracture has also become a more common practice (Nelson and Lewandowski, 2009; Edwards et al., 2010; Thompson et al., 2009). Probabilistic methods have been used in several areas to help predict human motion (Horiuchi et al., 2006).

Cloutier et al. (2012) reported the sensitivity analysis for one hand reach tasks with only joint angle uncertainty. The aim of this study is to extend this method for two hand reach tasks and both joint angle and link length uncertainty. A 30-DOF digital human model of the spine, right arm, and left arm will be used to demonstrate reach tasks. The sensitivities of all joint angles and link lengths for each reach task will be obtained. This will lead to conclusions about the relative importance of each joint angle and link length to reach tasks.

This paper is organized as follows: Section 2 gives the problem definition and briefly reviews the 30-DOF human model used in this work. Section 3 describes the probabilistic theory applied to this study. Section 4 shows results. Section 5 discusses the results and limitations. Section 6 provides concluding remarks.

2   PROBLEM DEFINITION AND HUMAN MODEL

Subjects are instructed to reach nine target points, as shown in Figure 1. Each point has the same displacement along the z-axis but varies along the x- and y- axes. For a reach task, the left arm of the model reaches for one of the three points to the left while the right arm reaches for one of the remaining six points. Using all combinations of target points creates a total of eighteen reach tasks. The goal is to determine the sensitivity of joint angle and link length variation on the reach tasks.

The model for the upper body is composed of 30 DOF as seen in Figure 2 (Yang et al., 2004), and the Denavit-Hartenberg (DH) notation is used to define the transformation matrices between the local frames (Denavit and Hartenberg, 1955).

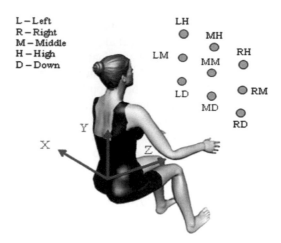

Figure 1 Location and naming syntax for nine target points

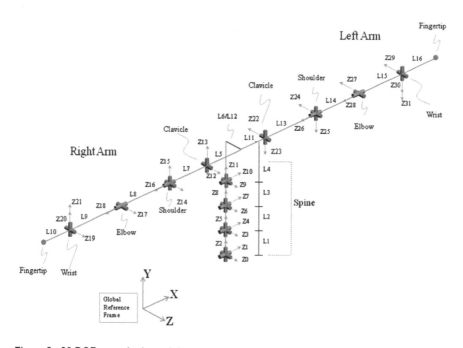

Figure 2 30 DOF upper body model

3   PROBABILISTIC THEORY

In a probabilistic analysis, each input parameter containing uncertainty is considered as a random variable with an associated probability distribution. Both the joint angles and link lengths are assumed to have normal (Gaussian) distributions, which is a reasonable assumption due to the Central Limit Theorem (Haldar and Mahadevan, 2000). A normal distribution is defined by the random variable's mean and standard deviation. For the posture prediction problem, the mean of each joint angle's distribution is taken as the value of the joint angle in the deterministic solution. The coefficient of variation, the ratio of the standard deviation to the mean, is specified as $\delta=0.02$ (2%) for all joint angles. A small coefficient of variation is assumed because varying the joint angles too much can significantly alter the reach task.

The mean values for each link length's normal distribution were determined based on the average link lengths obtained from motion capture experiments used to validate the results of posture prediction problems. Male and female link lengths were averaged separately using a total of forty subjects. A standard coefficient of variation for limb length anthropometric data is roughly 3% to 6% (McDowell et al., 2008). Therefore, a coefficient of variation of $\delta=0.05$ (5%) is used for all link lengths.

With all joint angles and link lengths considered as random variables, the threshold between a succeeding and failing result is represented by the limit state equations in Eq. (1) and Eq. (2).

$$z_1 = g(\overline{q}, \overline{L})_1 = \varepsilon - \left[\left(x_{21} - TP_{x1} \right)^2 + \left(y_{21} - TP_{y1} \right)^2 + \left(z_{21} - TP_{z1} \right)^2 \right] \tag{1}$$

$$z_1 = g(\overline{q}, \overline{L})_1 = \varepsilon - \left[\left(x_{21} - TP_{x1} \right)^2 + \left(y_{21} - TP_{y1} \right)^2 + \left(z_{21} - TP_{z1} \right)^2 \right] \tag{2}$$

In the limit state equations, (x_{21}, y_{21}, z_{21}) and (x_{30}, y_{30}, z_{30}) are the positions of the two end-effectors, $(TP_{x1}, TP_{y1}, TP_{z1})$ and $(TP_{x2}, TP_{y2}, TP_{z2})$ are the target points for the right and left hand, respectively, and ε is the desired level of precision. For this problem, a value of $\varepsilon = 0.1$ was used. If the value of either limit state equation for a certain position of the end-effectors is less than zero, i.e. either hand falls outside of the precision limit, the result is considered failing. The probability of failure (P_f) is defined by Eq. (3).

$$P_f = P(z(\overline{q}, \overline{L})_1 \cup z(\overline{q}, \overline{L})_2) < 0) \tag{3}$$

The safety index, β, is the ratio of the mean of the limit state equation to the standard deviation. It is defined by Eq. (4), and it is used to calculate the sensitivity factors. The sensitivity factors, α_i, are the partial derivatives of the safety index with respect to each random variable, and the partial derivatives are taken with respect to both the mean and standard deviation. They are calculated using Eq. (5). All sensitivity factors are calculated using NESSUS software.

18

$$\beta = \frac{\mu_z}{\sigma_z} = -\Phi^{-1}(P_f) \tag{4}$$

$$\alpha_i = \frac{\partial \beta}{\partial \mu_i} \, or \, \frac{\partial \beta}{\partial \sigma_i} \tag{5}$$

4   RESULTS

The primary results of this analysis are gained from the sensitivities for each joint angle and link length. Figure 3 shows the sensitivities for all thirty joint angles for three reach tasks, and Figure 4 shows the sensitivities of all sixteen link lengths for six reach tasks. Note that the reach tasks are named according to the syntax in Figure 1; the right hand reach point is specified, then the left reach point is specified.

The sensitivity levels in Figures 3 and 4 represent the effect of the mean of the joint angles and link lengths on the probability of failure. In almost all cases, the sensitivity levels are positive. A positive sensitivity factor indicates that the probability of failure would increase if the mean were raised; if the sensitivity factor is negative, raising the mean will lower the probability of failure.

Several general trends throughout the data can be observed by comparing the magnitudes of the sensitivity levels throughout reach tasks. Knowing the sensitivities of each joint angle and link length provides information about which parameters' variability affect a particular posture the most. Also, certain patterns in the data can indicate specific zones of reach for which certain joint angles and link lengths can be expected to have higher sensitivities.

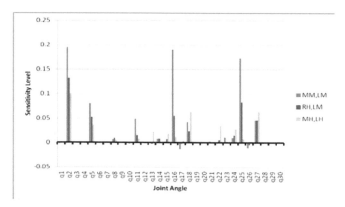

Figure 3   Sensitivities of the joint angles for selected reach tasks

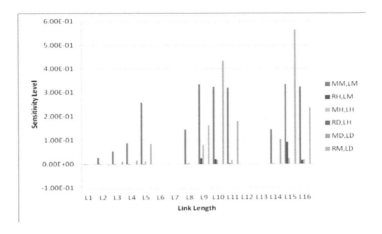

Figure 4 Sensitivities of the link lengths for selected reach tasks

In general, the sensitivity levels of the link lengths are much higher than the sensitivities of the joint angles. This is partially due to the larger amount of variability allowed in the link lengths, which have much higher standard deviations than the joint angles. However, several other trends exist in the observed data. In all cases, the sensitivities of q_2, q_5, q_8, and q_{11} are much higher than the sensitivities of the other eight spine joints (refer to Figure 2). This pattern exists because those four joint angles all correspond to the bending motion of the four spine joints. Because all of the reach tasks require the subject to lean forward more than moving or twisting from side to side, the joint angles which represent a bending spine motion contribute more to the final posture. Naturally, their sensitivities would be higher. The joint angles corresponding to elbow bend (q_{18} and q_{27}) and wrist tilt (q_{20} and q_{29}) always have small sensitivities, and the link lengths L_5, L_6, L_{11}, and L_{12} always have zero sensitivity. In reach tasks for which the right hand reaches for a point to the far right and for which both hands are on the same vertical plane, the sensitivities of the spine joint, with the exception of q_2, q_5, q_8, and q_{11}, are zero. This is due to the symmetry of the reach task for the right and left arms; in these three cases, the spine requires no rotation or tilting to complete the reach task, and therefore, the probability of failure is not sensitive to these joint angles.

Sensitivity levels are typically high for link lengths representing the upper arm, L_8 and L_{14}, the lower arm, L_9 and L_{15}, and the hand, L_{10} and L_{16}, and for the joint angles corresponding to the forward and backward motion of the shoulder, q_{16} and q_{25}. This shoulder sensitivity is generally higher for central reach points than for reach points farther to the right.

5   DISCUSSION

The results of the sensitivity analysis provide valuable information about the effects of joint angle and link length variability on the likelihood of completing reach tasks. The generally higher values of the link lengths' sensitivities indicate that it is important to include considerations of uncertainty into posture prediction. These high importance levels to the probability of failure show that variations in human anthropometry may significantly affect a person's ability to complete a reach task. Additionally, it shows that models simply based on an average height or stature may result in a loss of accuracy when predicting human posture; for instance, a person of average height does not necessarily have average link lengths, and due to high sensitivity levels in the arms, variability in arm link lengths may vary the posture a person ultimately chooses for a reach task.

There are several limitations associated with this study. First, a model of only the upper body is used. Including the lower body in the sensitivity analysis might cause some variation in the results. This work also examines only a small portion of reach tasks. Extending the number of reach points to include more spine rotation and varying points along the z-axis may provide additional useful information. Finally, the link lengths were determined based on a relatively small number of subjects through motion capture data. Because motion capture experiments determine joint center locations based on the positions of markers on the skin, there is inherent uncertainty associated with joint center locations. To account for this uncertainty, it would be helpful to draw data from a larger pool of subjects. These limitations will be addressed in future work. Future work will also include incorporating considerations of uncertainty into the posture prediction model itself.

6 CONCLUSION

A sensitivity analysis has been performed on the results of a deterministic posture prediction problem for generalized reach tasks in order to quantify the influence of uncertainty in the joint angles and link lengths toward the likelihood of completing a reach task. A 30-DOF human upper body model was used to demonstrate postures. Joint angles and link lengths were both considered as random variables in the analysis. The sensitivity results for eighteen different reach tasks for both male and female models indicate that it is important to consider anthropometric uncertainty in posture prediction problems. It is particularly important to include variability in the link lengths because of high sensitivities in the arm.

The results of this study also show that sensitivities of joint angle and link lengths are dependent on the nature of the reach task; for example, because the reach tasks in this work required the subject to bend the spine forward more than rotate it, the sensitivities which correspond to the bending movement were much higher than all other spine sensitivities. Future work will include extending the results of this study to include sensitivities for a full body posture prediction problem and considering variability in posture prediction problems themselves (as opposed to applying sensitivity analysis to a deterministic result).

ACKNOWLEDGMENTS

This work was partly supported by the Undergraduate Research Fellowship at The Honors College, TTU, and National Science Foundation (Award #0926549 and 1048931).

REFERENCES

Barratt, D. C., Chan, C. S. K., and Edwards, P. J. et al. 2008. Instantiation and registration of statistical shape models of the femur and pelvis using 3D ultrasound imaging. *Medical Image Analysis* 12: 358-374.

Browne, M., Langley, R. S., and Gregson, P. J. 1999. Reliability theory for load bearing biomedical implants. *Biomaterials* 20: 1285-1292.

Cloutier, A., Gragg, J., and Yang, J. 2012. Sensitivity analysis of achieving a reach task within a vehicle considering joint angle variability. In. *SAE World Congress*. Detroit, MI.

Dar, F. H., Meakin, J. R., and Aspden, R. M. 2002. Statistical methods in finite element analysis. *Journal of Biomechanics* 35: 1155-1161.

Denavit, J. and Hartenberg, R. 1955. A kinematic notation for lower-pair mechanisms based on matrices. *Journal of Applied Mechanics* 22: 215-221.

Dopico-Gonzalez, C., New, A. M., and Browne, M. 2009. Probabilistic analysis of an uncemented total hip replacement. *Medical Engineering and Physics* 31: 470-476.

Dopico-Gonzalez, C., New, A. M., and Browne, M. 2009. Probabilistic finite element analysis of the uncemented hip replacement—effect of femur characteristics and implant design geometry. *Journal of Biomechanics* 43: 512-520.

Easley, S. K., Pal, S., and Tomaszewski, P. R. et al. 2007. Finite element-based probabilistic analysis tool for orthopaedic applications. *Computer Methods and Programs in Biomedicine* 85: 32-40.

Edwards, W. B., Taylor, D., and Rudolphi, T. J. et al. 2010. Effects of running speed on a probabilistic stress fracture model. *Clinical Biomechanics* 25: 372-377.

Fitzpatrick, C. K., Baldwin, M. A., and Rullkoetter, P. J. et al. 2010. Combined probabilistic and principal component analysis approach for multivariate sensitivity evaluation and application to implanted patellofemoral mechanics. *Journal of Biomechanics* 44: 513-521.

Galbusera, F., Anasetti, F., and Bellini, C. et al. 2010. The influence of the axial, antero-posterior and lateral positions of the center of rotation of a ball-and-socket disc prosthesis on the cervical spine biomechanics. *Clinical Biomechanics* 25: 397-401.

Haldar, A. and Mahadevan, S. 2000. *Probability, reliability and statistical methods in engineering design*. New York, NY: John Wiley & Sons, Inc.

Holden, J. P. and Stanhope, S. J. 1998. The effect of variation in knee center location estimates on net knee joint moments. *Gait and Posture* 7: 1-6.

Horiuchi, T., Kanehara, M., and Kagami, S. et al. 2006. A probabilistic walk path model focused on foot landing points and human step measurement system. In. *2006 IEEE International Conference on Systems, Man, and Cybernetics*. Taipei, Taiwan.

Langenderfer, J. E., Rullkoetter, P. J., and Mell, A. G. et al. 2009. A multi-subject evaluation of uncertainty in anatomical landmark location on shoulder kinematic description. *Computer Methods in Bomechanics and Biomedical Engineering* 12: 211-216.

Laz, P. J. and Browne, M. 2009. A review of probabilistic analysis in orthopaedic biomechanics. *Engineering in Medicine*, 224: 927-943.

Li, K. and Zhang, X. 2010. A probabilistic finger biodynamic model better depicts the roles of the flexors during unloaded flexion. *Journal of Biomechanics* 43: 2618-2624.

McDowell, M. A., Fryar, C. D., and Ogden, C. L. et al. 2008. Anthropometric reference data for children and adults: United States, 2003-2006. *National Health Statistics Reports* 10.

Morton, N. A., Maletsky, L. P., and Pal, S. et al. 2007. Effect of variability in anatomical landmark location on knee kinematic description. *Journal of Orthopaedic Research* 25: 1221-1230.

Nelson, E. S., Lewandowski, B., and Licata, A. et al. 2009. Development and validation of a predictive bone fracture risk model for astronauts. *Annals of Biomedical Engineering* 37: 2337-2359.

Nicolella, D. P., Thacker, B. H., and Katoozian, H. et al. 2006. The effect of three-dimensional shape optimization on the probabilistic response of a cemented femoral hip prosthesis. *Journal of Biomechanics* 39: 1265-1278.

Pal, S., Haider, H., and Laz, P. J. et al. 2008. Probabilistic computational modeling of a total knee replacement wear. *Wear* 264: 701-707.

Rajamani, K. T., Styner, M. A., and Talib, H. et al. 2007. Statistical deformable bone models for robust 3D surface extrapolation from sparse data. *Medical Image Analysis* 11: 99-109.

Rao, G., Amarantini, D., and Berton, E. et al. 2006. Influence of body segments' parameters estimation models on inverse dynamics solutions during gait. *Journal of Biomechanics* 39: 1531-1536.

Reinbolt, J. A., Haftka, R. T., and Chmielewski, T. L. et al. 2007. Are patient-specific joint and inertial parameters necessary for accurate inverse dynamics analyses of gait? *IEEE Transactions on Biomedical Engineering* 54: 782-793.

Rohlmann, A., Boustani, H., and Bergmann, G. et al. 2010. Effect of pedicle-screw-based motion preservation system on lumbar spine biomechanics: A probabilistic finite element study with subsequent sensitivity analysis. *Journal of Biomechanics* 43: 2963-2969.

Santos, V. J. and Valero-Cuevas, F. J. 2006. Reported anatomical variability naturally leads to multimodal distributions of Denavit-Hartenberg parameters for the human thumb. *IEEE Transactions on Biomedical Engineering* 53: 1555-163.

Thompson, S., Horiuchi, T., and Kagami, S. 2009. A probabilistic model of human motion and navigation intent for mobile robot path planning. In. *4th International Conference on Autonomous Robots and Agents*. Wellington, New Zealand.

Valero-Cuevas, F. J., Johanson, M. E., and Towles, J. D. 2003. Towards a realistic biomechanical model of the thumb: The choice of kinematic description may be more critical than the solution method or the variability/uncertainty in musculoskeletal parameters. *Journal of Biomechanics* 36: 1019-1030.

Yang, J., Marler, R., and Kim, H. et al. 2004. Multi-objective optimization for upper body posture prediction. In. *10th AIAA/ISSMO Multidisciplinary Analysis and Optimization Conference*. Albany, NY.

Zhang, L., Wang, Z., and Chen, J. et al. 2010. Probabilistic fatigue analysis of all-ceramic crowns based on the finite element model. *Journal of Biomechanics* 43: 2321-2326.

CHAPTER 3

Probabilistic and Simulation-Based Methods for Study of Slips, Trips, and Falls-State of the Art

Jared Gragg, Jingzhou (James) Yang*, and Daan Liang

Texas Tech University
Lubbock, TX 79409, USA
james.yang@ttu.edu

ABSTRACT

Slips, trips, and falls (STF) can cause serious harm or death. The biomechanics of slip have been extensively studied in the literature. Understanding the biomechanics of slip is the beginning of developing prevention strategies for STF and allows one to develop slip-resistant footwear and other technologies as well as understanding the difference between slips that lead to recovery and those that result in falls. It is generally understood that there are certain key factors that can contribute to slips. Ground reaction forces at the shoe-floor interface have been determined to be a critical biomechanical factor in slips (Redfern et al., 2001). Other biomechanical factors known to influence slip include: the kinematics of the foot at heel contact, human responses to slipping perturbations, the mental state of the individual, the individual's perception of the environment, the individual's age, walking speed, compensatory stepping, trained exposure to slips, and gait stability. One property of the shoe/floor interface that is closely associated with slip is the coefficient of friction (COF). Often, the COF is compared to the ratio of shear to normal foot forces generated during gait, which is known as the required coefficient of friction (RCOF). In early simplistic STF studies, it was held that the condition for slip was when the RCOF was less than the COF. However, it was proven that this is not always the case. Now there are more complicated expressions that predict the probability of slip under certain conditions. In addition to biomechanical factors that influence slip, there are environmental factors that influence slip probability.

24

Environmental factors include: temperature; presence of rain, ice, and snow; presence of obstacles or ramps; and surface characteristics of the floor/surface such as roughness and waviness. In addition to studies on the biomechanics of slip, biomechanical factors of slip, and environmental factors of slip there are numerous studies on STF as it relates to occupational hazards. Survey-based studies have been conducted for occupations with a high risk of STF including: hospital and healthcare workers, postal workers, construction workers, and fast-food/restaurant workers. Other non-occupational survey-based studies on STF include: STFs from ladders, STFs and nursing homes, and STFs during bathtub/shower ingress and egress. While there is a great deal of literature on the biomechanics of slip and STF detection and prevention, most of the knowledge on STF has been obtained from conducting surveys and experiments that test various conditions and environments where STF is common. In contrast to the amount of experimental work done on STF there are limited studies available that propose probabilistic or simulation-based methods of STF detection and prevention. This paper aims to present a comprehensive literature review about probabilistic or simulation-based methods for STF. Based on the literature review, we would like to develop a more efficient simulation-based STF method.

Keywords: slips, trips, and falls (STF); simulation-based STF detection and prevention

1   INTRODUCTION

Slips, trips, and falls have serious impact on humans at work, at play, outdoors, and indoors. From pedestrian accidents on walkways, to postal workers delivering mail, to nursing home residents, STF have major impact on everyday life. It is estimated that the annual direct cost of occupational injuries due to STF in the USA was more than $37 billion from 1993 to 1998 or over $6 billion annually (Courtney 2001). There are data from various sectors listing recorded falls from slips and trips, as well as the severity of the fall. The literature is full of various studies focused on STF including: STF detection, STF statistics, biomechanics of slip, STF simulation, measurement of slipperiness, classification of STF, occupational and non-occupational STF, STF and the elderly, fall prevention once slip has occurred, injury prevention after a fall has occurred, etc. Redfern et al. (2001) presented a review of the biomechanics of slip citing the leading theories at the time. This reference is a useful starting point when beginning to study STF, but has become somewhat outdated. In addition, its focus was on the biomechanics of slip and in general the cause of STF. This paper expounds on the literature review presented in Redfern et al. (2001) by presenting a literature review of current STF literature focusing on those studies that deal with the probabilistic determination of STF and simulation-based methods for the study of STF. The conclusion provides a forecast for the direction of simulation-based studies which include probabilistic methods.

2   PROBABILISTIC DETERMINATION OF STF

Experiment-based studies dominate the literature on STF. There are numerous experiments listed in the literature to determine slipperiness on various floor surfaces under various conditions (Aschan, 2009; 2005; Chang, 2003; 2004b;, 2006; 2008a; 2008b; Courtney, 2006; 2010; Li, 2007; Marpet, 2001; 1996). The traditional theory of slip and fall considers a floor with an average coefficient of friction (COF) between the floor and a given footwear material of $\bar{\mu}$. The theory states that no slip, and thus no fall, will occur when $\bar{\mu} > \mu_c$, where μ_c is the critical COF for the given footwear material. Often times the critical COF has been determined through a set of experiments aimed at measuring slips for a given floor and footwear type. Other times, the critical COF has been established for a certain condition and expanded to include other conditions. In general, the concept of a critical COF criterion for determining the probability of slip is not helpful. A deterministic approach such as this is dangerous in that it gives one outcome regardless of variability in the input conditions. In contrast, a probabilistic approach allows one to include input variability into the simulation or prediction and determine a probability of failure. In the case of STF, there is variability in the input conditions for experimental measurement of the floor COF. Some of the variability is due to human gait dynamics; the slipmeter, a tribometric device employed to measure the COF, effectiveness; floor variability and wear; and footwear material variability and wear. In the case of STF the probability of failure would be the probability that one slips. For problems where it is more interesting or important to study the probability of success, the probability of failure is recast in terms of the reliability. The reliability is the probability of success, calculated as one minus the probability of failure. Thus, the reliability for the case of STF is the probability that a floor is safe, i.e. there are no slips. In summary, it is better to characterize the probability of slip for a given floor condition and footwear material than to determine a critical COF. If a slip and fall were to occur on a floor deemed safe from the measurement of the mean COF, traditional theory does not have a way to explain the phenomenon. Incorporating a probabilistic approach to the prediction of slip and fall allows one to include variability and prevent liability in the case of a slip and fall.

There are several studies in the literature that incorporate probabilistic approaches to the determination of slip. Hanson et al. (1999) investigated the relationship among measurements of the COF, the biomechanics of gait, and actual slip and fall events. Five subjects walked down a ramp at angles of $0°$, $10°$, and $20°$ with either a tile or carpeted surface under dry, wet, or soapy conditions. The dynamic COF for each instance was recorded and the required COF, heretofore referred to as the critical COF, was assessed by examining the foot forces during trials in which no slips occurred. The data was then fit to a logistic regression model to estimate the probability of a slip based on the difference between the available and required COF. The available COF, or sometimes referred to as the available friction, is the actual COF recorded between the shoe-floor interface. The required

COF, or required friction, is the minimum friction needed to support human activity. Thus if the available friction is lower than the required COF, a slip is likely. The drawback of this study was that mean values instead of a stochastic distribution were used to estimate the probability of a slip, as pointed out by Barnett (2002) and Chang (2004).

Fathallah et al. (2000) estimated the probability of slip for drivers exiting commercial vehicles equipped with steps and grab-rails under assumed icy conditions. The study measured the required COF of 10 male subjects as they exited the vehicles. The logistic regression model detailed in Hanson et al. (1999) was then used to estimate the probability of a slip when employing the grab-rails and not employing the grab-rails. As expected, the probability of slip was greatly decreased when exiting the vehicles with the aid of the grab-rails. This study represents a useful application of probabilistic methods to STF study, but as will be discussed later, the methods developed in Hanson et al. (1999) for predicting the probability of slip have some drawbacks.

Marpet (2002) acknowledged the inherent variability of tribometric slipmeter test results for measuring the COF of a test surface and recommended two techniques to accommodate for the variability. There are limitations when accepting that a single-point characterization of the required COF is sufficient in preventing a slip. Typically, the performance of a floor surface is characterized by the sample mean for a number of tests. This study presents scenarios where the single-point characterization of the required COF is dangerous when employing the sample mean. Since typically it is the lower COF values that result in slip and falls, employing the sample mean can underestimate the safety of the floor. The study recommends using a lower percentile value, say 10% of the sample mean, for characterizing floor performance. The study also recommends applying multi-point characterization of the required COF. While this study moves the slip and fall analysis in the right direction, there are better ways to characterize floor performance as listed hereafter.

Barnett (2002) reformulated the classical slip and fall analysis to a probabilistic approach in order to account for the stochastic nature of friction. Through the incorporation of extreme order statistics, it was possible to determine a relationship between the probability of slipping, the required COF criterion, distance traveled by the subject, and the distribution of friction coefficients. The developed theory also revealed that short walks lead to fewer falls and that in some cases lower friction floors are better than high friction ones. The first conclusion is a natural one, but the second conclusion is significant in that classical theory could not predict lower slip probability for a floor with a lower COF.

Chang (2004) studied the effect that the stochastic nature of the required COF had on the probability of slip. For the study in Hanson et al. (1999) mean values were used instead of stochastic distributions. For the study in Barnett (2002), the required COF was assumed to have a deterministic value, or a single value. In reality, both the required COF and the available COF are stochastic by nature. Chang (2004) extended the probabilistic approach in Barnett (2002) to include stochastic variations in the required COF and compared the results to previous studies where the required COF was taken as a single, deterministic value. The results showed that the study in Barnett (2002) underestimated the probability of slip and Hanson et al. (1999) overestimated the probability of slip. In addition a friction adjustment coefficient, $\Delta\mu$, was introduced in order to accommodate the situation where humans were able to regain their balance after a slip. Compared to the study in Barnett (2002) and Hanson et al. (1999) it is more realistic to include the stochastic nature of the required COF in addition to the available COF.

Barnett and Poczynok (2005) applied the probabilistic approach to classical slip and fall analysis to calculate floor reliabilities of various asphalt tile floors under straight walking scenarios and for various lengths of walk. For a typical asphalt tile floor deemed safe under classical slip and fall theory the reliability of the floor was practically 100%, that is less than three slips occurred during five million miles of walking. Thus, the probabilistic results were similar to the classical theory. The additional benefit of the analysis was that a classical theory would not predict even a single slip. While the probabilistic approach still deems the floor safe, it is interesting that there is still a remote possibility of a slip. This is more realistic than the classical approach and explains why a slip may still occur on a "safe" floor. For an asphalt tile floor with a COF of 0.366 (unsafe floor according to classical theory), the results also match intuition. Walks of 10 steps resulted in an 80% reliability and walks of 1000 steps resulted in a 3% reliability. These results are more intuitive than the classical theory which deems the floor unsafe. The probabilistic approach still deems the floor unsafe, but allows for the slight possibility of a successful walk. One of the main drawbacks of the probabilistic approach is that extreme care must be taken when identifying the statistical parameters involved in the simulation. Careful judgment must be made to the number of experiments that are necessary to calculate the mean and standard deviation of the COF during floor slipperiness measurements. Also, careful judgment must be taken when deciding what level of floor reliability is acceptable.

Barnett et al. (2009) applied a probabilistic approach to slip and fall theory to explain how a woman slipped on a concrete walkway, a floor deemed safe by the classical approach to slip and fall. The study proclaims that typically a walking surface is deemed safe if the average COF between the surface and a footwear material is higher than 0.5. In the case of concrete, this is typically true. The study rejects the classical theory and analyzes the concrete that the woman slipped on with a probabilistic approach similar to Barnett (2002) and Barnett and Poczynok (2005). A safety analysis of the concrete in question yielded an average COF of 0.51. The fifty measurements of the local COF (which average to 0.51) were recast in terms of a Weibull probability density distribution, using extreme value statistics. The reliability of the concrete was calculated and it was estimated that the concrete in question gave rise to an average of 29.4 slips per year. The average COF of the concrete in question was noted as being considerably lower than normal and only slightly higher than the required 0.5. An alternative asphalt tile was suggested as an alternative to the concrete.

Chang (2004) introduced a method to estimate the probability of slip and fall incidents by comparing the available and required COF considering the stochastic nature of each. Chang et al. (2008c) introduced a methodology to quantify the stochastic distributions of the required COF for level walking. In this study, a walkway with a layout of three force plates was designed in order to capture a large number of walking strides under various walking conditions. The required COF was recorded for a single participant. Three distributions, a normal (Gaussian), log-normal, and Weibull distribution, were fitted to the data and a Kolmogorov-Smirnov (K-S) goodness-of-fit test was used to determine an appropriate distribution for the data. According to the K-S tests, each of the distributions fit the data well, but the normal distribution was chosen for its ease of application. The limitation of this study was that it only presented results for one walker. However, the methodology could be applied to more subjects in the same manner.

In summary, it is better to model the stochastic nature of the available and required COF for a given shoe-floor interface when estimating the likelihood of slip. The method developed in Chang (2004) provides a sound methodology for predicting the probability of slip including the stochastic nature of both the available and required COF. The methodology developed in Chang et al. (2008c) details the steps to determining the appropriate stochastic distribution parameters to be used for the prediction of the probability of slip. Including the stochastic nature of friction into the prediction of slip enhances the results. Classical slip and fall theory cannot predict why slips may occur on a floor deemed safe, while probabilistic methods can. In addition, probabilistic methods enhance the predictions by: allowing the inclusion of variability of human gait, pedestrian walking style, age, gender, health, speed, etc; accounting for distance walked; explaining why lower friction surfaces sometimes produce fewer slips; addressing the lowest COF encountered, not the mean; and incorporating the notion of traffic patterns and duty cycles on a walking surface (Barnett et al., 2009).

3   SIMULATION-BASED METHODS FOR STUDY OF STF

In the literature, there is little in the way of simulation-based methods for the study of STF. However, a series of studies by Yang (Yang et al., 2008a; 2008b; Yang and Pai, 2010) incorporate simulation models for the determination of backwards loss of balance (BLOB) and slip in gait. Other simulation-based studies of STF include Beschorner et al. (2008), Mahboobin et al. (2010), Park and Kwon (2001), Smeesters et al. (2007), Yang et al. (2007), Yang et al. (2009), and Yang et al. (2008c).

Yang et al. (2008a) employed a 7-link bipedal model and forward dynamics simulation integrated with a dynamic optimization to derive threshold values of the center of mass (COM) velocity relative to the base of support (BOS) in order to prevent BLOB. The COM velocity was consistently higher under slip conditions than under nonslip conditions. The findings were verified by experimental data collected during walking experiments. Limitations in the study included the model only considered motion in the sagittal plane, the pelvis was assumed to have no horizontal rotation, and the relative position in the sagittal plane between the two hips was held constant.

Yang et al. (2008b) employed a 7-link, moment-actuated human model to predict the threshold of the COM velocity relative to the BOS required to prevent BLOB during single stance recovery from a slip. Five dynamic optimization problems were solved to find the minimum COM velocities that terminated with the COM above the BOS for different initial conditions of the COM based on averaged data from experimental trials. The forward velocities of the COM necessary to prevent BLOB were compared to experimental results from 99 adults during 927 slips. The 7-link model was shown to provide more accuracy than a 2-link model. Similar limitations were present as in Yang et al. (2008a).

Yang and Pai (2010) employed the forward-dynamics model developed in previous studies, including Yang et al. (2008a; 2008b), to determine the impact of reactive muscular response from individual lower limb joints on regaining stability control after a slip. Ten young adults' resultant moments at three lower limb joints, derived from an inverse-dynamics approach from empirical data, were applied as inputs to the forward-dynamics model. By systematically altering the moments of each joint, participants were able to improve the COM stability. The model simulation revealed that the systematic altering of the joint moments had little effect on the COM velocity, but had substantial impact on the BOS velocity reduction.

Simulation-based methods have significant advantages over experiment-based methods. Some of the advantages for the study of STF include: simulation-based methods have no limitations for the number of trials; simulation-based methods are not influenced by participant fatigue; experiment-based methods typically do not reproduce a real "fall" since the subject is protected by a harness; dynamic COF measurements can be difficult to obtain during slip scenarios; and simulation-based methods pose no risk of injury. Thus, it is important to develop simulation-based methods for the study of STF. Current literature is lacking in the depth and breadth of simulation-based methods. The methods presented above have limitations as listed and also include the human model employed in the study. The 7-link model ignores the effect of arms and arm-swing in the simulation. More accurate human models must be employed in future slip and fall simulations and the limitations of the studies must be addressed in the future.

4   CONCLUSION

There are numerous experiment-based and survey-based studies in the literature that deal with STF. In contrast, there are precious few that employ probabilistic methods or simulation-based methods. Probabilistic methods have several advantages over classical slip and fall theory. These advantages include, but are not limited to: the incorporation of the stochastic nature of friction; classical slip and fall theory cannot predict why slips may occur on a floor deemed safe, while probabilistic methods can; allowing the inclusion of variability of human gait, pedestrian walking style, age, gender, health, speed, etc; accounting for distance walked; explaining why lower friction surfaces sometimes produce fewer slips; addressing the lowest COF encountered, not the mean; and incorporating the notion of traffic patterns and duty cycles on a walking surface. One of the main drawbacks of the probabilistic approach is that extreme care must be taken when identifying the statistical parameters involved in the simulation. Careful judgment must be made to the number of experiments that are necessary to calculate the mean and standard deviation of the COF during floor slipperiness measurements. Also, careful judgment must be taken when deciding what level of floor reliability is acceptable.

Simulation-based methods have several advantages over experiment-based methods. These advantages include, but are not limited to: simulation-based methods have no limitations for the number of trials; simulation-based methods are not influenced by participant fatigue; experiment-based methods typically do not reproduce a real "fall" since the subject is protected by a harness; dynamic COF measurements can be difficult to obtain during slip scenarios; and simulation-based methods pose no risk of injury. Current limitations of the simulation-based STF methods include the accuracy of the human model employed in the simulations and the depth and breadth of the studies available in the literature.

It is proposed that a combination of the probabilistic approach and simulation-based approach be taken for the future study of STF. Limitations of the probabilistic method include the number of experiments necessary to provide adequate distribution data. One of the advantages of simulation-based methods is that the number of trials is only limited by the allotment of time and computation power. One of the limitations of the current simulation-based methods is that they do not account for the stochastic nature of friction. Including the stochastic nature of friction is important for calculating the probability of slip. It is better to calculate the probability of slip rather than calculate limiting or terminal values for slip or BLOB. Furthermore, simulation-based methods must be developed that allow for repeated simulations of slip and fall without risk to subjects in experimental settings. Combining the advantages of probabilistic methods and simulation-based methods will allow for a robust model capable of predicting the probability of slip under a wide range of scenarios. Future work will be dedicated to developing such a model.

ACKNOWLEDGEMENTS

This work was partly supported by AT&T Chancellor's Fellowship, the Whitacre College of Engineering Dean's Fellowship, and the Graduate School, Texas Tech University.

REFERENCES

Aschan, C., Hirvonen, M., Rajamäki, E., Mannelin, T., Ruotsalainen, J., and Ruuhela, R., 2009. Performance of slippery and slip-resistant footwear in different wintry weather conditions measured in situ. *Safety Science*, 47: 1195-1200.

Aschan, C., Hirvonen, M., Mannelin, T., Rajamäki, E., 2005. Development and validation of a novel Portable Slip Simulator. *Applied Ergonomics*, 36: 585–593.

Barnett, R., 2002. "Slip and fall" theory- extreme order statistics. *International Journal of Occupational Safety and Ergonomics*, 8 (2): 135-159.

Barnett, R., and Poczynok, P., 2005. Floor reliability with respect to "slip and fall". In: *Proceedings of the 2005 ASME International Mechanical Engineering Congress and Exposition,* Orlando, FL, USA, Nov. 5-11.

Barnett, R., Ziemba, A., and Liber, T., 2009. Slipping on concrete: a case study. In: *Proceedings of the 2009 ASME International Mechanical Engineering Congress and Exposition,* Lake Buena Vista, FL, USA, Nov. 13-19.

Beschorner, K., Higgs, C., Lovell, M., and Redfern, M., 2008. A mixed-lubrication model for shoe-floor friction applied to pin-on-disk apparatus. In: *Proceedings of the STLE/ASME International Joint Tribology Conference*, Miami, FL, USA, Oct. 20-22.

Chang, W., Cotnam, J., and Matz, S., 2003. Field evaluation of two commonly used slipmeters. *Applied Ergonomics*, 34: 51–60.

Chang, W., 2004. A statistical model to estimate the probability of slip and fall incidents. *Safety Science*, 42: 779–789.

Chang, W., Li, K., Huang, Y., Filiaggi, A., and Courtney, T., 2004b. Assessing floor slipperiness in fast-food restaurants in Taiwan using objective and subjective measures. *Applied Ergonomics*, 35: 401–408.

Chang, W., Li, K., Huang, Y., Filiaggi, A., and Courtney, T., 2006. Objective and subjective measurements of slipperiness in fast-food restaurants in the USA and their comparison with the previous results obtained in Taiwan. *Safety Science*, 44: 891–903.

Chang, W., Li, K., Filiaggi, A., Huang, Y., and Courtney, T., 2008a. Friction variation in common working areas of fast-food restaurants in the USA. *Ergonomics*, 51 (12): 1998-2012.

Chang, W., Huang, Y., Li, K., Filiaggi, A., and Courtney, T., 2008b. Assessing slipperiness in fast-food restaurants in the USA using friction variation, friction level and perception rating. *Applied Ergonomics*, 39: 359–367.

Chang, W., Chang, C., Matz, S., and Lesch, M., 2008. A methodology to quantify the stochastic distribution of friction coefficient required for level walking. *Applied Ergonomics*, 39: 766–771.

Courtney, T., Sorock, G., Manning, D., Collins, J., and Holbein-Jenny, M., 2001. Occupational slip, trip, and fall-related injuries- can the contribution of slipperiness be isolated? *Ergonomics*, 44 (13): 1118-1137.

Courtney, T., Huang, Y., Verma, S., Chang, W., Li, K., and Filiaggi, A., 2006. Factors influencing restaurant worker perception of floor slipperiness. *Journal of Occupational and Environmental Hygiene*, 3: 593–599.

Courtney, T., Verma, S., Huang, Y., Chang, W., Li, K., and Filiaggi, A., 2010. Factors associated with worker slipping in limited-service restaurants. *Injury Prevention*, 16 (1): 36-41.

Fathallah, F., Grönqvist, R., and Cotnam, J., 2000. Estimated slip potential on icy surfaces during various methods of exiting commercial tractors, trailers, and trucks. *Safety Science*, 36: 69-81.

Hanson, J., Redfern, M., and Mazumdar, M., 1999. Predicting slips and falls considering required and available friction. *Ergonomics*, 42 (12): 1619-1633.

Li, K., Hsu, Y., Chang, W., and Lin, C., 2007. Friction measurements on three commonly used floors on a college campus under dry, wet, and sand-covered conditions. *Safety Science*, 45: 980–992.

Mahboobin, A., Cham, R., and Piazza, S., 2010. The impact of a systematic reduction in shoe–floor friction on heel contact walking kinematics- A gait simulation approach. *Journal of Biomechanics,* 43: 1532–1539.

Marpet, M., 2001. Problems and progress in the development of standards for quantifying friction at the walkway interface. *Tribology International,* 34: 635-645.

Marpet, M., 1996. On threshold values that separate pedestrian walkways that are slip resistant from those that are not. *Journal of Forensic Science,* 41 (5): 747-755.

Park, J., and Kwon, O., 2001. Reflex control of biped robot locomotion on a slippery surface. In: *Proceedings of the 2001 IEEE International Conference on Robotics & Automation,* Seoul, Korea, May 21-26.

Redfern, M., Cham, R., Gielo-Perczak, K., Gronqvist, R., Hirvonen, M., Lanshammar, H., Marpet, M., Pai, C., and Powers, C., 2001. Biomechanics of slips. *Ergonomics,* 44 (13): 1138-1166.

Smeesters, C., Hayes, W., and McMahon, T., 2007. Determining fall direction and impact location for various disturbances and gait speeds using the Articulated Total Body model. *Journal of Biomechanical Engineering,* 129: 393-399.

Yang, J., Jin, D., Ji, L., Zhang, J., Wang, R., Fang, X., and Zhou, D., 2007. An inverse dynamical model for slip gait. In: *First International Conference on Digital Human Modeling, ICDHM,* Beijing, China, July 22-27.

Yang, F., Anderson, F., and Pai, Y., 2008a. Predicted thresholds of dynamic stability against backward balance loss under slip and nonslip bipedal walking conditions. In: *2008 Asia Simulation Conference- 7th Intl. Conf. on Sys. Simulation and Scientific Computing,* Beijing, China, Oct. 10-12.

Yang, F., Anderson, F., and Pai, Y., 2008b. Predicted threshold against backward balance loss following a slip in gait. *Journal of Biomechanics,* 41: 1823–1831.

Yang, F., Passariello, F., and Pai, Y., 2008c. Determination of instantaneous stability against backward balance loss: Two computational approaches. *Journal of Biomechanics,* 41: 1818–1822.

Yang, F., Espy, D., and Pai, Y., 2009. Feasible stability region in the frontal plane during human gait. *Annals of Biomedical Engineering,* 37 (12): 2606–2614.

Yang, F., and Pai, Y., 2010. Role of individual lower limb joints in reactive stability control following a novel slip in gait. *Journal of Biomechanics,* 43: 397–404.

CHAPTER 4

Helmet Risk Assesment for Top and Side Impact in Construction Sectors

James Long, Zhipeng Lei, Jingzhou (James) Yang*, and Daan Liang

Texas Tech University, Lubbock, TX 79409, USA
james.yang@ttu.edu

ABSTRACT

In recent years there has been a push for greater job safety in all industries. Personnel protective equipment (PPE) has been developed to help mitigate the risk of injury to humans that might be exposed to hazardous situations. The human head is the most critical location for impact to occur on a work site. A hard enough impact to the head can cause serious injury or death. That is why industries have adopted the use of an industrial hard hat or helmet. The objective of this safety device is to reduce the risk of injury to the head. There has only been a few articles published that are focused on the risk of head injury when wearing an industrial helmet. A lack of understanding is left on the effectiveness of construction helmets when reducing injury. The scope of this paper is to determine the threshold at which a human will sustain injury when wearing a helmet. Complex finite element, or FE, models were developed to study the impact on construction helmets. The FE model consists of two parts the helmet and the human model. The human model consists of a brain, enclosed by a skull and an outer layer of skin. The level and probability of injury to the head is determined using both the Head Injury Criterion (HIC) and tolerance limits. The HIC has been greatly used to assess the likelihood of head injury while in a vehicles. The tolerance levels are more suited for finite element models, but lack wide scale validation. Different cases of impact were studied using LSTC's LS-DYNA. This study assesses the risk of injury for wearers of construction helmets or hard hats.

Keywords: impact simulation, construction/industrial helmet, injury prediction

1   INTRODUCTION/RESEARCH MOTIVATION

The head is the most critical area of the human body. Severe trauma to the head can lead to death or long term disability. In fact, emergency rooms treat and release one million Americans a year for traumatic brain injury, TBI. Out of these patients 50,000 die; 80,000 to 90,000 experience long term disability; and 230,000 are hospitalized and survive. The major causes of TBI include vehicular accidents, violence, falls, sports and industrial incidents (Goldsmith 2001). A reduction in head injury could save thousands of lives and prevent injury in thousands of others that survive, but are living with disabilities. Much research has been devoted to the topic of head injury. Understanding the mechanics and biomechanics of head injury is vital for engineers and scientist to prevent injury. The study of head injury biomechanics can be divided into experimental, analytical, numerical and regulatory information (Goldsmith 2001). An understanding of experimental and regulatory information is key to utilizing numerical results to aid in the design process. Experimental results aid researchers in validating analytical and numerical solvers. Experimentation is vital to obtain the behavior of the materials simulated in finite element, or FE, solvers. Most devices designed to reduce head injury need to meet approval of the regulatory body governing the use of a particular device. An important device in mitigating head injury has been the helmet. Helmets not only prevent the skull from being perforated, they can also dampen the force of the impact object transmitted to the wearer. Hard hats first saw wide spread use in 1931 on the Hoover Dam Project. In the United States employers must follow Occupational Safety and Health Administration, or OSHA, regulation and ensure that their employees wear head protection if any of the following conditions apply: "objects might fall from above and strike them on the head; they might bump their heads against objects such as exposed pipes and beams; or there is a possibility of accidental head contact with electrical hazards (OSHA 3151-12R, 2003). Hard hats are also regulated by the American National Standards Institute, ANSI. The "American National Standard for Industrial Head Protection" or ANSI Z89.1-2009 is referenced in the research conducted. The ANSI standards for impact only dictate two types of helmets, Type I and Type II. Type I helmets must reduce the force of impact on only the top of the head and Type II must reduce the force of impact on the top and sides of the head. ANSI creates a minimum amount of protection that an industrial helmet should provide to the wearer. Not all helmets provide the same amount of protection and designs are constantly evolving to make more protective and comfortable helmets. Designers need tools that can predict the mitigation of injury to optimize and improve helmet design. The FE model and injury prediction methods explained, in this article, attempt to accurately predict injury and provide a useful tool for helmet designers.

A basic understanding of the human anatomy is required for the set up of the FE model. This article will start with a brief overview of human head anatomy. Next the proposed FE model will be explained in great detail. This includes initial set up and the material models utilized to accurately predict the reaction of the helmet and the head under impact. After the FE model is explained, the different injury criterion models will be listed. Each individual responds differently to head impact. The values collected to determine injury equate to a probability to a various level of head injury. Once the various levels of injury are made clear, the results from various impact scenarios will be discussed. This paper will end with a conclusion.

1.3   Human Head Anatomy

There are three main sections or layers surrounding the human brain. These layers include the scalp, skull and the meninges. The scalp is stretched over the outer layer of the skull. The scalp has an average thickness of 3 to 6 mm and is composed of five anisotropic layers. These layers from descending order are; "(a) the skin with hairy coverings, (b) the layer of tela subcutanea, a loose, fiberous connective tissue that attaches the skin to the deeper structures; (c) the aponeurotic layer, a fiberous membrane constituting flattened tendon connecting the frontal and occipital muscles; (d) a loose subaponeurotic layer of connective tissue; and (e) the pericranium, a tough vascular membrane, also designated as the subpericranial layer proximate to the skull" (Goldsmith, 2001). The next main layer, the skull is a more uniform and rigid structure. The skull has an average thickness of 9.5 to 12.7 mm. The skull has an outer and inner layer of calcified compact bone. Sandwiched in between the inner and outer layer of the skull is a vesicular layer that resembles a honeycomb. The skull encloses the entire brain except for an opening at the bottom for the spinal cord. The final layer, the meninges, consists of three sub layers. It has an average thickness of 2.5 mm. The first sub layer, the dura, which is located below the skull is tough, dense, inelastic and an anisotropic membrane consisting of connective tissue. Between the dura and second layer, the arachnoid, is a space. This space is referred to as the subdural space. The arachnoid is a delicate nonvascular membrane if interconnected trabecular fibers. The arachnoid trabecular fibers connect to the next the final layer, the pia. The pia is a layer of white fibrous tissue that is attached to the surface of the brain. There is another space in between the arachnoid and pia layers. This space is referred to as the subarachnoid space. This space is occupied by water like fluid known as the cerebrospinal fluid (CSF). The CSF provides damping and cushions the brain in impact situations. The CSF is also produced in cavities of the brain and circulates through the spinal canal and perivascular space.

The brain is divided into three main sections. The largest fraction being the cranium is divided into two convoluted hemispheres. The section that connects the brain the spinal cord is the brain stem. The final section the cerebellum is where higher level functions are concentrated, see figure 1. All of the sections of the brain are separated by dura mater and coated with pia and anachnoid layers. The human head is complex structure that needs to be simplified in order to be simulated by a numerical solver.

1.4   Finite Element Model

The model shown in Figure 2 and 3 is the 3D representation of a human head. The model is reconstructed from cross section images of the Visible Human Dataset (http://www.nlm.nih.gov/research/visible/visisble_human.html). The volumetric meshing was performed in CFD-GEOM (Version 2009). The entire head model is modeled with solid elements. The outer layer of human skin which includes the scalp, consist of 128,061 elements and 25,798 nodes. The skull is modeled as a single layer with 44,938 elements and 11,828 nodes. The brain is also modeled as a homogenous structure with 33,786 elements and 6,009 nodes.

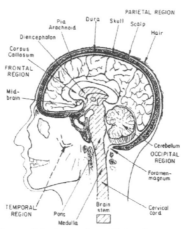

Figure1 Further illustrates the layers of the human head. Take of anatomical geometry and compare to the proposed FE model. (From Goldsmith. 2001)

Yan and Pangestu (2011), in "A modified human head model for the study of impact injury", assumed the behavior of the scalp was elastic. Yan and Pangestu's material model of the skin is utilized, and is listed in Table 1. In order to maximize the effectiveness of the tolerance limits proposed by Deck and Willinger, this model will use the material models employed by Deck and Willinger for the skull and brain. Deck and Willinger assume the skull is a single elastic layer, see Table 1 for further details. The brain is modeled as a viscoelastic material. The viscoelastic response due to shear behavior is modeled by the following equation:

$$G(t) = G_\infty + (G_0 - G_\infty)\exp{(-\beta t)}$$

Where G_0 is the dynamic shear modulus and has a value of 528 kPa. G_∞ is the static shear modulus and has a value of 168 kPa. The final variable β, is a decay constant and has a value of $0.035 \mathrm{ms}^{-1}$. The constants of the viscoelastic shear behavior are applied to the viscoelastic material card in LS-Dyna.

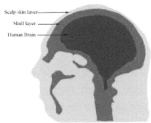

Figure 2 Displays the cross sectional view of the FE model on the left compared to the cross sectional view of the human head. The FE model is simplified but still remains anatomically accurate.

The remaining part of the FE model is the industrial hard hat. The hard includes an outer shell along with two straps that are in place for the suspension system. The suspension system is crucial to the effectiveness of the helmet. According to ANSI Z89.1 – 2009, the suspension is connected to the harness and acts an energy-absorbing mechanism. Also this harness should leave a 1.25 inch gap between the suspension and the inner helmet shell. The shell and strap pieces are modeled as shell elements. The helmet shell consists of 3,752 elements with 1,878 nodes. Each strap contains 2,916 elements and 1,708 nodes.

Two more different material cards, in LS –Dyna, are employed to model the plastic response of the helmet shell and straps, see Table 2. Industrial hard hats are usually molded from high density polymers or thermoplastics. The material properties of current hard hats are proprietary to the manufactures. Sabic's Ultem ATX 100, a common thermoplastic for impact and a popular addictive for hard hat construction, was chosen to represent the helmet shell. The material responses of plastics are dependent on strain rate. To model this phenomena the Piecewise Linear Plasticity card was chosen in LS-Dyna. With this material option users input several true stress and true strain curves, at different strain rates, to more accurately model the effects of plasticity in materials. Three stress strain curves at different strain rates are provided at from the manufacture of Ultem, Sabic, at (http://www.sabic-ip.com/gep/en/Home/Home/home.html). These are then imputed into a table in LS-Dyna. There is even less published information on the materials used for the suspension system. Some common additives include nylon and plastics, such as Ultem 1000. Assuming that the material properties of the suspension is similar to the additives, a plastic kinematic material card was developed from known material properties of Ultem 1000, once again provided by Sabic, see Table 2.

Human Head Material Property Table

	Material Type	Density (kg m^{-3})	Youngs Modulus (MPa)	Poisson's ratio	Bulk Modulus (MPa)
Skin	Elastic	1130	16.7	0.42	N/A
Skull	Elastic	2100	6000	0.21	N/A
Brain	Viscoelastic	1140	0.675	N/A	5.625

Table 1 The material models utilized in LS-Dyna to simulate the human head. The skin material properties are from Yan and Pangestu's "A modified human head model for the study of impact head injury". The Skull and Brain material properties are from Willinger et al. "Three_Dimensional Human Head Finite_Element Model Validation Against Two Experimental Impacts.

Helmet Material Table

	Material Type	Density (kg m^{-3})	Youngs Modulus (MPa)	Poisson's ratio	Yield Stress (MPa)	Failure Strain (m m^{-1})
Helmet Shell	Piecewise Linear Plastic	1210	3000	0.3	68	0.8
Helmet Straps	Plastic Kinematic	1270	3580	0.3	110	0.6

Table 2 The material models utilized in LS-Dyna to simulate the industrial helmet. The materials properties can be downloaded from (http://www.sabic-ip.com.html) with approved membership. Sabic informs that these material properties are for selection purposes only and that the user is responsible for their own material testing.

The first step of simulating the model is to properly place the helmet on the head. An initial simulation is run to achieve the proper gap between the suspension and the helmet shell. This step also molds the suspension to the human head and represents the wearer placing the helmet on his head. As a side note the straps of the suspension system are constrained to the sections of the helmet that house the connectors for the suspension system.

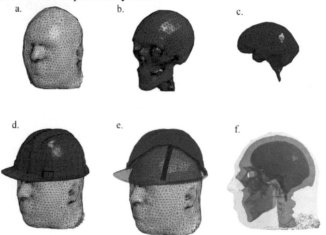

Figure 3 Different parts of the FE model are displayed. The following images are as follows: a. is the outer layer of the human head, the skin and scalp; b. is the middle layer of the head, the skull; c. is the solid elements of the brain; d. is the human head and helmet model, after placement of the helmet; e. the helmet shell is set to transparent to show the suspension system or straps; f. the skin and skull parts are set to transparent to show the anatomical positions of the human head parts.

1.5   Injury Criteria

Two different injury criterions are utilized in this article. The first the HIC is widely accepted measure of the likelihood of head injury. There is several disadvantages when relying on the HIC. The HIC was developed for use in crash test dummies, not FE models. The biggest disadvantage is that the HIC is based only on translational acceleration. Critics argue that rotational acceleration influences head injury as well. Also the HIC doesn't distinguish between specific mechanisms of injury in the head. Despite the limitations of the HIC it is the most validated head injury criterion to date. A HIC score correlates to a probability for a level of injury. The HIC levels of injury are as follows: *Minor Head Injury* is a skull trauma without loss of consciousness; fracture of nose or teeth; superficial face injuries, *Moderate Head Injury* – Skull trauma with or without dislocated skull fracture and brief loss of consciousness, *Critical head injury* – Cerebral contusion, loss of consciousness for more than 12 hours with intracranial hemorrhaging and other neurological signs, recovery is uncertain (Prasad and Mertz, 1985). See figure 4 for likelihood of injury at each level.

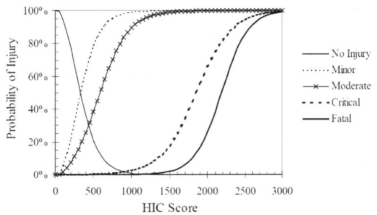

Figure 4 Probability of Specific Head Injury Level for a given HIC Score, downloaded from the Canadian Playground Advisory Inc at (http://www.playgroundadvisory.com)

Deck and Willinger (2008) proposed head injury criteria based on tolerance limits for the separate parts of the human head. This criterion was designed for ULP head model, a FE model developed by Willinger. Deck and Willinger reconstructed 68 known head impact conditions that occurred in motorcyclist, American football and pedestrian accidents. The study concluded with proposed limits for injury mechanisms. These mechanics include Moderate and Severe Diffuse Axonal Injury (DAI), skull fracture and SubDural Hematoma. The proposed limits for 50% chance of injury are as follows: For a mild DAI the value for von Mises Stress recorded in the brain is 26 kPa, for Severe DAI the value for von Mises Stress recorded in the brain 33kPa, for skull fracture the value is 865 mJ of Skull strain energy and the minimum amount of CSF pressure for a subdural heamatoma is -135kPa. It is important to note these values are from the max value calculated for any element within the part involved in the injury mechanism (Deck and Willinger, 2008).

2 RESULTS & DISCUSSION

For studying top impact two falling objects where chosen. The first object was a cylindrical bar, with similar material properties of steel. The steel bar has a weight of 2 kg. The second object chosen was a wood board. The board was set to dimensions similar to that of W. Goldsmith's boards he used when impacting a construction helmet in, "Construction Helmet Response Under Severe Impact". The boards used weighed around 5kg and had the dimensions of (90mm x 90mm x 160mm) (Goldsmith, 1975). For the cases of side and front impact a rigid wall was employed.

The results for front and side impacts did not show any significant risk of injury. The worst possible case considered was if the human was moving 5 m/s that is 11 mph and is much faster than most individuals are capable of running. The highest HIC value was less than one, the highest von Mises stress calculated in the brain is 3.77 kPa and the highest strain energy in the skull was 2 mJ. As stated by OSHA, a hard hat is designed to protect employees from bumping their heads. If a worker is wearing properly wearing a hard hat, it is safe to assume they should be protected from the most typical injuries. Fall injuries were not considered since they are out of the scope of design requirements for most hard hats.

For top impact the only limitation on speed is what height an object may fall. Simulations for top impact were conducted until an HIC score of 1000 was reached or in the case of the 5kg board until the helmet experienced a catastrophic failure. For a summary of the results see figure 5,6, and 7.

The HIC scores predict less injury than the tolerance level of Deck and Willinger. An HIC score of 500 means there is about 20% chance of no injury, 80% chance of a minor injury, and a 40% chance of a moderate injury. The tolerance levels proposed by Deck and Willinger, show that for around an HIC score of around 500, the threshold for a 50% chance of a DAI has already been passed and is either close to or past the 50% chance of a skull fracture.

The 2kg steel cylinder has a 50% chance of causing mild DAI around an impact speed of8.5 m/s, a severe DAI around a impact speed of 11 m/s, skull fracture at 15 m/s and receives and HIC score of 1000 at 18.5 m/s. The 5 kg wood board has a 50% chance of causing mild DAI around an impact speed of 6 m/s, a severe DAI around an impact speed of 8 m/s, skull fracture at 13 m/s and never quite reached the HIC score of 1000. The helmet fails within the first few milliseconds of impact, at an impact speed of 18 m/s. This castrophic failure cause the simulation to abort an HIC score cannot be tabulated. The 5 kg does cause an HIC score of 903 at 17 m/s.

3 CONCLUSION

With the data collect using a FE model and utilizing the tolerance levels proposed by Deck and Willinger and also tabulating an HIC score it is possible to predict injury in wearers of industrial hard hats. Not only can the threshold of injury be calculated, but also the mechanism of injury can be predicted thanks to the work completed by Deck and Willinger. The results from the two criteria's for head injury are not an exact match. This could be caused by the limitations of the HIC and/or the limited amount of cases reconstructed by Deck and Willinger.

Future work and considerations for the FE model proposed will be model validation with known forces transmitted to a head form. The FE model will need to be constructed to mimic the test conducted by Werner (Werner, 1975). Also the FE model needs a layer of CSF around the brain to predict when a significant risk of a subdural hematoma might occur.

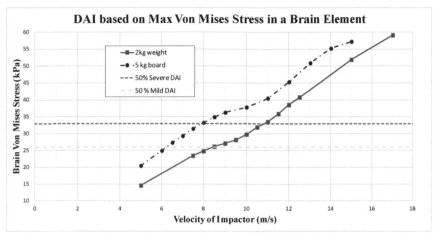

Figure 5. Based on injury tolerance limits proposed by Deck and Willinger

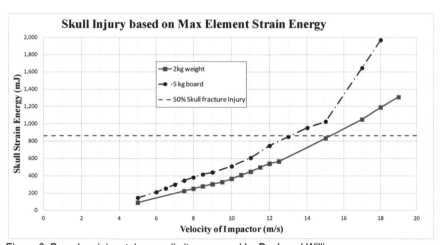

Figure 6. Based on injury tolerance limits proposed by Deck and Willinger

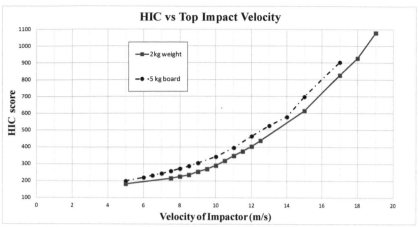

Figure 7. HIC scores from the various cases of top impact

REFERENCES

American National Standards Institute, Inc., 2009, *Z89.1 – 2009 American National Standard for Industrial Head Protection*, Arlington, Virginia: International Safety Equipment Association.

Canadian Playground Advisory Inc. "Risk of head injury and HIC scores."On-line. Available from Internet, http://www.playgroundadvisory.com, accessed 8July 2009.

Deck,C., and Willinger, R., "Improved head injury criteria based on head FE model." *International Journal of Crashworthiness* Volume 13 Issue 6 (Dec. 2008): 667-678.

Occupational Safety and Health Administration, 2003, *OSHA 3151-12R 2003 Personal Protective Equipment.*

Pinnojil, K., and Puneet M.. "Finite element modeling of helmeted head impact under frontal loading." *Sadhana* Volume 32 Part 4 (August 2007): 445–458.

Prasad, P., and Mertz, H.J. "The Position of the United States Delegation to the ISO Working Group on the Use of HIC in the Automotive Environment." *SAE Technical Paper No. 851246*, 1985.

Sabic Innovative Plastics, Material Data Sheets [online] Available at:
http://www.sabic-ip.com/gepapp/eng/datasheetinter/dswinter [Accessed 10 January 2012]

Werner, G., "Construction Helmet Response Under Severe Impact" Journal of the Construction Division Volume 101 Issue 2 (1975): 335-343.

Werner, G., "The State of Head Injury Biomechanics: Past, Present, and Future: Part1" *Critical Reviews in Biomedical Engineering* Volume 29 Issue 5 (2001): 441:600.

Willinger, R., Kang, H-S, and Diaw, B. "Three-Dimensional Human Head Finite-Element Model Validation Against Two Experimental Impacts." *Annals of Biomedical Engineering* Volume 27 Issue 3 (May 1999):403–410.

Yan, W. And Pangestu, O., "A modified human head model for the study of impact head injury." *Computer Methods in Biomechanics and Biomedical Engineering* Volume 14 Issue 12 (May 2011): 1049-1057.

CHAPTER 5

Stochastic Optimization Applications for Robotics and Human Modeling - A Literature Survey

Qiuling Zou, Jingzhou (James) Yang*, and Daan Liang

Texas Tech University, Lubbock, TX 79409, USA
james.yang@ttu.edu

ABSTRACT

In digital human modeling, optimization-based methods have been developed to simulate postures and motions due to the ability to predict posture and motion according to different criteria. In robotics area, optimization algorithms have also been used for path planning. Current available optimization methods in either human modeling or robotics are deterministic approaches. However, human anthropometry varies from one person to another and all parameters have uncertainties. Robotics has similar characteristics in terms of interaction with the environment. It is important to take account into these uncertainties in the optimization formulation. Stochastic optimization has been developed to consider uncertainties in the optimization problems. Reliability based design optimization (RBDO) and stochastic programming (SP) address the stochastic optimization problems. RBDO methods include most probable point (MPP)-based approaches and sampling-based approaches. Three policies can be found in the MPP-based approaches: nested RBDO, sing-looped RBDO, and sequential RBDO. For the sampling-based methods, factorial sampling, Monte Carlo sampling, importance sampling, and constraint boundary sampling (CBS) are developed. And some meta-models are often combined in the sampling: response surface models (RSM), kriging model, and artificial neural network (ANN). Three kinds of stochastic programming: stochastic programming with recourse, chance-constrained programming, and stochastic dynamic programming. Stochastic optimization has been applied in many fields in robotics and digital human modeling. This paper attempts to have a literature review on stochastic optimization applications in digital human modeling and robotics.

Key words: Uncertainty, Stochastic optimization, RBDO, robotics, digital human modeling

1.   INTRODUCTION

Deterministic optimization has gained great efforts in last decades. The key idea is to minimize objective functions while certain constraints must be satisfied. Considerable literatures can be found for both optimization algorithm and applications in robotics and digital human modeling (Tlalolini et al., 2010; Kim et al., 2009; Arnold et al., 2007; Ren et al., 2007, Marler and Arora, 2004).

In reality, uncertainties prevail in real world, such as in robotics, most engineering analysis and design optimization applications. Moreover, each person has different anthropometric parameters. These uncertain factors can impact the accuracy and reliability of results. Two types of uncertainties exist: inherent uncertainties and the uncertainties in modeling, decision making and simulation. For the first one, inherent difference in humans is easy to be understood. In robotics, manufacturing, assembling tolerances, and errors in the joint actuators and controllers are very difficult to avoid (Rao and Bhatti, 2001). For the second type of uncertainty, due to the random nature of some input data, although the boundaries are known exactly, but it is impossible to provide an exact values for variables, even a reasonable probability distribution function for human modeling. In addition, lack of enough information often leads to subjective uncertainties. Some errors are always with numerical models (Agarwal, 2004).

Stochastic optimization works to model the optimization problems under uncertainties, in which, uncertainties are often characterized with probability theory, generally. Stochastic optimization was first introduced as early as 1950's by George Dantzig, and then, it experienced a rapid development in algorithms. By now, stochastic optimization has been one of the most powerful tools for optimization problems under uncertainties. Generally, RBDO and stochastic programming address the uncertain design problems.

In RBDO approaches, objective functions are minimized with a reliability level for various sources of uncertainties. Two types of algorithms are included in this approach: MPP-based methods and sampling-based methods. The general MPP-based methods include nested RBDO (Hohenbichler and Rackwitz, 1986; Nikolaidis and Burdisso, 1988), single-loop RBDO (Liang et al., 2008) and sequential RBDO (Du and Chen, 2004). Based on the general algorithms, many customized approaches have been developed.

The research on stochastic programming often focuses on decision-making processes, in which, the uncertainty can be gradually reduced as more information is available. The goal is to find values of the initial decision variables and functions to update these variables when additional information about the problem evolves. Stochastic programming includes programming with recourse (Sahinidis, 2004), chance-constrained programming (Prékopa, 1995), and stochastic dynamic programming (Bellman, 1957).

Stochastic optimization has many applications in digital human modeling, for example, in motion control, object recognition, robot system design, dynamics and kinematics, etc. Comparing with the large amount of methods, the algorithms used in robotics are very limited by now.

The rest of the paper is organized as the follows: the second section will give an overview of the theories and algorithms of stochastic optimization; the third section is an introduction of the applications of stochastic optimization in robotics and digital human modeling; in the last section, a conclusion and future research work will be described.

2. STOCHASTIC OPTIMIZATION METHODS

2.1 Reliability based design optimization

RBDO is to obtain an optimal solution with a low probability of failure. A general formulation is:

$$\text{Minimize: } f(\mathbf{d})$$

$$\text{Subject to: } P[g(\mathbf{d}, \mathbf{x}) \geq \bar{z}] \leq \bar{p}_f$$

Where, the objective function f is a function of only the deterministic design variables $\mathbf{d}$, and the response function in the reliability constraint g is a function of $\mathbf{d}$ and $\mathbf{x}$. $\mathbf{x}$ is a vector of random variables defined by known probability distributions. The probability of failure p_f is defined as the probability that the response function $g(\mathbf{d}, \mathbf{x})$ exceeds some threshold value $\bar{z}$ and is calculated as:

$$p_f = \int\limits_{g>\bar{z}} \ldots \int h_{\mathbf{x}}(\mathbf{x})d\mathbf{x}$$

Where, $h_{\mathbf{x}}$ is the joint probability density function of the random variables $\mathbf{x}$. Generally, $h_{\mathbf{x}}$ is impossible to obtain and even available, evaluating the multiple integral is impractical. So, approximations of this integral are used. The common approximations include two kinds of approaches: MPP- based and sampling-based method.

MPP is defined as a particular point in the design space that is most probable to satisfy the limit-state function. In a standardized normal space, MPP is often on the limit-state that has the minimum distance to the origin. This distance is called reliability index β or safety, used to estimate the probability of system failure (Li et al., 2010). There are three types of MPP-based RBDO methods: a) nested RBDO.

The upper level optimization loop generally involves optimizing an object function subject to reliability constraints. The lower level optimization loop computes the probabilities of failure corresponding to the failure modes that govern the system failure. Reliability index approach (RIA) (Hohenbichler and Rackwitz, 1986) and performance measure approach (PMA) (Tu et al., 1999) are two examples of this method. b) Single-looped RBDO. This method simultaneously optimizes the objective function and searches for the MPP, satisfying the probabilistic constraints only at the optimal solution. First order reliability constraints are formulated to deterministic constraints with KKT conditions. c) Sequential RBDO. The deterministic optimization and reliability analysis is decoupled, and the iterative process is performed between optimization and uncertainty quantification by updating the optimization goals based on the latest probabilistic assessment results. Reliability estimation and deterministic optimization are performed sequentially. Table 1 lists the advantages and disadvantages for three kinds of MPP-based RBDO.

Table 1 Advantages and disadvantages for three kinds of MPP-based RBDO

MPP-based RBDO	Advantages	Disadvantages
Nested RBDO	Simple and direct	Possible to fail to locate MPP
Single loop RBDO	Far more efficient than nested RBDO	The accuracy depends on the accuracy of the first order assumption
Seqeuntial loop RBDO	More efficient than nested RBDO and more accurate than single-llo RBDO	Computationally expensive and inaccurate if approximation is poor

In order to improve the efficiency in reliability assessment, some advanced methods were developed, for example, sequential optimization and reliability assessment (SORA) (Liu et al., 2003), enhanced SORA (Cho and Lee, 2010), efficient global reliability analysis (EGRA) (Bichon, 2009).

Comparing with MPP-based methods, sampling-based methods are independent of the shape of limit state, which often leads to more accuracy. In this approaches, meta-models are commonly used in RBDO to reduce the computation effort.

Meta-models are often used to approximate the relationship between response and design variables within design space. There are local meta-models, such as kriging model (Choi et al., 2008), and global models, such as neural network (Rumelhart et al., 1986), and response surface model (Bucher and Bourgund, 1990).

Sampling points exert a direct impact on the quality of meta-models. Usually, three sampling methods are employed. The first is the space-filling sampling. Sampling points are distributed evenly in the design space. This is a direct way, but this concept cannot provide accurate optimum or satisfy constraints if highly nonlinear factors exist in objective functions and/or constraints (van Dam et al., 2007; Viana et al., 2009; Crombecq et al., 2011).

The second way is Monte Carlo sampling (MCS) (Metropolis and Ulam, 1949). The key idea is to apply numerical sampling technique to simulate a two-step process: find the random variables and determine if a particular event occurs for each simulation. With MCS, the cumulative distribution function (CDF) and the probability density function (PDF) of a system output will be built based on the sampling data. Among these sampling methods, MCS is often used. The accuracy of MCS does not depend on the dimension of the random input variables. And it is a reliable approach for reliability analysis for the problems which have multiple failure regions. However, the obvious weakness is intensive computation for any general case. To deal with this issue, Papadrakakis and Lagaros (2002) proposed the application of neural networks in the MCS.

The last method is the importance sampling. The importance sampling (Anderson, 1999) is an unbiased sampling method derived by modifying MCS methods to improve the computational efficiency. Random variables are sampled from different densities than originally defined. These densities are built to pick important values to improve the estimation of a statistical response of interest. This approach is attractive in some cases, for example, when more samples need to be generated in a failure region when estimating failure probability. However, the importance sampling cannot be implemented efficiently for a general framework to provide solutions for many classes of problems. Royset and Polak (2004) presented an implementable algorithm for stochastic programs using sample average approximations and discussed its applications. This method resulted in a better accuracy at the expense of expensive computation.

In addition, parallel asynchronous particle swarm optimization (Koh et al., 2006) has been developed to compensate the computational cost and improve accuracy. Genetic algorithm (Deb et al., 2009) and simulated annealing (Kirkpatrick et al., 1983) are used improve accuracy and efficiency. As a classical optimization technique, simulated annealing is performed globally, and it fits for a wide range of optimization problems, for example, the annealing stochastic approximation Monte Carlo algorithm (ASAMCA) (Liang, 2007) showed an outperformance in both training and test errors, and great computation cost at the same time.

2.2 Stochastic Programming

Stochastic programming often focuses on decision-making processes, in which uncertainties can be reduced gradually. Three main approaches have been developed:

The first is stochastic programming with recourse. It is a two-stage problem and focuses on the minimization of expected recourse costs. The key idea is that: in the first stage, before realizing the random variables ω, the decision variables x should be chosen to optimize the expected value of an objective function which depends on the optimal second stage objective function. In the second stage, variables are interpreted as corrective measures or recourse against any infeasibilities arising due to a particular realization of uncertainty. So the first-stage variables can be regarded as proactive and are often associated with planning issues; and second-stage decision variables as reactive and often associated with operating decisions. Stochastic linear programming (Kall and Wallace, 1994), integer programming (Dempster et al., 1981) and non-linear programming (Bastin, 2001) are based on this concept. The second is chance-constrained programming. Different from stochastic recourse programming, chance-constrained programming focuses on the reliability of a system. That is to say, the system's ability must satisfy feasibility in an uncertain condition. In this approach, there must be at least one certain probability constraints (chance constraints).

The last is stochastic dynamic programming (SDP). This is to deal with multi-stage decision processes. From the start, uncertainty was considered as an integral part of the dynamic environment. A discrete-time evolves over N time periods. For a time point k, we can describe a system with the current state (x_k), a control action (u_k) and a random parameter (w_k):

$$x_k = f_{k-1}(x_{k-1}, u_{k-1}, w_{k-1}), \quad k = 1, ..., N$$

Where, u_k is selected from a set of feasible actions based on the present state, $u_k \in U_k(x_k)$; the uncertainty w_k follows a distribution depending on the current state and control action: $P_{w_k}(x_k, u_k)$.

Finite element method with the parallel computing was used in stochastic dynamic programming (Chung et al, 1992), to improve the efficiency to the large-scale problems.

3. APPLICATIONS

Stochastic optimization has been applied in robotics and DHM. In this section, several applications are summarized.

3.1 Control

In robotics control, some stochastic optimization algorithms were used together with controlling method. The particle swarm optimization (PSO) with Gyro Feedback and phase resetting was developed to control the walking pattern of a humanoid robot (Faber and Behnke, 2007). This method improved the computation efficiency and as a result the walking speed was increased significantly.

Simulated annealing with central pattern generator (CPG) has several examples in motion control, such as walking gaits generation of a simple robot composed of lower limb (Itoh et al., 2004), head motion stabilization control (Costa et al., 2010), bipedal robot motion generation (Taki et al., 2004). The results proved the efficiency and the stability in walking.

Morgan (2008) focused on learning CPG controllers for 3- and 4- link swimming robots using a hybrid optimization algorithm combining simulated annealing and genetic algorithm. By this way, with or without a previous notion of correct swimming, the stochastic optimization could be performed effectively. But this research was limited to steady state swimming, without consideration of velocity and acceleration.

Brown and Whitney (1994) applied stochastic dynamic programming to plan robot grinding tasks. By selecting the volumetric removal and feed-speed for each pass, a grinding sequence planner was implemented. SDP is used to generate stage-by-stage control decisions based on a quasi-static model of grinding process. The process model verified to be sufficiently accurate for this purpose.

3.2   Motion planning

Monte-Carlo localization method for soccer robot self-localization was investigated (Li et al., 2008). The environment information, such as lines, goals, etc. was retrieved and processed, and then the position and pose of robot could be simulated by a group of different weighted samples. Experiments proved Monte-Carlo method a fast and stable approach of robot self-localization.

Simulated annealing technique was also used to perform stochastic optimization for grasping task (Bounab et al., 2010). Based on central-axis theory and friction cone linearization, FC condition was reformulated as a linear programming problem. Simulated annealing was employed for synthesizing suboptimal grasps. This method can explore the global surface and lead to a suboptimal result.

Marti and Qu (1998) formed the problem of adaptive trajectory planning for robots under stochastic uncertainty as a chance constrained problem. This problem was solved by means of spline approximation and neural network models. Through this way, the optimal control can be calculated in real-time.

3.3   Dynamics and Kinematics

Kastenmeier and Vesely (1996) combined stochastic methods and molecular simulation to solve the inverse kinematic problem, for redundant robots when meeting an obstacle. They borrowed SHAKE algorithm (Ryckaert et al., 1977) to compute the constrained kinematics, and performed path optimization with simulated annealing. The results proved it an efficient and flexible algorithm for the inverse kinematic problems.

Bowling et al. (2007) applied RBDO strategy in the prediction of the dynamic performance of a robotic system, along with the dynamic capability equations (DCE). PMA method was used to perform RBDO, which contributed to achieve the desired performance level for the robotic system design. Because the DCEs were at the core of the actuator selection problem, a more reliable final design could be generated when handling with the problem.

In working space optimization, Stan et al. (2008) used genetic algorithm in stochastic optimization for a six degree of freedom micro parallel robot. A number of solutions were allowed to be examined in a single design cycle, instead of searching point by point. Another advantage is its robustness and good convergence. Newkirk et al. (2008) also used DCE in conjunction with a RBDO strategy to predict the workspace of a robotic system.

3.4   Objects recognition

It is challenging to recognize the right object among multiple objects and to estimate the right pose. Stochastic optimization has been employed in this field as a promising solution, especially for high dimension human motion capture.

Choo and Fleet (2001) used hybrid Monte Carlo (HMC) to obtain samples in high dimensional spaces. In this method, multiple Markov chains using posterior gradients were adopted to improve the speed of exploring the state space. HMC sampling scales could fit the increase in the numbers of the degree of freedoms (DOFs) of models and the computation time would not increase significantly for high DOF models.

Kehl et al. (2005) presented a novel stochastic sampling approach for body pose tracking. From silhouettes in multiple video images, a volumetric reconstruction of a person was extracted. Then, an articulated body model is fitted to the data with stochastic meta descent optimization. This is a fast and robust tracking method, but its accuracy need to be improved.

Gall et al. (2007a) proposed global stochastic optimization for 3D human motion capture. This is also called interacting simulated annealing (ISA). And it is a general solution for markerless human motion capture. It comprised a quantitative error analysis comparing the approach with local optimization and particle filtering. Based on particle-based global optimization, Gall et al. (2007b) proposed a clustered stochastic optimization. The particles were divided into clusters and migrated to the most attractive cluster after detecting potential bounded subsets of the searching space. This method did not require any initial information about the position or orientation of the object.

3.5   Robot System Design

In robot system design, Shi (1994) developed a reliability analysis and synthesis procedure for robot manipulators. All the kinematic parameters of a manipulator were assumed as independent random variables with normal distribution. The reliability was defined as the probability of the end-effector position and orientation within a reliable region. It provides a novel way to compute optimal assignments of manufacturing tolerances to individual kinematic parameters which make sure the manipulator reliability while minimizing the overall cost.

3.6   Biomechanics

Simulated parallel annealing with a Neighborhood (SPAN) (Higginson et al., 2005) was provided for biomechanics community. This method consists of three loops: inner-most control loop making performance evaluations, search-radius loop executing the control loop several times before altering the size of search neighborhood, and temperature-reduction loop executing the search radius loop some times before reducing the temperature. This algorithm was implemented on two sites simultaneously. The communications in SPAN was accomplished through message passing interface. A simple parabola test and a complex pedaling simulation showed a large speed-up without compromising the heuristic of this algorithm.

4. CONCLUSION AND FUTURE WORK

Comparing with deterministic optimization, stochastic optimization problems contain uncertainties which are from inherent factors or from design model and simulations. From the literature, stochastic optimization obtained great development in theory and algorithm in last decades. Stochastic optimization has been applied in robotic control, motion planning, robot dynamics and kinematics, system design and object recognition. From the literatures, the methods applied in robotics are very limited in the piles of stochastic optimization methods.

Our future research will focus on how to apply stochastic optimization in DHM, including posture prediction and human dynamic simulation. The first step is to perform uncertainty analysis, in order to determine probability distributions about the uncertain parameters. Then, with the uncertainties, stochastic optimization will be performed. Accuracy of the optimum results and the efficiency in reliability analysis will be investigated. For a special human model, several stochastic optimization methods will be performed to compare the properties of each approach.

REFERENCES

Agarwal, H., 2004. "Reliability based design optimization: formulations and methodologies", PHD Dissertation, University of Notre Dame.
Anderson, E., 1999. "Monte Carlo methods and importance sampling", *Simulation*, 10:1-8.
Arnold, A., D. Thelen, M. Schwartz, F. Anderson, and S. Delp, 2007. "Muscular coordination of knee motion during the terminal-swing phase of normal gait", *Journal of Biomechanics*, 40(13): 3314-3324.
Bastin, F., 2001. "Nonlinear stochastic programming", Ph.D thesis, Department de Mathematique, Faculte des Sciences, Facultes Universitaires Notre-Damede la paix, Namur, Belgium.
Bellman, R., 1957. "Dyanmic Programming", Princeton, PA: Princeton University Press.
Bichon, B., S Mahadevan, and M. Eldred, "Reliability-based design optimization using efficient global reliability analysis", 50th AIAA/ASME/ASCE/AHS/ASC Structures, Structural dynamics, and Materials Conference, Palm Springs, California, USA, 2009.
Bounab, B., A. Labed, and D. Sidobre, 2010. "Stochastic optimization based approach for multifingered grasps synthesis", *Rototica*, 28:1021-1032.
Bowling, A., J. Renaud, J. Newkirk, and N. Patel, 2007. "Reliability-based design optimization of robotic system dynamic performance", *Journal of Mechanical Design*, 129(4): 449-455.
Brown, M., and D. Whitney, 1994. "Stochastic dynamic programming applied to planning of robot grinding tasks", *IEEE Transactions on Robotics and Automation*, 10(5):594-604.

Bucher, G., and U. Bourgund, 1990."A fast and efficient response surface approach for structural reliability problems", *Structural Safety*, 7(1):57-66.

Cho, T., and B. Lee, 2010. "Reliability-based design optimization using convex approximations and sequential optimization and reliability assessment method", *Journal of Mechanical Science and Technology*, 24:279-283.

Choi, K., G. Lee, S. Yoon, T. Lee, and D. Choi, "A sampling-based reliability-based design optimization using kriging metamodel with constraint boundary sampling", 12th AIAA/ISSMO Multidisciplinary Analysis and Optimization Conference, Victoria, British Columbia, Canada, 2008.

Choo, K., and D. Fleet, 2001. "People tracking using hybrid Monte Carlo filtering", *IEEE Computer Society*, 2:321-328

Chung, S., F. Hanson, and H. Xu, 1992, "Parallel stochastic dynamic programming: finite element methods", *Linear Algebra and its Applications*, 172:197-218.

Costa, L., A. Rocha, C. Santos, and M. Oliveira, "A global optimization stochastic algorithm for head motion stabiliazation during quadruped robot locmotion", 2nd International Conference on Engineering Optimization, Lisbon, Portugal, 2010.

Crombecq, K., E. Laermans, and T. Dhaene, 2011. "Efficient space-filling and non-collapsing sequential design strategies for simulation-based modeling", *European Journal of Operational Research*, 214:683-696.

Deb, K., S. Gupta, D. Daum, J. Branke, A. Mall, and D. Padmanabhan, 2009. "Handling uncertainties through reliability-based optimization using evolutionary algorithms", *IEEE Transactions on Evolutionary Computation*, 13(5):1054-1074.

Dempster, M., M. Fisher, L. Jansen, B. Lageweg, J. Lenstra, and A. Rinnooy Kan, 1981. "Analytical evaluation of hierarchical planning systems", *Operations Research*, 29:707-716.

Du, X., and W. Chen, 2004. "Sequential optimization and reliability assessment method for efficient probabilistic design", *Journal of Mechanical Design*, 126(2):225-233.

Elhami, N., M. Itmi, and R. Ellaia, 2011. "Reliability-based design and heuristic optimization MPSO-SA of structures", *Advanced materials Research*, 274:91-100.

Faber, F., and S. Behnke, "Stochastic optimization of bipedal walking using gyro feedback and phase resetting", IEEE-RAS 7th International Conference on Humanoid robots, Pittsburgh, USA, 2007.

Fouskakis, D., and D. Draper, 2002. "Stochastic optimization: a review", *International Statistical Review*, 70(3):315-349.

Gaivoronski, A., 1995. "A stochastic optimization approach for robot scheduling", *Annals of Operations Research*, 56:109-133.

Gall, J., B. Rosenhahn, T. Brox, and H. Seidel, 2010. "Optimization and filtering for human motion capture", *International Journal of Computer Vision*, 87(1-2):75-92.

Gall, J., T. Brox, B. Rosenhahn, and H. Seidel, 2007a. "Global stochastic optimization for robust and accurate human motion capture", Technical Report, MPI-I-2007-4-008, Max-Planck-Institut für Informatik, Germany.

Gall, J., B. Rosenhahn, and H. Seidel, 2007b. "Clustered stochastic optimization for object recognition and pose estimation", *29th Annual Symposium of the German association for Pattern Recognition*, 4713:32-41.

Higginson, J., R. Neptune, and F. Anderson, 2005. "Simulated parallel annealing within a neighborhood for optimization of biomechanical systems", *Journal of Biomechanics*, 38:1938-1942.

Hohenbichler, M., and R. Rackwitz, 1986. "Sensitivity and importance measures in structural reliability", *Civil Engineering Systems*, 3:203-209 .

Itoh, Y., T. K. Take, S. Kato, and H. Itoh, "A stochastic optimization method of CPG-based motion control for humanoid locomotion", Proceedings of the 2004 IEEE Conference on Robotics, Automation and Mecharonics, singapore, 2004.

Kall, P, and S. Wallace, 1994. "Stochastic programming", New York, NY: Wiley.

Kastenmeier, T., and F. Vesely, 1996. "Numerical robot kinematics based on stochastic and molecular simulation methods", *Robotica*, 14(3):329-337.

Kehl, r., M., Bray, and L. van Gool, 2005. "Full body tracking from multiple views using stochastic sampling", *IEEE Computer Society Conference on Computer Vision and Pattern Recognition (CVPR 2005)*, 2:129-136.

Kim, J., Y. Xiang, R. Bhatt, J. Yang, H. Chung, J. Arora, and K. Abdel-Malek 2009. "Generating effective whole-body motions of a human-like mechanism with efficient ZMP formulation", *International Journal of Robotics and Automation*, 24(2):125-136.

Kirkpatrick, S., C. Gelatt, and M. Vecchi, 1983. "Optimization by simulated annealing", *Science*, New Series, 220(4598):671-680.

Koh, B., A. George, R. Haftka, and B. Fregly, 2006. "Parallel asynchronous particle swarm optimization", *International Journal for Numerical Methods in Engineering*, 67(4):578-595.

Li, F., T. Wu, M. Hu, and J. Dong, 2010. "An accurate Penalty-based approach for reliability-based design optimization", *Research in Engineering Design*, 21:87-98.

Li, W., Y. Zhao, Y. Song, and Z. Yang, "A Monte-Carlo based stochastic approach of soccer robot self-localization", IEEE Conference on Human System Interactions, Krakow, Poland, 2008.

Liang, F., 2007. "annealing stochastic approximation Monte Carlo algorithm for neural network training", *Machine Learning*, 68:201-233.

Liang, H., "Possibility and evidence theory based design optimization: a survey", 2010Seventh International Conference on Fuzzy systems and Knowledge Discovery, 264-268, 2010.

Liang, J., Z. Mourelatos, and J. Tu, 2004. "A single-loop method for reliability-based design optimization", *International Journal of Product Development*, 5(1-2):76-92.

Liu, H., W. Chen, J. Sheng, and H. Gea, "Application of the sequential optimization and reliability assessment method to structural design problems", 2003 ASME Design Automation Conference, Chicago, IL, USA, 2003.

Marler, R., and J. Arora, 2004. "Survey of multi-objective optimization methods for engineering", *Structural Multidisciplinary Optimization*, 26:369-395.

Marti, K., and A. Aurnhammer, 2003. "Robust feedback control of robots using stochastic optimization", *Proceeding in Applied Mathematics and Mechanics*, 3:497-498.

Marti, K., and S. Qu, 1998. "Path planning for robots by stochastic optimization methods", *Journal of Intelligent and Robotic Systems*, 22:117-127.

Metropolis, N., and S. Ulam, 1949. "The Mont Carlo method", *Journal of the American Statistical Association*, 44(247):335-341.

Morgon, B., 2008. "Stochastic optimization for learning swimming robot gaits", Washington University Technical Notes, Washington University in St. Louis.

Newkirk, J., A. Bowling, and J. Renaud, "Workspace characterization of a robotic system using reliability-based design optimization", 2008 IEEE International Conference on Robotics and Automation, Pasadena, CA, USA, 2008.

Nikolaidis, E., and R. Burdisso, 1988. "Reliability based optimization: a safety index approach", *Computers and Structures*, 28(6):781-788.

Papadrakakis, M., and N. Lagaros, 2002. "Reliability-based structural optimization using neural networks and Monte Carlo simulation", *Computer Methods in Applied Mechanics and* Engineering, 191:3491-3507.

Prékopa, A., 1995. "Stochastic Programming", Dordrecht, The Netherlands: Kluwer Academic Publishers.

Rao, S., and P. Bhatti, 2001. "Probabilistic approach to manipulator kinematics and dynamics", *Reliability Engineering and System Safety*, 72:47-58.

Ren, L., R. Jones, and D. Howard, 2007. "Predictive modeling of human walking over a complete gait cycle*", Journal of Biomechanics*, 40(7):1567-1574.

Royset, J., and E. Polak, 2004. "Reliability-based optimal design using sample average approximations", *Probabilistic Engineering Mechanics*, 19:331-343.

Rumelhart, D., G. Hinton, and R. Williams, 1986. "Learning internal representations by error propagation", *Parallel Distributed Processing: Explorations in the Microstructures of Cognition*, 1:318-362.

Ryckaert, J; G. Ciccotti and H. Berendsen, 1977. "Numerical integration of the Cartesian equations of motion of a System with constraints: molecular dynamics of n-Alkanes". *Journal of Computational Physics*, 23(3):327–341.

Sahinidis, N., "Optimization under uncertainty: state-of-the-art and opportunities", *Computers and chemical Engineering*, 28, 971-983

Shi, Z., 1994, "Reliability analysis and synthesis of robot manipulators". Proceedings of Annual Reliability and Manitainability symposium, 201-205.

Stan, S., V. Maties, and R. Balan, "Stochastic optimization method for optimized workspace of a six degree of freedom micro parallel robot", 2008 Second IEEE International Conference on Digital Ecosystems and Technologies, 502-507, 2008

Tadoloro, Y., T. Ishikawa, T. Takebe, and T. Takamori, "Stochastic prediction of human motion and control or robots in the service of humans", IEEE International Conference on Systems, Man and Cybernetics, Le Touquet, France, 1993.

Taki, K., Y. Itoh, S. Kato, and H. Itoh, 2004. "Motion generation for bipedal robot using neuro-musculo-skeletal model and simulated annealing", Proceedings of the 2004 IEEE conference on robotics, Automation and Mechatronics, Singapore, 699-704

Tlalolini, D., Y. Aoustin, and C. Chevallereau, 2010. "Design of a walking cyclic gait with single support phases and impacts for the loco-motor system of a thirteen-link 3D biped using the parametric optimization", *Multibody System Dynamics*, 23:33-56.

Tu, J., K. Choi, and Y. Park, 1999. "A new study on reliability-based design optimization", *Journal of Mechanical Design*, 121(4):557-564.

van Dam, E., B. Husslage, D. den Hertog, and H. Melissen, 2007. "Maximin latin hypercube design in two dimensions", *Operations Research*, 55:158-169.

van den Berg, J., P. Abbeel, and K. Goldberg, 2011. "LQG-MP:optimizaed path planning for robots with motion uncertainty and imperfect state information", *The international Journal of Robotics Research*, 30(7):895-913.

Viana, f., G. Venter, and V. Balabanov, 2009. "an algorithm for fast optimal latin hypercube design of experiments", *International Journal for Numberical Methods in engineering*, 82:135-156.

Section II

Problem Solving Applications

Service Management System Based on Computational Customer Models Using Large-scale Log Data of Chain Stores

Hitoshi KOSHIBA, * *Takeshi TAKENAKA, Yoichi MOTOMURA*

National Institute of Advanced Industrial Science and Technology
Tokyo, JAPAN
*hitoshi.koshiba@aist.go.jp

ABSTRACT

Human-related factors such as customer behavior, employee skills, managerial decision-making, and social interactions are receiving greater attention for their relevance to the improvement of existing services and design of new services. This paper discusses a service management system that employs computational customer models constructed from large-scale log data of retail or restaurant chains. It introduces a demand-forecasting approach using a probabilistic model and large-scale customer log data. Then, it introduces a management support system for service industries on this basis.

Keywords: management support system, services, large-scale data, user model

1   INTRODUCTION

Over the past few decades, a considerable number of studies have been conducted for the improvement of business efficiency, in the fields of operations research, decision-making, game theory, behavioral economics, and others. In the context of service industries today, cost reduction and risk prediction are important issues. In the restaurant industry, for example, great difficulty arises in the prediction of customer behaviors because many unspecified customers visit irregularly and some customers will never come again if they don't like the restaurant (Takenaka, 2011b). In other words, it is very difficult to determine customer needs or satisfactions using simple purchase data. Therefore, modeling of customer behaviors considering environmental factors should an important aspect of reducing those risks.

This paper discusses a customer modeling method using large-scale purchase data without identification acquired from a restaurant chain. It introduces some modeling technologies for customer behaviors that have been developed in service engineering and proposes an integrated method embedding those technologies as a practical support system.

Then, it introduces the challenge of embedding those technologies in the real service field. In actual services, because a great part of managerial decision making depends on managers' experience and intuitions, we must know their decision-making processes in their routine practice. Because a prominent technology is not always suitable for practical use, we have to pay attention to the design of a useful system in the field. Therefore, we must think about both modeling methodology and system development.

2   COMPUTATIONAL CUSTOMER MODELING USING LARGE-SCALE LOG DATA

When we construct customer models, there should be two approaches. One is a direct approach of interviewing customers and asking about their purpose, motivation, situation, context, etc. This approach can achieve more accurate results than the other method, but it is inefficient, coming at high time cost to employees and customers. The other approach is indirect. It involves making customer models using log data. This approach is useful because it use existing business data (i.e., purchase data, survey data, and behavioral data of customer, as well as possibly other sources.).

In this paper, we use Probabilistic Latent Semantics Indexing (PLSI) for customer classification on the basis of large-scale purchase data. PLSI is similar to the method of LSI, which is used as a topic classification method in Natural Language Processing (NLP). The LSI method's basic concept is simple: similar topics use similar words, and one document includes one topic. Currently, LSI/PLSI is used in various fields. For example (Ishigaki, 2010), one approach that has been tried is PLSI-based customer-behavior modeling using ID-POS data. It is possible that PLSI can make computational customer models from large-scale log (i.e., simple purchase) data.

Herein, we consider a dataset used for computational customer modeling. We use simple purchase data provided by a Japanese restaurant chain (Ganko Food Service Co., Ltd.). The data was gathered from September 2, 2007 to December 6, 2011, from transactions at 44 stores in Japan. The dataset contains 34,423,914 transactions, including 7,395 item IDs and 6,919,664 receipt IDs. Although purchase data doesn't include customer IDs, it records the date and time of a given purchase transaction, the name and price of the purchased item, the store where the item was purchased, and other information; each item is given two hierarchical code, one large and one small. For example, tofu salad is assigned the following two codes: "Food" as a large classification and "Salad" as a small classification. The head office of this restaurant chain generates these hierarchical classifications manually.

In this case, the co-matrix for PLSI is a 7,395 x 6,919,664 matrix, which is large and supers. Thus, data categorization is difficult. So we try compels item data using hierarchical small codes. We remove some categories and items (e.g., the category of "Soft drink" includes a lot of complimentary drink items (service items for set, course, or party menus), and receive a lot of orders. Similarly, some categories are only on the kids' menu, but the kids' menu is not ordered without accompanying orders from the regular menu) And we summarize some categories; for example, "Sushi" has several types, ("Nigiri," "Mori," "Maki," "Chirashi," etc.), which we summarized as "sushi." Thus, we reduced the original 40 categories to 24. See Table 1.

In addition, we classified receipt data by situation: Day-Normal (Day_Nrm, 2,436,786 receipts), Day-Reserve (Day_Rsv, 284,433 receipts), Night-Normal (Night_Nrm, 4,060,444 receipts), Night-Reserve (Night_Rsv, 138,001 receipts).

Table 1 Item Categories for Customer Modeling using PLSI

Lunch set	Sushi	Fried	Vinegared	Beer
Other set	Nabe	Broiled	Soup	Sake
Traditional Japanese set	Rice	Sautéed	Salad	Shochu
Course menu	Noodle	Boiled	Dessert	Other alcohol
Party menu	Sashimi	Steamed	Appetizer	

We classified customers into 19 categories. In Figure 1, "Night_Nrm: Drink with à la carte" category items and probability are as follows: Beer 16.6%, Fried 15.7%, Broiled 11.9%, Sashimi 11.6%, Japanese distilled spirit 11.3%, etc., while "Day_Nrm: Lunch set" is 100.0% Lunch set.

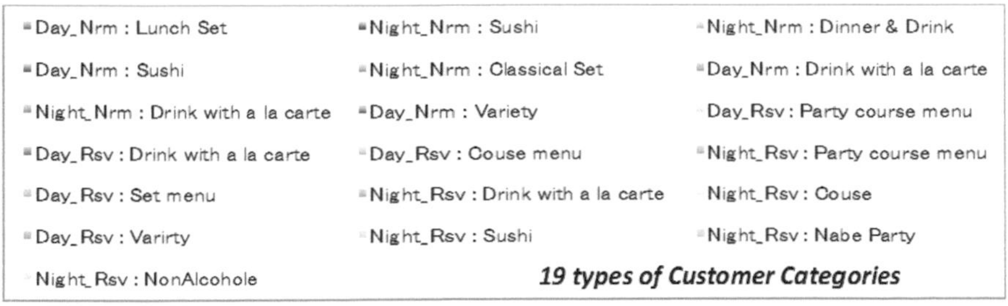

Figure 1 Results of Customer Categorization

Category component percentages and average spending per customer are different from location to location, as shown in Figure 2. That suggests that this customer model can draw out characteristics specific to specific shops.

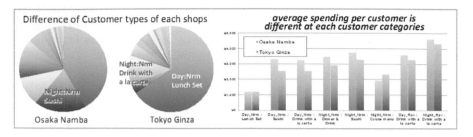

Figure 2 Differences in Customer Category in Each Location

Next, we chose a large location and tried to forecast demand for each category and bundle of categories using large-scale log data such as purchase data and behavioral data for customers. Details of our demand-forecasting methods and techniques are given in Takenaka (2011a). For the present analysis, we made a model using purchase data to forecast demand from October 1, 2009 to November 31, 2010 and October 1 to October 30, 2010.

The results are shown in Figure 3. The bundle of all categories indicates a higher recall ratio, and some categories indicate similar recall ratios. This result suggests that our customer model can be of real service.

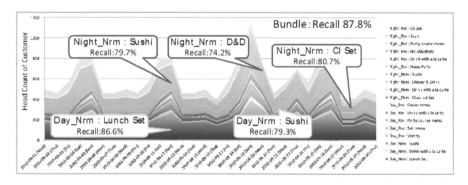

Figure 3 Sales Forecasting for Each Category at the Case Location

3   A MANAGEMENT SUPPORT SYSTEM, "APOSTOOL": DEVELOPMENT AND IMPLEMENTATION

In this section we develop a customer-clustering method based on PLSI, sales estimation using various datasets, etc. These methods can call on WEB-API. And some group functions divide to components. The system framework is shown in Figure 4.

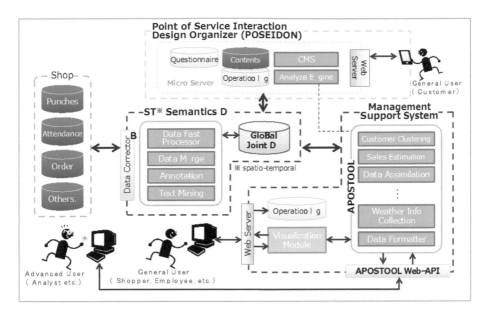

Figure 4 The system framework of Management Support System (APOSTOOL) and Related Components

The system has three components, as follows; 1. A service-log data collection and integration component. We call that the "Spatio-Temporal Semantics Database." 2. A data analysis toolkit and visualization. We call that the "Management Support System," and it is explained in this paper. 3. A point-of-service interaction design organizer. We call that "POSEIDON."

Service-log data collection and integration component: This component collects and integrates data from several databases including some SQL DBMSes. For example, purchase data and customer questionnaire data could be joined to client IDs to analyze customer attitudes and behavior, POS data could be summarized by shop and time, etc. This is an important component of system imprecation, because database schema already in use are different in each location. Therefore, our system (which includes a management support system) has to support these various environments. This component allows our system to adapt.

Data analysis toolkit and visualization component: This component includes an analysis toolkit for customer models, PLSI toolkit for customer categorization, Bayesian network (BN) toolkit for customer-behavior causal-structure analysis. These toolkits, which we collectively call "APOSTOOL," can be used with web API, so expert users can make a custom system themselves. For other users, we provide a basic visualization system based on a GUI, explained in this paper. Since system developed by researchers may not always meet the practical needs of employees and customers, we take a co-creation approach, combining field interviews with behavioral observation of employees and a lifestyle-survey method to elucidate the customer decision-making process, followed by agile development and trial.

Point of service interaction design organizer component: This is like a questionnaire system for the customer at the point of service. The details are reported in the other paper (Takenaka, 2012).

In light of the wide range of research in this area, it is worth remembering the valley of death and Darwinian Sea metaphors: laboratory results cannot be implemented directly in the real world. Methodologies such as human-centered design and user-centered design reflect users' needs, improve usability, and facilitate modification by means of custom design. In addition, services and especially restaurants are a difficult field depending heavily on contexts such as the people involved, the type of operation, the equipment used, and the environment.

In the present case, we try to develop a stakeholder analysis–based system using interviews and agile software development. We identified two kind of user at this restaurant chain: regional managers and front-line managers. We have interviewed these managers (as shown in Figure 5) and identified their main needs, which are order in the workplace and lack of absenteeism. They are also interested in our research findings regarding the characteristics of their restaurants.

Figure 5 Interviews with stakeholders

We customized the management support system's user interface and functions using APOSTOOL with WEB-API to satisfy these needs in a way congruent with the plan-do-check-act (PDCA) cycle. We suggested four functions, as follows: 1. Indicate actual sales information (e.g., sales, customer headcount), and compare it with 1-2 years ago and with results for other locations. 2. Run demand-forecasting and order-/work attendance-support functions using the forecasted results. 3. Indicate function of a shop's busyness by time (15-minute units) and day of the week. 4. Employ the journal-input function.

The managers (one regional manager and three frontline managers) used this system in their jobs, as shown in Figure 6.

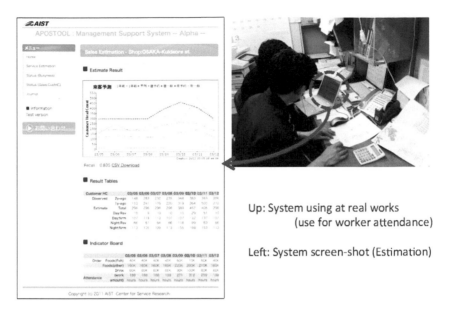

Up: System using at real works
(use for worker attendance)

Left: System screen-shot (Estimation)

Figure 6 Use of APOSTOOL in the workplace

4   CONCLUSION

In this paper, we develop a service-management support system and suggest way that it can be applied and further improved. The service industry depends heavily on uncertain contexts such as the person involved, the type of operation, the equipment used, and the environment.

We constructed a methods for a computational customer model, demand forecasting, etc., developed a support system and applied it in a real workplace. Our "develop/improve" approach is human-/community-based, and needs to be further supported based on tenacious interview-based information-gathering and agile PDCA cycling.

The system is already being used by managers in their real jobs. In the future, we will analyze the system log data and improve the customer model and the system. In particular, we will expand the use of Bayesian networks. In addition, we must synthesize data from managers, customers, and employees, for understanding services from all relevant perspectives.

ACKNOWLEDGMENTS

This paper is supported by projects on service engineering from the Japanese Ministry of Economy, Trade, and Industry (2009-2012). The authors also appreciate the contribution of Ganko Food Service Co., Ltd, which kindly made sites and data available for our research.

REFERENCES

Ishigaki, T., Takenaka, T., and Motomura, Y. (2010). Customer-item category based knowledge discovery support system and its application to department store service. *In Proceedings of the 2010 IEEE Asia-Pacific Services Computing Conference*, pages 371-377, Washington, DC: IEEE Computer Society.

Takenaka, T., Ishigaki, T., Motomura, Y., and Shimmura, T. (2011a). Practical and interactive demand forecasting method for retail and restaurant services. *In Proceedings of the International Conference on Advances in Production Management Systems (APMS2011)*, Stavanger: APMS Organizing Committee.

Takenaka, T., Koshiba, H., and Motomura, Y. (2012). Computational modeling of real services toward co-creative society. *In Proceedings of the 2nd International Conference on Applied Digital Human Modeling (ADHM2012)*, San Francisco:

Takenaka, T., Shimmura, T., Ishigaki, T., Motomura, Y., and Ohura, S. (2011b). Process management in restaurant service: A case study of Japanese restaurant chain. *In Proceedings of the International Symposium on Scheduling 2011*, Osaka: JSME.

CHAPTER 7

Development of Database on Daily Living Activities for Realistic Biomechanical Simulation

Hiroyuki Kakara[1,2], Yoshifumi Nishida[2], Sang Min Yoon[3]

Yusuke Miyazaki[4], Hiroshi Mizoguchi[1,2], Tatsuhiro Yamanaka[5]

hiroyuki.1842.kakara@aist.go.jp, y.nishida@aist.go.jp, sangmin.yoon@yonsei.ac.kr
y-miyazaki@t.kanazawa-u.ac.jp, hm@rs.noda.tus.ac.jp, tatsuhiro-yamanaka@nifty.com

[1]Department of Mechanical Engineering, Faculty of Science and Technology
Chiba, Japan

[2]Digital Human Research Center
National Institute of Advanced Industrial Science and Technology
Tokyo, Japan

[3]Yonsei Institute of Convergence Technology, Yonsei University
Incheon, Korea.

[4]Department of Mechanical Science and Engineering, Kanazawa University
Ishikawa, Japan

[5]Ryokuen Children's Clinic
Kanagawa, Japan

ABSTRACT

This paper describes the development of the fall database for a biomechanical simulation. First, data of children's daily activities were collected at a sensor home, which is a mock daily living space. The sensor home comprises a video-surveillance system embedded into a daily-living environment and a wearable acceleration-gyro sensor. Then, falls were detected from sensor data using a fall detection algorithm developed by authors, and videos of detected falls were extracted from long-time recorded video. The extracted videos were used for fall motion analysis. A new computer vision (CV) algorithm was developed to automate fall motion analysis. Using the developed CV algorithm, fall motion data were accumulated into a database. The developed database allows a user to perform conditional searches of fall data by inputting search conditions, such as a child's attributes and fall situation. Finally, a biomechanical simulation of falls was conducted with initial conditions set using the database.

Keywords: Childhood Injury Prevention, Fall Database, Biomechanical Simulation

1   INTRODUCTION

Unintentional injuries are the top and second cause of death in Japan among children aged 1 to 19 years old according to the 2010 Vital Statistics of Japan (MHLW, 2010). The number of unintentional injuries can be decreased from engineering approach. One promising approach consists of 1) collecting data of real living activities, 2) conduct the realistic biomechanical simulation by using real data to analyze injury risk in daily living spaces (Miyazaki, 2008)(Miyazaki 2009), 3) provide information of injury risk to improve environments or cause behavioral changes. Until now, there were no data of real dynamics data on living activities, therefore we couldn't conduct the simulation realistically.

This paper describes the development of the database on childhood daily living activities, especially we focus on fall motion, because fall is a leading cause of unintentional injuries (AIST, 2010) and the cost of medical care due to injuries (WHO, 2008). Finally, we will show an example of application of the database to the simulation.

2   DEVELOPMENT OF THE LIVING ACTIVITIES MEASURING SYSTEM AND THE MEASUREMENT

2.1   The behavior measuring system

We developed the behavior measuring system (shown in Fig. 1) in the ordinary apartment to obtain unintentional fall data. The system consists of 12 video cameras and the Bluetooth-compatible acceleration-gyro sensor attached to the child's body trunk. Video cameras and a Bluetooth adapter are connected to the PC in the monitoring room. Sensor data and video data are synchronized by adding frame numbers to the end of sensor data at the moment.

We conducted measurements under following conditions:

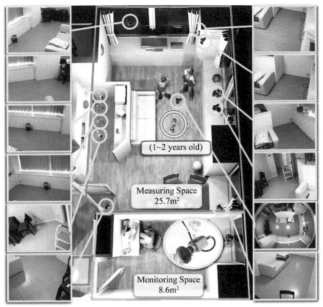

Figure 1 Configuration of the daily activities measuring system

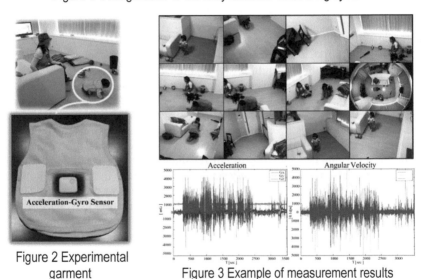

Figure 2 Experimental garment

Figure 3 Example of measurement results

1) Age of the subject: approximately 1 ~ 4 years old
2) Measurement time: approximately 60 minutes
3) Subject number per a measurement: maximum 2 person

The age group of children was selected considering that children at this age fall frequently. The measurement time was set in consideration of child's fatigue. All parents signed a consent form for the recording and release of data prior to the experiment. The height and weight of each child (i.e., database search conditions) were first measured and then each participant was dressed in the experimental garment shown in Fig. 2. To make wearing the sensor less cumbersome for the children, the sensor was placed in the back pocket of the garment.

In the measuring space, there were many toys to stimulate lively behavior. During the measurement, we never ask children to fall intentionally but ask them to play with their accompanying person as usual with using those toys.

So far, we've conducted measurements for 19 children between 11 and 50 months of age, and measured 103 falls. Fig. 3 is an example of the measuring result.

3   FALL VIDEO EXTRACTION

We developed the fall detection algorithm shown in Fig. 2, to extract fall videos from long-duration videos automatically.

Rapid changes in acceleration or angular velocity occur during falls. Particularly, Gy (acceleration of a body's lengthwise direction), Gz (acceleration of a body's anterior-posterior direction), and ωx (angular velocity about body's abscissa axis) show fall feature more obviously. Our measuring results indicated that fall features distribution among these three data varied substantively depending on the development stage of the child. For example, the fall features of toddling children tend to appear in Gy because they tend to fall assuming a seating action. On the other hand, fall features of more developed children tend to appear in Gy or ωx because they tend to fall forward. Thus, these three data (Gy, Gz, and ωx) should be considered for detection of falls by both toddling and active children. We developed the algorithm that fulfills this requirement.

Earlier studies have developed fall detection algorithms (Bourke, 2006)(Zhang, 2006)(Hazelhoff, 2008). However they can't be applied to this case because of following reasons. First, child's fall postures differs from adult's ones largely, so it is difficult to detect falls by a simple threshold processing of acceleration data. Fall features don't appear at the completely same instant in sensor data. To deal with this point, the algorithm developed by us utilizes integration of acceleration and angular velocity. Second, child's body is very small and its figure is quite different from that of adult, so it is difficult to detect falls using only image processing. Thus, we developed a new fall detection algorithm using the acceleration-gyro sensor.

As the first step shown in Fig. 5, the algorithm calculates dimensionless Gy, Gz, and ωx by dividing absolute values by sensor maximum values (Gy$_{max}$, Gz$_{max}$, and ωx$_{max}$). Each of the sensor maximum values are shown in Fig. 5.

Next, the algorithm adds three dimensionless values in order to calculate the comprehensive scale of F. Moreover, considering that there is a maximum time lag of 200 [msec] among the three values, the algorithm integrates 40 data (for 240 [msec]) and adds the integration to the comprehensive scale. In order to normalize the comprehensive, the algorithm divides F by "3N". The "3" indicates the added data number (Gy, Gz, and ωx), and "N" indicates the integrated data number (40). As a result of this operation, the maximum value of F becomes 1.

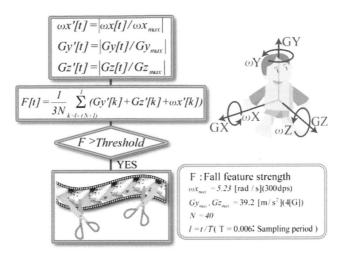

$$\omega x'[t] = |\omega x[t] / \omega x_{max}|$$
$$Gy'[t] = |Gy[t] / Gy_{max}|$$
$$Gz'[t] = |Gz[t] / Gz_{max}|$$

$$F[t] = \frac{1}{3N} \sum_{k=l-(N-1)}^{l} (Gy'[k] + Gz'[k] + \omega x'[k])$$

$F > Threshold$

YES

F : Fall feature strength
$\omega x_{max} = 5.23$ [rad / s](300dps)
$Gy_{max} , Gz_{max} = 39.2$ [m/s²](4[G])
$N = 40$
$l = t / T$ (T = 0.006: Sampling period)

Figure 4 Fall detection algorithm

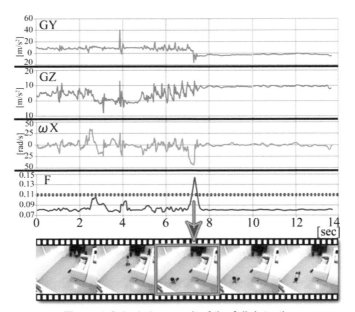

Figure 5 Calculation result of the fall detection

Finally, the algorithm performs threshold processing. If F exceeds a given threshold, the system determines that a fall has occurred and extracts fall videos automatically by referring to the frame numbers data added after the sensor data. The example of calculation result is shown in Fig. 3.

70

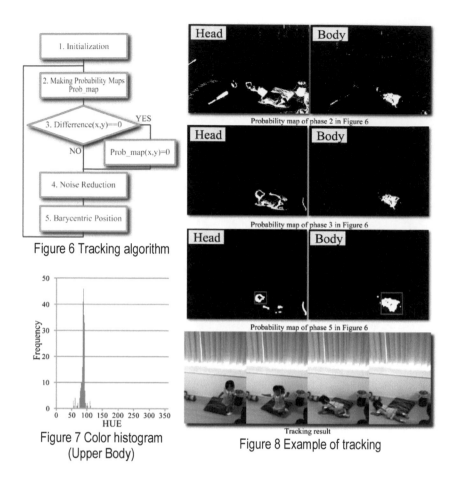

Figure 6 Tracking algorithm

Probability map of phase 2 in Figure 6

Probability map of phase 3 in Figure 6

Probability map of phase 5 in Figure 6

Tracking result

Figure 7 Color histogram
(Upper Body)

Figure 8 Example of tracking

4   FALL MOTION ANALYSIS

Speeding up a motion analysis process is required in order to develop the large database. So, we developed the system which extracts body trunk's motion automatically using an image processing technology. In our previous report (Kakara, 2011), we extracted that manually by clicking the corresponding point of the stereo vision. The developed tracking system doesn't need these manual procedures. It tracks head and upper body with no markers by using the algorithm based on color and motion information as shown in Fig. 6. Some examples of image processing are shown in Fig. 8.

Initialization We specify a head area and an upper body area in the first frame of the fall video by clicking those areas, to create histograms as shown in Fig. 7. The system creates luminance histogram of the head area to utilize the fact that the head area is characterized by lower luminance, and color histogram of upper body area that is characterized by green.

Creating probability maps The system calculates the event probability map of color or luminance using a current frame and the quantized color histograms. In the top image of Fig. 8, a white area indicates high similarly which means that the area has the color or luminance similar to head or upper body.

Similarity measure Our proposed methodology measures the similarity between previous frame and current frame in each pixel. If the difference becomes 0, the system determines that there is no moving object in that pixel and the corresponding pixel in the probability map is changed to 0 (the second image of Fig. 8).

Noise reduction The system converts the probability map to binary image, and expands white pixels by morphological operation to reduce noise.

Barycentric position calculation The system calculates the barycentric position of white pixels inside the searching region*. And the system recognizes this barycentric position as a tracking target's position.

* The search region is determined by the number of pixels within a square, $5/3 \times \sqrt{count}$ on a side, and its center is a tracking target's position in the previous frame ("*count*" means a total count of white pixels in the previous frame).

5   THE DATABASE ON DAILY LIVING ACTIVITIES

We developed the database on fall motion. and includes the following information: 1) fall dynamics data, 2) children's attributes such as age, height, weight, and 3) fall situation such as landing body site, action immediately before the fall, and cause of fall. The first two sets of data are used as search conditions. Furthermore, we developed the database browser shown in Fig. 9. We explain the browser below.

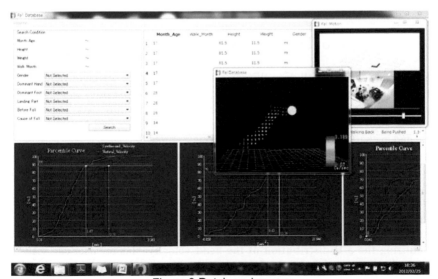

Figure 9 Database browser

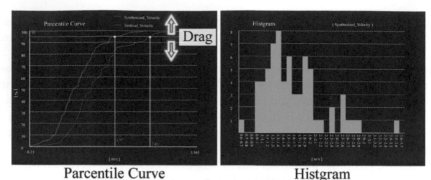

Parcentile Curve   Histgram

Figure 10 Probabilistic distributions

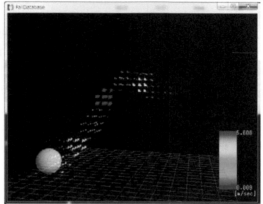

Figure 11 Browser for 3-D distribution of head velocity

5.1   Fuctions of the database browser

Conditional search of fall data We can conduct conditional search of fall data. We can set conditions in the left of the window. Searching conditions include child's attribution (age, height, weight, etc.) and fall situation (landing part, cause of fall, action immediately before the fall). The corresponding data will be shown in the table and in charts in the bottom of the window.

Visualizing the probabilistic distribution This browser can visualize probabilistic distributions of fall dynamics and situations in the bottom of the window. Charts are updated automatically after the conditional search. This browser can visualize both histograms and percentile curves as shown in Fig. 10, in order to enable the user to conduct not only worst-case analysis, but also analysis based on percentile or histogram.

Visualizing the 3-D distribution of the head velocity This browser can visualize the 3-D distribution of the head velocity by selecting a fall in the right-hand-table as shown in Fig. 11. Specifically, it divides space surrounding a child into a reticular pattern, and visualizes velocities in each cell. Velocities are described by circular cones. The cone's direction indicates the velocity vector, and color and size indicates the scalar velocity.

Displaying individual fall videos and motion analysis videos This browser can show the fall video and the resultant video of the motion analysis by selecting a cell in the right-hand table.

The database and its browser are opened to the public from following: https://www.dh.aist.go.jp/projects/kd/cocreation2011.php. Currently, we deal with only fall data. But data on many other activities have been obtained. So we will include those into the database and develop the database on daily living activities.

5.2 Application for the biomachanical simulation

Floor Material	Concrete	Wood	Cushion	Shock absorbing
Maximum principal strain [10^{-3}]	4.66	4.14	2.70	1.48
Event probability of cranial bone fracture	12%	10%	5%	2%

Table 1 Event probability of cranial bone fracture

This database has already been used in the biomechanical simulation. We extracted the maximum velocity among hand landing falls from the database, and used it as an initial condition of the simulation. Results are shown in table 1.

5 CONCLUSION

This paper describes the development of the database on daily living activities and the biomechanical simulation. We developed the daily activities measuring system and conducted measurement. Fall videos were extracted automatically by fall detection algorithm developed by authors. Next, we developed the system which extracts body trunk's motion automatically using an image processing technology. Finally, we developed the database and its browsing software, and conducted the realistic biomechanical simulation using the database. We will develop the database on whole daily activities in the future.

ACKNOWLEDGMENTS

This research was supported in part by the "Program for Supporting Kids Design Product Development" of the METI (The Mnistry of Economy, Trade, and Industry), Japan, and the MKE (The Ministry of Knowledge Economy), Korea, under the "IT Consilience Creative Program" support program supervised by the

NIPA(National IT Industry Promotion Agency) (NIPA-2010-C1515-1001-0001).

REFERENCES

Ministry of Health, Labour and Welfare (MHLW), "The 2010 Vital Statistics of Japan", 2011

Y. Miyazaki, Y. Murai, Y. Nishida, T. Yamanaka, M. Mochimaru and M. Kouchi. "Head Injury Analysis In Case of Fall from Playground Equipment Using Child Fall Simulator"; The impact of Technology on Sport, Vol. III, pp.417-421, 2009

Y. Miyazaki, Y. Nishida, Y. Motomura, T. Yamanaka and I. Kakefuda. "Computer Simulation of Childhood Head Injury Due To Fall From Playground Equipment"; The 2nd Asia Pacific Injury Prevention Conference, November 2008

National Institute of Advanced Industrial Science and Technology (AIST). "Project of Building a Safety-Knowledge-Recycling-Based Society 2009"

World Health Organization (WHO). "World Report on Child Injury Prevention," (edited by M. Peden, K. Oyegbite, J. Ozanne-Smith, A. A. Hyder, C. Branche, A. F. Rahman, F. Rivara and K. Bartolomeos). 2008

A. K. Bourke, J. V. O'Brien and G. M. Lyons. "Evaluation of a threshold-based tri-axial accelerometer fall detection algorithm"; Gait and Posture, Vol. 26 Issue 2, pp.194-199, November 2006

T. Zhang, J. Wang, P. Liu and J. Hou. "Fall Detectionby Embedding an Accelerometer in Cellphone and Using KFD Algorithm"; IJCSNS International Journal of Computer Science and Network Security, Vol. 6 No. 10, pp. 277-284, October 2006

L. Hazelhoff, J. Han and P. H. N. de With. "Video-Based Fall Detection in the Home Using Principal Component Analysis"; ACIVS 2008, pp298-309, October 2008

H. Kakara, Y. Nishida, Y. Miyazaki, H. Mizoguchi, T. Yamanaka, "Development of Database of Fall Dynamics Probabilistic Distribution for Biomechanical Simulation," The Proc. of The 2011 International Conference on Modeling, Simulation and Visualization Methods (MSV'11), pp. 82-88, July 19 2011 (Las Vegas)

CHAPTER 8

Social System Design Using Models of Social and Human Mind Networks- CVCA, WCA and Bayesian Network Modeling

T. Maeno[1]*, T. Yasui[2], S. Shirasaka[1] and O. Bosch[3]
[1]Graduate School of System Design and Management, Keio University
[2]Keio Advanced Research Center, Keio University, Yokohama, Japan
[3]University of Adelaide Business School, Adelaide, Australia
E-mail*: maeno@sdm.keio.ac.jp

ABSTRACT

Examples of CVCA, WCA and Bayesian network modeling are shown in which human minds (mental models) and behavior are modeled. First, the fundamental idea of CVCA and WCA are explained. This is followed by an example of using CVCA and WCA for modeling cause-related marketing such as the "Drink one, give ten campaign" by Volvic. A student workshop has been used to demonstrate the effectiveness of WCA to visualize and understand the needs and wants of humans. Second, an example of how Bayesian Network Modeling can be used effectively through a participatory process to design policies for child safety. Both results (CVCA/WCA and the Bayesian Network Modeling) have shown that these are useful tools for workshop-based education of service/products system design.

Keywords: system design, wants chain analysis, Bayesian network

1 INTRODUCTION

Innovation has been moved from technology-oriented to human-needs-oriented. Hence, human-centered product and service system design are becoming increasingly important. For example, causal-loop diagrams are often used in system thinking, analyzing the relationships among behaviors of humans. However, human models are often not sufficient to fully understand the relationships and interactions involved.

First, humans are only thought to be input-output systems for developing products/service systems. It has become important in future system design, especially those that are human-centered, that human minds and behavior models will become more and more precise and reliable. For this reason CVCA (customer value chain analysis) and WCA (wants chain analysis) were used to analyze the minds and behavior of humans, with a special focus on wants and needs. In the development of products and services these analyses are useful for qualitative visualization of the wants and needs of humans.

Causal-loop diagrams are also highly useful in describing the relationships and interactions between components of the system. However, for these models to be more precise and to convert them into dynamic models, quantitative information and data are required. This is often very difficult in the social sciences, especially when the components that are involved include human behaviors, feelings, ethical considerations, cultural differences, etc. Bayesian network modeling has been found highly useful for the analysis of the minds of humans and their different behaviors.

In this paper, CVCA, WCA and Bayesian network modeling are utilized to explore their effectiveness for social system design.

2 CVCA AND WCA FOR SOCIAL SYSTEMS

2.1 CVCA/WCA

Customer Value Chain Analysis (CVCA) is one of the methods developed by Donaldson as a tool for engineering design (Donaldson, 2006). In the CVCA, the relationship of stakeholders and flow of various values (including money and products) are shown in Figure 1, using the "Drink one give ten campaign" by Volvic as an example.

It is an effective tool for understanding existing technological and social systems as well as designing new types of systems. Its usage is targeted for

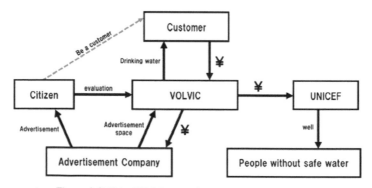

Figure 1 CVCA of "Drink one, give ten campaign" by Volvic

product/service development teams engaged in group discussions to express their thoughts and generate ideas during meetings such as design reviews and presentations. While CVCA focuses on "how" the stakeholders are related to each other, it does not explicitly focus on "why" the system structure (relationships among stakeholders) are formed.

"How and Why": Human behavior is motivated by various needs that manifest consciously or non-consciously. Hence, people's needs in their mind should be analyzed and visualized, even though it is not easily quantifiable.

Various psychologists including Murray (1938) and Maslow (1987 and 1998) have classified human needs. However, since psychologists' interests have been on the minds of persons, they seldom pay any attention to utilizing their results for analyzing or designing social systems where human relationships (between persons and groups of persons) play an important role.

From a classification point of view, "Maslow(1987)'s "hierarchy of needs" can be used. Maslow assumed five levels of needs - physiological; safety; belongingness and love; esteem; and self- actualization needs from the bottom to the top. These five steps of needs can be used as categories. Maslow (1998) also described two other needs outside the above hierarchy of five needs, called "basic cognitive needs". They include the desires to know and understand and to fulfill aesthetic needs. Hence, from a classification point of view, we utilize seven categories of human needs as shown in Figure 2.

Another important issue is that the subject and object of those needs are not usually focused on in psychological studies. For example, when humans are discussing "safety" needs, they usually think of their own safety. Some may state that the object (or subject) of personal needs is usually him or herself because humans are fundamentally selfish. Others may say that there exist altruistic needs for others. The following example of "safety" indicates that if the relationships among people are considered, there are four safety needs involved:

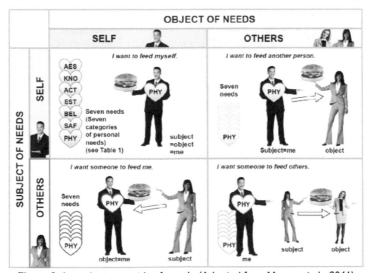

Figure 2 A two-by-two matrix of needs (Adapted from Maeno et al., 2011)

1.   want to protect myself.
2.   want someone to protect me (e.g. I want car companies to produce safe cars for me).
3.   want to make someone safe (e.g. I want to donate to the flood victims).
4.   want someone to protect someone else (e.g. I want my son's teacher to provide him with a good education).

Those four situations can be described by a two-by-two matrix based on the subject and the object of needs as shown in Figure 2. In the figure physiological needs (needs of food) are illustrated. Symbols in the hearts represent the seven categories of needs described above. They will be used in WCA figures. Colored hearts indicates that the subject of needs is himself/herself. On the other hand, a white heart with a colored outline indicates that the subject of needs is someone else. The red color indicates that the object of needs is "he or she", whereas the green color indicates that the object of needs is someone else. The four different features of marked hearts will be important for analyzing the relationship of stakeholders using WCA.

It is important that for designing systems, the various stakeholders and the structure of their personal needs should be considered. That implies that the seven categories shown in the two-by-two matrix in Figure 2 should be taken into account to visualize stakeholders' needs.

2.2 Case study: Analysis of social systems using CVCA/WCA

Figure 3 shows the result of WCA for the "Drink 1, give 10 campaign". Figure 4 shows the example of the workshop by students using CVCA and WCA.

Comparing Figure 1 and Figure 3 reveal clear differences. By adding the heart marks we can clearly see the humans' needs/wants. The color of the hearts is of special importance. The red color represents "self" needs, implying they are the "selfish" needs. On the other hand, the green hearts represent needs of others. Hence, they are relatively altruistic needs. That is why CRM is accepted by people in our current world and regarded as sufficient/good for satisfying Corporate Social Responsibility (CSR) as well. We can say that WCA is a visualizing tool of selfish or altruistic actions made by various people.

Of particular importance is that the two white hearts with the green outline in "Customer" and "Volvic" are connected by green arrows to the green heart in "UNICEF". This means that the need of "want someone to save African people without safe water" is realized by "UNICEF" through the green arrows chain. The reason why we use the heart figure is that people's hearts are connected by arrow chains to relay their wish until it is realized by a heart filled with green color.

WCA is useful in our current systems, because social values such as social safety, security, health, welfare, sustainability of environment, food, water and materials are becoming increasingly more important. For example, climate change problems cannot be solved if all people are selfish. Everybody has to get together to save the earth. However, not everybody's moral standards are high enough to try to save the earth as a first priority. It means that people's personal needs are not at the "Self-Actualization" level. Many people want to fulfill various other needs to live a

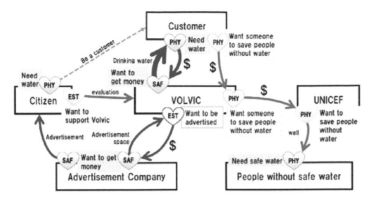

Figure 3 WCA of "Drink one, give ten" campaign by Volvic (Adapted from Maeno et al., 2011)

Figure 4 Example of workshop by use of WCA

life that is balancing with morality. That's why we should systematically construct business and social systems taking into account the various needs of people. Volvic's case implies that it is important to construct sustainable and robust systems by adding a little green heart even if the majority of people are driven mostly by a red heart.

After the workshop of CVCA/WCA, we have asked students their opinions of the outcomes and interpretations. The main positive comments included:

(1) WCA can visualize humans' wants/needs even though it is usually difficult

(2) Red and green hearts/arrows are useful to see clearly the selfish/altruistic needs

(3) Thinking of classification of Maslow's needs lead me create new ideas of service system

It can be concluded that expected opinions can be obtained by students as users of WCA. Negative opinions are that WCA is much more difficult to use compared with CVCA. However, it might not be a problem because the purpose of CVCA and WCA are different. CVCA is a simple tool for visualizing the flow of money, products and information among stakeholders and its simplicity has clear benefits. On the other hand, WCA is for visualizing the minds of humans (their needs/wants). It is not possible to visualize human minds perfectly and therefore should only be regarded as an assumption. However, it is important to think like others in order to understand others. WCA is for helping the analyzer to think like others even though it is difficult.

3 BAYESIAN BELIEF NETWORK MODELING OF SOCIAL SYSTEMS

3.1 The Bayesian network modeling

Bayesian Belief Network Modeling is a method to formulate a visual network with calculated conditional probabilities. This modeling method is applied with the use of causal relation diagrams to achieve a particular goal for a complex problem (Nguyen *et al.*, 201; Bosch *et al.*, 2007; Checkland and Scholes, 1990). The process to develop a BBN is an excellent mechanism for "Participatory Systems Analysis" (PSA) (Smith *et al.*, 2007) since diverse groups of stakeholders can participate in collaboratively making a decision to solve the problem of their system or achieving a particular goal (Ames and Neilson, 2001).

A key motivation for stakeholders during is the visualized co-designing process (PSA) under the theory of value co-creation (Sanders and Stappers, 2008) with the Bayesian Network software. Previous studies using BBN modeling include a wide variety of applications, e.g. policy-creation targeting specific sectors or communities with particular problems (e.g., water management in a region (Ames and Neilson, 2001), sustainability development in UNESCO defined biospheres (Nguyen *et al.*, 2011), the engineering world, medical diagnosis, etc.). This paper explores the usefulness of the model when it is applied to a national-level macro policy in Japanese society.

3.2 Case study: policy design for child safety in Japan

This subsection is to explore the usefulness of BBN modeling as a tool for designing a public policy to enhance the requirements of stakeholders. The authors selected the policy issue to prevent children from unexpected casualties by accidents. In OECD member countries, more than 125,300 children died from injuries, which amounts to 39% of all deaths in 1991-95. Japan was ranked as a medium risk performer in deaths by drowning, fire, falls and intent, whereas deaths by cars accidents were significantly lower than in other countries (UNICEF, 2001). Japanese society often regards parents as the only people responsible for child safety. Japanese parents tend to be isolated and frustrated, because there is a clear lack of a coordinated approach with other stakeholders in the society to help prevent their children from injuries (Kakefuda *et al.*, 2008). A PSA for child safety could therefore be regarded as an essential and most urgent needed approach for the policy agenda to realize a more safe and secure society.

3.3 Sequential Approach

The PSA for child safety was carried out on 22 September 2011 on the Keio University Campus near Yokohama.

Figure 5 Experts at the Focus Group Meeting and CLD Result

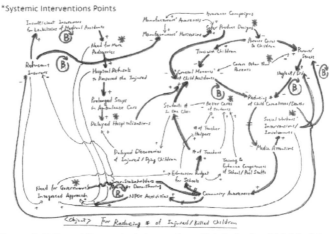

Figure 6 CLD and Systemic Interventions Points for the Child Safety

This approach was used to develop a Bayesian network model for child safety. This was done in two main steps:

Constructing a Causal Loop Model

A focus group meeting was convened to hold a brainstorming session to identify and visualize all factors related to child injuries as shown in Figure 5. The participating stakeholders collaboratively draw a causal loop diagram to identify loops, that is the interactions and relationships between the components of the model. This was followed by the identification of possible leverage points and systemic interventions by the stakeholders. This was carried out through "visual" scenario testing (observing the potential degree of change that will be caused by changes to particular components of the system). The leverage points were identified as the most effective points in the model that will help achieving the policy goal (Senge, 1990) and are indicated in Figure 6 for the case of child safety. Seven systemic interventions points were identified: safer product designs, caring volunteers to support frustrated parents, social worker's involvement, more integrated approach by government, more pediatricians, shortened time before hospitalization, and better care of students in schools.

3.4 Policy Implications by the Bayesian network modeling

Constructing a Bayesian network Model

The focus group experts collaborated to establish a Bayesian Network Model, using the Causal Loop model as a basis. This included the identification of possible root causes affecting the unintentional injuries of children in Japan.

Participating experts used the seven systemic interventions points to structure the Bayesian belief network model for designing policies on child safety (Figure 7). To populate the model, they jointly decided the probabilities of how the parent nodes will determine the probability that a child node would be achieved. For example, what are the probabilities that more scholarships and better insurance policies will increase the probability that there will be more pediatricians (see Fig. 7, right hand side). Through this co-designing process, the stakeholders recognized that their populated Bayesian model indicated that there is under current policy only a 21.5% probability to reduce the child injury ratio (Figure 8).

A sensitivity analysis of the model indicated that the most effective parameter to reduce child injuries was the policy to increase the number of volunteer nursing councilors. The populated model showed that the probability of decreasing child injuries will rise from 21.5% to 46.7% if the number of volunteer nursing councilors is set at 100%. Therefore this policy was determined by the participating experts to be the leverage point for improving child safety.

By completing the above sequential steps, the Bayesian network model proved to be highly effective for creating a public policy with stakeholders' involvement.

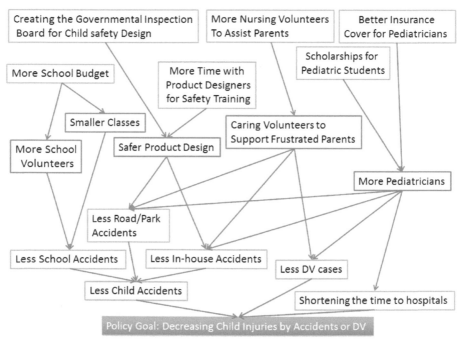

Figure 7 Factors for Injury-free children: The descriptive Bayesian Model

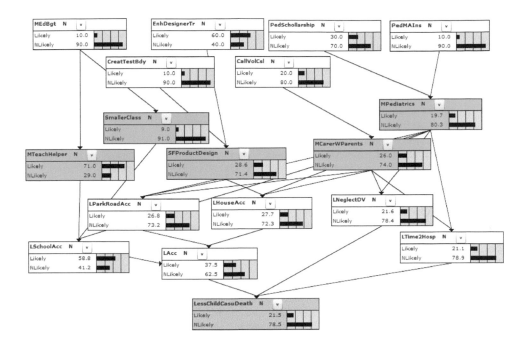

Figure 8 Populated Bayesian Model for Child Safety:

4 CONCLUSIONS

It is clear that CVCA/WCA and BBN modeling are in all cases useful tools for studying human needs/wants in a workshop-based situation. These methods should not only be used for education but are highly valuable in practice and for generating innovative ideas in the real world.

ACKNOWLEDGEMENTS

This study was financially supported by the KAKENHI (23611038), the Grant-in-Aid for Scientific Research (C) through the Japan Society for the Promotion of Science (JSPS) and the Ministry of Education, Culture, Sports, Science and Technology (MEXT). The Center for Education and Research of Symbiotic, Safe and Secure Systems Design of the Keio University Advanced Research Center also financially supported this study by the MEXT Global COE Program (Keio University GCOE H-10).

84

REFERENCES

Adkins, S. 1999. *Cause Related Marketing*, Oxford: Butterworth-Heinemann.

Ames, D.P. and B.T. Neilson 2001. 'A Bayesian Decision Network Engine for Internet-Based Stakeholder Decision-Making', *American Society of Civil Engineers (ASCE), World Water and Environmental Resources Congress 2001, Conference Proceedings*, Accessed September 23, 2011,
http://ascelibrary.org/proceedings/2/ascep/111/40569/169_1

Bosch, O.J.H., C.A.King, J.L. Herbohn, I.W.Russell, and C.C.Smith 2007. 'Getting the Big Picture in Natural Resource Management: Systems Thinking as 'Method' for Scientists, Policy Makers and Other Stakeholders', *Systems Research and Behavioral Science*, Syst. Res. 24, pp.217-232.

Checkland, P. and J. Scholes 1990. *Soft Systems Methodology in Action,* Chichester, UK: John Wiley & Sons, Ltd.

Donaldson, K., Ishii, K., and S. Sheppard 2006. Customer Value Chain Analysis, *Research in Engineering Design*, Vol. 16, pp. 174-183.

Ishii, K. Olivier de Weck, Shinichiro Haruyama, Takashi Maeno, Sun K. Kim, and Whitfield Fowler 2009. 'Active Learning Project Sequence: Capstone Experience for Multi-disciplinary System Design and Management Education', *Proceedings, International Conference on Engineering Design*, pp. 57-68.

Kakefuda, I., T. Yamanaka, L.Stallones, Y.Motomura, Y.Nishida 2008. Child Restraint Seat Use Behavior and Attitude among Japanese Mothers, *Accident Analysis and Prevention*, 40, pp.1234-1243.

Kotler, P. and K. Keller 2008. *Marketing Management* 13th Edition, New York: Prentice Hall.

Maeno, T., Yurie Makino, Seiko Shirasaka, Yasutoshi Makino and Sun K. Kim 2011. 'Wants Chain Analysis: Human-Centered Method for Analyzing and Designing Social Systems', *Proceedings, International Conference on Engineering Design*, August 2011, Copenhagen, Demark, pp. 302-310.

Maslow, A. H. 1987. *Motivation and Personality*, Harpercollins College Div., 3rd Edition.

Maslow, A. H. 1998. *Toward a Psychology of Being*, New York: Wiley, 3rd Edition.

Ministry of Economy, Trade and Industry, Japan 2009. *Selected fifty five social businesses* (in Japanese), Accessed February 21, 2012,
http://www.meti.go.jp/policy/local_economy/sbcb/sb55sen.html.

Murray, H. A.1938. *Explorations in Personality*, New York: Oxford University Press.

Nguyen, N.C., O.J.H. Bosch, K.E. Maani 2011. 'Creating 'Learning Laboratories' for Sustainable Development in Biospheres: A Systems Thinking Approach', *Systems Research and Behavioral Science*, Syst. Res. 28, pp.51-62.

Project Management Institute 2008. *PMBoK Guide 4th Edition*, Newton Square, PA: Project Management Institute.

Sanders, E. B. and P.J. Stappers 2008. 'Co-Creation and the New Landscapes of Design', *CoDesign*, Vol.4, No.1, March 2008, pp.5-18.

Senge, P. 1990. *The Fifth Discipline: The Art and Practice of the Learning Organization*, New York: Doubleday.

Smith, C., L.Felderhof, O.J.H.Bosch 2007. 'Adaptive Management: Making it Happen Through Participatory Systems Analysis', *Systems Research and Behavioral Science*, Syst. Res. 24, pp.567-587.

UNICEF 2001. *A League Table of Child Deaths by Injury in Rich Nations*, Innocenti Report Card, Issue No.2, February 2001.

CHAPTER 9

Promotion of Social Participation among the Elderly based on Happiness Structure Analysis using the Extended ICF

Mikiko Inoue, Koji Kitamura, Yoshifumi Nishida

National Institute of Advanced Industrial Science and Technology (AIST)
Tokyo, Japan
mikiko-inoue@aist.go.jp

ABSTRACT

Social participation is key to maintaining health and contributes to a high level of quality of life in the elderly. However, finding the motivation for social participation is a challenge because motivation varies widely among individuals. A new technology is needed to reveal what promotes or impedes social participation among the elderly. This study attempts to identify factors that facilitate active social participation based on past experiences. In this paper, we examine a new approach to canonicalizing and understanding the relationships between the senior's happy feelings and daily-living components and visualizing it as a network structure. To canonicalize daily life data, we used the WHO's International Classification of Functioning, Disability, and Health (ICF) codes to describe the senior's daily life. We analyzed how a specific daily-living component affects one's life using degree centrality. This paper also discusses a new methodology of community design that describes how one can achieve self-fulfillment and keep one's happiness structure by integrating the happiness structures of others.

Keywords: Happiness analysis, Social participation, Elderly, Quality of life, International Classification of Functioning, Disability, and Health (ICF)

1   INTRODUCTION

A paradigm shift has occurred in the field of public health and medicine in that acute infectious diseases are no longer major causes of death and disability in most industrialized countries. In the 21st century, lifestyle-related diseases place the heaviest burden on our society, accounting for 70% of all deaths in the United States (Kung et al, 2008). This change has put great emphasis on disease prevention through behavioral changes. Although researchers in many disciplines around the world have tried to develop a new technology that would help people live the type of lifestyle that combats the burden of chronic diseases, such technology is not yet available.

Social participation is defined as involvement in life situations (World Health Organization, 2001) and is considered one of the best ways to promote health, especially in seniors. Researchers have a prominent role in developing new technology that brings out one's abilities or incorporates one's experience in order to increase willingness to take part in social activities, a so-called Human Involving Technology. Human Involving Technology should not replace human activities of daily life but should coexist with them by helping people to lead fulfilling lives. Helping people identify problems and find solutions, forming new ties between people, supporting one another, and giving people more pleasure and satisfaction through social participation are examples of Human Involving Technology.

Thanks to technology advancements, Human Involving Technology allows us to solve real life problems. Foldit, an online game that people with no or little science background try to solve protein-structure problems, is one such example. It creates an opportunity for players to compete and collaborate with each other to find an answer (Khatib et al., 2011). This is a technology that shares and integrates people's knowledge and skills. In the field of robotics, Matsumoto et al. have proposed a new method of describing the "benefit of a robot" and "needs of a user". They described a person's abilities to perform a certain task and a robot's abilities to assist the person. This is a person-robot matching technology to help people with disabilities perform necessary tasks while preventing disuse syndrome (Matsumoto et al., 2011). Population health technology surely improves the public's health by detecting health threats, exchanging information, and spreading the word, which was not possible 20 years ago (Eng, 2004 & Eysenbach, 2003).

In this study, we tried to develop Human Involving Technology focusing on promoting social participation and vitalizing a society. To design an individualized lifestyle based on the whole aspect of one's life, we need to understand not only whether one suffers from a certain disease but also whether one is satisfied with everyday life. The International Statistical Classification of Diseases and Related Health Problems (ICD) allows us to visualize disease status by indicating whether or not one has a disease, but lacks a way to visualize one's life status. We need a new way to visualize life status to develop Human Involving Technology.

In this paper, we suggest a new approach to understand the structure of daily life. We make use of the WHO's International Classification of Functioning, Disability, and Health (ICF) codes to record daily-living components including 1) one's daily activities such as *go out*, *eat breakfast*, *talk to friends*, and 2) environmental factors such as *family members, consumer goods,* and *natural events*. Because the current version of ICF is developed for objective evaluation of one's health and is unable to deal with one's feelings and subjective experience, we developed additional ICF codes (extended ICF) to describe them. We discuss how to canonicalize data on one's life using the extended ICF. This allows us to describe daily life quantitatively and accumulate reusable life data. We visualized one's daily life as a network and analyzed how a specific daily-living component affects one's life. Then, we focused on events during which people felt happy and developed a happiness network. We compared the structural differences of two individuals' happiness networks using degree centrality. At the end of this paper, we discuss how we apply our suggested approach to promotion of social participation by integrating one's functioning into another's.

2 DEVELOPMENT OF THE EXTENDED ICF AND A METHOD OF DAILY LIFE VISUALIZATION

Establishing the method of daily life visualization is one of the most important themes for synthesiology of functioning. Functioning is defined in details in the following subsection, but simply stated, it is people's abilities to live. Synthesiology of functioning is the study of everyday life design by combining people's functioning and artificial objects. When the method of describing our daily life is fully established, it would be possible to share data on people's living (life data) among different stakeholders and to scientifically predict or take control over one's life. In this section, we discuss the development of the extended ICF as a fundamental language to describe a daily life structure.

2.1 Development of the extended ICF

In this study, we used ICF codes to describe daily life. ICF is the WHO family of international classifications, focusing on health and functioning rather than on disability. It is a standard language to describe one's health. There are more than 1400 codes available, which are categorized into three levels: level 1, body functions & structures; level 2, activities; and level 3, participation. Examples of ICF codes are provided as follows:
- a445: hand and arm use
- a510: washing oneself
- p910: community life
- p950: political life and citizenship
- e130: products and technology for education
- e410: individual attitudes of immediate family members

The prefix characters "a" represents "activities", "p" represents "participation", and "e" represents "environmental factors". These three levels are called "functioning", an umbrella term for neutral or positive aspects of human life, in ICF (WHO, 2001).

One's state of health or functioning is considered to be an outcome produced by interactions between health conditions (having high blood pressure or suffering from heart disease) and contextual factors (friends and family members, or being female) as shown in Figure 1.

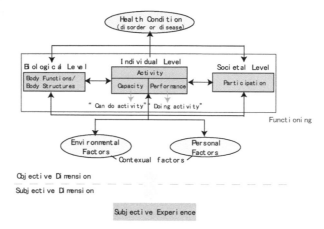

Figure 1. Model of Subjective and Objective Dimensions of Human Functioning and Disability

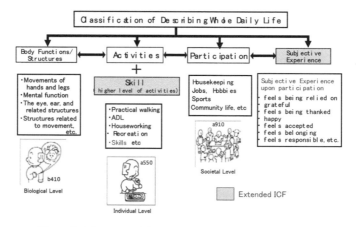

Figure 2. A Conceptual Framework of the Extended ICF Model

While the ICF focuses on the classification of objective health state, Ueda and Okawa, advisers to the Japanese Society for Rehabilitation of Persons with Disabilities, stress that people's state of health cannot be defined without considering the subjective experience of their health condition (Ueda and Okawa, 2003). These authors developed a tentative framework of a classification of subjective dimension of functioning and disability for future inclusion into ICF. We used *Subjective experience 6* through *Subjective experience 9* of the proposed classifications and coded each construct. We consider the subjective dimension, for example, a category of *Individual's Value and Meaning of Life*, as the fourth level in ICF (the extended ICF) (Figure 2). When it is difficult to categorize people's subjective experience using the classification of Ueda and Okawa, we developed additional categories of subjective experience and constructs as necessary. The extended ICF codes allow us to canonicalize life data not only on objective health but also on subjective health and to conduct scientific analyses.

2.2 Method of describing daily life structure using the extended ICF

In this subsection, we discuss the method of describing daily life structure using the extended ICF. We collected life data using a questionnaire and extracted five types of data from responses to an open-ended question [*ex. Did you have any happy events in the past 2 weeks? If you did, please specify.*]: 1) feelings, 2) subjective experience, 3) an event, 4) activities around an event, and 5) personal and environmental factors related to an event. Then, we converted 2) subjective experience, 4) activities around an event, and 5) personal and environmental factors into the extended ICF codes. For instance, when the response regarding an event that made one happy was a hot spring trip with one's son and his wife, we recorded 1) "happy" for feelings, 2) "ex218.1: enjoys life" for subjective experience, 3) "a hot spring trip" for an event, 4) "a920 (recreation and leisure): going on a trip" and "a3: communication" for activities around an event, and 5) "e310: immediate family" for personal and environmental factors. "ex" represents "subjective experience".

3. COLLECTING LIFE DATA

We conducted surveys of 20 seniors who did not have any difficulties in activities of daily living (ADL) from April 2011 to March 2012. These 20 seniors were selected by members of the local government office. There were two types of surveys. One was on instrumental ADL conducted once a month, and the other was on experiences in daily life conducted once every two weeks. The method of data collection was home visits or telephone interviews by a nurse.

4. ANALYZING HAPPINESS STRUCTURES

In this section, we discuss the method of network analysis focusing on the relationships between events that made one happy and one's functioning. This research is a trial to promote a deeper understanding of how to increase the motivation for social participation among seniors. In the present study, we used life data collected from the survey on experiences in daily life.

90

4.1 Feelings and experience generating model

We constructed a happiness network model according to the feelings and experience generating model (Figure 3). By following the rule of a node connection and describing relationships between nodes, it is possible to quantitatively analyze one's daily life. It is important to note that our research findings themselves are not necessarily important products for our study. Rather, this present study is an opportunity to examine the possibilities of applying our proposed method of describing one's daily life as a network structure to the development of new Human Involving Technology that can prevent people from lowering their level of functioning and that can motivate people to engage in social participation based on their subjective experience.

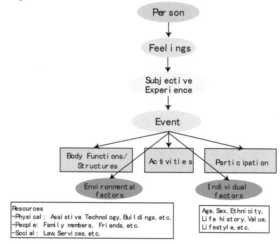

Figure 3. Feelings and Experience Generating Model

4.2 Comparison anlaysis of happiness structures

We focused on events related to one's happiness and developed a happiness network. In this subsection, we compare the structural differences using degree centrality. The degree centrality is defined as the number of direct connections that a node has. As shown in Figure 4, the score of degree centrality for happiness was 5 for senior A and 3 for senior B. This means that senior A encountered more subjective experiences than senior B. The same trend can be observed for a subjective experience. The degree centrality for a subjective experience was 8 for senior A and 6 for senior B, indicating that senior B did not have as many various activities as senior A to enjoy subjective experiences. In this case, senior B had fewer opportunities to use functioning and, therefore, it is possible to say that we should provide supports to senior B first based on the result of the analysis. Furthermore, to support senior B in having have the subjective experience of *enjoys life*, having events related to ground golf or a concert will effectively motivate senior B to participate. In addition, because we collected data related to people's happiness, it might be possible to predict events that senior B enjoys by comparing other people's happiness networks whose structures are similar to senior B's. This way, the method of visualizing and analyzing one's happiness structures makes it possible to help develop personalized motivation strategies for participation.

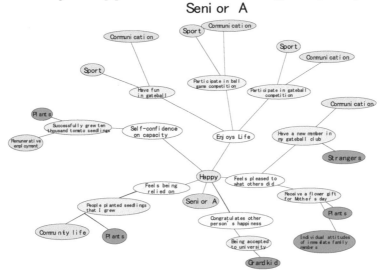

Degree Centrality for Happiness: 5
Degree Centrality for Subjective Experience: 8

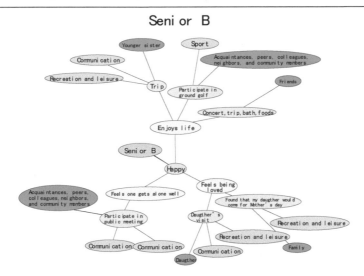

Degree Centrality for Happiness: 3
Degree Centrality for Subjective Experience: 6

Figure 4. Structural Difference for Happiness Network

4.3 Case study of functioning integration for self-fulfillment and promoting social participation

In this subsection, we discuss a case example of achieving self-fulfillment and social participation by integrating functioning of others into one's functioning. We found this case example from an interview of a handyman who worked in a town where we collected life data.

The senior, a 90-year-old lady, has suffered from physical pain since a few years ago and had hip replacement surgery in 2011. She lived in a rural area and enjoyed sending homegrown vegetables to her son who lived in Tokyo every year. Tokyo was far from her residence, and she always looked forward to his phone calls when she sent vegetables. Her happiness network is shown in Figure 5.

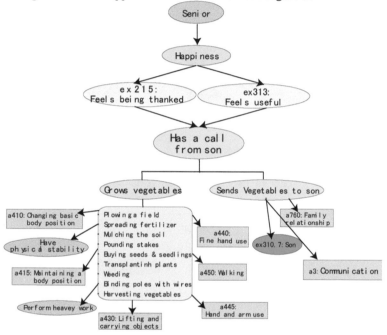

Figure 5. Happiness Network Structure

She used to maintain her happiness structure by herself. However, because of a lower level of functioning, she was no longer able to perform farm work. This work consisted of multiple tasks such as plowing a field, spreading fertilizer, mulching the soil, and binding poles with wire, and these tasks also consisted of a combination of multiple functions. In short, when the level of a particular functioning became weakened, it was not possible for her to perform complex tasks like farm work. If she could not perform this one farming task, that is, growing vegetables, and subsequently send them to her son, she would not be able to reach happiness.

Figure 6 indicates that tasks of plowing a field and mulching the soil were completed by getting offers for three types of functioning from others (a functioning giver); "a410: changing basic body position," "having physical stability," and "performing heavy work". In this way, she, a functioning taker, successfully kept her happiness using her own functioning as much as possible while having support from a functioning giver for the tasks that she had difficulties performing. Considering this case example from the standpoint of a functioning giver, the giver needed to use not only functioning that was expected to provide but also all other necessary functioning to complete each aspect of the farm work. Moreover, the giver will have the subjective experience of *feeling relied on, feeling thanked,* or *having high self-confidence* regarding his/her farming skills by engaging in social participation. The way of describing one's life structure and integrating multiple individuals' functioning like this case example can be the starting point for the development of new technology that will help promote social participation and enhance one's quality of life.

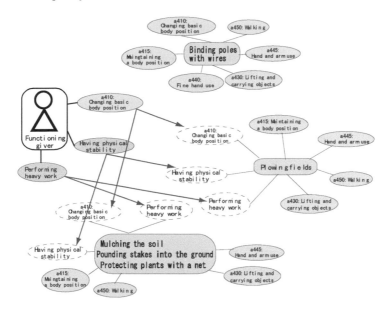

Figure 6. Example of Social Participation by Integrating Functioning and Skills

4.4   Discussion of functioning integration based on a happiness network analysis

From the result of the happiness structure analysis and a case example of functioning integration, we revealed that it was possible to analyze one's life structure using the extended ICF. Visualizing life structure allows us to understand the relationships between one's subjective experience and functioning, or more broadly, between one's happiness and daily life activities. In this subsection, we discuss the application possibilities of our suggested method for social participation promotion based on one's motivation. The method of visualizing one's life structure using the extended ICF makes it possible to find individuals who have a similar life structure to identify events in daily life related to their happiness, and develop effective strategies for social problems such as social isolation and loneliness among the elderly. In the case example, we found a way of keeping one woman's happiness structure by integrating her functioning with others. Creating social events based on the result of happiness analysis will help individuals naturally engage in social participation. Consequently, a society will be vitalized. Thus, the development of a method for describing one's life and analyzing happiness structures can be applied to health promotion and social vitalization and will lead to solutions for societal health problems.

5. CONCLUSION

To develop a new life design technology that promotes social participation, we developed the extended ICF to canonicalize life data and suggested a new method to analyze our daily life. To examine the applicability of our suggested method, we constructed a happiness network using the extended ICF and found that visualizing one's daily life structure was possible. From the case example, we revealed that it would be possible to develop individualized strategies for social participation based on one's happiness structure by integrating functioning from others.

In the future studies, we are trying to conduct community-based participatory research to advance the life design technology, taking one's subjective experience into account. Although our research is only a preliminary step in understanding our daily life structure, it is fundamental to the development of Human Involving Technology. We hope to gather more life data and increase our understanding of the complex structure of our daily life using the extended ICF.

REFERENCES

Eng, T. 2004. Population health technologies: emerging innovations for the health of the public. *American Journal of Preventive Medicine* 26(3):237-242.

Eysenbach, G. 2003. SARS and population health technology. *Journal of Medical Internet Research* 5(2):e14.

Khatib, F. 2011. Crystal structure of a monomeric retroviral protease solved by protein folding game players. *Nature Structural & Molecular Biology* 18(10): 1175-1177.

Kung, H., Hoyert, D., Xu, J., Murphy, S. 2008. Deaths: Final Data for 2005. *National Vital Statistics Reports.*

Matsumoto, T., Nishda, Y., Motomura, Y., Okawa, Y. 2011. A concept of needs-oriented design and evaluation of assistive robots based in ICF. *Proceedings of IEEE International Conference on Rehabilitation Robot*:1-6.

Ueda, S., Okawa, Y. 2003. The subjective dimension of functioning and disability: what is it and what is it for. *Disability and Rehabilitation* 25, No.11-12:596-601.

WHO. 2001. International Classification of Functioning, Disability and Health.

CHAPTER 10

Barcode Scanning Data Acquisition Systems in Temporary Housing Area to Support Victim Community

Takuichi Nishimura*1, Hitoshi KOSHIBA*1, Yoichi Motomura*1*2, and Kohtaro Ohba*3

*1 Center for Service Research,
National Institute of Advanced Industrial Science and Technology (AIST), JAPAN
*2 The Institute of Statistical Mathematics (ISM), JAPAN
*3 Intelligent Systems Research Institute,
National Institute of Advanced Industrial Science and Technology (AIST), JAPAN
Email address taku@ni.aist.go.jp

ABSTRACT

We are conducting a community-based participatory study called the "Kesennuma Kizuna" project. The aim, current status, and research issues are introduced in this paper. There are still many victims trying to recover from the Northeast Japan earthquake, which hit March the 11th, 2011. Temporary houses have been built and new life has started. But people in those houses in each district are gathered from different district and have to establish new social network. We have been trying to contribute to whose new communities with local organizations and nonprofit organizations. People will use a "Thank You Card" which is printed a barcode to participate charities meeting or to get coupon and so on. The system helps the local support members to find inactive people or needs by analyzing the log data. This work will present detail results of the system and efficiency to support the community.

Keywords: Community-based participatory research, Service engineering, Behavior observation, Large-scale data, Life care technology

1   INTRODUCTION

Due to the Great East Japan Earthquake and the ensuing tsunami, many have lost their homes and jobs and have been forced to move into makeshift residences. Many local communities have also been devastated as a result of the disaster. Kesennuma is one of many municipalities severely affected by the disaster whose main industry was marine products industry. Engineering technologies are expected to support community reconstruction through general rebuilding efforts to solve social problems, to create living environments and jobs. It is an object of urgent social concern to find solutions through ongoing research.

Members of the National Institute of Advanced Industrial Science and Technology entered the actual makeshift residences and workplaces of people in Kesennuma, and they began an experimental study of engineering technology development. This research takes a community-based approach [1] that supports the livelihoods of community members and fosters bonds (called *kizuna* in Japanese) between people. This paper reports the current status and the issues encountered through the "Kesennuma Kizuna Project."

2   OBJECTIVE

This project aims to develop engineering technologies and smart community technologies that revitalize human connections, and to implement such technology at reconstruction sites to prevent disuse syndrome and even the deaths of isolated individuals. In sociological terms, *kizuna* can be defined as a concept that emphasizes the value of social capital, conveying the belief that the efficiency of a society can be increased by the practice of cooperative behavior. Social capital is considered to be impossible to quantify directly, and previous studies in sociology inferred levels of social capital from subjective evaluations, such as voting rates, surveys on basic social activities, and residents' sense of accountability to the local communities. The ultimate goal of the project is to propose an engineering solution for the improvement of social capital. The researchers intend to accomplish this by using technology to create a model to assess social capital based on quantifiable values, and by conducting simulations using such a model.

In order to execute this project, the target community must be defined, and devices that observe the behaviors of community members must be installed into locations where they engage in everyday activities. The installation of such devices faces many obstacles, including human and economic costs, as well as psychological costs such as privacy concerns. Therefore, it is important to analyze and to understand the prospective users in order to design the devices in a way that increases residents' willingness and motivation to use them. This process is deeply related to studies concerning dissemination of products and technologies in marketing and innovation theories. Service engineering methodologies [2, 3] can also be applied for the observation, design, and analysis of service activities, which are accumulations of social interactions within the community.

3   COMMUNITY-BASED APPROACH

Studies on the revitalization of human bonds need to focus on the intangible concept of "livelihood," which is embodied by the totality of human interactions rather than the development of life support technologies. Actual participation in the community is necessary in order to make observations on livelihood as objects of study. In the case of the disaster area, participation in rebuilding efforts is expected. The authors of this paper stayed in a trailer house installed next to the makeshift homes, and participated in the community by living there. Many events were hosted in the multi-purpose room of the trailer house, where equipment (e.g. advanced healthcare equipment) was installed for the local residents to use. It is hoped that habitual use of the equipment by residents will increase data collection opportunities.

Because the application process to move into makeshift homes was conducted urgently in order to move residents out of temporary shelters, the neighborhood assignment process was often unresponsive to the will of applicants and communities, and the process severed many preexisting local community ties. Thus it became imperative that the researchers participate in activities as members of the community in order to support the development of a new community in the makeshift neighborhood. To this end, the researchers organized numerous events in cooperation with many nonprofit and aid organizations, and participated in events such as tea times and dinner parties in order to foster a cooperative community atmosphere.

Fig. 1: Trailer house installed next to makeshift homes

4   PROJECT CHALLENGES

Three trailer houses were installed on January 28, 2012, next to the makeshift residences in Goemon-gahara, and the occasion was celebrated with a "kickoff event." The trailer houses quickly became established as places for daily life activities in the new community through the use of their multi-purpose rooms for activities such as participatory events, eating, drinking, and product sales. The behavioral patterns of community members within the trailer houses become observable through the installed devices. The devices will continue to be developed and deployed in the trailer houses, and it is necessary to promote activities that encourage residents to utilize the devices while revitalizing the community by involving as many residents as possible.

On the other hand, behaviors within the trailer house make up only a fraction of the community's activities. The next step is to observe behavior in a wider area. It is important that observations of behavior be accompanied by the consent of and a sense of agency for the users. Instead of using devices such as cameras, this project uses barcode scanners (fig. 2) that require voluntary actions on the part of users for the observation of outdoor behaviors.

Unmanned station with coupon-issuing features

- Equipped with high spec scanner
- Equipped with a printer
- Built in Wi-Fi, mobile card support

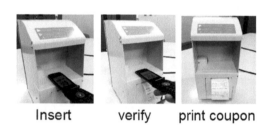

Insert   verify   print coupon

Fig. 2: Barcode scanner (Unmanned Type)

There is a need not only to develop barcode scanner systems, but also to design services that engender motivation to use the systems. The latter will be accomplished through a partnership with a local makeshift shopping mall known as "Recovery Stalls Village" (*Fukkou Yatai Mura*), which houses the community's eateries and a product sales service. The researchers plan to bring barcode scanner service to the mall, allowing visitors to scan barcodes upon arrival in order to receive information and a gift as a reward. In order to realize such a service, and in order to increase the appeal of the barcode scanner stations, the researchers are designing a system that encourages service utilization and repeat visits from visitors from outside the area.

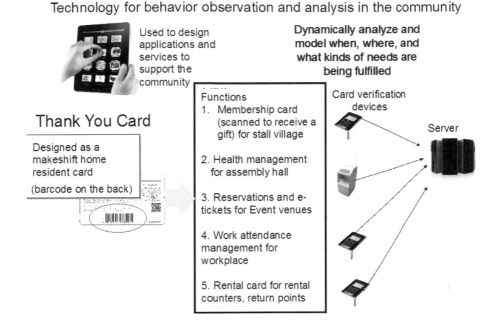

Fig. 3: Behavior observation and analysis system using "Thank You Cards"

A variety of projects using the scanners are related to rebuilding efforts that improve service delivery in disaster areas. In other words, observation and modeling of life activities are advanced as by activities that support reconstruction efforts. Partnerships and coordination with community organizations also become important as means to secure the continuation of system use.

5   CONCLUSION

This paper has discussed the progress and the issues of the community-based livelihood support technology development project in Kesennuma. In order to study people's mutual interactions during their everyday lives, it is essential that we collect and analyze observations of daily activities. Lewin proposed an approach called "research as a service," which unifies public service with research on everyday life [4]. This research shares a common framework with research as a service; i.e., investigators create new value for the community by bringing research into the field.

ACKNOWLEDGMENTS

This study was supported by the Project of Service Engineering Research in 2011 from Japanese Ministry of Economy, Trade and Industry (METI). And, the authors would like to express their thanks to Kesennuma city members, *Fukkou Yatai Mura* members, and those who support this project.

REFERENCES

[1] B. Israel, et al.: Methods in Community-Based Participatory Research for Health, Jossey-Bass (2005).

[2] K. Naito: Introduction to Service Engineering, University of Tokyo Publishing (2009).

[3] Y. Motomura, et al.: Large-Scale Data Observation, Modeling, and Service Design for Service Innovation, Journal of the Japanese Society for Artificial Intelligence, 23-6, pp. 736-742 (2008).

[4] Y. Motomura: Everyday Life Behavior Prediction Modeling from Large-Scale Data, Synthesiology, 2-1, pp. 1-11 (2009).

Section III

Information Processing and Intelligent Agents

Analysis of the Effect of an Information Processing Task on Goal-directed Arm Movements

Thomas Alexander

Fraunhofer-FKIE
Neuenahrer Str. 20
53343 Wachtberg
thomas.alexander@fkie.fraunhofer.de

ABSTRACT

Today's Digital Human Models (DHMs) are primarily used for designing new processes, workplaces, or products. Most of these applications have in common that they focus on the spatial design. The DHMs allow the inclusion of anthropometric dimensions, biomechanics and their variability. Available functions include sight, reach, range, and posture analyses. But modern work and workplaces often involve information technology and human information processing. According to literature research information processing and cognitive workload affect human movement behavior and, thus, require additional modifications of spatial workplace design.

It is hypothesized that the interrelationship primarily affects motion planning and less motion execution. N=60 participants (male, 19 – 27 yrs) took part in the experiment to investigate this interrelationship. A console workplace was selected because users experience a vast amount of cognitive workload at similar workplaces (e.g. surveillance workplaces or radar system controllers). The setup facilitated an in-depth analysis of different goal-directed movements of the upper extremities towards a target area. During the experiment the participants reached to 14 different target positions which were assigned in a random order. In parallel, they performed a mathematic processing task for information processing. Results were analyzed using ANOVA and subsequent post-hoc tests. The ANOVA of subjective workload ratings confirmed that an additional information processing task resulted into significantly longer time to prepare a goal-directed movement (F=33.55; p<0.01 and F=37.58; p<0.01). The measurement of effectiveness o^2 was larger than the effect of a spatial shift of the target location (o^2=0.11 and o^2=0.19). However, the additional task did not result into longer movement times (F=6.8; p<0.01 and F=2.1; p=0.13). A subsequent in-depth analysis of the temporal spatial trajectory revealed that a significant effect is only observed for the initial posture.

The results support the effect of an information processing task on motion behavior. It affects primarily motion planning but not motion execution. However, in order to exclude effects on the spatial movement and posture, subsequent analyses should follow.

Keywords. Digital Human Models, Human Information Processing, Multiple Information Processing Ressources

1   INTRODUCTION

Many working processes in today's industrial life are characterized by both: spatial movements and information processing. In facts there are hardly any actions with just one of them. But most workplace designers and industrial scientists consider these aspects in an isolated way and separately from each other. Spatial design of workplaces is nearly exclusively based on anthropometry and biomechanics, and hardly refers to effects of information processing and cognitive workload. Examples for this are manifold and can be found in workplace design, car interior design, or design of production workplaces. Today, there are many software tools available to support the spatial design of workplaces, e.g., Digital Human Models (DHMs). But nearly none of them includes information processing. On the other hand, designing information presentation and user interaction hardly refers to the overall spatial workplace design. Instead, it addresses primarily usability, human information processing capabilities, and perceptual and cognitive workload. Publications in the domain of human-computer interaction provide many examples for this.

But an ergonomic design requires a comprehensive approach, taking both aspects into consideration. This is supported by the results from a literature research.

1.1   Physical and mental work

In general, physical or energetic work can be divided into static and dynamic muscular work. With static work the muscular tension is in balance with external forces and no movement takes place. Despite, dynamic muscular work includes muscular contraction and movements. In most cases a mixture between both types of work can be observed (Bubb, 1992). The high amount of complexity of modern machines has led to a large portion of information processing work compared to purely physical work. Information processing is characterized by perceptive and mental attention requiring mental and cognitive resources. The resulting workload is influenced by information density and by the complexity of the input/output interface between user and system. Schmidtke (1973) already pointed out that a separation between muscular and mental workload is primarily based on purely operational principles. Both types of workload are still closely connected because an external stress always affects the whole organism.

1.2   Human information processing

Human information processing is a complex process which can be divided into various subtasks. Hart & Wickens et al. (1990) proposed that human information processing requires processing during three different phases: perception, cognition and motoric reaction It is not a single linear chain. Instead, a complex interrelation between different components takes place.

Different types of work or tasks, however, require different resources of information processing. According to the available resources work performance can be increased or decreased. If two tasks are performed in parallel, they might either compete for the same resources or require different resources. In case of shared resources performance of either of the both tasks will be decreased. Wickens et al. (1998) argue in favor of a three-dimensional separation of resources. Its dimensions are stage, modality, and codes/response. Figure 1 shows the general structure of the model.

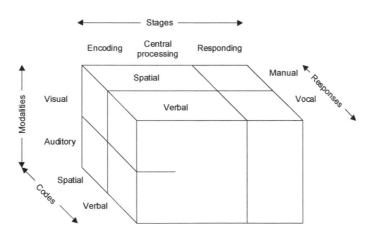

Figure 1. Model of information processing resources (based on Wickens et al., 1998)

Stage separates an early (encoding, central processing) from a late (responding) information processing. *Modality* describes the sensual modality of perceiving information. Wickens et al. (1998) proposes to divide it into visual and auditive information input. *Coding* the information differs between spatial and verbal in an early processing stage. In later stage, it differs between manual and vocal response.

1.3   Movement analysis

As stated before, spatial design and analysis often happens without considering information processing and cognitive workload. A movement analysis has to address the temporal as well as the spatial aspects of a movement for a comprehensive analysis of expected interrelationships between both. Temporal measures like reaction time and movement time only allow a first analysis. Such measures are frequently used because they can be measured easily with minimal efforts. For industrial science and planning of production processes they are frequently applied to calculate required times. Examples of this are Work Factor (WF) or Methods-Time Measurement (MTM) (Barnes, 1980; Kracht & Lorenz, 1971). But reducing the complexity of a movement to movement time seems unsatisfying and insufficient for spatial workplace design. Neither spatial positions of limbs nor motion trajectories are considered.

Other characteristics of a movement refer to the spatial positions and trajectories. They include the position of different body limbs for characteristic postures. A typical one is the maximum lateral position of the elbow. It defines spatial freedom requirements to both sides. Joint angles are also important for estimating comfort and discomfort at that point of time.

1.4   Cognitive processes and motion planning

In addition to the anatomical and physiological basis of human movements, cognition and human information processing are relevant for human movements. This includes preparatory sets as planning, attention and motivation (Ferner, 1995; Keele, 1986). The sets are carried out during motion planning and –programming. Bennet & Castiello (1995) and Schmidt & Thewes (2000) show that the preparation of a movement can be linked to activating neuronal processes in different cortical areas. The internal planning leads to straight motion trajectories of the executing body limb (Flash, 1990, 1994). A subsequent movement is initiated and the motor program is executed.

The movement itself is divided into two phases: In a first, ballistic phase no control exists and the effecting limb, e.g. the fingertip, moves towards the target region. In this phase a visual control takes place until the target is reached. Another basic psychophysical law describing the interrelationship between an index-of-difficulty of a movement and movement time was specified by Fitts' (1954). These examples show that there is a close interrelation between human information processing and human movements. This motivated an investigation of the effect of human information processing on human movements.

1.5   Functional data analysis

An analysis of potential effects requires a comprehensive description of human movements. Functional data analysis, as introduced by Ramsay & Silverman (1992) is a powerful method for it. It has also been applied for analysis of climatic data and repetitive motion data. It has also become a tool for motion analysis in workplace design (Chaffin et al., 2000; Reed, Flanagan et al., 2001). In principle, FDA is based on an exchange of the discrete (constant) values by (changing) functions. This way it facilitates a continuous analysis of functional data.

When used for the analysis of goal-directed human movements, a continuous functional description of the trajectories is required before FDA can be used. Alexander (2002) has combined this approach with a functional regression analysis for specifying an alternative, biokinematic approach for a continuous description of goal-directed motions. With a clever definition of the regression function it is possible to use the basic principles of FDA for a continuous, functional analysis of variance. For this purpose a polynomial function suit well as input function.

2   METHOD

The results of the literature research described in the previous chapter gave evidence for an interrelation between information processing and motor action. The hypothesized interrelation is analyzed with the experiment described in this chapter.

2.1   Hypothesis

It is hypothesized that information processing and, thus, cognitive workload affects goal-directed human movement. For the experiment information processing is induced by a secondary task of varying task difficulty. It is expected that an effect would primarily occur during motion preparation and planning. This will result into longer reaction times (i.e. time between stimulus and motion start). No effects are expected during actual motion execution. This requires an in-depth analysis of human movement. By applying FDA it is possible to investigate the occurrence of a potential effect during this phase throughout the total movement execution.

2.2   Participants

Sixty male subjects aged from 19 to 27 years volunteered to participate in the experiment. Pre-experimental screening considering 44 anthropometric dimensions and their interrelations showed no significant differences to the male German population of a similar age. Therefore, body dimensions are considered as representative for a larger sample of that age group.

2.3   Independent and dependent variables

Because of the goal of the study, information processing and spatial position of targets served as independent variables. The information processing task included a mathematic processing task as it will be described in section 2.5. The level of difficulty of the task was varied from none, to medium and difficult. In parallel the position of the target of the goal-directed movement was varied as the second and third independent variable. The horizontal position was the second, the vertical position the third independent variable. Details about the position are given in the following section 2.4.

The dependent variables used in the experiment were time measurements, position / trajectory data of the effecter (fingertip) and subjective workload rating. Time measurements included reaction time as a variable for the amount of cognitive processing and pre-programming. Movement time served as a variable for motor action. Because time measurements allowed no in-depth analysis, motion trajectories were also analyzed.

2.4 Apparatus

An experimental workplace of a radar controller was taken as main application. It allowed structured investigation of goal-directed motion. The experimental layout was designed according recommendations and existing console layouts (Lohr, 1989). The setup allowed a vertical orientation of the panel and also an orientation of 55° against the horizontal. A matrix of 3 (rows) x 4 (columns) switches were positioned on the panel as targets. The horizontal and vertical distance between them was constantly 20 cm. Two additional switches were positioned at the right side of the panel to get more detailed results in this area. This is because interaction elements are often positioned at that position. An IR-optical system was used for motion capture (Qualisys, 2000). The system uses spherical markers (25 mm diameter), which were attached to 15 anthropometric landmarks of the participants. Their positions were captured with a measurement rate of 60 Hz.

2.5 Experiment

Participants sat at an adjustable seat and lay the right hand onto a hand rest table in front of them. One of the switches was lighted and participants reached for it as fast as possible. After operating the switch they returned their hand back to the initial rest position and waited for the next switch to light. After each trial the angle of the panel was adjusted (90°, 55°) and the next trial started. The first two trials served as exercise and the according movements were not captured. The following two trials were captured and used for the analysis.

In parallel to the motor task the participants performed a cognitive task. Stimuli were presented acoustically; the subject's response was also by speech. In order to induce cognitive workload we used a mathematic processing task as described by Schlegel & Shingledecker (1985), Schlegel & Guilland (1990) and Shingledecker (1984). The task is standardized and established. It consists of two or three mathematic operations (+, -) which connect three or four one-digit numbers. Participants answered whether the result was larger or smaller than five. Restrictions specified by the authors were considered for the task.

A complete test took about 60 minutes, and 10.080 movements were captured.

2.6   Analysis

The analyses of reaction and movement times were performed by using Analysis of Variances (ANOVA). Because such times are usually lognormal distributed, the logarithmic values were used for the analysis.

The in-depth analysis of the spatial movement was more complicated to allow a detailed analysis. At first, a regression analysis with a polynomial function of 5th order was carried out for each of the trajectories. As shown by Alexander (2010), this is sufficient for a functional description of the trajectory of the fingertip with goal-directed movements. Subsequently, functions for mean and standard deviation were calculated for each movement under each condition. Following the principle of functional data analysis, functional ANOVAs were carried out afterwards.

To merge the different functions of the trajectories the time was normalized from $t'=0$ (start) and $t'=1$ (end). This allowed for continuously analyzing the trajectories. If a significant effect was revealed, subsequent calculation of the measurement of effectiveness o^2 followed.

3   RESULTS

The results refer to workload and the temporal effects of the additional cognitive task. The main results were expected for the spatial trajectories, so that these results require a more detailed description..

3.1   Reaction time

Temporal measures are usually lognormal distributed. This is supported for the measured data by performing a Kolmogoroff-Smirnoff test with Lilliefors-modification (Sachs, 1992). Two three-way ANOVA (task difficulty, horizontal and vertical target position) reveals a significant effect of each of the three factors on reaction time for a panel angle of 55° (F=33.55; p<0.01) and 90° (F=37.58; p<0.01). The measurement of effectiveness o^2 shows that a large fraction of the overall variance is explained by task difficulty ($o^2=0.111$ (55°) and $o^2=0.190$ (90°)). The explained variance of vertical and horizontal target position are considerably smaller (horizontal. $o^2=0.046$ at 55°, and $o^2=0.018$ at 90°; vertical. $o^2=0.017$ at 55° and $o^2=0.009$ at 90°).

3.2   Movement time

With regard to movement time, ANOVA reveals that there is a significant effect of task difficulty at 55° panel angle (F=6.80; p<0.01). At 90° there is no significant effect (F=2.10; p=0.13). The measurement of effectiveness $o^2=0.009$ is small. It is larger for horizontal target position ($o^2= 0.124$ at 90° and $o^2=0.204$ at 55°), or vertical target position ($o^2= 0.049$ at 90° and $o^2=0.168$ at 55°). The results show that task difficulty affects reaction time, and there is no effect on movement time.

110

3.2 Trajectories of fingertip

The results of the functional ANOVA used for the analysis show significant effects of each of the three factors on the movement to the front and to the side. However, the effect of the cognitive task diminishes over time and is not significant at the end of the movement (55°: t'>0.55; 90°: t'>0.75). The vertical movement is just affected at the start of a movement (t'<0.1).

The measurement of effectiveness o²(t') show that the additional cognitive task has just a minimal effect compared to the spatial position of a target. It shows that the cognitive task has a minor effect during the initial phase of a movement. As it is shown in figure 2 it is always below o²=0.05.

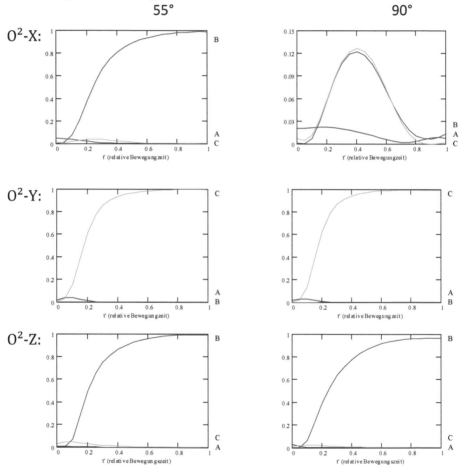

Figure 2. Measurement of effectiveness (o²(t')) for x-, y-, z-coordinates (top, middle, bottom) and for 55° (left) and 90° panel surface incline. Functions show the characteristics of task difficulty (A), horizontal (B) and vertical (C) target position from the start (t'=0) to end of the movement (t'=1).

4   CONCLUSION AND OUTLOOK

The expected effect of an information processing task on a goal-directed movement was found only for the initial phase of a movement. This is explained by a different posture, which was more upright and stretched. As a matter of facts it resulted into a different position of the hand and, thus, the fingertip. With ongoing movement this effect diminishes and the effect of the target position increased.

This result also supports the initial hypothesis which expects significant effect primarily for motion planning and not for motion execution. Our hypothesis is also supported by longer reaction times. Because of additionally required perceptive and cognitive resources there are fewer resources available for motion planning. Consequently, longer reaction times occurred. It also resulted into no effect of cognitive task on movement time. During motion execution less cognitive resources are required and the effect of a cognitive task is reduced.

The detailed analysis of the trajectories and the spatial motion behavior also supports this finding. Similar to the results described before, this effect can be neglected during actual motion execution.

LITERATURE

Alexander, T. (2010). *Funktionale Analyse und Modellierung der zielgerichteten Armbewegung.* Zeitschrift für Arbeitswissenschaft (ZArbWiss) 2/2010

Barnes, R.M. (1980). *Motion and Time Study. Design and Measurement of Work.* New York. Whiley.

Bennet, K.M.B.; Castiello, U. (Eds) (1995). *Insights into the Reach to Grasp Movement.* (Advances in Psychology - 105) Amsterdam. Elsevier Science B.V.

Bubb, H. (1992). *Menschliche Zuverlässigkeit. Definitionen - Zusammenhänge - Bewertung.* Landsberg/Lech. Ecomed.

Chaffin, D.B., Faraway, J.J., Zhang, X., Woolley, C. (2000). Stature, Age, and Gender Effects on Reach Motion Posture. Human Factors, Vol. 42 (3), pp. 408-420.

Faraway, J.J.; Hu, J. (2001). *Modeling Variability in Reaching Motions.* SAE-Paper 2001-01-2094. Proceedings of the 4rd SAE Digital Human Modeling for Design and Engineering Conference in Arlington, VA, USA. Warrendale, PA. Society of Automotive Engineers.

Ferner, M. (1995) *Empirische Untersuchung von Greifpräferenzen.* Studienarbeit an der Technischen Hochschule Darmstadt. TH Darmstadt.

Fitts, P.M. (1954). *The Information Capacity of the Human Motor System in Controlling the Amplitude of Movement.* Journal of Experimental Psychology, 1954, 47 (6), 381-391.

Flash, T. (1990). The Organization of Human Arm Trajectory Control. In. Winters & Woo (Eds). *Multiple Muscle Systems. Biomechanics and Movement Organization.* New York. Springer-Verlag.

Flash, T. (1994). *Trajectory learning and control models. From human to robotic arms.* Proc. of the IFAC/IFIP/IFORS/IEA Symposium on Analysis, Design and Evaluation of Man-Machine Systems at the Massachusetts Institute of Technology, Cambridge, USA. Cambridge, USA. Massachusetts Institute of Technology.

Hart, S.G.; Wickens, C.D. (1990). Workload Assessment and Predicition. In. Booher (Ed). *MANPRINT. An Approach to Systems Integration.* New York. Van Nostrand Reinhold, S. 257-296.

Keele, S.W. (1986). Motor Control. In. Boff, Kaufman, Thomas (Eds). *Handbook of Perception and Human Performance. Volume II. Cognitive Processes and Performance.* New York. John Wiley and Sons.

Kracht, R.; Lorenz, J. (1971). *Grundzüge des MTM-Verfahrens. Programmierte Unterweisung der Deutschen MTM-Vereinigung.* Stuttgart.

Lohr (1989). *Ergonomische Methodik Sitzarbeitsplatz Marine.* Wehrtechnischer Bericht. Kiel. Wehrtechnische Dienststelle 71.

Qualisys (2000). Product information for system MacReflex.

Ramsay, J.O.; Silverman, B.W. (1997). Functional Data Analysis. New York, Berlin. Springer-Verlag.

Sachs, L. (1992). *Angewandte Statistik. Anwendung statistischer Methoden.* Berlin. Springer.

Schlegel, R.E.; Gilliland, K. (1990). *Evaluation of the Criterion Task Set.* Technical Report AAMRL-TR-90-007. Wright-Patterson-AFB, Dayton, OH. Air Force Aerospace Medical Research Laboratory.

Schlegel, R.E.; Shingledecker, C.A. (1985). *Training Characteristics of the Criterion Task Set Workload Assessment Battery.* Proceedings of the Human Factors Society 29th Annual Meeting in Baltimore, MD. Santa Monica, CA. Human Factors and Ergonomics Socity.

Schmidt, R.F.; Thews, G. (Eds.) (2000). *Physiologie des Menschen.* Berlin, Heidelberg, New York. Springer.

Schmidtke, H. (Hrsg.) (1973). *Ergonomie 1. Grundlagen menschlicher Arbeit und Leistung.* München. Carl Hanser.

Shingledecker, C.A. (1984). *A Task Battery for Applied Human Performance Assessment Research.* Technical Report AFAMRL-TR-84-071. Wright-Patterson-AFB, Dayton, OH. Air Force Aerospace Medical Research Laboratory.

Silverman, B.W. (1992). Function Estimation and Functional Data Analysis. In Joseph (Ed). *Proceedings of the First European Congress of Mathematics.* Basel, Boston, Berlin. Birkhäuser.

Wickens, C.D.; Gordon, S.E.; Liu, Y. (1998). *An Introduction to Human Factors Engineering.* New York, Reading. Addison Wesley Longman.

CHAPTER 12

Cognitive and Affective Modeling in Intelligent Virtual Humans for Training and Tutoring Applications

Robert Sottilare and John Hart

U.S. Army Research Laboratory
Human Research & Engineering Directorate
Orlando, Florida
{robert.sottilare, john.hartiii}@us.army.mil

ABSTRACT

This chapter reviews current and emerging trends in cognitive and affective (e.g., emotions, motivational) modeling within virtual humans and their application to training and tutoring domains for individuals and small groups. Virtual humans have become commonplace in computer games and other digital entertainment applications, but their use for training and one-to-one tutoring applications is evolving and remains primarily focused on well-defined training/tutoring domains (e.g., procedural tasks and rule-based domains including mathematics and physics). In order to support viable self-regulated learning environments, future training and tutoring systems will require virtual humans with enhanced cognitive and affective capabilities that are adaptive, engaging and motivating in ill-defined domains.

Keywords: virtual humans, training, computer-based tutoring

1. INTRODUCTION

In 2002, Egges, Kshirsagar and Magnenat-Thalmann proposed a goal to create virtual humans that can interact spontaneously using a natural language, emotions and gestures in a manner similar to and even indistinguishable from real humans. Toward this goal, this chapter compares and contrasts the capabilities of virtual humans and reviews their current roles and limitations in entertainment and training. Potential roles and capabilities for future virtual humans in training and computer-based tutoring contexts are also discussed along with recommended design goals and areas for future/continued research.

To engage trainees and gain their confidence as credible actors in training environments, future virtual humans will need to be able to understand the trainee's language and interpret their behaviors to assess the trainee's states (e.g., cognition and affect) and then respond appropriately. They will also need to recognize context in the training environment (e.g., increases or decreases in progress toward training objectives). Using this trainee and training environment information, future virtual humans will be able to adapt their interaction (e.g., direction or support) to influence and even optimize learning. By affiliating with and motivating learners, virtual humans will develop rapport and trust as credible and supportive mentors.

2. COMPARING AND CONTRASTING VIRTUAL HUMANS

There is a great fascination with the creation of artificial humans. It is not a recent idea and can be seen in literature and entertainment throughout many years. In Mary Shelley's *Frankenstein*, Dr. Frankenstein desired to create an artificial human. Modern science fiction introduced us to many artificial humans such as in the Terminator films, and the human clone replicants in *Blade Runner*. There are many other examples of artificial humans developed to perform roles to assist humans or perform dangerous missions as in weapons of war in literature and entertainment. Now, imagine our world today with similar artificial humans. What about an artificial human that could act as a virtual receptionist or assist in training as a tutor or teammate. What if you could deal with a character that was unscripted, had knowledge and could reason about environment, understood and expressed emotion, communicated both verbally and nonverbally, and could play different roles as needed. Recently, renewed interest in artificial humans is making this a reality. This quest to build an artificial human is becoming a reality due to developments in virtual human technologies. It can be seen that there is a role for virtual humans in our world today and into the future.

Just as all men are not created equal, so it is true for virtual humans. Virtual humans can be thought of as "software entities [that] look and act like people and can engage in conversation and collaborative tasks, but they live in simulated environments." (Gratch, et. al., 2002). They can also include different cognitive states to include beliefs, desires, goals, intentions, and attitudes (Rickel and Johnson, 1999; Traum, Swartout, Gratch, and Marsella, 2008). Many virtual humans have been developed over the years, but each one was developed for a specific purpose as in to support training and to provide a new interface for the delivery of information. Forms of virtual humans are being used today as web-based airline reservationists as on Alaska Airlines on *www.alaskaair.com*, building receptionists as in MicroSoft Research's "Situated Interaction" project (Bohus, 2008), museum guides (Swartout, et al., 2010), Army recruiting (Artein, 2009), and training applications (Rickel and Johnson, 1999; Hill, et. al., 2006). In other areas, virtual humans can also be helpful in the medical and social sciences fields that include diagnosis, treatment, and therapy skills.

Differences in virtual humans start with how they are controlled. Basically virtual humans are controlled by either a human via a keyboard and mouse, joystick, and other interfaces such cameras and motion sensors. Human or user controlled characters are popular in computer games (ie., Call of Duty, Battlefield, Halo), massively multiplayer online games (ie., World of Warcraft, Sims Online), and virtual worlds (ie., Second Life, Active Worlds). The user controls the actions of the character in these environments. There environments do allow for some characters to be controlled via the computer and they are known as "bots". The computer controlled characters have become more than just "bots" with the development of computational algorithms in natural language, emotions, and behaviors. These developments have allowed virtual humans to perceived and respond to and within the environment without human intervention (Johnsen, Beck, & Lok, 2010; Kenny, et al., 2007; Swartout, 2010).

Johnsen (2008) describes four categories that affect the human-computer interaction when dealing with virtual humans. Those categories are the virtual human, the simulation system, the user environment, and the user. In recent years, there has been a large body of research focused on the virtual human development in the areas of anthropomorphism (Dehn & van Mulken, 2000; Yee, Bailenson, &, Rickersten, 2007); appearance (Garau, et al., 2003; Bailenson, et al., 2005; MacDorman, Coram, Ho, & Patel, 2010); and behaviors (Garau, et al., 2003; Bailenson, et al., 2005; Gratch, et al., 2007). The simulation system includes the human to computer interface. There are a number a methods for interacting with virtual humans that include speech and natural language (Traum, et al., 2007; Johnsen & Lok, 2008); text and natural language (Rizzo, et al., 2010, Sproull, et al., 1996); and menu systems (Hill, et al., 2006; Lester, Stone, & Stelling, 1999).

2.1 Roles of virtual humans in games and digital entertainment applications

As one participates in a game, virtual world, or virtual training environment, it can be analogous to an interactive drama where the user experiences the story first hand (Bates, 1992; Mateas & Stern, 2002). Virtual humans are then considered the actors within the virtual environment. In screenwriting and other entertainment, actors typically take on primary or supporting roles within the story (Stout, 2011). Primary roles include both the protagonist and antagonist. One could develop a single user interactive play model for the virtual environment where the user assumes the role of the protagonist. In this model, the antagonist and other supporting roles are played by virtual humans. Using the interactive play model, a more appropriate description of the interaction might be an improvisational play where the actors are free to an act or react to other actors within the limits of the story or script (E. H. LeMasters, personal communications, Sept. 1, 2011). Supporting roles for virtual humans include that of a team mate, adversary, instructor, tutor, or mentor.

In developing the improvisational play model, the roles a virtual human can play can be divided into either roles as an actor or as a supporting role. The supporting role provides realism for the scenario and may cause the actors to act or react to the supporting members. An example may be a crowd of people in the street, where they may become angry based on the action or decision of the primary actors. The actor roles in the improvisational play help to "drive" the story by taking actions based on the current state of the scenario. A virtual human could participate as a team member or ally, an adversary, or as a mentor or instructional role.

The training domain makes use of virtual humans in instructional roles. Frenchette (2008) describes some of these roles as supplanting agents, scaffolding agents, demonstrating agents, modeling agents, coaching agents, and testing agents. These types of agents provide different forms of instructional support to the training environment. Steve, a half-bodied human-like agent, was an early implementation of an instructional agent (Johnson, Rickel, & Lester, 2000). Steve shared the virtual environment with the student and aided them in learning how to operate shipboard equipment. In the environment, Steve would first start with a demonstration of how to perform an activity opposed to just explaining the activity. The student could ask questions of Steve, and even ask to complete an activity he had started. Steven monitored the student's performance and provided assistance when needed. Another agent was Herman the Bug (Lester, et al., 1997). Herman helped students understand biological processes for plants and provided advice in response to students' problem solving activities. Lester, et al. (1997) proposed that virtual humans can play a critical role in the motivation of students through the way the agent interacts with the student.

2.2 Roles of virtual humans in training and tutoring domains

This section examines the roles of virtual humans in sample training and tutoring environments. There are several working examples of embodied conversational agents in both training and tutoring contexts (Rickel & Johnson, 1999; Traum, et al, 2007; Person, Graesser, Kreuz, Pomeroy & the Tutoring Research Group, 2001). In the training domain, virtual humans are used to take the place of human role-players and thereby reduce the labor required to support training exercises. In the tutoring domain, virtual humans are the interface for providing feedback (e.g., direction, support, information) to the trainee.

Examples of the use of virtual humans in the training domain include virtual patients used for both training clinical therapists (Kenny, Parsons, Gratch, Leuski and Rizzo, 2007) and medical doctors performing military sick call (Kenny, Parsons and Garrity,

2010). In the tutoring domain, both AutoTutor (Graesser, Chipman, Haynes and Olney, 2005) and the Cognitive Tutor (Ogan, Aleven, Kim & Jones, 2010) use virtual human interaction to guide instruction and provide feedback to the trainee.

2.3 Cognitive and affective modeling of virtual humans in training and tutoring domains

"Emotion pulls the levers of our lives, whether it be by the song in our heart, or the curiosity that drives our scientific inquiry." (Picard, 1995). In order to fulfill the goal of being indistinguishable if one is interacting with a virtual human or human, emotion is needed. Emotion is important to the social interaction. The Institute for Creative Technologies of the University of Southern California has been researching and developing virtual human technologies over the last decade and has found that an emotional model can have an impact on the virtual human's cognitive processing. The emotional model can provide insight to assist in natural language processing such as the disambiguation of ambiguous references. The model can also inform the decision making process (Swartout, 2010). de Melo, Carnevale, and Gratch (2010) developed a virtual human to play games of prisoner's dilemma and found indications that emotions in the virtual human players had influence on the human participant's decision making. As one considers the development of a computational model for a virtual human, one quick finds that it involves a multi-disciplinary approach. The theory emotions are rooted in the science of psychology. As one begins to use facets of artificial intelligence, the development of a computational model can unveil assumptions and hidden complexities with the theory of the emotional model (Marsella, Gratch, and Petta, 2010). Some of the more popular models are based on appraisal theory, which is based on an individual's judgment concerning the relationship between events and the individual's beliefs (Lazarus, 1991). Dimensional theories characterize emotion not as discrete events but a points in a continuous multi-dimensional space. Other models include anatomic theories which attempts to reconstruct neural links and process that control emotions; rationale approaches that take a more abstract approach looking at the purpose of the adaptive function of emotion; and communicative approaches in which one views the emotional processes as a communicative system (Marsella, Gratch, and Petta, 2010).

3. DESIGN GOALS FOR EFFECTIVE VIRTUAL HUMANS IN TRAINING AND TUTORING

This section addresses constraints, learning-related design goals and desired attributes for virtual humans to be effective training and tutoring tools. Constraints focus on barriers to real-time interaction with virtual humans while learning-related design goals highlight the potential influence of virtual humans on engagement and motivation during training/tutoring sessions. In conclusion, we address virtual human perception as a gateway to providing adaptable training and tutoring.

3.1 Constraints in training and tutoring domains

As in all human-technology interactions, there is an expectation of real-time (or near real-time) natural language and gesture interaction to maintain engagement. The process of determining what the trainee says/does, what a particular phrase/gesture means and then selecting an appropriate response takes time. It would be frustrating for trainees to wait several seconds for virtual human responses. This delay can be compounded when we are thinking about applying virtual human technology in: training domains with a large volume of phrases; and in mobile and/or other distributed training/tutoring applications. Until understanding natural language technologies become more efficient, the authors recommend that application of virtual human technologies in very specific and localized contexts. More specific training applications will limit the size of the corpus that must be understood by the virtual human and reduce the search time. Localizing virtual humans to run in the same geographic location as the trainee and training application will minimize delays due to data transport on wide-area networks. Next, we will explore learning-related design goals for virtual humans to enable them to support training and tutoring.

3.2 Learning-related design goals

In assessing the design goals of virtual humans as tutors, we examined validated studies of "expert" human tutors and came across a set of studies (Lepper & Chabay, 1988; Lepper, Aspinwall, Mumme & Chabay, 1990; Lepper, Woolverton, Mumme & Gurtner, 1993; Lepper, Drake & O'Donnell-Johnson, 1997) that provided an analysis of factors contributing to successful tutoring outcomes and resulting in the INSPIRE model of tutoring success (Lepper, Drake & O'Donnell-Johnson, 1997). INSPIRE is an acronym of the attributes of a successful human tutor: intelligent, nurturing, Socratic, progressive, indirect, reflective and encouraging.

If we expect to use virtual humans as coaches, mentors and tutors, we should have design goals that include the capability for them to influence learning by influencing engagement and motivation. Below we posit how virtual humans might be designed to allow them to manage trainee engagement and motivation.

3.2.1 Virtual human influence on trainee engagement

In implementing the INSPIRE model within virtual humans to influence trainee engagement, key attributes to consider in the virtual human design are intelligence, Socratic interaction, reflection and indirect feedback. Virtual humans in the role of the tutor must be knowledgeable of the subject matter and pedagogy to demonstrate credibility and maintain the trust/engagement of the trainee. The virtual human must be able to ask leading questions and provide hints instead of directions/answers that could result in trainees losing interest in the subject matter.

Klein & Baxter (2006) assert that advanced problem solving on the part of trainees requires the recognition of flaws in their existing mental models in a process called the Cognitive Transformation Theory (CTT). CTT links learning objectives to the person's current mental models and promotes reflective processes for shedding flawed mental models for less flawed models. Virtual humans that are designed to support discovery and reflective processes per the INSPIRE model and CTT are anticipated to be more effective tutors/training partners.

3.2.2 Virtual human influence on trainee motivation

In implementing the INSPIRE model within virtual humans to influence trainee motivation, key attributes to consider in the virtual human design are the capabilities to nurture and encourage. Nurturing virtual humans should be able to develop rapport with trainees which indicates a persistent trainee model that is maintained by the virtual human (or in its cognitive architecture or in a tutoring architecture) and used to demonstrate a history of "shared experiences" and mutual trust. Lepper & Malone (1987) and Malone & Lepper (1987) noted five complementary sources of motivation for learning under the heading of encouragement. Virtual humans should be able to encourage trainees and affect their motivation by: enhancing the trainee's feelings of competence and mastery; challenging the trainee to accomplish more; piquing the trainee's sense of curiosity; providing trainees with a sense of control over the learning process; and by providing context and relevance for training content.

3.3 Desired virtual human attributes for training and tutoring

In addition to the influence on trainee engagement and motivation, it is desirable for virtual humans to be able to perceive their environment including the trainee and the training environment (e.g., virtual world, game or other simulation). Perception is a prerequisite for the virtual human to interpret the trainee's state (e.g., cognition and affect) and then use this trainee state data along with training context to formulate optimal training strategies (e.g., feedback, direction, questions, support).

Adapting training strategies based on individual differences (traits), changes in state, context and progress allows for tailored training that is optimized for the individual trainee. A tutoring architecture will provide the capability for virtual humans to recognize the state of the trainee and the training environment. Computer-based tutoring systems have concept maps to describe the trainee state and training context. Markov decision processes might be used to weight and optimize the reward resulting from movement from one state to another. For example, the tutor might determine that a trainee's learning outcome might be optimized by selecting a reflective training strategy over a directive strategy based on the trainee's current motivational level, engagement level, progress and competency. To optimize this decision, it is essential for the tutoring architecture to be able to sense the behaviors, interactions and even the trainee's physiology to determine the trainee's cognitive and affective states.

4. DISCUSSION

Finally, our discussion leads us to understanding of general research domains for modular and integrated virtual humans, their capabilities and limitations and recommendations for future research. In general, virtual humans are difficult to author and modify, they have limited understanding of their environment, limited natural language understanding and prescriptive cognitive and affective responses. Virtual humans have almost no ability to retain and use knowledge gained during a training/tutoring scenario in subsequent scenarios. So, how can we improve the adaptability of virtual humans to be more like their human counterparts?

A major objective is to enable more adaptive interaction between virtual human agents, their live human counterparts, and their environment the following research domains are on the critical path for enhanced virtual humans: natural language understanding and generation, sensing technologies, cognitive modeling, affective modeling and value modeling. The ability for virtual humans to perceive and interpret voice and gestures of multiple live trainees is limited. The ability for virtual humans to identify spoken words/phrases is improving, but the corpus is generally limited to a small number of phrases from a single user that can be recognized and interpreted for a rationale response. Research is needed to improve the capability to recognize, interpret and respond in near real-time to multi-sided conversations that include human and virtual human participants and larger topical domains to support social interaction.

Research is needed to enhance a virtual human's capability to recognize and interpret its environment including the behaviors of human participants. To support a tutoring context, virtual humans must be able to recognize changes in the cognitive state (e.g., engagement) and affective state (e.g., frustration, boredom) of trainees. Research is needed to improve the capabilities of unobtrusive physiological and behavioral sensing technologies to this end.

The modeling of how virtual humans perceive and judge their environment is tied to their cognitive and affective modeling. These models are in turn closely tied to value modeling and affect the decision-making processes of virtual humans. Incorporating value models that include ethics and personality preferences would have a significant impact on the variability of actions taken by virtual humans and would move them from prescriptive beings to true decision makers.

Finally, another major objective is to facilitate the development and authoring of new virtual human characters. Interface standards and a framework for virtual humans would go a long way toward making virtual humans easier to construct and move them from the purview of computer scientists to application domain experts (e.g., training and tutoring developers) who could use virtual human tools and standards to create virtual humans specific to their application needs.

REFERENCES

Cassell, J., Bickmore, T., Campbell, L. and Vilhjalmsson, H. (2000). Human Conversation as a System Framework: Designing Embodied Conversational Agents. In Cassell, J., Sullivan, J. Prevost, S. and Churchill, E. eds. Embodied Conversational Agents. MIT Press, Cambridge, MA.

de Melo, C. M., Carnevale, P., and Gratch, J. (2010). The Influence of Emotions in Emobdied Agents on Human Decision-Making. In J. Allbeck, N. Badler, T. Bickmore, C. Pelachaud, and A. Safonova (Eds.) Intelligent Virtual Agents. 10th International Conference, IVA 2010. Springer.

Egges, A., Kshirsagar, S. and Magnenat-Thalmann N. Imparting individuality to virtual humans. In First International Workshop on Virtual Reality Rehabilitation (Mental Health, Neurological, Physical, Vocational) (November 2002).

Graesser, A.C., Chipman, P., Haynes, B.C. and Olney, A. (2005). AutoTutor: An Intelligent Tutoring System With Mixed-Initiative Dialogue. IEEE Transactions on Education, VOL. 48, NO. 4, November 2005.

Johnsen, K. (2008). Design and Validation of a Virtual Human System for Interpersonal Skills Education. (Doctoral dissertation, University of Florida, 2008) Dissertation Abstracts International, DAI-B 69/10. (UMI No. 3334476).

Johnsen, K., Beck, D., and Lok, B. (2010). The impact of a mixed reality display configuration on user behavior with a virtual human. In J. Allbeck et. al. (Eds.), Intelligent Virtual Agents 2010. (pp. 42-48). Springer-Verlag Berlin Heidelberg 2010.

Johnson, W. L., Rickel, J. W., & Lester, J. C. (2000). Animated pedagogical agents: Face-to-face interaction in interactive learning environments. International Journal of Artificial intelligence in education, 11(1), 47–78.

Kenny, P., Parsons, T., Gratch, J., Leuski, A., and Rizzo, A. (2007). Virtual Patients for Clinical Therapist Skills Training. In C. Pelachaud et al. (Eds.), Intelligent Virtual Agents, pp. 197-210. Springer-Verlag Berlin Heidelberg 2007.

Kenny, P., Parsons, T., Pataki, C., Pato, M., St-george, C., Sugar, J. and Rizzo, A.A. (2008). Virtual Justina: A PTSD Virtual Patient for Clinical Classroom Training. Review Literature And Arts Of The Americas, 6(1), 113-118.

Kenny, P., Parsons, T. and Garrity, P. Virtual Patients for Virtual Sick Call Medical Training. In Proceedings of the Interservice/Industry Training, Simulation, and Education Conference (I/ITSEC) 2010 (November 2010).

Klein, G. and Baxter, H. (2006). Cognitive Transformation Theory: Contrasting Cognitive and Behavioral Learning. Presented at the Interservice/Industry Training Systems and Education Conference, Orlando, Florida, December 2006.

Lazarus, R. (1991). Emotion and Adaptation. NY, Oxford University Press.

Lester, J. C., Converse, S. A., Kahler, S. E., Barlow, S. T., Stone, B. A., & Bhogal, R. S. (1997). The persona effect: affective impact of animated pedagogical agents. Proceedings of the SIGCHI conference on Human factors in computing systems (p. 359–366).

Lester, J. C., Towns, S. G., & Fitzgerald, P. J. (1999). Achieving affective impact: Visual emotive communication in lifelike pedagogical agents. International Journal of Artificial Intelligence in Education, 10(3-4), 278–291.

Marsella, S., Gratch, J. and Petta, P. (2010). Computational Models of Emotion. In in Scherer, K.R., Bänziger, T., & Roesch, E. (Eds.) A blueprint for a affective computing: A sourcebook and manual. Oxford: Oxford University Press.

Ogan, A., Aleven, V., Kim, J., & Jones, C. (2010). Intercultural negotiation with virtual humans: The effect of social goals on gameplay and learning. In V. Aleven, J. Kay, & J. Mostow (Eds.), Proceedings of the 10th international conference on intelligent tutoring systems, ITS 2010 (Vol. 1, pp. 174-83). Berlin: Springer.

Person, N. K., Graesser, A. C., Kreuz, R. J., Pomeroy, V., & the Tutoring Research Group (2001). Simulating human tutor dialogue moves in AutoTutor. International Journal of Artificial Intelligence in Education, 12, 23-39.

Picard, R. W. (1995). Affective Computing. (Technical Report No. 321). Cambridge, MA: M.I.T Media Laboratory Perceptual Computing Section. Retrieved from http://affect.media.mit.edu/pdfs/95.picard.pdf on February 26, 2012.

Rickel, J. and Johnson, W.L.: Virtual humans for team training in virtual reality. In: Proceedings of the Ninth International Conference on Artificial Intelligence in Education, IOS Press (1999) 578-585.

Rizzo, A. a, Lange, B., Buckwalter, J. G., Forbell, E., Kim, J., Sagae, K., Williams, J., et al. (2010). An intelligent virtual human system for providing healthcare information and support. Proceeding of 8th International Conference on Disability, Virtual Reality & Associated Technologies (Vol. 163, pp. 503-9). Vina del Mar/Valparaiso, Chile. Retrieved from http://www.ncbi.nlm.nih.gov/pubmed/21335847

Swartout, W. (2010). Lessons Learned from Virtual Humans. AI Magazine. Spring 2010.

Swartout, W., et al. (2010). Ada and Grace: Toward Realistic and Engaging Museum Guides. In J. Allbeck, N. Badler, T. Bickmore, C. Pelachaud, and A. Safonova (Eds.) Intelligent Virtual Agents. 10th International Conference, IVA 2010. Springer.

Traum, D., Roque, A., Leuski, A., Georgiou, P., Gerten, J., Martinovski, B., Narayanan, S., Robinson, S., & Vaswani, A. (2007). Hassan: A Virtual Human for Tactical Questioning. Proceedings of the 8th SIGdial Workshop on Discourse and Dialoque, pages 71-74 (Antwerp, Belgium, September 2007).

CHAPTER 13

Intelligent Agents and Serious Games for the Development of Contextual Sensitivity

*Alberto Tremori[1], Claudia Baisini [2], Tommy Enkvist[3],
Agostino G. Bruzzone[4,] James M. Nyce[5]*

[1,4] University of Genoa,
Genoa, Italy
alberto.tremori@simulationteam.com, agostino@itim.unige.it
[2] Swedish National Defence College
Stockholm, Sweden
Claudia.Baisini@fhs.se
[3] Swedish Def. Research Agency, Dep. of Psychology, Uppsala University, Sweden
tommy.enkvist@foi.se
[5] Ball State University,
Muncie, IN, USA
jnyce@bsu.edu

ABSTRACT

This paper suggests an alternative approach to develop the necessary competences to handle decision making in complex environments, stemming from the challenges that the military has faced in the past ten years of conflicts. However, the authors believe, the ideas suggested here can be employed and benefit any other field requiring agile and adaptive thinking. The conflicts in which our Armed Forces are engaged are largely characterized by Interactive Complexity: the system is nonlinear, its proportions unstable and cause-effect patterns ambiguous. Large civilian presence and involvement, difficulties in identifying possible threats, high tempo, and dense terrain are typical features of the so called "three block war", introduced by Gen Krulak, which requires the capability of making a broad range of decisions in little or no time at the micro tactical level. In order to obtain optimal situational awareness, it is necessary to provide the necessary skill set (not tools, not rules) that allows to 'read' the operational environment and understand its regulating rules, rather than applying frames of reference that are derived from the Domestic environment. The capability of learning from the Context, stretching the dominant mental models and transcend the obvious is crucial. Research suggests that visual orientation can be an important feature for a group leader especially in urban scenarios; what one sees and how he interprets this can be decisive. Furthermore, he must do it fast, which is why the visual dimension and Intuition emerged as so critical: he must get that "Coup d'Oeil" that was considered crucial by Napoleon, and by many after him. This paper suggests a training framework to develop effective and innovative solutions by Modeling & Simulation (M&S) and specifically by Serious Games, to establish a solid ground for re-framing and creative decision making. The proposed approach moves away from traditional Simulation and Gaming products, where the tendency is to represent reality. We suggest a training that focuses on developing cognitive skills to recognize what is salient in an Operational Context characterized by a high level of Interactive Complexity.

Serious Games and Agent Driven Simulation based on Intelligent Agents and Human Behavioral Models Libraries are the proposed M&S methodologies. Serious Games can provide an opportunity to improve performances with reduced efforts and great attention to story and emotional involvement; Agent Driven Simulation based on Intelligent Agents and Human Behavioral models allows to create complex scenarios, considering even the behavior of individual agents, and, at the same time, can be developed rapidly and cost efficiently. This paper suggests an "educational path" from portable serious game solution to more complex and immersive synthetic environments which can educate the user's intuitive thought in recognizing key patterns that are contextually salient. At the same time we want to encourage user involvement so as to support some of the key factors in tactical decision making process: speed and stress. The idea here is to develop an "educational path" to create a set of tools that move from general cultural models to very specific contextual elements using scenarios with validated notions of context "to drive" the acquisition of expert knowledge by different scalable architectures, representations, and methodologies for decision makers' education. The use of Behavioral Models can support this cultural and technological "scalable" approach by providing common, well validated, basic models to re-create agents' actions that could be applied to different scenarios. This paper will also sketch out some different solutions for ergonomic issues: from the simple hand-held device to immersive environment with all senses involved: sight, sound and smell.

Keywords: Intelligent Agents Computer Generated Forces, Serious Games, Contextual Learning

1 «LE COUP D'ŒIL», INTUITION AND CATEGORIZATION

Napoleon referred to the intuitive capability to rapidly assess a situation and make a fast decision as "Coup d'Oeil" or "strike of the eye": he believed it was a gift of nature. In fact, behavioral psychologists have identified the creative-intuitive personality as being "alert, confident, foresighted, informal, spontaneous and independent. Not afraid of its experiences, accepts challenges readily, unconventional yet comfortable in its role, able to live with doubt and uncertainty, and not afraid of exposing to criticism" (Mrazek, 1972). However, the military believes that although heredity and personality certainly play a role here, intuition can be cultivated and developed.

One definition of intuition is: "intuition is a developed mental faculty which involves the automatic retrieval and translation of subconsciously stored information into the conscious realm to make decisions and perform actions. Organized databases of knowledge gained through education – experiences, memorization, sensations and relationships – are the building blocks for intuitive thought" (Reinwald, 2000, p.86).

While MDMP (Military Decision Making Process) has demonstrated its effectiveness in long term planning, it carries some risks, identified in the literature as "bounded rationality" (Wolgast, 2005). An alternate approach is the one named as Intuitive (or Naturalistic) Theory of decision making based on the premise that individuals often use less formal but much faster decision making strategies (Gigerenzer et al. 2004, 1999; Gilovich et al., 2002) in real time situations (Bryant et al., 2003).

The "Coup d'Oeil" refers to the instant, global understanding of a situation. This refers to what the eye seizes, both literally and metaphorically. It is also the ability to see the whole and also to see what is not there, and act.

Stemming from Klein's definition of intuition as based on "experience to recognize key patterns that indicate the dynamic of a situation" a problem arises: a soldier has no experience of the local environment in which he is deployed, therefore his intuition would be based on experiences, patterns and dynamics that emerge from, and are applicable to, his prior context and circumstances. Such patterns, however, not necessarily are applicable to the context in which he is deployed and often can be deceptive. These issues are key to "understanding" and "learning" about the operational environment. The way in which an individual sees the world is the product of the individual's personal history, experiences, upbringing, personality, and of his social context. However, these frames not necessarily apply to the context in which he is to conduct his operation; in this case they can be misleading, as illustrated in the example below. Both Figure 1 and 2 could be a main road to a rural community anywhere but Figure 2 looks exotic to most Westerners.

126

Figure 1: Private Photo   Figure 2: Private Photo

It represents the most important road in the State of Malawi. We know that because there is always someone walking aside the road, any time of the day or night. However, as read through our Western experience, this would never be considered as a main road. This is because, according to what is normal for us, on a main road there would be no one walking along it and there would be a lot of car traffic. In fact, most of us would suspect that something is wrong if driving on an Interstate we would see many people walking aside it. Hence, if I found myself in Malawi and saw the road as illustrated in Figure 1, it would be the most relevant indicator for local people (or an intuitively skilled war fighter) that there **is** something wrong. This raises the question of to what extent can we generalize knowledge about local conditions.

2   THE CHALLLENGE FROM INTUITION AND CATEGORIZATIONS

When we face a new situation, the brain sets it in relation to existing knowledge and experiences (not only knowledge we are aware of, but also the so-called tacit knowledge and unconscious memories) (Duggan, 2005). The two paradigms illustrated below give examples of how experience drives perception (in this case Visual Perception).

<u>Contextual Cueing:</u> the contextual cueing paradigm states that "visual context can assist localization of individual objects via an implicit learning mechanism" (Chun & Olson, 2002)

<u>Categorizations</u>: categorization works best in familiar environments in which decision makers can rely on habitual associations. In novel environments, previously learned categories may not be appropriate and may in fact hinder adaptive decision-making, leading to unfavorable outcomes.

The results illustrated above challenge traditional training and simulation paradigms, especially those that attempt to mimic reality (the context) as perfectly as possible, to train the soldier in a sort of automatic "internalized and reflexive response". We need to take these representations one step further and address elements which cannot be handled well in photorealistic simulations as presently implemented. What is crucial here is to be able to see what we are not conditioned to see. To achieve this, we need to move education, training, and simulation technology beyond a concern with detecting and reinforcing certain rules of behavior or by producing "better" and "better" reproductions of reality.

We suggest approaching the operational environment as a system characterized by Interactive Complexity. These systems are *not proportional, replicable, or additive, and the link between cause and effect is ambiguous. They are inherently unstable, irregular, and inconsistent.* Reductionism and analysis are not as useful with interactively complex systems because they can lose sight of the dynamics between the components (Baisini, 2009).

We also rely upon theories of Single and Double Loop Learning; learning can reflect in a change in behavior (single loop learning), or in a change in behavior and of the criteria/paradigms that were regulating it (double loop learning). Learning is defined as a process of "cognitive restructuring", "frame breaking" or "reframing". (Argyris & Schön, 1996; Schein, 1992).

The approach suggested here builds upon these learning theories and focuses on generating the disconfirmation necessary to bring about re-thinking and re-framing and, with continuous use, can increase mental agility.

Another big challenge is due to redirect attention: research reveals that decreased attention, due to time pressure or distraction, increases the impact of implicit effects of cues on judgment (Greenwals & Banaji, 1995).

3   THE CONCEPT: SERIOUS GAMES AND HUMAN BEHAVIOR MODELS

Moving away from traditional Simulation and Gaming products, where the tendency is 'to represent' reality, we suggest a training that focuses on developing cognitive skills to recognize what is salient in an Operational Context characterized by a high level of Interactive Complexity. The task should be to trigger disconfirmation that generates rethinking, increase awareness of implicit effect of categorizations in order to develop the ability (mental agility) to move beyond what is obvious to them. This leads individuals to re-frame perception according to the context in which one is immersed, in an adaptive manner.

The methodology and the training tools we propose are based on the usage of simulation and in particular on Serious Games (Abt., 2002; Michael and Chen, 2005; Bergeron, 2006; Iuppa and Borst, 2006). According to David Kolb Experiential Learning Model (Kolb and Fry 1975) the learning process is composed of four main steps: (1) concrete experience, (2) observation of and reflection on that experience, (3) formation of abstract concepts based upon the reflection, (4) testing the new concepts. Based on this model, Serious Games can be designed to facilitate iteratively proceeding all the above four steps. Serious Games can be a very powerful tool for developing skills and analytical capacity for decision-making, in knowledge development and also to change individual and personal attitudes.

For the purpose of developing Contextual Sensitivity it might be necessary to add realism by including Human Behaviors through Artificial Intelligence and to add active entities in the game.

The authors represent a very good sample of interdisciplinary approach

128

including experts in learning theories, anthropology, psychology, and engineers with a long experience in Modeling and Simulation. This part of the team has a long and consolidated experience in research and development of Computer Generated Forces managed by Intelligent Agents based on Human Models. The experience of the M&S team extends from project PIOVRA (sponsor EDA) and MIAC (Private Industry) to the on-going CAPRICORN (EDA) and CGFC4IT (Italian MoD) where behavioral models have been studied and developed (fear, stress, fatigue…) for reproducing complex, non conventional scenarios with a strong involvement of civilian. (Bruzzone et al. 2007, 2009, 2011, 2012).

Based on these experiences we plan an extensive use of Intelligent Agents Computer Generated Forces (IA_CGF) for managing numerous actors in the game with autonomous and "intelligent" behaviors. The IA_CGF are a family of behavioral libraries, units and non-conventional frameworks based on authors previous researches and developed by Simulation Team (MAST and DIME-University of Genoa) (Bruzzone, Tremori at al. 2007, 2009, 2011). The intelligent agents can manage the behavior of the different human actors in the scenario reproducing normal and anomalous behaviors based on human factors and social, political, and economic parameters for that specific area. In Figure 4 we have a sample of chart summarizing the Fear Conceptual Model used for IA_CGF Human Behavioral Libraries.

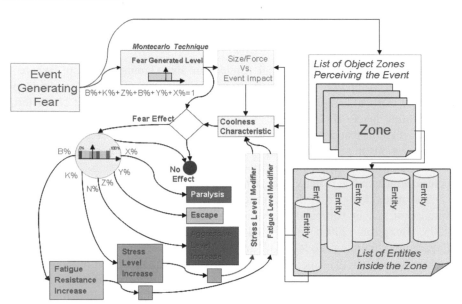

Figure 4: Fear Conceptual Model

For the purpose of training Coup d'Oeil we suggest an "educational path" that starts from a simple, portable serious game solution and can move to more complex and immersive synthetic environments. Throughout, the intention is to encourage intuitive thought and the recognition of key patterns mainly based on the behavior of humans in the scenario. The goal is also to engage users and reproduce the key factors in tactical decision making process: speed and stress.

The idea in the "educational path is to create a set of tools that move from general models to the very specific local habits and uses.

A deployed soldier has to be aware of local habits, what it is normal and what is not. But at the same time we suggest that his skills can be extended to include a contextual sensitivity that goes beyond rough approximations of what is normal or not, to notice shades and nuances. The end result is that users can then approach totally new environment(s) with enhanced capabilities to analyze and understand (see Figure 5).

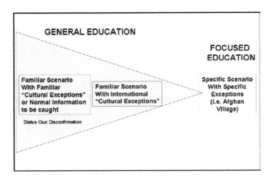

Figure 5: Educational Path from general to specific training

From a technological point of view this strategy could support different kinds of architectures and technologies: from the simple hand-held device where the serious game is running (see next paragraph "Prototype") to immersive environments where different senses are stimulated, such as sight, sound and also smell. With what we call a "Sensory Box" (see sample in figure 6) we can stimulate the users with different types of non-coherent stimuli (i.e. the synthetic environment representing an afghan village with the noise of New York and the smell of grandma's cookies) to create disorientation and support different key patterns discrimination.

 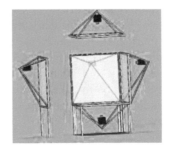

Figure 6: Sketch of a Sensory Box

The next step is to provide less familiar scenarios: for instance for US users moving in a European town. In this new and "less familiar" scenario we could start provide exceptions to be recognized with characteristics of the environment we are reproducing.

These training tools can also support light portable gaming.

4   PROTOTYPE: A SERIOUS GAME

We suggest the development of a prototype that could include still shots and animated scenes, to represent logics that that do not correspond to familiar categorizations or common sense. Such scenes/images would be created in order to:

◊   Identify and enhance the effects of existing categories and stereotypes.

◊   Generate "disconfirmation"

◊   Require the user to increase attention, observation skills, and interaction with the context in order to learn and identify clues relevant to understanding the logic underlying the scene.

◊   Relate to the context as a living system, one of constant movement and change rather than having a static picture as often found in Handbooks and briefs.

◊   Combine the identified cues in multiple ways building different plots in order to challenge common sense and stretch the boundaries of own categories to re-define them (Re-think/Re-frame).

◊   Identify what is happening in relation to the context.

As an example, imagine a very simple portable application such as a game running on a hand-held device (i.e. tablet pc or i-pad) in which the user is involved in a competition in a synthetic environment representing a familiar scenario. Two main issues here are to be 'worked through':

◊   Status Quo Disconfirmation (or Unfreezing): the rules of the competition are not explained and they are not consistent with one's common sense and normal rules (i.e. in a car race where you do not score if you go faster than others, but rather you win when minimize travel (distance not time). Further, one is not confined to using roads). The user soon learns that he cannot increase his score by applying the usual frame of reference. In order to win then, he has to find the right way to score and so has to re-think his customary behaviors (and the logics underlying them).

Recognize Key Patterns: while the user is focusing attention on the contest, due to the unusual rules (that create cognitive confusion), we start to inject in the scenario incongruent elements. This would lead the user to re-create experience in ways that demonstrate the value of re-thinking 'common sense'. One strategy might be to have users become familiar with the game environment, assume that this is a 'stable' reality, and then slowly destabilize the 'taken for granted' order of things.

5   CONCLUSIONS

Most existing research attempts to address problems related to Situational Awareness by trying to build models that reproduce the reality of the operational environment (for critiques of this strategy in a maritime environment see Dekker, 2009). Often training is focused on notional or normative models designed to teach "how things are" or "how they are supposed to be". This project takes an alternative path – one that can enhance users' cognitive systems by teaching them to learn by observing and engaging with operationally relevant field indicators, in more fundamental adaptive ways. This will be based on Serious Games where Human Behavioral Models will be widely used to reproduce complexity of scenarios where humans (military but in particular civilians) are involved. The intention is to support users to become more receptive to the signals sent by specific and generic environments. Further, it will help users "boot strap" from one environment to another without the need to return to first base each time. Most learning and training scenarios work off of a set of pre-defined, deterministic categories. This is true even with sense making models. How to reframe and train for beyond the obvious is an issue that has relevance not only in relation to the Modern Operational Environment, but to all branches of Decision Making. Understanding complex environments and to be able to extrapolate from one context to another would strengthen the war fighter's cognitive assets and help him or her meet the challenges of today's warfare, providing the ability to succeed also, and particularly, when reality turns out to be different from 'how it was supposed to be' (according to the handbook).

REFERENCES

Abt C.C. (2002) Serious Games, University Press of America

Argyris, C. and Schön, D. (1996) Organizational learning II: Theory, method and practice, Reading, Mass: Addison Wesley

Baisini C., Tremori A., Enkvist T., Bruzzone A.G., James M. N. (2011), White Paper "Agile Intuition An innovative approach for educating Context Sensitive Coup d'Oeil"

Baisini, C and Deinlein, E (2002) Knowledge in Knowledge Intensive Organinzations. The case of Crime Investigation and Consulting Firms. Masters Thesis. Gothenburg School of Economics and Law, Graduated Business School.

Baisini, C. (2009) USMC and US Army at the Aftermath of OIF and OEF, Internal Report for Swedish National Defense College, Department of Command & Control Studies

Baisini, C. and Nyce, JM (2010) Lethal Modeling, Military Intelligence Professional Bulletin, June-July

Baisini, C. Bjurström, E, Gemeinhardt, D. (2010) Incorporating Contextual Sensitivity and Metacognition into Law Enforcement and Intelligence Activity, FBI Terrorism Research and Analysis Program, Vol I, US Government Publication

Bergeron B, (2006) Developing Serious Games, Charles River Media

Bruzzone A.G. (1996) Object Oriented Modelling to Study Individual Human Behaviour in the Work Environment: a Medical Testing Laboratory, Proc. of WMC'96, San Diego, January

Bruzzone A.G. (2009) Intelligence and Security as a Framework for Applying Serious Games, Proceedings of Serixgame, Civitavecchia, November

Bruzzone A.G. Tremori A., Massei M. (2011) Adding Smart to the Mix, Modeling Simulation & Training: The International Defense Training Journal, 3, 25-27, 2011

Bruzzone A.G., Cantice G., Morabito G., Mursia A., Sebastiani M., Tremori A. (2009) CGF for NATO NEC C2 Maturity Model (N2C2M2) Evaluation, Proceedings of I/ITSEC2009, Orlando, November 30-December 4

Bruzzone A.G., Elfrey P.,Cunha G., Tremori A. (2009) Simulation for Education in Resource Management in Homeland Security Proceedings of SCSC2009, Istanbul, Turkey, July

Bruzzone A.G., Frydman C., Tremori A. (2009) CAPRICORN: CIMIC And Planning Research In Complex Operational Realistic Network MISS DIPTEM Technical Report, Genoa

Bruzzone A.G., Massei M. Tremori A., Bocca E., Madeo F., Tarone, F. (2011) CAPRICORN: Using Intelligent Agents and Interoperable Simulation for Supporting Country Reconstruction, Proceedings of DHSS2011, Rome, Italy, September 12 -14

Bruzzone A.G., Massei M., Tremori, A. (2009) Serious Games for Training and Education on Defense against Terrorism - NATO MSG-069 Symposium Use of M&S in: Support to Operations, Irregular Warfare, Defence Against Terrorism and Coalition Tactical Force Integration", Bruxelles, Belgium October 15, 16

Bruzzone A.G., Tremori A., Madeo F., Tarone F, (2012) "Intelligent agents driving Computer Generated Forces for simulating human behaviour in urban riots", Under Publication on International Journal of Simulation and Process Modelling (IJSPM).

Bryant Dr. D. J, Webb Dr. R. D.G. and McCann C. (2003) Synthesizing two approaches to decision-making in Command and Control Canadian Military Journal Spring

Chun, M and Marois, R. (2002) The dark side of visual attention Current Opinion in Neurobiology 12.XXX

Chun, M and Olson, I. (2002) Perceptual Constraints on Implicit Learning of Spatial Context Visual Cognition 9 (3) pg. 273-302

Chun, M. (2000) Contextual Cueing of Visual AttentionTrends in Cognitive Sciences Vol. 4, No. 5 May pp.170-177

Dekker, Dahlstrom, van Winsen, Nyce (2009) Fidelity and Validity of Simulator Training, in Theoretical Issues in Ergonomics Science, Taylor & Francis Ed, 29 January

Duggan, W. (2005) Coup d'Oeil: Strategic Intuition in Army Planning, SSI Publication, US Army War College

Gigerenzer, G. (2004). Fast and frugal heuristics: The tools of bounded rationality. In D. Koehler & N. Harvey (Eds.), Blackwell handbook of judgment and decision making. Oxford: Blackwell.

Gigerenzer, G., Todd, P. M., & the ABC Research Group (1999) Simple heuristics that makes us smart. Oxford: Oxford University Press.

Gilovich, T., Griffin, D., & Kahneman, D. (2002). Heuristic and biases. The psychology of intuitive judgment. Cambridge: Cambridge Univeristy Press

Greenwals, A. & Banaji, M. (1995) Implicit Social Cognition: Attitudes, Self-Esteem, and Stereotypes, Psychological Review, Vol. 102, No.1, 4-27

Iuppa N, Borst T (2006) Story and Simulations for Serious Games: Tales from the Trenches Focal Press

Klein, G. (1998) Sources of Power: How People Make decisions MIT Press Cambridge MA

Klein, G. (2007) Flexecution as a Paradigm for Replanning, IEEE Computer Society Vol. 22, No. 5

Kolb. D. A. and Fry, R. (1975) Toward an applied theory of experiential learning. in C. Cooper (ed.) Theories of Group Process, London: John Wiley

Krulak Gen , C. (1999) Cultivating Intuitive Decision Making Marine Corps Gazette, May

Massei M., Tremori A., Pessina A., Tarone F (2011) Competition and Information: Cumana a Web Serious Game for Education in the Industrial World, Proceedings of MAS2011, Rome, Italy, September 12 -14

Michael D, Chen S (2005) Serious Games: Games That Educate, Train, and Inform, Course Technology PTR

Mrazek Col, J. (1972) Intuition: an instantaneous backup system?Air University Review January- February

Reinwald Maj, B. R. (2000) Tactical Intuition Military Review September-October p. 88

Schein, E. H. (1992) Organizational Culture and Leadership. 2d. Ed. San Francisco, CA.: Jossey Bass

Wolgast, K. (2005) Command Decision Making: Experience Counts, U.S Army War College Carlisle Barracks

CHAPTER 14

Outdoor Augmented Reality for Adaptive Mental Model Support

Jan A. Neuhöfer, Thomas Alexander

Fraunhofer Institute for Communications,
Information Processing and Ergonomics
55343 Wachtberg, Germany
jan.neuhoefer@fkie.fraunhofer.de

ABSTRACT

Initially, this paper gives a brief introduction to the principles of Outdoor Augmented Reality, its impact on mental model buildup in a military context and an overview over a new system developed at Fraunhofer FKIE. The system features differential GPS for translational tracking, inertial sensors for rotational tracking and a modular, helmet-attachable vision mockup with a digital camera as source for the reality context. To assure always up-to-date information feed, a wireless information infrastructure has been designed, in conjunction with an executive command information system and an advanced blue force tracking approach. With special focus on enhanced situation through Augmented Reality, details of an empirical study are conducted, focusing on the interplay of colors, symbols and quantity information as components of NATO tactical signs, used as additional information in the open field. Twenty male subjects have been tested on their capabilities of situation awareness creation by utilizing a laboratory setup of the AR system. Results are discussed, and as a conclusion, an adaptive information layer approach is proposed to handle information overflow.

Keywords: Outdoor Augmented Reality, Situation Awareness, Mental Model, Information Layers, Tactical Signs

1 INTRODUCTION

The future dismounted soldier will be confronted with high requirements on physical and mental capabilities. Strong emphasis on personal accountability as well as direct confrontation with unpredictable threats (asymmetric warfare) demands a new type of personal equipment, granting an always up-to-date level of the information status. Since additional equipment brings about additional weight, an ergonomic integration into the soldier's personal equipment is of major importance. At this stage, digital human models can be utilized for early geometric and anthropometric workload evaluation.

Since information sources are manifold (mission briefing, radio, reconnaissance in advance, updates via network etc.), new technologies must be investigated to merge, filter and prepare these information streams. Outdoor Augmented Reality (OAR) is such an upcoming technology, displaying all mission-relevant information and latest enemy movements directly into the operator's field of view. In order to realize OAR, the reality is captured with a digital video camera and merged with virtual content which relates to the environment the user is moving in (Figure 1). This can be geographical data, but also any other information referring to any dynamic or static objects of interest. As video-based solution, a combination with night vision and thermal imaging can generate an additional operational advantage.

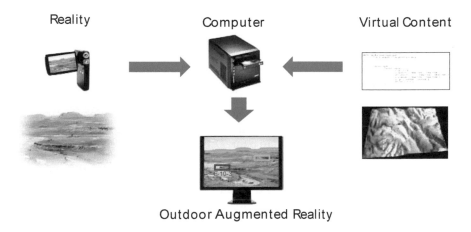

Figure 1 Principle of video-based outdoor Augmented Reality

The foremost precondition for spatially accurate placement of the virtual information into reality is that the soldier's position as well as his viewing direction (known as "outdoor tracking", e.g. by differential GPS, inertia sensors etc.) are determined with appropriate accuracy. For the virtual content, crucial conditions especially in military scenarios are reliability, accuracy and relevance. Thus, the OAR system must be embedded in an information infrastructure.

2 SITUATION AWARENESS AND AUGMENTED REALITY

The process of mental model creation directly depends on the soldier's situation awareness. Many scientific models of situation awareness have been brought forward in the last decades amongst which the approach of Mica Endsley (1995) is one of the most popular and accepted ones. The model shown in Figure 2 is based on that approach, comprising situation awareness to include three levels: perception, comprehension and projection. This cascade finally leads to a decision as a precondition for the action, impacting the situation. Of course, besides the situation in back-loop, there are individual factors as well as environmental factors influencing the outcome of perception, comprehension and projection.

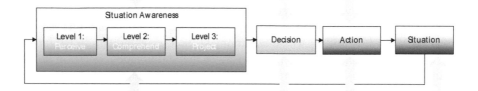

Figure 2 Situation Awareness model, based on Endsley (1995)

OAR is supposed to influence the soldier's situation awareness in all three levels of situation awareness. For example, regarding perception, bright colors and highlighting can draw the user's attention to enemy appearances and movements early before he or she would have notices without any augmentation. In terms of comprehension, color coding is useful for better comprehension of a scene, especially when vision conditions are poor. Different colors (blue, red or green) can be used for quick differentiation between object affiliations (friendly, enemy or neutral). And finally, OAR technology is capable of visualizing the invisible, e.g. what is inside a building or behind a mountain. Thus, planning and projection of actions of both own and enemy actions become clearer.

3 RESEARCH GOALS

Examples for indoor Augmented Reality are widely spread and reach from robot programming support (Chong et al., 2007) to maintenance and overhaul (Kleiber & Alexander, 2011), but examples of functional high-accuracy OAR are comparably rare. One major reason is that indoor Augmented Reality can rely on a steady infrastructure and a static environment which facilitates tracking, e.g. with infrared video technology. For OAR, higher effort is required for viable results since the working volume is much bigger, and the usability depends on many more side conditions like weather and temperature. Examples like ARVISCOPE (Behzadan, 2011), TINMITH (Avery et al., 2008) or VIDENTE (Schall et al., 2008) use GPS, combined with additional sensors like inertia, laser distance measuring etc. for localization. So, high-accuracy outdoor tracking is the most critical challenge, followed by general ergonomic usability in the open field.

Besides the technologic view, research focuses on cognitive aspects because especially for military applications, situations can get confusing and mentally demanding. This is why the AR system's human-computer-interface needs to adaptively support mental model creation, differentiating between primary (important) and secondary (negligible) information and setting information density appropriately. As a standard, augmented information may be constructed from a predefined set of signs. In this case, official NATO tactical signs serve best to transmit information compatible to standard information systems. Still, the question is: How much and which kind of information chunks may be presented to optimally support the user's mental model of the situation for an adequate level of situation awareness?

4 OUTDOOR AUGMENTED REALITY SYSTEM

4.1 Components

A personal computer is responsible for merging reality with virtual content, so a ruggedized tablet PC is the centerpiece of the functional demonstrator. It is worn attached to a harness at the back of the soldier. For translational positioning, a differential GPS device is applied to the harness, sending the data to the ruggedized tablet PC via Bluetooth. The device is capable of receiving both GPS and (Russian) GLONASS satellite signals on 72 channels and two frequencies (L1 and L2), but requires an additional antenna which has been placed on a modular carrier on the helmet of the soldier for undisturbed reception. The differential GPS operates with the help of a correction signal received via GPRS and reaches centimeter accuracy under ideal conditions. The carrier is equipped with a flap, so that the vision unit connected to the flap can be utilized as needed by flapping it down, but does not disturb the natural view when flapped up. The vision unit itself consists of a monocular, head-mounted display (HMD), an inertial sensor with three degrees of freedom (3DOF) including a digital compass and a digital camera. Effectively, the camera is placed as close as possible to the HMD to reduce cyber sickness. The integration has been done with a digital human model for easy experimentation with different attachment setups. The result is shown in Figure 3. Since the tablet PC comes with an integrated, non-reflective display, not only a single soldier shall benefit the augmented view, but the whole team. Thus, the role of a soldier equipped OAR can be considered as a specialist, like a radio operator. Strong emphasis has been put on the system's modularity and exchangeability: In case that one component drops out, it can easily be replaced and when the soldier is hit, the system can be transferred to another team member.

138

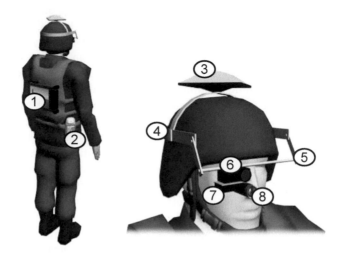

Figure 3 Components of the integrated outdoor Augmented Reality vision system.
(1): Ruggedized UMPC, (2): Differential GPS device, (3): External GPS antenna,
(4): Detachable Carrier. (5): Flap, (6): Monocular HMD, (7): 3DOF Inertial Sensor,
(8): Digital camera

Merging the live video stream with virtual content on the software side is done with FKIE's framework "Open Active World" (2011), a software development toolkit based on a generic software development approach that allows the implementation VRML/X3D browser components. In this way, generic 2D as well as 3D objects can be placed into the real environment. Generally, Open Active World works with Cartesian coordinates. Additionally, as a precondition for operation in conjunction with a command executive system, WGS84 coordinates as well as UTM coordinates are accepted.

4.2 Infrastructure

As actuality of data is important for an effective usage of OAR in the field of military scenarios, the surrounding infrastructure needs further description. Taking house takeover as an example, so-called Blue Force Tracking is needed to track all soldiers participating in operation. As Blue Force Tracking is not topic of this paper, is should only be mentioned that a combination of an inertial measurement unit and light detection and ranging with a laser scanner has been applied (Neuhöfer et al., 2012). Figure 4 shows the communication layout and information flows. During the house takeover, the soldiers transmit their position to the executive command system (ECS) via directed wireless LAN which used to overcome walls and other obstacles. The ECS stores their information as well as all other information gathered by reconnaissance like the layout of the house. The data is redirected to the Augmented Reality vision system.

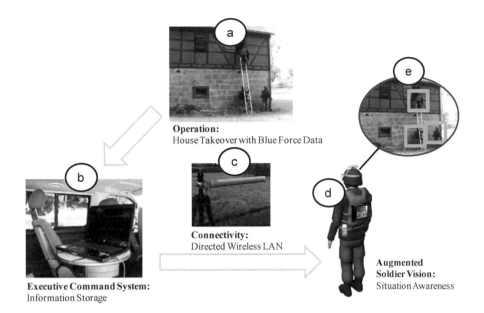

Figure 4 Communication layout and information flows for the outdoor Augmented Reality system. The soldiers (a) transmit their positions to the ECS (b) via directed wireless LAN (c) to the soldier (d) who gets an augmented (e) vision of the scene

5   EXPERIMENTAL STUDY

An experimental study has been initiated to follow the question brought up in chapter 3 on how much and which kind of information should be presented to the soldier to optimally support his or her mental model creation for optimized situation awareness. As basis for the augmented information, a subset of the NATO tactical signs has been used. The research questions are:

1. How much information is perceived within a time frame of five seconds?
2. What kind of information is perceived in case of information overflow?

5.1   Setup

A sample of 20 male participants aged between 20 and 41 years volunteered for the experiment. They sat down in front of the ruggedized tablet PC (size: 10.4", resolution: 1024x768px, color depth: 24 Bit) (Figure 5 a). It is the same as used for the OAR system described in chapter 4.1. The distance between head and display was 40 centimeters in average. For participation, a minimum acuity of 80% was presumed as well as color vision. Written instructions were given to the subjects before the experiment took place. In the experiment, a digital image of a landscape was shown to the subject. It was presented twenty times with duration of five seconds each time. In this way, a brief tactical situation check-up was simulated. The landscape was overlaid with one to four rectangular tactical signs, each sign as a combination of affiliation (colors red, green or blue), type (symbol within: semi-circle, circle or fork) and quantity (one to three marks above the rectangle). These enriched landscapes are referred to as a "scene" in the following. See Figure 5 (b) for an example of a sign and Figure 5 (c) for an example of a scene with two signs.

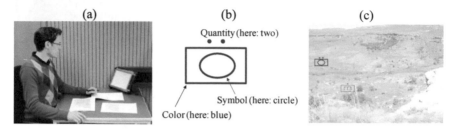

Figure 5 Setup (a), an example for a sign (b) a scene displayed to the subject (c) for five seconds

After showing a scene for five seconds, it was faded out. The subjects were then prompted to reproduce all information on a printout of the landscape with the same dimensions as the PC screen. The reproduction procedure was done according to a specific scheme: Firstly, a sign's position had to be checked on the landscape. Then, a connecting line to an arbitrary free "attribute block" had to be drawn. Finally, all information still present in the subject's memory had to be checked: Affiliation (color), type (symbol) and quantity (number of quantity marks). For each scene, a new sheet was used and the participant had to await scene fading before starting.

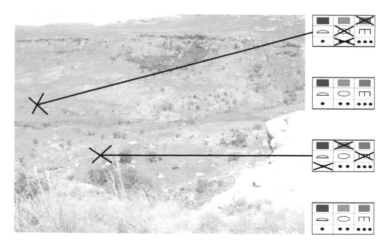

Figure 6 Information reproduction for two signs.

Each participant had a brief training with five scenes beforehand with an increasing level of difficulty (one sign = easy, four signs = hard). For the twenty scenes assessed, the level of difficulty then was uniformly distributed and balanced. The composition of the signs (affiliation + type + quantity) was random. To prevent sequence effects, two sequences were generated and each participant worked on one of them. They were requested not to guess. The duration of the experiment was 15 minutes for each subject.

5.2   Results

As the first step of the reproduction procedure was checking the position on the landscape, Table 1 shows a comparison of the mean difference and standard deviation of difference between where the position actually was and where it has been checked on the paper sheet. When looking at the mean difference, we see that it more than doubles when showing two signs compared to one sign, and then remains at that level with increasing number of signs.

Table 1 Means and standard deviations of the difference for position checking

Number of signs displayed in scene [int]	One	Two	Three	Four
Mean of difference [mm]	6.7	14,22	14,71	15,76
Standard deviation of difference [mm]	2.87	7,31	3,81	4,87
Omission rate [%]	0	0	3.5	10.75

The variance analysis shows that the number of signs significantly effects the error in position checking when switching from one to two signs ($F = 17.593$, $p < 0.05$), but no more significant effect when adding more signs. Here, it is important to mention that especially when four signs were displayed at a time, not all of them were checked on the sheet, so there are deviation values missing for higher number of signs. Instead of introducing a penalty error which would obfuscate the results, the omission rate is given separately: No omission occurred with one or two signs, but scenes with three signs had an omission rate of 3.5% and scenes with four signs even 10.75%. Effectively, one out of ten signs were completely left out when four signs were displayed.

Simultaneously, the success of information reproduction also depended on the number of signs displayed. Figure 7 shows the results. Obviously, when looking at a scene for 5 five seconds, a 100% probability for full information reception was granted with one sign only. Adding more signs led to an information overflow and an increasing loss of information. Then, color was constantly that part of the information which was perceived most reliably. For two signs, a success rate of over 80% was still given for affiliation, type and quantity marks. For three signs, the success rate dropped below 80% for all parts. Displaying three or four signs additionally led to an inferior perception of symbols in comparison to the quantity marks.

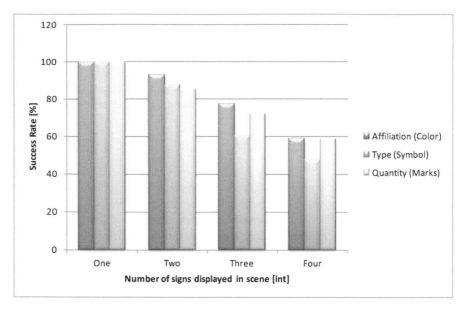

Figure 7 Success rate when specifying affiliation (color), type (symbol) and quantity (marks above the rectangle) against the number of signs displayed in a scene

As shown above, a general limit of two signs should not be exceeded to ensure an 80% perception rate when working with signs comparable in complexity to those presented. Fortunately, no overflow-related information loss was identified: the more information was shown (up to four signs at a time), the more information was reproduced by the participants overall.

6   CONCLUSIONS

Although outdoor Augmented Reality means an elaborate technical effort, it is worth it. Direct information placement into the real view provides clear advantages for the creation of situation awareness, so in particular for situation perception, comprehension and projection. But even when all technology is set up, information overload may lead to a loss of all benefits gained. In context of military missions, the approach followed in this research program is close to NATO standards, classifying all objects by their affiliation, type and quantity. But the results of our experiments show that the borderline to information overflow is reached quickly. One way out of this misery would be an introduction of information layers. These would allow the user to switch between "views", e.g. by showing units of certain affiliations only. The system has already been modified accordingly. Of course, most situations will demand more differentiated information selection and filtering. This will require more intelligent data handling and more sophisticated field-suitable interaction paradigms.

ACKNOWLEDGMENTS

We gratefully acknowledge funding of the project by the German Federal Office of Defense Technology and Procurement (BWB).

REFERENCES

Avery, B., B. H. Thomas, and W. Piekarski 2008. User Evaluation of See-Through Vision for Mobile Outdoor Augmented Reality, Proceedings of the 7th International Symposium on Mixed and Augmented Reality ISMAR 2008, pp. 69 – 72.

Behzadan, A. H. 2011. ARVISCOPE: Georeferenced Visualization of Dynamic Construction Processes in 3D Outdoor Augmented Reality. Proquest, Umi Dissertation Publishing, September 2011.

Chong, C. W. S., S.K. Ong, and A. Y. C. Nee 2007. Methodologies for Immersive Robot Programming in an Augmented Reality Environment, The International Journal of Virtual Reality, Vol. 6, pp. 69 – 79.

Endsley, M. R. 1995. Toward a Theory of Situation Awareness in Dynamic Systems. In: Human Factors, 37 (1), pp. 32-64.

Kleiber, M. and T. Alexander 2011. Evaluation of a Mobile AR Tele-Maintenance System, Proceedings of the 14th International Conference on Human-Computer Interaction HCI 2011, 9. – 14. July 2011, Orlando, USA.

Neuhöfer, J. A., F. Govaers, H. Elmokni, and T. Alexander 2011. Adaptive Information Design for Outdorr Augmented Reality. In: The 18th World Congress on Ergonomics, 12.-16. February 2012, Recife, Brazil.

Open Active World 2011. Open Augmented Reality and Virtual Reality Toolkit. Fraunhofer FKIE, http://open-activewrl.sourceforge.net/

Schall, G., E. Mendez, E. Kruijff, E. Veas, S. Junghans, B. Reitinger, and D. Schmalstieg 2008. Handheld Augmented Reality for Underground Infrastructure Visualization, Journal on Personal and Ubiquious Computing.

Section IV

Human Surface Scan, Data Processing and Shape Modeling

CHAPTER 15

Topology Free Automated Landmark Detection

Chiayuan Chuang, John Femiani, Anshuman Razdan

Department of Engineering
College of Technology & Innovation
Arizona State University Polytechnic
Mesa, AZ
john.femiani @ asu.edu

Brian Corner

WarSTAR
US Army Natick Soldier RDEC
Natick, MA

ABSTRACT

Given a corpus of 3D laser scan data (geometry and luminance) of head and shoulder region of a subject with preplaced circular labels that mark the location of facial landmarks, we have developed a novel approach to automatically locate the preplaced markers. Accurate landmark location then allows for morphometric modeling and bio statistical analysis of facial form. We achieve this by proposing a topology independent shape matching algorithm that allows for deformable/non-rigid alignment of a template based on a modified ICP discrepancy measure. The measure involves weighted geometry and luminance components, so that markers are aligned when present, and geometrically-based landmark definition is used when markers are unavailable. The algorithm allows for per feature variance to accommodate local tolerances. We ran our algorithm on a sample of 110 scans (92 males and 18 females) with 94.2% success overall with approximately less than 23

148

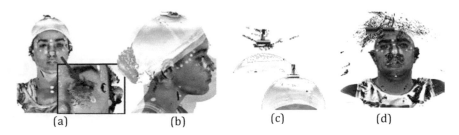

(a)   (b)   (c)   (d)

Figure 1: Model pointsets drawn superimposed over an input pointset; (a) an input model with the local errors for a single landmark highlighted in red (in the callout); (b) The same mesh while another landmark is being aligned, viewed from the left; (c) examples of a head stabilizer; (d) a pointset with a fair amount of scanning artifacts, landmarks are successfully identified in green.

seconds for 24 landmarks search including I/O time. We plan to extend the algorithm to include a global warp from the model to the target mesh and thus creating a topology and pre-processing free method to establish a dense correspondence between two meshes which will have significant impact in the area of shape matching.

Keywords: Shape matching, topology, landmarks, morphometrics, non-rigid transformations, ICP, alignment, laser scanned data.

1   INTRODUCTION

We are interested in an application driven by the US Army's effort to digitize and model variation of the faces of their personnel for the purpose of understanding bio-statistical/anthropological distribution of facial form. The process includes 3D laser scanning of the subject's head (Cyberware PX scanner, 360 degrees and includes some portion of the shoulders). Prior to scanning, a human operator places a 1/4" size circular marker (Avery part #5793) on a previously determined set of facial. Landmarks are located using standard techniques that involve palpation of underlying bony surface to establish the position of anatomical features (Farkas 1994). Markers are then placed at landmark position. This process is subject to skill/knowledge of the operator and there is some variation in both the consistency of the position and placement of the markers. The subject wears a nylon wig to minimize intrusion of hair. The marker color is not always distinguishable from the cap, clothing, specular reflections, and other digitization artifacts. The markers are sometimes absent, either not placed or fell off during/just prior to scanning.

Specifically, we are given:
  a) A set of points $v(x, y, z, lum^*)$ and connectivity as a typical triangular mesh supplied by 3D scanners and associated with each subject. Each scan is approximately million vertices. In this case Cyberware PX head

scanner is being used. The *lum* or color values (r, g, b)* are associated on a per vertex basis. We are given a corpus of such data.

b) A list of semantic landmark names and locations (see **Figure 2**). Semantically meaningful landmarks rely on prior knowledge of landmark points and their name; we represent these as a subset from a statistical model built from a 1988 survey of landmark points (Annis and Gordon 1988).

We aim to do the following:
1) For a target mesh surface/subject find approximate center of each marker and associate it with the corresponding semantic landmark point from **Figure 2** irrespective of expressions, noise, artifacts, etc. and with no pre-processing required.
2) Determine which of the markers may be missing or could not be identified and estimate missing landmarks associated with geometry such as back of the head. Even if the markers are correct and can be identified, we must bring them into correspondence to a canonical set of landmarks so that we can determine the semantic identity of each landmark.
3) Remove undesirable geometry (such as below neck and head support) as well as fill geometric information. For anthropological consistency the aim is to get the surface of the head (specially back of the head) which is not *distorted* by hair sticking out (see Figure 1).

1.1   Approach and Claims

There are many challenges that prevent a direct segmentation classification approach in solving the problem at hand. These are:
a) The markers are not separable just on luminance alone because of specular reflections, presence of other materials such as head caps and clothing that have similar luminance and the luminance value of the markers is not uniform across scans.
b) There is noise both in the geometry of the 3D scan and in the luminance. For example a device to stabilize the head and reduce movement artifacts is used to ensure subject heads are scanned in as consistent a posture as possible. Also, there is hair and other artifacts (Figure 1).
c) The markers themselves are applied to the face surface by a technician who is skilled but since these features represent the bony surface underneath, marker placement is still subjective to interpreting where these lie on the skin. This creates some inconsistency in placement of the markers. Additionally, without notice of the operator, markers may fall off or are missed in the first place.

d) Many of the challenges described above require significant pre-processing such as bringing the models into consistent number of vertices and/or create correspondence with a model mesh (Blanz, et al. 2004). Our aim is to require no pre-processing of the data.

Our approach is based on a deformable template idea (see section 3 for details). We start by taking an arbitrary subject from the sample and identifying all the markers on the model mesh in **Figure 2**. We mark additional features on the model mesh such as tip of the nose. We can specify any number of additional features based on surface geometry (i.e. not labeled with the markers) and add semantic labels. This is done as an additional soft constraints and helps in deciphering the relative positions of the markers. Then for any target mesh we do a rigid coarse alignment using ICP. The geometrically defined features described above play an important role at this step. Then each marker on the model mesh, based on a specific sequence, is allowed to move to the closest marker on the target surface based on a discrepancy measure that includes luminance. Each marker has its own spatial variance (based on experimentation) used to determine the sizes of geometric features. Some markers tend to have close neighbors such as *Ectoorbitale, Zygofrontale* and *Frontotemporale*, and therefore require a tighter set of tolerance than others (Figure 2). Each marker is allowed to morph locally irrespective of how it *pulls* the global model. The local morphing is a key feature of our algorithm. Once all markers from the model mesh find corresponding markers in the target mesh we can specify which of the markers may be missing from the target mesh. This process also creates a correspondence in terms of a non-linear transform between the model and target mesh. Specifically our solution is robust to missing landmarks, hair, changes in orientation, scale, expression of the subject, and also the head support.

In addition to solving an important applied problem, we make the following contributions:

- We present a method to align a sparse set of landmark points to dense models exploiting prior knowledge of the distributions of landmark points.
- In the process create a dense correspondence between meshes. This correspondence then can be used for variety of other applications.

The remainder of this document is organized as follows: section 2 presents prior art; section 3 presents our novel approach in detail; in section 4 we present an evaluation of our approach. Section 5 concludes the document. In order to validate our approach we have created a demo application; it will be made available online through http://i3dea.asu.edu.

2 PRIOR ART

We review prior art relating to shape alignment and registration. This is an exciting research area with many applications; the interested reader can refer to the recent survey (Heimann and Meinzer 2009) for additional material.

2.1 Template Alignment and Registration

Alignment algorithms attempt to minimize a discrepancy term over all transforms T in a set $\mathcal{T}$, such as Euclidean or similarity transforms, so that a template pointset A is as close to another as possible to a target pointset B according to some measure of discrepancy E. Alignment results in a correspondence μ of point-pairs in A and B. The classical algorithm is due to Besl and McKay (Besl and McKay 1992) and is called Iterative Closest Points (ICP). It alternates between finding point-correspondence $\mu \subset A \times B$ that minimizes E while holding T fixed, and then choosing a transform T to minimize E with μ fixed. The method is often fast but it can get stuck in local minima. It has been extended in many ways to address this; the survey (Rusinkiewicz and Levoy 2001) discusses many of its variants. TPS-RPM (Chui and Rangarajan 2003) accomplished non-rigid alignment using thin plate splines (TPS). They use a robust point matching (RPM) method similar to that used in EM-ICP (Granger and Pennec 2006) . Thin plate splines were also used by Brown and Rusinkiewicz (Brown and Rusinkiewicz 2007) for nonrigid alignment. In order to construct a warp they identify landmarks, however their landmarks are automatically located based on geometric features in the model such as curvature. The process is aimed at correcting small warps caused by scanning issues rather than larger morphometric differences between models.

2.2 Landmarks & Shape

Bookstein (Bookstein 1997), Kendall (Kendall 1984), and Goodall (Goodall 1991) provide theoretical frameworks for understanding shape distributions. Prior knowledge of a shape distribution can be exploited for non-rigid alignment problems, but much work relies on datasets with a consistent set of point landmarks that correspond across all shapes. However these can be extremely difficult to identify. Active shape and active appearance models have been extensively explored by Cootes, Taylor, and others at the Imaging Sciences research group at the University of Manchester (Cootes, et al. 1995) . The method has been employed to register full body scans of human subjects (Allen, Curless and Popović 2003), and also for reconstructing faces from undersampled meshes (Blanz, et al. 2004). Active shape and active appearance models use a template with fixed topology. Our approach to alignment uses landmarks on a template mesh, but the appearance away from the landmark is captured only as a distribution of points centered near each mark, unlike active appearance models. This allows us to align point-sets without parametrizing them or mapping them to a simpler domain; in fact we ignore connectivity and treat out samples as pointsets.

2.3 Motion Capture

Our problem bears some similarity to marker-based motion capture (MoCap) used for performance driven facial animation. A classic work in this area is due to Williams (Williams 1990), who represents a model-mesh using cylindrical coordinates with manually-placed landmarks. In order to capture a performance Williams placed Scotchlight® reflective tape from 3M, and a hole punch in order to create markers that were easily located in the image. The landmarks would be manually located based in an initial frame, and tracked between subsequent frames exploiting temporal coherence of the video sequence. The cylindrical facial model was warped based on the locations of the markers in order to recreate an actor's performance in 3D. Today very similar approaches are used in commercial software such as Vicon, PhaseSpace, etc. Twenty-plus years of commercial development have produced robust systems, but unlike our approach these systems can exploit temporal coherence, allowing them to reduce the search space after an initial interactive initialization process. They start from a set of 2D images captured under controlled conditions, and use markers that are separable based on luminance because they are made of retro-reflective material, or even radiant LED markers.

3   METHOD

Our approach is as follows:
1) One dataset (mesh) out of the corpus is selected as a *model* mesh. All datasets conform to the following coordinate systems. From the subject's point of view, z is up, the subject facing the scanner is y direction, and x direction is towards the subject's left shoulder. Landmarks are manually placed on the model mesh using an interactive tool.
2) Landmarks are given attributes including: (a) whether a marker is used at the landmark, or it is entirely based on geometry; (b) the size of the geometric feature around the landmark to consider, expressed as the standard deviation σ for a normal distribution centered at the landmark; (c) the size of the marker if one was used, e.g. ¾ of an inch, and (d) a weight (called the *lumweight* or w_{lum}) to balance the cost of aligning the underlying geometry against aligning the marker when seeking out a correspondence. The landmarks are also prioritized, so that landmarks that may be easier to identify are listed first (see the discussion in section 3.3).
3) For each target mesh:
   a.   The model mesh is rigidly aligned to the target mesh using a weighted ICP algorithm. We call this *global* alignment because we attempt to minimize the geometric discrepancy

Figure 2: Landmarks used in this work.

between the features associated with all landmarks at once. The aim of global alignment is to position each landmark close to its corresponding point on the target mesh.

b. We iterate through each landmark, in order, and for each landmark we sample points from the model mesh according to a normal distribution centered at the landmarks location with a standard deviation based on the feature size of the landmark.

c. We align the landmarks feature to the target mesh using a modified version of ICP that favors alignment to bright (high luminance) points on the target mesh for model points that are within a certain distance of a marker. The *lumweight* parameter controls the degree to which the expected marker drives the alignment process.

d. Once a landmark is aligned, we *erase* the marker from the target mesh by overwriting the luminance values with zeroes (black) within a disk that is twice as large as the marker. The aim of erasing markers is to prevent the multiple landmarks from being aligned to the same marker, which can occur when multiple markers are placed close together in a featureless area of the mesh (Figure 3). This works best when landmarks with distinct features or fewer nearby landmarks are aligned first; for instance the landmark *Zygion* has one neighbor within 2cm, whereas the *Ectoorbitale* landmark has two, so we align *Zygion* before *Ectooribitale*.

3.1 Modified ICP For Markers

The ICP method minimizes a discrepancy measure between two pointsets by alternating between (1) finding a corresponding target point for each model point to minimize a discrepancy measure, and (2) solving for a similarity transform (translation, rotation, and uniform scale) that minimizes the discrepancy, holding the correspondence fixed. Generally the target point for ICP can be anywhere on a *surface*, e.g. a model point could be aligned to a point on a face in a polygonal target mesh. However, since our data is sampled so densely we limit our correspondence search to vertices of the meshes. In order to drive the local alignment search to align landmarks we use the following discrepancy measure:

$$D(M,T) = \sum_{a \in M} w_a d(a, c(a)),$$

$$d(a,b) = \|\mathbf{x}_a - \mathbf{x}_b\|^2 + w_{\text{lum}} \times \begin{cases} 0 & \text{if } r_a > r_{\text{mark}} \\ \lambda_b & \text{else,} \end{cases}$$

where M and T are the model and target pointsets (after sampling the model mesh near the landmark), a is a vertex of the target mesh and $b = c(a)$ is a *corresponding* point in the target mesh. The points $\mathbf{x}_a$ and $\mathbf{x}_b$ are the spatial positions at two vertices, λ_b is the luminance at vertex b, r_a is the distance between the point at vertex a and the location of a landmark on the model mesh, r_{mark} is the radius of a marker, w_{lum} is the weight of luminance relative to geometry for a given landmark, and w_a is the weight given to vertex a based on the feature size of the landmark. We choose weights according to a Gaussian function

$$w_{\text{a}} = \begin{cases} 0 & \text{if } r_a > 3\sigma \\ \exp\left[-r_a^2 \big/ 2\sigma^2\right] & \text{else,} \end{cases}$$

where σ is a parameter controlling the feature size of a landmark point.

This modified ICP is solved using a traditional method, such as Horn's method (Horn 1987), to solve for the optimal transform. Various existing methods of weighted ICP have been published and we refer the reader to (Rusinkiewicz and Levoy 2001) for details. We use a KD tree with vertices at the leaves in order to accelerate the search for corresponding points that minimize D. The luminance λ_b only has an influence for model-vertices whose position is within r_{mark} of the landmark. For those points we can treat $-w_{\text{lum}} \times \lambda_b$ as a fourth coordinate of each target point, we use zero as the fourth coordinate of the model point, and we use a four dimensional search through the KD tree to find the nearest-neighbor. A three dimensional nearest-neighbor search through the KD tree for other points can be accomplished by always visiting every sub tree when we encounter a node that is split on the fourth dimension during the nearest-neighbor search.

3.2 Coarse Alignment

The local landmark alignment process is iterative and the discrepancy function may have several local minima. To improve the odds that each landmark converges to the desired minimum location, we use a *coarse* alignment step to determine a starting point for each landmark. Both the model and target pointsets may include many samples that are not near any landmark, so in order to do coarse alignment we use a *weighted* version of ICP where the weight assigned to each model point is the sum of the weights given to that point by each landmark:

$$w_a^{\text{coarse}} = \sum_{k \in K} w_a^k$$

where K is the set of landmarks and a is a vertex in the model mesh, and the superscript k indicates which landmark's position was used to determine w_a.

3.3 Landmark Order

Even with the global alignment step, initial positions of some landmarks can put them close to the wrong marker in the target mesh. This is most likely to happen when multiple landmarks are located in close proximity – such as the *Frontotemporale*, *Ectoorbitale*, and *Zygion* landmarks. Each Avery marker becomes a basin that can attract a landmark potentially to the wrong location. Since the *Ectoorbitale* is positioned in-between two other landmarks it is difficult to identify; if the local landmark alignment process approaches that marker from below it will be pulled into the *Zygion* marker instead, and if it approaches from above it will be pulled into the *Frontotemporale* mark or to the subject's cap. The *Zygion* landmark, however, can be identified easily if the model landmark starts anywhere at, below, or even slightly above the proper location on the target mesh. This makes the landmark more robust, and easier to identify. The robustness can be further improved by deliberately placing the landmark below its desired position in the model mesh (so that its vector of approach comes from below). Once a marker has been assigned to a landmark, we do not wish any other landmark to be assigned to the same marker. This is accomplished by replacing the luminance values of the target mesh with zeros at every vertex within $2 \times r_{\mathrm{mark}}$ of an aligned landmark's position. The new luminance values are usually darker than a subject's skin, so they serve to slightly repel rather than attract landmarks during the local alignment process. After the *Zygion* is erased, the *Ectoorbitale* can be identified robustly if it starts anywhere below its true location; it will no longer be confused with the *Zygion* landmark. This suggests the following approach for listing landmarks; first, list the landmarks that seem the most robust. Landmarks that are clustered near several others should be listed last. Second, if one observes that landmark A is often moved to the marker for B, one should consider listing B prior to A in the list of landmarks.

4   RESULTS AND EVALUATION

We have run the proposed approach for landmark identification on a sample of 110 laser-scanned head and shoulder regions collected by the US Army Soldier Research Center in Natick, MA. The dataset is composed of 92 males, and 18 females. Prior to scanning each subject had orange-colored circular markers, applied by a skilled technician, to their face to indicate the location of skeletal landmarks (identified by palpation) that cannot always be determined from a laser scan alone. Each subject had a head-brace to keep them from moving during the digitization process. Some subjects presented with a shaved head, the remaining subjects wore wig-caps that have similar luminance values to the circular markers. Many of the female subjects had long hair that was pulled back into a bun; which is important because the hair digitizes poorly. The digitized scans had approximately

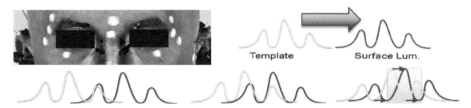

Figure 3: (top-left) Three dots a row make convergence to best solution difficult; (top-right) A one dimensional example of the problem; (bottom-left) as the template moves right, it gets stuck in a false minimum; (bottom-center) yet another false minimums; (bottom-right) there is a small "Goldilocks" zone of initial points where we can succeed with alignments using ICP.

Figure 4: Examples of landmark alignment from our corpus of pointsets. The landmarks are indicated by green dots.

one million vertices. Figure 4 is a snapshot of a few scans within our corpus, illustrating the range of artifacts in the scans.

We implemented out approach in ANSI-C on a laptop with an Intel P8600 CPU and 4Gb of RAM running Windows 7. We recorded the amount of time per-scan to align the landmarks. Our implementation saves images of the aligned landmarks superimposed over each scan in order to determine which landmarks were aligned and which failed, and it also crops the shoulders using a clipping plane oriented relative to the model mesh. We consider a landmark to be aligned when a marker is present, and the final position of the landmark is within the marker. We aligned 24 landmarks; the average time to process a mesh was 22.84 seconds, including reading the model and target meshes, global alignment, local landmark alignment, cropping the mesh, and saving the cropped mesh and reference image. We processed the scans using a luminance threshold of 0.7 (on a scale of 0-1, the landmarks had a variety of geometric variances ranging from 0.005 to 0.03 meters. Our results for a subset of the landmarks are summarized in Table 1 below. In summary, the results show fairly consistent and robust landmark detection using our approach with an average success rate of 94.2% per landmark. Errors are distributed such that 3 meshes contributed 2% of the overall errors; and our model mesh was chosen arbitrarily so that some asymmetry in the model mesh (particularly where the bald-cap impinges on the ears) introduced some failures for the *Tragion* landmarks.

Table 1: Summary of landmark alignment; Landmarks without markers are omitted, landmark numbers are out of full set of 24.

		Lum weight (w_{lum})	Geometric deviation (σ).	Percent with a Marker
	Overall	*2.55*	*0.012m*	*68.2%*
1	Orbitale (right)	1	0.01m	89.0%
2	Orbitale (left)	1	0.01m	97.2%
3	Frontotemporale (right)	1	0.01m	94.5%
4	Frontotemporale (left)	1	0.01m	94.5%
5	Zygioin (right)	1	0.01m	92.7%
6	Zygioin (left)	1	0.01m	94.5%
11	Glabella	5	0.015m	100.0%
12	Tragion (right)	0.01	0.01m	96.3%
13	Tragion (left)	0.01	0.01m	92.7%
14	Menton (gnathion)	10	0.02m	96.3%
16	Gonion (right)	10	0.01m	100.0%
17	Gonion (left)	10	0.01m	99.1%
18	Zygofrontale (right)	10	0.01m	96.3%
19	Zygofrontale (left)	10	0.01m	96.3%
22	Ectoorbitale (right)	0.01	0.01m	98.2%
23	Ectoorbitale (left)	0.01	0.01m	100.0%
24	Sellion	0.01	0.03m	98.2%

5 CONCLUSION AND FUTURE WORK

We have presented a topology free approach for detection of features that uses a new discrepancy measure that involves geometry and luminance. This shape matching method allows for a locally defined variance and hence a non-rigid/deformable method which can be used in other shape matching applications. The topology free is an important component thereby skipping the pre-processing step that most other state of the art algorithms require (such as same number of vertices, triangles, etc.). Future work includes adding a global warp function such as using thin plate splines to combine the local deformations on the model mesh to create a global transformation of the model mesh to the target mesh. This will allow for filling holes or missing geometry and correcting noise in the target mesh and will also create a dense correspondence between the model and the target mesh.

158

REFERENCES

Allen, B, B Curless, and Z Popović. "The space of human body shapes: reconstruction and parameterization from range scans." *ACM Transactions on Graphics (TOG)* (ACM) 22, no. 3 (2003): 587-594.

Annis, J. F., and C. C. Gordon. "The development and validation of an automated headboard device for measurement of three-dimensional coordinates of the head and face." Technical Report, Natick, Massachusetts, 1988.

Besl, P J, and N D McKay. "A method for registration of 3-D shapes." *IEEE Transactions on pattern analysis and machine intelligence* (Published by the IEEE Computer Society), 1992: 239-256.

Blanz, Volker, Albert Mehl, Thomas Vetter, and Hans-Peter Seidel. "A Statistical Method for Robust 3D Surface Reconstruction from Sparse Data." *3D Data Processing Visualization and Transmission, International Symposium on.* Los Alamitos, CA, USA: IEEE Computer Society, 2004. 293-300.

Bookstein, F L. "Landmark methods for forms without landmarks: morphometrics of group differences in outline shape." *Medical Image Analysis* (Elsevier) 1, no. 3 (1997): 225-243.

Brown, Benedict, and Szymon Rusinkiewicz. "Global Non-Rigid Alignment of 3-D Scans." *ACM Transactions on Graphics (Proc. SIGGRAPH)* 26, no. 3 (August 2007).

Chui, H, and A Rangarajan. "A new point matching algorithm for non-rigid registration." *Computer Vision and Image Understanding* (Elsevier) 89, no. 2-3 (2003): 114-141.

Cootes, T F, C J Taylor, D H Cooper, J Graham, and others. "Active shape models-their training and application." 1995: 38-59.

Farkas, Leslie G. *Anthropometry of the head and face.* Raven Press: NY, 1994.

Goodall, C. "Procrustes methods in the statistical analysis of shape." *Journal of the Royal Statistical Society. Series B (Methodological)* (JSTOR) 53, no. 2 (1991): 285-339.

Granger, S, and X Pennec. "Multi-scale EM-ICP: A fast and robust approach for surface registration." *Computer Vision—ECCV 2002.* Springer, 2006. 69-73.

Heimann, T, and H P Meinzer. "Statistical shape models for 3D medical image segmentation: A review." *Medical Image Analysis* (Elsevier) 13, no. 4 (2009): 543-563.

Horn, B.K.P. "Closed-form solution of absolute orientation using unit quaternions." *Journal of the Optical Society of America A* (OSA) 4 (1987): 629--642.

Kendall, D G. "Shape manifolds, procrustean metrics, and complex projective spaces." *Bulletin of the London Mathematical Society* 16, no. 2 (1984): 81.

Rusinkiewicz, S, and M Levoy. "Efficient variants of the ICP algorithm." *3dim.* 2001. 145.

Williams, Lance. "Performance-Driven Facial Animation." *Computer Graphics / SIGGRAPH* 24, no. 4 (August 1990): 235-242.

CHAPTER 16

Geometric Analysis of 3D Torso Scans for Personal Protection Applications

Jeffrey A. Hudson, Gregory F. Zehner, Brian D. Corner

InfoSciTex, Dayton, OH
Jeffrey.hudson@wlpafb.af.mil

AFMC 711 HPW/HPS
WPAFB, OH
Gregory.zehner@wpafb.af.mil

US Army Natick Soldier RDEC
Natick, MA
Brian.corner@us.army.mil

ABSTRACT

Optimization of body armor is desired to ensure that military personnel are provided with the most advanced, state-of-the art protection available. Part of the goal in optimizing body armor is to improve the fit and contour of the body armor, while maintaining the required area of protective coverage. Although current body armor systems have been developed to accommodate both sexes, the impression from female warfighters is that they are not optimally accommodated. Several efforts are underway by the US Army to redesign ballistic body armor to better fit males and female warfighters, thereby improving protection, maneuverability, and comfort. The goal of the current effort is to characterize meaningful size and shape variation in the female torso to aid in the development of ballistic armor design forms. A geometric method employing Principal Component Analysis (PCA) was applied to three-dimensional torso scans from the Joint Strike Fighter female subsample (JSF CAESAR, n= 722, Hudson et al. 2003) of the CAESAR (Civilian American European Surface Anthropometry Resource) database. The torsos were aligned to an anatomical coordinate system defined by their superior iliac spine landmarks (anterior and posterior), cervicale, and suprasternale. Four hundred and fourteen vectors (spaced regularly around the vertical axis at nineteen vertical levels) were generated from the anatomical origin outward to the torso scan surface. PCA was applied to the vector distances representing the anterior torso (217 vectors) resulting in three unique components which explain more than 70% of the total variance. Using this three principal component solution, cases were plotted in the multivariate distribution which defined the torso shapes for generating prototype torso forms for armor design.

Keywords: 3D Human Shape Analysis, Body Armor, Principal Components Analysis

1   BACKGROUND

Principal Components Analysis (PCA) is a statistical approach that is helpful in both understanding the relationships between application-relevant anthropometric measurements, and identifying the most important variation within a large set of dimensions as a whole. This is traditionally an exploratory statistical technique used to reduce the complexity of large amounts of data. It has been used effectively by the USAF to define the anthropometry of boundary cases in cockpit and workstation design to ensure a minimum percentage of users are physically accommodated (Zehner et al 1993 and Zehner 1996). In the context of workspace accommodation it is assumed that, if the boundary cases are accommodated, so too are all potential cases within the boundary. In this study, PCA is used to identify and quantify size and shape variation in the anterior female torso represented by 217 vector lengths sampled from each three-dimensional (3D) anterior torso scan. The torso cases, or forms, generated using the PCA solution are located throughout the sample distribution and represent potential size roll centers and not levels of accommodation.

2   METHODS

Torso body segments from the female JSF CAESAR whole body scan sample (n=722) were used in this study. However, this geometric approach to torso size and shape characterization is ultimately intended for application to the next generation US Army anthropometric survey (ANSUR II). A total of 414 vector length measurements were made on each torso segment of a standing whole body scan, and of these, 217 were used to represent the anterior half of the torso. These 217 vector lengths served as input for the torso PCA.

2.1   Alignment of Torso Scan Using Anatomical Landmarks

Before the torso vector lengths were generated, a coordinate system was constructed using anatomical landmarks on the torso and then aligned to the world coordinate system. Below in Figure 1, the scan associated landmarks (Anterior and Posterior Superior Iliac Spines (ASIS and PSIS), cervicale, and suprasternale) were used to define the vertical torso axis which was aligned with the Z axis. The Z axis was constructed between a point midway between cervicale and suprasternale, and a centroid representing the four landmarks: right and left ASIS and right and left PSIS. The origin for the vector dimensions is located at the midpoint of this vertical torso axis. The torso is rotated around the vertical axis until the line connecting both anterior iliac spines is parallel to the YZ plane. The resulting X axis projects through the torso front while the Y axis projects through the left torso side.

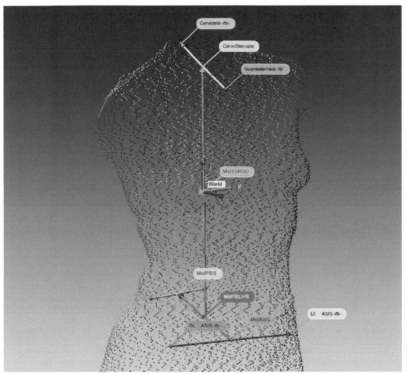

Figure 1: Example of constructed torso coordinate system and its position within the torso scan.

2.1 Vector Extensions: Extraction of Size and Shape Data

Below in Figure 2, the spacing of the vectors in a regular fashion is illustrated as they originate from the torso center, as well as an example torso with all vectors stopping their extension at the scan surface. To minimize the possibility of a vector travelling through arm, neck, and leg openings in the scan, "stop planes" passing through the acromion at the shoulders, cervicale at the neck, and mid-anterior superior iliac point at the pelvis, provided a block for the extension of the vectors in these directions. The stop planes also served to define the torso segment. In addition, a hole patching algorithm was used on the entire scan before vector extension. The resulting 217 vector lengths served as input for the torso PCA.

162

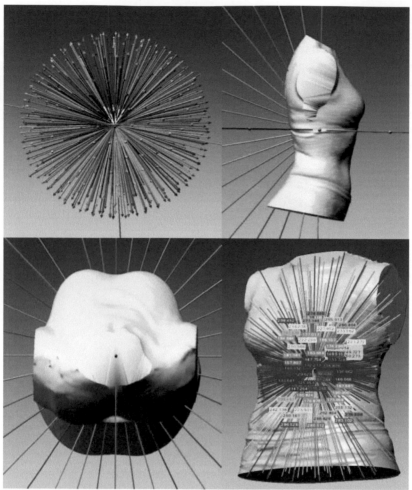

Figure 2: 414 regular spaced vector lengths originate from torso center seeking the scan surface (Top Left). Vector spacing though vertical plane (Top Right). Vector spacing around horizontal plane at mid level (Bottom Left). Example of vectors stopping on a 3D scan surface with stop planes not visible (Bottom Right).

2.2   Principal Component Analysis

Generally, for any design problem the important human dimensions have some relationship with each other. For example, Sitting Height and Sitting Eye Height are highly correlated. The paired relationships within a set of dimensions can be expressed as either a correlation or a covariance matrix. PCA can use either matrix to create a new set of variables called "principal components" which can be envisioned as a new coordinate axis system. All of its axes are orthogonal and rotated with respect to the original anthropometric dimension axis system and retains the same origin. In this study, instead of traditional anthropometric dimensions, the torso vector lengths are used. The total number of principal components is equal to the number of original variables, and the first Principal Component (PC) will always represent the greatest amount of variation in the multivariate distribution. The second PC is orthogonal to the first and describes the second greatest, and so on. The change in size and shape, or proportional change, can be visualized as you slide up and down the axes. For example, sliding along the first principal component, usually interpreted as describing *overall size*, is characterized by an increase or decrease in the original variable values of about the same relative magnitude.

Below in Table 1, the explained variance offered by the first 10 PCs (of the total 217) are presented both as the PC axis length (Eigenvalue) and percentage of total variance in the multivariate distribution. Note that the first three PCs explain a combined 71.5% of the total variation exhibited by the torso vector lengths. Figure 3 plots the explained cumulative variance from PC 1 to PC 30. An important question is: *how many PCs should be used in a solution to represent relevant total torso variation for armor design?* A solution with fewer PCs keeps the number of representative torso forms or model points down, but also ignores the variation explained by the higher PCs.

Table 1 Results of PCA on 217 anterior torso vector Lengths. First 10 of 217 total principal components shown only.

Principal Component	Eigenvalue	% Total Variance	% Cumulative Variance
1	102.8	47.4	47.4
2	36.6	16.8	64.2
3	15.7	7.2	71.5
4	10.3	4.8	76.2
5	6.7	3.1	79.3
6	6.2	2.8	82.1
7	4.4	2.0	84.2
8	2.7	1.3	85.4
9	2.7	1.2	86.7
10	2.5	1.2	87.8

After examining and interpreting the variation expressed by PCs 1 to 10 (below in Table 2), a three PC solution, PC 1 to PC 3 explaining 71.5% of the total variation, was chosen. However, a close examination was conducted of PCs 4 to 10 to determine if they might possibly represent important variation in armor applications. If so, torso design forms representing their variation will be included to augment the model points offered by the 3 PC solution.

164

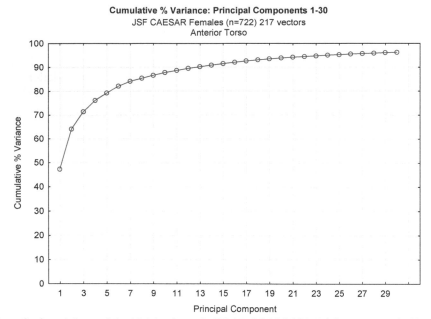

Figure 3: Cumulative explained total variance for PC 1 to PC 30 (of 217 PCs). Note that 71.5% is explained by PC 1 to 3, while 87.8% is explained by PC 1 - PC -10.

Table 2 Interpretation of Principal Components 1 to 10 based on the differences observed between model points along each axis. The first 3 PCs (red) represent the core solution and offer the torso forms representing 71.5% of the variation. PCs 4 to 10 are examined for possible relevance to armor design.

PC	% Total Variance	Interpretation of Primary Torso Variation Represented
1	47.4	Robusticity
2	16.8	Short/Wide/Round vs. Tall/Narrow/Elliptical
3	7.2	Girth Ratio: Upper to Lower Torso
4	4.8	Asymmetry: Right vs. Left Shoulder Height
5	3.1	Neck Length
6	2.8	Size Ratio: Breast to Belly
7	2.0	Hour Glass Shape vs. Straight
8	1.3	Flatter Sternum vs. Cleavage (fabric span?)
9	1.2	Indentation vs. Protrusion: 10th Rib and Sternum
10	1.2	Asymmetry: Right vs. Left Breast

165

3 RESULTS

3.1 Representative Female Torso Forms (Model Points)

Model points, or torso forms, are located in PC space by first defining a shell surrounding the percent of sample included (e.g., 50%, 90%). The model point is located at the intersection of the boundary shell and the PC axis. Using matrix algebra, the vector lengths of the model points are found from their location in PC space. Below, Figure 4 provides an example showing PC 1 vs. PC 2 and 30% and 90% inclusion boundaries. Determining where and how many model points to represent is associated with subsequent work involving fit testing prototypes to determine the most efficient size roll. The 3 PC solution is represented here by 4 models points on each axis plus the center or mean shape, for a total of 13 model points, or torso forms.

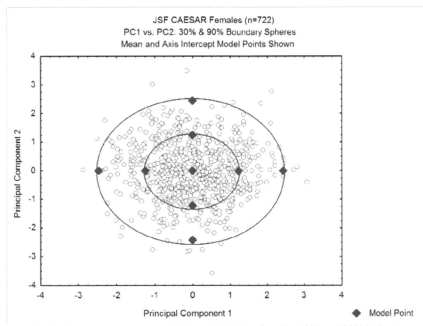

Figure 4: Factor Scores for sample on PC1 vs. PC 2 (Top Graph). 30% and 90% inclusion boundary shells define the location of torso forms or model points at the shell - axis intercepts.

Figures 5, 6, and 7, below present the 12 model points for PCs 1, 2, and 3 respectively. The mean is shown at center in each Figure. The changing size and shape evident in the model points, left to right, is analogous to sliding along the PC axis. The torso forms are intentionally not smoothed so the polygon vertices, representing the end of a vector length, are apparent.

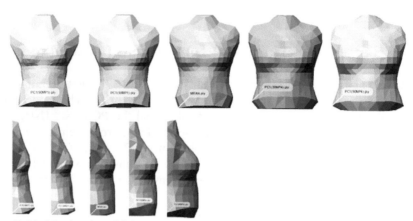

Figure 5: Model points from PC 1, showing variation in overall "Robusticity" or change in overall girth. Front view at top, with associated side view below. The mean shape is located at center.

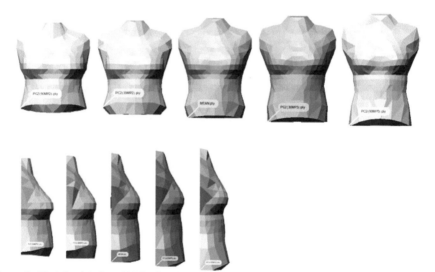

Figure 6: Model points from PC 2, showing variation contrasting torso height with width and depth: i.e. "Short, Wide, Round vs. Tall, Narrow, Elliptical." Front view at top, with associated side view below. The mean shape is located at center.

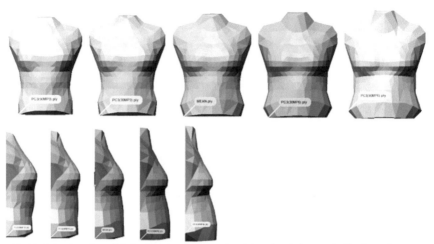

Figure 7: Model points from PC 3, showing variation in girth ratio of chest to belly circumference. Front view at top, with associated side view below. The mean shape is located at center.

4   DISCUSSION: THE "IGNORED" VARIATION, PC 4 - PC 10

Upon examination of the 90% model points offered by a 10 PC solution, some of the variation was excluded as not relevant to armor design but some is kept for possible relevance. (Scans of "nearest neighbors" to the model points were also helpful in the examination.) PC 4 (4.8% of total variation) represents an asymmetry of right to left shoulder height related either to a real asymmetry in anatomy or perhaps an odd posture during the scan. In any case armor will not be designed with asymmetries in mind. This also eliminates PC 10 (1.2% of variation), which represents variation in a postural difference (torso turned right or left) as well as differential right to left breast size. PC 5 (3.1% of variation) represents an apparent contrast in neck length, which in this study is not relevant to torso armor design. The model points and nearest neighbors for PC 8 (1.3% of variation) suggest a contrast between flatness of the sternum and breast cleavage. However, much of this is attributed to the presence of scanned fabric spanning the cleavage of the breasts and subsequent odd shapes from the hole filling procedure. At this time it is ignored. Model points representing variation described by the remaining PCs (PC 6, 2.8% of variation; PC 7, 2.0% of variation; and PC 9, 1.2% of variation) are of possible relevance to armor design and will augment the core 3 PC solution model points. PC 6 is clear contrast between breast size and belly size, while PC 9 suggests a protrusion vs. an indentation of both the sternum and 10th rib areas. PC7 suggests an hourglass shape vs. a vertically straighter torso. Figure 8, 9 and 10 below, illustrate the model points for these additional PCs. While the 12 torso forms offered by the first 3 PCs serves as the core solution to characterize female torso variation, the ad hoc addition of model points from the higher PCs may represent potentially important shape variation. Prototype testing on subjects will determine if this is the case.

168

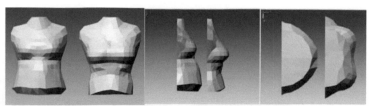

Figure 8: Front, right, and bottom views of the 90% model points on PC 6, Breast Size vs. Belly Size.

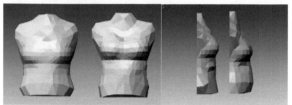

Figure 9: Front, and right views of the 90% model points on PC 9, protrusion of 10th rib and sternum (left model point) vs. indentation of those sites.

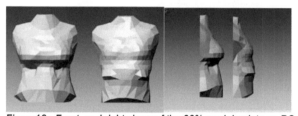

Figure10: Front, and right views of the 90% model points on PC 7, "hour glass" shape (left model) vs. a vertically straighter shape (right model). (The apparent protrusion and indentation at 10th rib may be representing a poor hole filling routine under the breasts.)

REFERENCES

Bittner, A.C., F.A. Glenn, R.M. Harris, H.P. Iavecchia, R.J. Wherry. 1987. CADRE: A family of manikins for workstation design. In Asfour, S.S. (ed.) *Trends in Ergonomics/Human Factors IV*. North Holland: Elsevier, pp. 733-740.

Hudson, J.A., G.F. Zehner, K.M. Robinette. 2003. JSF CAESAR: Construction of a 3-D Anthropometric Sample for Design and Sizing of Joint Strike Fighter Pilot Clothing and Protective Equipment. (Publication No.: AFRL-HE-WP-TR-2003-0142). AFRL/HECP, Wright-Patterson AFB, OH.

Zehner, G.F., R.S. Meindl, and J.A Hudson. 1993. A Multivariate Anthropometric Method for Crew Stations Design: Abridged, (Publication No.: AL-TR-1992-0164), Armstrong Laboratory, Air Force Systems Command, Wright Patterson Air Force Base, OH.

Zehner, G. F. 1996. Anthropometric Accommodation and the JPATS Program, *SAFE Journal*, 26(3).

CHAPTER 17

Shape Description of the Human Body Based on Discrete Cosine Transform

Peng Li, Brian Corner, Steven Paquette
US Army Natick Soldier Research, Development & Engineering Center,
Natick, MA 01760, USA

ABSTRACT

This paper investigates the shape description of the human body based on the discrete cosine transform (DCT). The DCT is able to compress a dense three-dimensional (3D) surface into a small   vector but preserve shape information truthfully. In order to apply the DCT to a 3D body scan, we segment each scan into six regions and represent them in a cylindrical coordinate system. In this paper we discuss this method and its application to torso shape retrieval.

Keywords: Discrete Cosine Transform, Shape description, Human body

1   INTRODUCTION

Describing the shapes of the human body has a long history. Traditional shape descriptions of the human body are textual and based on a few anthropometric measurements such as body mass index (BMI) and waist-hip circumference ratio (WHR), etc. [Tovée, et al 2002, Simmons 2002]. The words to describe body shape, such as 'hour glass', 'pear', 'apple', 'diamond', etc. are highly descriptive and are easily interpreted as a shape in the human brain. They also naturally form a classification system that helps designers in the clothing industry.

Recently 3D scanning technology has been employed in a number of large scale anthropometric surveys in the UK [www.shapegb.org/home], USA [www.sizeusa.com/], Japan [www.dh.aist.go.jp/bodyDB/s/s-01-e.html] and China [www.sizechina.com, Zhang 2011] and abundant 3D body scans are available. Thus developing 3D shape description of the human body has become an active research area [Paquet, Robinette and Rioux 2000, Simmons 2002, Ben Azouz, Rioux & Lepage 2002, Godil & Ressler 2005]. Paquet and Rioux first applied a surface property based shape descriptor [Paquet and Rioux 1997] for shape retrieval from

3D body scan databases. Simmons explored high-level textual shape description based on traditional anthropometric measurements such as chest, waist, and hip circumferences taken from 3D body scans. Ben Azouz et. al. applied principal component analysis (PCA) to describe body shapes from a 3D body scan database.

Computer based 3D shape descriptors, developed for searching and retrieving 3D models by software programs, map dense 3D surface coordinates to a finite-dimensional feature vector [Zhang 2004]. These feature vectors are usually in a numerical form and difficult for humans to understand. Because the shape of an object is an intrinsic characteristic irrelevant to viewpoint, a good shape descriptor should be invariant to translation, rotation and scaling. The major efforts in this area have been devoted to extracting these affine transform invariant feature vectors, and they include the density based methods [Akgül, C.B., Sankur, B. Yemez, Y., & Schmitt, F., 2007], histogram based methods [Paquet and Rioux, 1997] and spherical harmonic methods [Kazhdan 2004].

Our objective is to develop a shape descriptor that is not only suitable for shape retrieval, but also suitable for shape comparison and blending. For the former purpose, the shape descriptor should be easy to calculate and have a high compression ratio relative to the original surface. For the latter purpose, the shape descriptor should be able to build correspondence between shapes and contain information to reconstruct a 3D surface. Our target group is the human body in a standing posture. The Discrete Cosine Transform (DCT) is a good candidate shape descriptor. In section 2, we briefly introduce how the DCT is used to describe 3D shape. Section 3 discusses how to apply DCT to 3D body scans. Section 4 describes preliminary results using the DCT shape descriptors to search similar shapes.

2   THE DISCRETE FOURIER TRANSFORM BASED SHAPE DESCRIPTION

The Fourier transform has been a primary tool for signal and image processing for data smoothing, enhancement etc. [Gonzalez]. Using the DFT to describe shape was developed in the early 1970's [Zahn & Roskies 1972]. In this application a closed 2D contour is parameterized as a waveform and transformed by the DFT. Its first few DFT coefficients are formed a shape descriptor as a highly compact representation of a 2D contour. Recently this approach has also been used for 3D shape description [Vranić and Saupe 2001], where a 3D surface is voxelized and encoded by the 3D DFT. Then they used first 16~172 real-valued components as shape descriptors. The spherical harmonic transform recently became another popular choice for describing 3D shapes [Kazhdan 2004, Shen 2009], where a 3D object is centered in a spherical coordinate system and then sampled.

The DCT is a real value DFT and has been used for 2D image compression [Gonzalez] due to its excellent de-correlation and energy compact characteristics. Notably it has been a standard algorithm for used in JPEG (Joint Photographic Experts Group) image compression. Similar to the DFT, the DCT is also used in 3D shape description by voxelizing of a 3D surface as presented in Lmaati [2010].

A commonly used discrete cosine transform has form is:

$$F(u) = \alpha(u) \sum_{n=0}^{N-1} f_n \cos\left[\frac{\pi}{N}\left(n + \frac{1}{2}\right)u\right] \qquad u = 0, \ldots, N-1. \qquad (1)$$

For the 2D DCT, the form is :

$$F(u,v) = \alpha(u)\alpha(v) \sum_{m=0}^{M-1}\left\{\sum_{n=0}^{N-1} f_{m,n} \cos\left[\frac{\pi}{N}\left(n + \frac{1}{2}\right)v\right]\right\} \cos\left[\frac{\pi}{M}\left(m + \frac{1}{2}\right)u\right],$$

$$u = 0, \ldots, M-1, \quad v = 0, \ldots, N-1 \qquad (2)$$

Where function $\alpha(.)$ is defined as:

$$\alpha(u) = \begin{cases} \frac{1}{N} & for\ u=0 \\ \frac{2}{N} & \\ & for\ u=1,2,\ldots,N-1 \end{cases} \qquad (3)$$

In (2) $\{f_{m,n}\}$ are a 2D data array either from an image or other data source. For $u = 0$, $v = 0$, $F(0,0) = \frac{1}{MN}\sum_{m=0}^{M-1}\{\sum_{n=0}^{N-1} f_{m,n}\}$ is the average value of the sample image. This value is called the DC Coefficient in the literature. It contains information about overall image intensity (or general dimension of a 3D surface). All other transform coefficients are called the AC Coefficients, which reflect relative variation within an image.

The original image can be reconstructed from its DCT coefficients as follows:

$$f_{m,n} = \sum_{u=0}^{M-1}\left\{\sum_{v=0}^{N-1} F(u,v) \cos\left[\frac{\pi}{N}\left(n + \frac{1}{2}\right)v\right]\right\} \cos\left[\frac{\pi}{N}\left(m + \frac{1}{2}\right)u\right] \qquad (4)$$

However the 2D DCT has not been widely used in 3D shape processing due to it restrictive data format.

3   IMPLEMENTATION OF A DCT BASED SHAPE DESCRIPTOR FOR 3D BODY SCANS

In this section we explore the DCT as a shape descriptor for the human body surface. As shown in the above section, the 2D DCT requires data in a 2D array format, which is difficult to generate for an arbitrary, mesh-based 3D shape. Often it requires a segmentation method to decompose the surface into parts so that each part can be properly parameterized into 2D space.

A cylindrical coordinate system is a 2D space to represent 3D points. In a cylindrical coordinate system, a 3D data point can be represented in longitudinal direction (z axis) , azimuth angle (θ) and the radial distance from the longitu-dinal axis to the point (r), as illustrated in Figure 1. In order to fit a surface into a cylindrical coordinate system at every point of the surface, its radius must be single-valued. Given the specific pose of a 3D body scan, this restriction can be met.

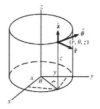

FIGURE 1. Cylindrical Coordinate System

Our implementation of the DCT shape descriptor for 3D body scans consists of the following steps: 1) pose normalization, 2) body segmentation, 3) normalized cylindrical sampling and applying the DCT with low pass filtering. This section presents a detailed description of these steps.

3.1 POSE NORMALIZATION

A typical 3D body scan in a standing pose is illustrated in Figure 2. The first step we take is to normalize the orientation of these standing scans. Although using the principal component analysis (PCA) to find major axes of an object is a popular method in the pose normalization [Vranić and Saupe 2001, Sfikas 2010], the directions of principal axes are not uniquely determined from the PCA method. In our case, it is essential to have all subjects facing the same direction with feet on the ground. To obtain such orientation our method is to identify the direction of the toes from a scan and set the toe direction to $\theta = 0$ degree.

In order to identify the toe direction we obtain a cross section of lower leg and a cross section of the feet from a 3D whole body scan as shown in Figure 3. We calculate the centroid of each section and then remove their height difference. A directional vector from the centroid A of the leg section to the centroid B of the feet section forms the toe direction.

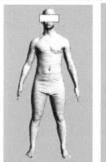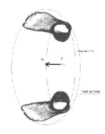

FIGURE 2. 3D scan with typical standing pose   FIGURE 3. Toe orientation vector $\overrightarrow{AB}$

3.2   BODY SEGMENTATION

To apply the 2D DCT to a 3D body scan we have to map the 3D surface to a 2D coordinate system or parameter space. This is difficult to achieve with the standing pose since the arms and legs are branched out from the trunk. Our approach is to segment a 3D body scan into six parts, namely head, torso, left and right arms, left and right legs as shown in Figure 4. In this way, each body segment can be fit into a cylindrical coordinate system. With the sampling method discussed in the next section the surface cab be transferred into a 2D array.

Since torso shape is the most prominent part of the whole body shape, the following discussions and implementation are on the torso shape only. But the methodology can be generalized to other body segments, and a whole body shape descriptor can be constructed from each partial shape descriptor.

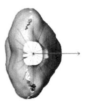

FIGURE 4. Body segments   FIGURE 5. Cylindrical sampling scheme on the torso surface

3.3   CYLINDRICAL SAMPLING AND DCT

Once the pose is segmented, we use the PCA to find the longest axis for a segment. This is the longitudinal axis of a cylindrical coordinate system and it passes through the centroid of the segment. Then we create a series of sampling planes along the longitudinal axis, where the longitudinal axis is the normal direction of the sampling planes. For each plane, we sample the body surface in equal angular space (the azimuth angle) and obtain a distance value from the longitudinal axis to the surface of the segment. This distance is the radius of each sampled point. Figure 5 illustrates the above sampling frame.

After the above sampling process a torso surface is represented in a 2D array, in which each row is a horizontal section at height h, each column is a vertical slice at a certain azimuth angle θ, and each element is a radial distance from the surface to the longitudinal axis. For a torso sampled as a 64x128 matrix, 64 is the number of slices and 128 is the number of points in each slice,. The resultant DCT coefficients are also a 64x128 matrix, which has 8192 elements. Figure 6 shows a torso surface in its original mesh and as a reconstructed surface from 128x128 and 64x128 DCT. Reconstruction of the surface from DCT coefficients is an inverse process.

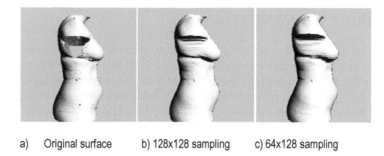

a)   Original surface   b) 128x128 sampling   c) 64x128 sampling

FIGURE 6. Torsos and their reconstructed surfaces from the DCT

As a comparison, we calculated surface to surface distances between the original 3D scan and reconstructed surfaces from 128x128 and 64x128 DCT coefficients for ten scans (five males, five females). Here we use $d(A, B)$ for the distance calculated from a vertex of surface A to surface B with a ray-face intersection algorithm. Each ray is at the surface normal direction of the vertex. Therefore $d(A, B)$ is different from $d(B, A)$. Table 1 shows mean surface distance and standard deviation results between the original scan and the reconstructed surfaces. The values are average of five subjects for each gender.

Table 1. Statistics of distance between the original scan and reconstructed surfaces -- 128x128 (16384 coefficients) and 64x128 sampling (8192 coefficients). Here unit = mm. and A is the original surface and B is the reconstructed surfaces

		$d(A,B)$		$d(B, A)$	
		Male	Female	Male	Female
128x128 DCT	Mean	0.105	0.086	-0.172	-0.168
	STD	0.767	0.700	0.694	0.641
64x128 DCT	Mean	0.279	0.204	-0.363	-0.332
	STD	1.315	1.233	1.238	1.107

In the above table, the mean distances count for positive and negative values. A negative distance in d(A, B) means that at the calculated vertex the surface A is bigger than surface B. Comparing the mean distances between d(A,B) and d(B,A), it shows that the reconstructed surface is slightly larger than the original surface consistently across all subjects. The comparison of standard deviations in the above two distances reveals that distance variation in d(A,B) is consistently larger than those in d(B,A). This is because the distance calculation originated from the raw surface has more vertices intersected to non-sampled areas. On the other hand, the difference between two corresponding standard deviations is very small (<0.2 millimeter). This means the reconstructed surface interpolates the space between sampled points smoothly.

By applying a low pass filter to the original DCT coefficient matrix, non-zero elements can be reduced significantly. Our experiment shows that the 3D torso surface (normally with more than 50k points) can be represented and reconstructed from a small set of DCT coefficients. We computed the reconstructed surface distances from the full DCT coefficient set (64x128 sampling) to filtered DCT coefficients at cutoff frequency 50, 30 and 20 respectively. The average of each gender group is presented in Table 2. It shows that even with 1:24 (335: 8192) reduction ratio the reconstructed surface distances between filtered DCT coefficients and full DCT coefficients are still very small. Therefore with 335 non-zero coefficients, the shape of the original body is well preserved.

Table 2. Distance statistics between reconstructed surfaces of 64x128 sampling (8192) and filtered DCT coefficients

Cutoff frequency		Male	Female
50 (2012 coeffs)	Mean	-0.007	-0.006
	STD	0.367	0.332
30 (736 coeffs)	Mean	-0.021	-0.019
	STD	0.908	0.812
20 (335 coeffs)	Mean	-0.066	-0.051
	STD	1.707	1.575

4   SHAPE RETRIEVAL BASED ON LOW PASS FILTERED DCT COEFFICIENTS

We explored shape retrieval with the filtered DCT coefficients. For shape description purpose, using over several thousand coefficients is still a high dimensional problem. We analyzed DCT coefficient spectra of 128x128 and 64x128 sampling, and found that 99% of the spectrum energy is concentrated in low frequencies (cutoff frequency < 20), regardless of the original sampling density as shown in Figure 7. Therefore, by applying an ideal low pass filter with a cutoff frequency equal to 20, we obtained a shape vector with much fewer elements than those used for shape reconstruction. With a cutoff frequency set to 20, the shape vector obtained has only 335 non-zero elements. The final shape descriptor is made up of the normalized DCT coefficients; that is, we divided all the DCT elements by the DC Coefficient. This shape descriptor is invariant to scale.

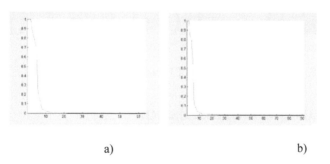

a)   b)

FIGURE 7. Energy distribution of torso's DCT coefficients, a) 64x128 DCT, b) 128x128 DCT

To define a distance for shape retrieval we consider cosine distance between two shape vectors [Qian 2004]. The "cosine" distance is defined as one minus normalized inner product of two shape vectors as in (5).

$$D(u, v)_c = 1 - \frac{\sum_{i=2}^N v_i * u_i}{\sqrt{\sum_{i=2}^N v_i^2 * \sum_{i=2}^N u_i^2}} \tag{5}$$

where $\{v_i\}$ and $\{u_i\}$ are two shape vectors.

Note that the subscript in the above formula starts at 2. This is to exclude the DC coefficient of DCT which is an amount far bigger than the rest of coefficients. Using a subset of scans (517 females and 614 males) from the CAESER database [http://store.sae.org/Caesar/], we retrieve similar shapes according to a given shape vector. The shape vectors of these scans are pre-calculated. The similar shapes are ordered in ascending distance value. For example, in Figure 8 we list two given shapes (a male and a female) and their eight nearest shapes under "cosine" distance.

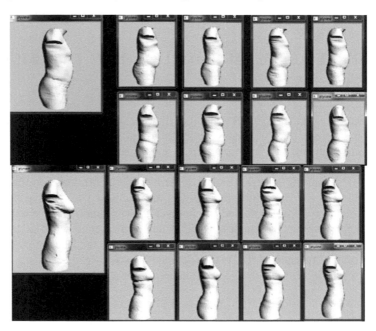

Figure 8. Retrieval results using 335 DCT coefficients as the shape vector

5   CONCLUSIONS AND DISCUSSIONS

Our research demonstrated that the DCT is an efficient way to encode torso shape into a small shape vector. The reconstructed surface from the filtered DCT coefficients has a predictable error tolerance with regards to the cutoff frequency.

Even at 1:24 compression ratio, the original surface shape can be reconstructed from only 335 coefficients with less than a standard deviation of three millimeters.

Using DCT to represent a surface benefits shape comparison and blending. Since our sampling scheme already builds a common parameter frame for all torso surfaces, shape comparison and blending can be done in DCT coefficients directly. For example, Figure 9 shows new surfaces reconstructed from blended DCT coefficients of two different shapes.

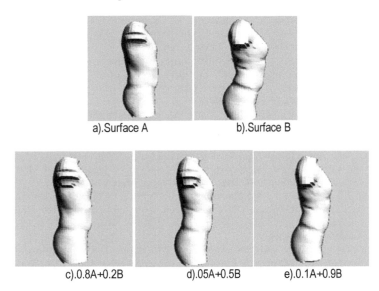

a).Surface A   b).Surface B

c).0.8A+0.2B   d).05A+0.5B   e).0.1A+0.9B

FIGURE 9. Surface blending with DCT

In future work, we will extend the DCT shape descriptor to the whole body. We will apply the DCT to each segment separately and integrate them into a 'big' shape vector. At this point, we may need a hybrid approach to describe the overall body shape, especially proportional relationship between limbs and trunk. The hybrid description may incorporate length measurements of each body segment.

One of important objective to describing 3D surfaces with a 'short' shape descriptor is to classify the shapes in a lower dimensional space. We will explore data clustering techniques to classify the body shapes with the DCT shape descriptor. In this direction, the classification using 335 variables may require further dimensionality reduction.

REFERENCES

Akgül, C.B., Sankur, B. Yemez, Y., & Schmitt, F., "Density-Based 3D Shape Descriptors", *EURASIP Journal on Advances in Signal Processing,* Volume 2007, Hindawi Publishing Corp.

Ben Azouz, Z., Rioux, M., Lepage, R., "3D Description of the Human Body Shape Using Karhunen-Loève Expansion", *International Journal of Information Technology*, Vol. 8, No. 2, September 2002

Gonzalez, Rafael C., and Woods, Richard E., *Digital Image Processing*, Second Edition, Prentice Hall, 2001

Godil, A., Ressler, S., "Retrieval and Clustering from a 3D Human Database based on Body and Human Shape" *Digital Human Modeling* , SAE, 2006.

Kazhdan, Michael M., "Shape Representations and Algorithms for 3D Model Retrieval", Ph.D. Dissertation, Department of Computer Science, Princeton University, 2004.

Lmaati, E.A., El Oirrak, A., Kaddioui, M. N., Ouahman, A. A., and Sadgal, M., "3D Model Retrieval Based on 3D Discrete Cosine Transform", *The International Arab Journal of Information Technology, Vol. 7, No. 3, July 2010*

Paquet, E., and Rioux, M., "Nefertiti: a query by content software for three-dimensional models database management", in *Proceedings of the 1st International Conference on Recent Advances in 3-D Digital Imaging and Modeling (3DIM' 97)*, pp. 345-352, Washington, DC, USA, May 1997.

Paquet, E., Robinette, K., and Rioux, M., "Management of Three dimensional and Anthropometric Databases: Alexandria and Cleopatra", Journal of Electronic Imaging, Volume 9(4). October 2000.

Qian, Gang, Sural, Shamik, Gu, Yuelong and Pramanik, Sakti, "Similarity between Euclidean and Cosine Angle Distance for Nearest Neighbor Queries", ACM Symposium on Applied Computing (SAC'04), Nicosia, Cyprus, March 14-17, 2004.

Sfikas, K., Theoharis, T., and Partikakis, I., "ROSy+: 3D Object Pose Normalization Based on PCA and Reflective Object Symmetry with Application in 3D Object Retrieval", International Journal of Computer Vision, VOL. 91, Issue 3, pp. 262-279, 2010, Kluwer Academic Publishers.

Shen, Li, Farid, Hany, and McPeek, Mark A., "Modeling Three-dimensional Morphological Structures Using Spherical Harmonics", *Evolution*, 63-4: 1003-1016, April 2009.

Simmons, Karla K. P., "Body Shape Analysis Using Three-Dimensional Body Scanning Technology" Ph.D., Thesis, North Carolina State University, 2002

Tovée, M. J., Hancock, P. J. B., Mahmoodi, S., Singleton, B. R. R., and Cornelissen, P. L., "Human female attractiveness: waveform analysis of body shape", Proceedings of the Royal Society, London, B (2002) Vol. 269, pp 2205-2213

Vranic, D. V., and Saupe, D.,"3D Shape Descriptor Based on 3D Fourier Transform", *Proceedings of the EURASIP Conference on Digital Signal Processing for Multimedia Communications and Services (ECMCS 2001)* (editor K. Fazekas), Budapest, Hungary, pp. 271-274, September 2001.

Zahn, Charles T., and Roskies, Ralph Z., "Fourier Descriptors for Plane Closed Curves", *IEEE transactions on Computer*, VOL. c-21, NO. 3, March 1972

Zhang, L, Manuel João da Fonseca, Alfredo Ferreira, "Survey on 3D Shape Descriptors", POSC/EIA/59938/2004.

Zhang Xin, Wang Yanyu, Ran Linghua, Fang Ailan, He Ketai, Liu Taijie and Niu Jianwei, "Human Dimensions of Chinese Minors", *Digital Human Modeling, Lecture Notes in Computer Science*, 2011, Volume 6777/2011, pp 37-45.

Section V

---

*Student Models in Adaptive
Modern Instructional Settings*

Personalized Refresher Training Based on a Model of Competency Acquisition and Decay

W. Lewis Johnson Ph.D. and Alicia Sagae Ph.D.

Alelo Inc.
12910 Culver Bl., Suite J
Los Angeles, CA 90066 USA
ljohnson@alelo.com; asagae@alelo.com

ABSTRACT

This chapter describes an approach for automatically adapting and personalizing world language curriculum materials based on a model of each learner's knowledge and skill in using the language. The model takes into account the learner's history of language use as well as periods of disuse. It identifies which language competencies are at risk of decay, so that the personalized curriculum can focus on helping learners quickly recover those competencies. The approach is particularly valuable for learners who wish to recover their language skills quickly prior to going abroad. It is equally applicable to instructional settings where learners come with varied backgrounds and experience using the language, learn at different rates, and have mastered the target language competencies to differing degrees, making them differentially susceptible to language attrition.

Keywords: learner modeling, language attrition, personalized learning

1 INTRODUCTION

Nearly everyone who has studied a foreign language has at some point experienced periods of disuse of the language, or stopped using the language altogether. In the United States, only 18.5% of students are enrolled in foreign language classes (ACTFL, 2010), and enrollments vary widely from grade year to grade year (Abbott & Wilcox, 2008). In college, foreign language enrollments decrease to 8.6 per hundred enrollments, and 80 percent of these are introductory enrollments, implying that most students do not continue their foreign language study (Furman et al., 2010). Once learners leave school, most find their opportunities to practice their foreign language skills to be diminished further.

Once people stop using a language they are vulnerable to *language attrition* (Köpke et al., 2007). Second language attrition is a very common problem both after and during the language learning process. The incipient effects of language attrition can sometimes be detected within hours after a learner has completed a language learning session (Johnson et al., 2012). Yet in spite of this the phenomenon of second language attrition has received relatively little attention. Second language acquisition researchers (e.g., Gass & Selinker, 2008) tend to overlook attrition effects. Conventional linear foreign language curricula implicitly assume that language skills are learned in succession and then are retained once learned.

Part of what makes language attrition difficult to account for in curriculum design is that its effects may depend upon each learner's previous learning behaviors and individual characteristics. For example, overlearning, i.e., continuing to practice a skill after it has been learned, can facilitate subsequent retention of the skill (Driskell et al., 1992). A learner's metalinguistic and metacognitive skills can affect how the learner mentally organizes, learns, and retains linguistic knowledge (Reilly, 1988). Over time these individual differences in learning behaviors can increase the differences between learners in terms of what they know and retain, making it difficult to design and apply a one-size-fits-all refresher training solution.

This chapter describes a method for automatically adapting and personalizing world language curriculum materials based on a model of each learner's knowledge, skills, abilities, and attitudes (KSAAs). This model, known as a *competency model*, takes into account the learner's history of language use, as well as periods of lack of use. This allows it to predict which of the learner's competencies may have been lost or may be at risk of decay. The competency model is used to automatically generate *personalized learning trajectories*, i.e., curricula that are tailored to help each learner learn the KSAAs that are new to them and recover the KSAAs that are in the process of decay. The approach is useful not just for learners suffering language attrition but for any group of learners who have diverse learning needs.

2   CONTEXT: ALELO LANGUAGE AND CULTURE COURSES

This model is being applied to Alelo language and culture courses. These utilize social simulation technology and methods to help learners acquire communicative competence. These are being used widely around the world for world-language training and education. The dialog activity shown in Figure 1 illustrates the approach. It is taken from a prototype Chinese course that Alelo developed and trialed with high school students in the Virtual Virginia Online School (VVOS), to help meet the high demand for Chinese instruction across the state of Virginia.

Figure 1. Virtual Virginia / Alelo Chinese course prototype

In Figure 1, the student's character (left) is arranging to meet with a friend named Zhang Li (right), a non-player character. The learner speaks into the microphone in Chinese, and Zhang Li responds in Chinese. Zhang Li has just asked the learner: "Wǒmen lǐbàijǐ jiànmiàn?" (What day should we get together?) The learner needs to respond by suggesting a day to meet. A help menu on the left lists a number of possible ways of responding, with different days of the week and choices of phrasing. Once the learner is able to converse without reliance on prompts and hints, the dialog becomes a realistic simulation of real-world task-oriented dialog.

Alelo develops its own speech recognition models, tailored to the language and speech of language learners, which improve learner performance better than off-the-shelf speech technology. Note that although this activity utilizes speech recognition, the objective of the activity is not to pronounce phrases, but to communicate meaning in context. The learner must perform and understand a number of communicative functions, including greetings, suggesting days and times to meet, and agreeing or disagreeing with proposals. These in turn require the ability to recall and construct phrases that are grammatically correct and have the right meaning and pragmatic connotations. These underlying KSAAs are taught and practiced in a variety of activities, many of which also employ speech recognition.

Evaluations by the Marine Corps Center for Lessons Learned (MCCLL) have empirically documented the effectiveness of the Alelo social simulation approach (Yates, 2011). Prior to deploying to Iraq in 2006, the 3rd Battalion 7th Marines assigned two members of each squad to train for forty hours using Alelo's Tactical Iraqi course. The 3/7 Marines then completed a year-long tour of duty without losing a single Marine, the first Marine battalion deploying to Iraq that could make that claim. The MCCLL interviewed the battalion's officers and surveyed the

Marines, and determined that the Marines had acquired enough communication skills to perform many missions without an interpreter and this contributed significantly to the unit's operational effectiveness.

We are currently working to extend Alelo products so that they incorporate automated learner assessment and personalization throughout. As a learner engages in an activity such as the dialog in Figure 1, the product will automatically update its estimate of the learner's mastery of the communicative functions, vocabulary, grammatical structures, and usage practices involved. It will provide teachers and learners with summaries of the learner's KSAAs across these categories. It will allow teachers to listen to examples of the learners' language use, and thereby draw their own conclusions about the learners' level of mastery.

3   INSTRUCTIONAL MODEL AND LEARNING MODEL

The instructional model underlying Alelo's social simulation approach is as follows. Learners first gain a knowledge-level familiarity with words, phrases, and cultural concepts via computer-based activities or textbooks and classroom activities. They then engage in social-simulation-based activities involving communicative functions that utilize these words, phrases, and concepts. Learners then practice applying communicative functions in extended dialogs in simulations of real-world situations. The similarity between the simulated situations and real-world situations facilitates transfer of learned KSAAs to the real world.

This approach is motivated by key principles in how people learn in general, and how they learn communicative skills in particular (Bransford et al., 2000). People employ language not for its own sake, but to convey meaning. The context of use affects both how communicative skills are learned and how they are recalled and applied. Social simulations provide learners with realistic contexts for learning and practicing communicative skills, which aid both learning and recall. We typically give learners opportunities to practice their communication skills in multiple social simulations. This results in variable practice (i.e., practicing a set of skills across a range of situations), which has been found to be beneficial across many skill domains (Ghosdian et al., 1997), and stimulus variability during training, which has been shown to increase skill retention and transfer (Gick & Holyoak, 1987).

Recent theories of language acquisition and decay model the process as a progression of phases from slow, cognitively involved, error-prone performance to successful performance with reduced error frequency, and then on to automated, accurate, effortless performance (e.g., see Langan-Fox et al., 2007). Moreover, they acknowledge that acquisition is likely to occur in a nonlinear bi-directional fashion, with phase slippage and skill decay (Towell & Hawkins, 1994). This view has implications both for language acquisition and language attrition. On the attrition side, one of the first indicators of language attrition is reduced performance speed, as learners lose automated fluency and engage more in cognitively involved processing. In the context of social simulations this is manifested as accurate but slow performance, as learners spend time thinking about how to express themselves in the target language. As attrition progresses, performance speed continues to decrease and errors increase in frequency. By tracking both learner response time and response accuracy in simulated dialogs, we can potentially detect the early stages of language attrition and provide early corrective remediation.

Many methods for promoting language skill retention draw on classical research on learning and forgetting, stemming from Ebbinghaus's seminal work (Ebbinghaus, 1885). This research shows that the ability to recall verbal items decreases exponentially over time, yielding what is known as a *forgetting curve*. The rate of decay of the forgetting curve depends upon the number of training episodes, so that after each training episode the rate of forgetting decreases. This research has led to the development of flash-card language memorization aids such as BYKI, in which learners practice vocabulary repeatedly in multiple sessions in an effort to commit them to memory and reduce the rate of forgetting.

In our view, the automated flash-card approach takes a simplistic view of language acquisition and attrition. Flash-card recall is a very different task from real-world communication, and impedes transfer. Flash-card recall is performed out of context, which impedes retention. Instead what is needed is an approach that promotes recall and minimizes forgetting-curve decay rates, and that also employs realistic communication tasks and reflects a realistic model of language acquisition, attrition, and transfer to real-world use.

4   PRELIMINARY STUDIES

To validate the above model and its application to Alelo courses, we collected and analyzed log data and speech recordings from trainees who had trained with Alelo's Tactical Iraqi course for extended periods of time at the Marine Corps Air Ground Combat Center (MCAGCC) 29 Palms. We retrieved a total of 294 separate trainee profiles from 29 Palms, out of which we selected 34 for analysis. They had trained for over periods of one to four months, often with gaps of a month or more during which language attrition could occur. Many of these trainees had conducted extensive training with Tactical Iraqi, at least 25 Skill Builder lessons. These learners were native English speakers. Training dates for data used in this study ranged between October 2007 and September 2008. The final data set consisted of 9,615 unique speech attempts made during the course of training. These profiles were assigned anonymous codes to protect the identity of the participants. More detail about this study may be found in (Johnson et al., 2012).

In this study we focused on learner performance in speaking Arabic phrases that appear on language instruction pages. Language instruction pages introduce the words and phrases that learners need to know in order to perform roles in the subsequent social simulation activities. Learners hear recordings of native speakers saying the phrases, read transliterations of the phrases, translations, and explanatory notes, and practice saying the phrases. Then the system presents translations of the phrases and the learner attempts to recall each phrase and speak it. With each speech attempt the speech recognizer matches the learner's speech against the target utterance and evaluates whether it is sufficiently close to be positively recognized. The accuracy threshold for recognition tends to increase as learners progress through the course. Success at this task demonstrates knowledge of the target phrases and the ability to speak them intelligibly.

Learners typically attempted each phrase multiple times, either because the phrase occurs in multiple places in the course or because the learner made multiple attempts to get it right. Successful speech attempts were positively related to the probability of future success. This relationship lessened as the number of successful attempts increased. For example, a second successful attempt at a particular utterance increased the probability of success on the next attempt by .015. A third successful attempt increased the probability on the next attempt by .011. Thus if a learner demonstrates multiple times an ability to speak a particular utterance, we can conclude that he or she will likely be able to recall it successfully in the future.

In contrast, unsuccessful speech attempts were generally negatively related to the likelihood of future success for a given target utterance. This relationship lessened as the number of unsuccessful attempts increased. For example, the first unsuccessful attempt at an utterance (after 1 prior correct attempt) was associated with a .029 decrease in the probability of the success on the next attempt. The next unsuccessful attempt was associated with a slightly smaller decrease (.024) in the probability of success on the next attempt. Thus repeated failures may be indicative of a persistent difficulty or technical problem requiring instructor intervention.

The amount of time since one's last correct attempt (on a given target) was negatively related to the probability of success on future attempts, an indication that language attrition was taking place. The largest decrease in performance occurred during the first 50 hours, followed by a more gradual decrease up until 150 hours. The amount of decrease depended upon the number of past successful attempts. If the learner performed an utterance correctly ten or more times in the past the decrease in probability of future success was relatively modest. If, however, the learner performed the utterance correctly just once, the probability of future success decreased substantially over time. These results suggest that repeated success results in a level of mastery that is relatively resistant to attrition.

These results are consistent with the model of language learning and forgetting described in the previous section. The target activity, recalling and speaking phrases, is closer to real language use than flash-card recall. The challenges now are to extend the analysis to more realistic language use in social simulations, and use the results in real-time to optimize instruction and maximize retention.

5   DOMAIN MODELS FOR LANGUAGE LEARNING

We have developed an extended domain-modeling framework called Nutopia that makes it possible to assess learning and customize curricula based on language use, in accordance with the acquisition and decay model described above. It links learner behavior data, competencies, learner models, learning activities, and learning outcomes. Data is captured and stored in a common data hub, to facilitate analyses and mapping between subdomains. The framework estimates learning across a network of individual knowledge, skills, abilities, and attitudes that comprise linguistic competencies. These include learning objectives that are of particular interest to instructors and learners, so that they can track learners' progress. The KSAAs are linked to the particular learning activities that cover them, making it possible to dynamically select and reorder learning activities to focus on KSAAs that the learner has not yet learned or are at risk of attrition.

The Nutopia domain model comprises the following subdomains:

- Purpose: the purposes in language-for-specific-purposes curricula (Basturkmen, 2010); also missions and mission-essential tasks.
- Language: linguistic forms (utterances, translations, grammatical forms, sentence patterns, syntactic categories) and communicative functions.
- Culture: social interaction frames such as conversational openings and host-guest relationships, and sociolinguistic practices in language use.
- Curriculum materials: learning activities and their organization into lessons and courses.
- Scenarios: encounters between learners and non-player characters, as well as the process through which agents interact with their environment, represented as an observe-orient-decide-act (OODA) loop (Osinga, 2007).
- Learner history: the history of the learner's interactions with the learning environment.
- Learner model: the learner's KSAAs and other individual characteristics.

The learner history subdomain captures information about each interaction session, the properties of the system and environment in which it occurred, and the history of interaction events within the session, as advocated by Mostow and Beck (2009). Individual interaction events are time-stamped and indexed by computer, training site, and learner, to facilitate data archiving, retrieval, and analysis. Student identifiers are anonymized to protect learner privacy. Interaction events include all learner actions that involve language use, such as speaking an utterance, listening to a recording, or choosing an answer. Learner speech recordings are captured, as well as the system's interpretation of the learner's speech. When the learner logging system is enabled, it captures learner history data during each session and imports it into the data hub, where it is used to update the learner model.

The centerpiece of our approach is a simple but powerful framework for modeling the range of world language competencies, captured in the language and culture subdomains. Competencies are grouped into *forms*, *functions*, and *practices*.

- Forms include various subcategories of linguistic forms and patterns, such as sentence structures, vocabulary, and phrases.
- Functions cover the use of language to communicate meaning and perform actions, including communicative functions and discourse structures and functions.
- Practices cover the cultural norms and practices governing communication and language use.
- Competencies can have attributes that characterize and circumscribe their use, e.g., whether the competency involves speaking, listening, reading, or writing.

These competencies are organized into subcategories (taxonomies) and part-whole hierarchies (meronomies). Conversely, they can be grouped into collections. The result is a rich lattice of interconnected KSAAs.

In this framework each individual interaction event in the learner history can relate to multiple KSAAs. For example, consider the utterance spoken by Zhang Li (the non-player character) in Figure 1: "Wǒmen lǐbàijǐ jiànmiàn?" (What day should we get together?). This phrase is a linguistic form made up of component parts (words and subphrases). It utilizes grammatical forms (e.g., the use ofi'the "j particle to form a question). It has a modality attribute (listening, since the learner is listening to Zhang Li). It is an expression of a communicative function (requesting a meeting date). It conforms to a cultural practice (the absence of face-threat mitigating strategies, as is customary for conversation among friends). If the learner is able to respond appropriately to Zhang Li in the context of this dialog, it provides evidence that the learner has mastered each of these KSAAs.

6 LEARNER MODELING

The Nutopia learner modeling subsystem uses the learner history data to derive from elements of the competency model *assessments* and *predictions* of competence. Assessments are estimates of learner mastery that are supported by evidence from learner behavior. Predictions are derived estimates of mastery that rely on a mixture of behavioral evidence and assumptions about the learner and/or the learning process. The learner model makes predictions across the learner competency hierarchy, e.g., using assessments of mastery of some competencies to predict mastery of other related competencies. It also makes predictions across time, e.g., predicting that previously learned competencies are now undergoing attrition.

For each modeled competency the system derives likelihoods of *success* and *mastery*, in accordance with the Langan-Fox et al. (2007) three-phase learning model. Success here refers to middle-phase successful performance, and mastery refers to final-phase automatized performance. The system assesses and predicts *learning* (increases in competency) and *attrition* (decreases in competency). For higher-level competencies the system may also derive estimates of *generality* (ability to apply the competency to novel situations) or *coverage* (application of the competency to a range of subcompetencies). Thus, for example, it is possible for a learner to demonstrate a high degree of mastery of a competency in specific situations, but fail to demonstrate generality of mastery.

The learner model update process begins by classifying each interaction event in terms of the features that provide good evidence for or against learning. These include level of performance (*unsuccessful, successful,* or *automatized*) and degree of reliance on hints (*assisted* or *unassisted*). A response is classified as automatized if it satisfies the accuracy requirements of a successful response and the response time is sufficiently fast to imply automatized skill. Some of these feature detectors look at the immediate history of interaction, e.g., asking for a hint before performing an activity.

The behavior classifications are used to compute overall performance scores for providing feedback to the learner. They are also used to update assessments of each

KSAA exemplified by the behavior, starting at the individual forms, functions, and practices and traversing up the competency hierarchy. For the higher-level competencies the assessment takes into account both the performance level and the degree of generality of performance, i.e., the range of contexts in which the learner has demonstrated the KSAA. The system reports these to the learners and teachers to give them a current snapshot assessment of performance.

The learner model maintains an evidence history relevant to each KSAA. At designated time points, such as before and after each learning session, the prediction algorithm generates estimated likelihoods of success and mastery for each KSAA. This takes into account the passage of time since the previous demonstration of the KSAA, as well as the learner's performance on similar and related KSAAs.

7 PERSONALIZING LEARNING TRAJECTORIES

The personalized learning trajectory algorithm generates recommendations of learning modules for the learner to focus on and tailors the sequence of activities within those learning modules, based on the learning objectives that those activities address and the learner's level of mastery of those learning objectives. To facilitate this, the Nutopia content authoring system links curriculum material elements (referred to as *interactivities*) to the KSAAs that they address. Linkage is performed automatically based upon the type of interactivity and the author's choice of linguistic forms, functions, and practices in the exercise.

This approach makes it possible to tailor and personalize each learner's learning experience based on a combination of learner preferences, learner performance, and teacher guidance. Learners or teachers can indicate which learning objectives they wish to focus on, and the system can propose learning objectives to recover KSAAs that are at risk of attrition. The personalization algorithm will then select learning activities that focus on the common subset of learning objectives.

8 CONCLUSIONS

This chapter has presented an approach for automatically tailoring learning materials for language and culture training, based upon a model of each learner's learning and pattern of language acquisition and attrition. The approach was originally developed to generate personalized refresher training courses, however it can be applied quite generally to provide learners with personalized learning trajectories to help them achieve their learning goals.

We are now in the process of iteratively validating and calibrating the approach, using archival data and in-house formative trial data. Field trials with learners are planned in the summer and fall of 2012.

ACKNOWLEDGMENTS

The authors wish to acknowledge Eric Surface and Aaron Watson for contributions to the literature review and theoretical framework in this chapter. The

190

Office of Naval Research funded some of the work presented here. Opinions expressed here are the authors' and not official positions of the U.S. Government.

REFERENCES

Abbott, M.G. and S.L. Wilcox. 2008. Status report: Foreign language enrollment in K-12 public schools. Pre-conference presentation at ACTFL 2008. Raleigh: SWA Consulting.

ACTFL (American Council on Teaching Foreign Languages). 2010. *Foreign language enrollments in K-12 public schools: Are students ready for a global society?* Alexandria, VA: ACTFL.

Basturkmen, H. 2010. *Developing courses in English for special purposes.* Basingstoke: Palgrave Macmillan.

Bransford, J.D., A.L. Brown, and R.R. Cocking. 2000. *How people learn: Brain, mind, experience, and school.* Washington, DC: The National Academies Press.

Driskell, J.E., R.P. Willis, and C. Copper, 1992. Effect of overlearning on retention. *Journal of Applied Psychology,* 77, 615-692.

Ebbinghaus, H. 1885. Memory: A contribution to experimental psychology. Translated by H.A. Ruger and C.E. Bussenius (1913). New York: Columbia University.

Furman, N., D. Goldberg, and N. Lusin 2010. *Enrollments in languages other than English in United States institutions of higher education, Fall 2009.* New York: MLA.

Gass, S.M and L. Selinker. 2008. *Second language acquisition: An introductory course.* New York: Routledge.

Ghodsian, D., R.A. Bjork, and A.S. Benjamin, A.S. 1997. Evaluating training during training: Obstacles and opportunities. In M.A. Quinones & A. Ehrenstein (Eds.) *Training for a rapidly changing workplace,* (pp 63-88). Washington, DC: APA.

Gick, M.L. and K.J. Holyoak, K.J. 1987. The cognitive basis of knowledge transfer. In S. M. Cormier & J.D. Hagman (Eds.), *Transfer of training: Contemporary research and applications* (pp 9-46). New York: Academic Press.

Johnson, W.L., L. Friedland, A.M. Watson, and E.A. Surface 2012. The art and science of developing intercultural competence. In P.J. Durlach and A.M Lesgold (Eds.), *Adaptive technologies for training and education,* 261-285. New York: Cambridge U. Press.

Köpke, B., M.S. Schmid, M. Keijzer and S. Dostert (Eds.) 2007. *Language attrition: Theoretical perspectives.* Amsterdam/Philadelphia: John Benjamins.

Langan-Fox, J., Grant, S., & Anglim, J. 2007. Modeling skill acquisition in acquired brain injury. *Australian Psychologist, 42,* 39-48.

Mostow, J. & Beck, J. 2009. What, how, and why should tutors log? *Proceedings of EDM,* 269-278.

Osinga, F.P.B. 2007. *Science, strategy, and war: The strategic theory of John Boyd.* Abingdon: Routlege.

Reilly, T. 1988. Retaining foreign language skills. Washington DC: ERIC Clearinghouse on Languages and Linguistics.

Towell, R. & Hawkins, R. .1994. *Approaches to second language learning.* Philadelphia, PA: Multilingual Matters LTD.

Yates, W. 2011. Tactical Iraqi Language and Culture Training Systems: Lessons learned from 3rd Battalion 7th Marines 2007. Presentation at Gametech. Orlando, FL.

CHAPTER 19

Modeling Student Arguments in Research Reports

Collin Lynch, Kevin D. Ashley

University of Pittsburgh
Pittsburgh, Pennsylvania, USA
collinl@cs.pitt.edu

ABSTRACT

Argumentation is a core skill for expert problem-solving in ill-defined domains and for participation in an informed democracy. Skills of argument comprehension and argument generation, however, are often difficult to teach and assess. In this chapter we highlight the issues associated with instruction in argumentation and discuss the use of diagrammatic models of argument to scaffold students' argument construction and comprehension. We then describe ongoing research on the diagnostic utility of argument diagrams and the application of graph grammar induction to automatic graph classification and rule induction.

Keywords: argument diagrams, ITS, graph grammar induction.

1   INTRODUCTION

Argumentation is a central skill for real-world problem solving in domains from science to public policy. Scientists advance hypotheses or research proposals and defend the novelty, validity and relevance of their work by presenting and defending arguments. Legal advocates make argument their profession while business people substantiate their proposals using not just a bald statement of facts and figures about trade and traffic but with an argument for why the new company will be viable and successful. And, once every election cycle citizens of a democracy are tasked with examining arguments for and against policies and candidates to decide whom to trust as a representative.

Apart from courses in rhetoric or philosophy, argument is rarely taught explicitly or in isolation. Students in various disciplines typically learn how to argue by example, by reading published research reports or studying judicial decisions, and practice argument by drafting proposals or debating in a moot court. As domain experts, faculty members are accustomed to reading, assessing, and critiquing these types of arguments, and inculcate students in the norms of the discipline as they do.

Students benefit from this authentic practice as they work with realistic arguments and real domain concepts. They become familiar with the forms of argument employed in their professional lives. Law students, for example, grapple with building an argument using legal concepts at the same time that they come to understand the concepts. In the process, the argument's complexity and nuances are highlighted in a domain-appropriate way. The arguments produced, while simpler than those of experts, are directly relevant to the students' professional practice. Students thus get experience with writing legal briefs and other forms of argument that are employed by professional advocates.

This focus on written and oral argumentation practice is not without its disadvantages. Realistic arguments are complex, often employing dense language or implicit constraints. It can be difficult for novices to pull the thread of claims and counterclaims from even a well-woven report. Moreover the act of producing novel arguments entails writing them down or presenting them before an audience, both of which are skills in their own right. Thus students' comprehension of the structure of arguments and their own ability to produce them are often masked by their oral and literary limitations. In both events this can limit the quality of the argument-related feedback they receive and hamper their ability to benefit from classroom exercises.

Significantly for the present discussion, arguments of this type are not readily amenable to automated analysis. While NLP research progresses, we can not yet automatically extract complex arguments from real prose or provide reliable advice about open-ended essays. Argumentation as texts remains largely out of the scope of intelligent tutoring systems. Our work aims at addressing these limitations.

In sum, argumentation skills are at the core of many educational domains and are essential for real-world problem solving and professional practice. Argumentation is commonly taught implicitly as students read and produce written or oral arguments. While some scaffolding occurs, as discussed below, it often does not scaffold the whole argument construction process. While reading and writing arguments are beneficial, this process does not always provide the structural support that students may need nor is it amenable to automated diagnosis and feedback.

In our view, diagrammatic models of argument can address these limitations. As discussed below, such models allow for complex real-world arguments to be made while scaffolding key structural relationships. Argument diagrams can also be used to diagnose students' comprehension of arguments and argumentation skills, and can be analyzed to provide a platform for feedback.

To show that diagrammatic models are useful we need to show that they can be: constructed by students; graded reliably by domain experts; used to diagnose students' skills concerning more traditional forms of argument (e.g. written essays); and used for manual or automatic diagnosis and feedback. Here we discuss our ongoing series of studies and present some encouraging preliminary results.

This process is not specific to diagrammatic models of argument but be used in developing other educational interventions. Rather than focusing exclusively on developing a computer model to recognize existing human practice (the NLP Problem) or forcing users to modify their practice to a completely tractable method (Logic Puzzles) one develops a middle ground that trades small constraints in user behavior for larger payoffs in educational utility and computational tractability. Other middle ground approaches include using simple worksheets that lead students to classify their contributions as in the Belvedere interface (Suthers in-press). Or, one may restrict users to a known set of references. so that attempts to process any text (e.g. when performing anaphora resolution) are limited and tractable.

2   ARGUMENTATION AND EDUCATION

Real-world problems are not *well-defined.* Biologists, for example, are not expected merely to memorize all parts of a cell but to advance our understanding of their function. Civil engineers are not hired merely to memorize the building codes of their home state or the tensile strength of steel but to apply that knowledge to the design of new bridges. The problems of this latter type, problems that make up the bulk of professional practice in most domains are *ill-defined* (Lynch, et al. 2012).

In order to solve these problems, experts are tasked with: identifying unstated or assumed constraints (e.g. how much traffic must the bridge carry); clarifying often vague goals (e.g. what are the acceptable settlement terms for the case); and most importantly, *justifying* their decisions through means of arguments that often build on, and respond to, the arguments of others. Engineers, for example, compare their work to previous designs and seek to convince planners that their designs are more economical, aesthetically pleasing, innovative and so on. Scientists similarly consider, and sometimes reject, the hypotheses and research of others from when designing their own experiments and explaining their findings.

This act of making and responding to arguments is an essential task of real-world problem solvers and citizens. Indeed, as research by Voss and others have shown, this process of generating explanations and arguments is an integral part of the problem solving process (Voss 2006; Voss et al. 1983). Voss and his colleagues examined the decision-making of policy experts tasked with solving the Soviet Union's low agricultural output. They found that expert problem solvers constructed arguments for their solutions as part of the solution process, for instance, considering whether a proposed solution, increased mechanization, could be justified to the decision-makers in the politburo. Subsequent work by Fernandes and Simon (1999) carried this further, indicating that the types of reasoning and justification employed vary from domain to domain.

While educators acknowledge the importance of argumentation skills, as noted above, argumentation is not always taught explicitly but by example. Often, students learn argumentation through a form of *authentic practice* in which, ideally they gain an understanding of argument and test their understanding by peer feedback and expert review. Students in law study argumentation chiefly by reading and writing legal notes, engaging in moot-court sessions and in class debates, "that ritual of fire charitably known as the Socratic Method" (Aldisert 1989). A similar route is taken in science courses where students read published scientific papers and reference them when reporting on their own work.

As noted, this process of instruction is not without scaffolding. Students in science courses can be guided in the process of articulating their research goals, refining their hypotheses, selecting their citations, and writing a linear outline of their proposed argument. Even with this support, however, students often risk getting "lost in the text," focusing on grammatical issues and their writing or speaking style, and ignoring the deeper structural relationships in arguments. Students risk missing the nonlinear structure of arguments, the way in which each citation must be separately related to the hypotheses and compared. This is a manifestation, in part, of the relative complexity of arguments even in short scientific works. In fact, much of the structure of an argument is implicit in (and often at odds with) the ordering of a paper, related to the arguer's emphasis and to the relative significance of the points, and encoded in domain-specific ways that must be taught to novices (Llewellyn 1951).

One approach to addressing this problem is an increased focus on the outline process. In "Writing in the Social Sciences" (Greene 2009) describes constructing goal-based argument plans and using graphical outlines such as bubble diagrams, concept maps, and trees. While this focus on rhetorical goals and graphical layouts may help more than standard topic-sentence outlines, it does not necessarily make the argument structure explicit.

3   ARGUMENTATION AND DIAGRAMMING

Diagrammatic models of argument have a long history in law and philosophy. The models represent persuasive arguments or debates using a graph structure with assertions commonly represented as nodes and argumentative relationships represented as arcs. Toulmin-style diagrams separate the argumentative statements into *claim* nodes, which assert basic assertions or conclusions, *data* nodes which provide the data that validates a claim and *warrant* nodes which explain why the data supports the claim. The use of typed nodes with semantic rules that control the accepted relationships scaffolds the construction of appropriate arguments while accommodating complex real world discussions.

Diagrammatic models of argument have gained renewed interest in recent years as scaffolding devices and knowledge representations in domains such as Intelligence Analysis (Lowrance 2007), and legal modeling (Gordon 2011). For a more complete survey of the uses of diagrammatic models in law, philosophy, evidence and AI see (Reed, Walton and Macagno 2007; Scheuer et al. 2012; 2010).

The potential for Diagrammatic models to scaffold students' argument making and to instruct students in the production of argumentation generally has led to their increased use in educational contexts (Scheuer et al. 2010; van Den Braak 2006).

Diagramming systems have been employed to: help students annotate existing arguments by highlighting key features such as legal tests and hypothetical cases (Pinkwart et al. 2009; Easterday 2009); support and structure collaborative discussions among large groups (Scheuer 2010); and structure students' formation of novel arguments (Chryssafidou 2002). These and other systems have been applied with the twin goals of teaching students to make arguments, *learning to argue*, and in imparting domain knowledge, *arguing to learn* (Scheuer et al. 2010).

Despite the intuitive appeal of argument diagrams for educational interventions, prior studies have not always shown success. In van Den Braak (2006) the authors surveyed four systems in domains ranging from law to empirical science, and considered a range of outcomes from improved critical thinking to more coherent arguments. Despite promising trends they found a number of methodological limitations. Scheuer et al. (2010) arrived at similar conclusions. Here the authors examined argumentation support systems generally. While they found support for the hypothesis that argumentation support systems help students to form more coherent arguments via a *scaffolding effect,* they did not find any evidence for the hypothesis that argument support systems help students to acquire domain knowledge (*arguing to learn*), and inconsistent evidence for the view that such systems help to acquire general argumentation skills (*learning to argue*).

Crucially, they found support for the contention that the format of the argument representation directly affects students' argumentation behavior and comprehension. Representations that provided more structure prompted students to use the structure and this in turn led to more elaborated arguments. Structuring the students' process can also encourage students to engage in better argumentation. Thus, the pedagogical utility of argument diagramming has not been completely settled.

4 DIAGRAMS AND DIAGNOSIS

While argument-support systems scaffold students' arguments it is not clear whether these systems can effectively assess students' argument-planning process. Can we, for example, grade student-produced arguments and provide effective remediation that will, in turn, result in better written or oral arguments? And, by extension, can we do this automatically? Our goal in the present series of studies is to address these questions. We begin by focusing on the problem of diagnosis.

4.1 Expert Grading

As stated above, one of the key advantages of authentic practice in argumentation is that it transfers directly to professional practice and can be graded by domain experts according to the same or similar rubrics that they would use in practice. In (Lynch et al. 2009) we reported on the results of an agreement study designed to test whether domain experts, namely law school faculty, could grade student-produced argument diagrams in a consistent way.

For that study we collected a set of student-produced argument diagrams taken from the LARGO system (Pinkwart 2009). LARGO is a tutoring system designed to guide students in understanding the use of rule-like tests and hypothetical cases in legal arguments. Students use the system to read and annotate a series of oral argument transcripts taken from the U.S. Supreme Court. Annotations are made using a graphical argument model with nodes representing legal tests (a kind of warrant in the Toulmin sense), hypothetical cases, and relevant facts. The nodes are connected using arcs that represent: modifications of the tests; responses to hypothetical cases; and comparisons. Students are guided in their use of the system by on-demand self-reflection prompts and peer feedback.

In this study we selected a total of 198 diagrams covering three transcripts taken from studies that spanned first and third-year students as well as different classes. We selected a pair of experienced law school faculty and collaborated with them to develop a grading rubric, which they then applied to the diagrams. The rubric assessed how the diagram represented salient features of the argument, such as the legal tests, key hypothetical cases, and the way in which an advocate modifies his or her test in response to a hypothetical case posed by one of the justices. Grading was conducted independently by both faculty members with the diagrams randomized to conceal class level or other identifying information. Our preliminary results show that the faculty achieved high agreement on the overall scores.

While this work is promising, however it is not complete. Firstly, the LARGO system provides students with on-demand help and, as such, scaffolds their construction. While this was ideal for the educational goals of LARGO, we risk biasing the results by having guided the students to produce well-structured diagrams. Secondly, the grading of the diagrams was not connected with any written or oral arguments produced by the *students*. Thus we cannot assess the extent to which use of the system helped students to plan future arguments.

We are presently conducting a second grading study as part of a psychological research methods (RM) course at the University of Pittsburgh. As part of the course students are tasked with designing a psychology research study, carrying out the data collection and analysis phases, and writing up their results in a research report. When writing up an empirical research report the author presents a coherent argument, largely in the introduction section, to convince the reader that their research is relevant and novel, that the stated hypotheses are testable, and that the study being conducted will test the hypotheses adequately. For the purposes of this study we assigned students to outline their arguments using the LASAD system. LASAD is an online argument diagramming toolkit developed at Clausthal University of Technology (Loll 2011).

LASAD is designed as an argument diagramming platform that supports multiple frameworks for diagramming arguments (i.e., ontologies, the nodes, arcs, and their associated semantics) as well as peer collaboration. For the present study the system was set to use an ontology that defined four types of argument nodes: *Claims* representing assertions of psychological facts such as "Handedness affects haptic sensitivity"; *Hypotheses* representing psychology research hypotheses such as "If a person is right-handed then they will better identify shapes with the right hand." *Citations* of prior work; and *Current-Study* nodes that highlight key features of the students' present study such as "Our work focused solely on right-handed college students." The ontology also defined four types of arcs. *Supporting* and *Opposing* arcs are used to represent an argumentative relationship where the content of one node validates or contradicts the other much like the data supporting the claim in the Toulmin diagram above. *Undefined* arcs are used when one node adds information to another or provides some context. Finally, the *Comparison* arcs are used when the author wants to highlight similarities or distinctions between two nodes such as drawing comparisons between the current study and cited works. A partial LASAD diagram taken from this study is shown in Figure 1.

Figure 1. Subset of a LASAD diagram drawn from the present study.

For our present study we have collected 238 essays and a comparable number of diagrams. Some of the essays are tied to a specific companion diagram while others are from a subsequent assignment that did not involve argument diagramming. We have engaged a pair of experienced teaching assistants from the RM course to grade the essays and diagrams according to a grading rubric that focuses on key features of the diagrams, such as the presence of adequate hypotheses and citations, the relationships between those features, and the overall coherence and quality of the arguments presented. The grading rubrics for the graphs and essays are aligned so that we can compare scores across the two types of features. Thus we can correlate the scores on a question-by-question basis.

Once this analysis is complete we will be able to examine the correlation between the students' diagrams and their essay grades. If the correlations hold then it will validate our hypothesis that argument diagrams, when used for argument outlines, may be used to diagnose students' subsequent argument performance. We will also be able to test the extent to which students' performance on the argument tasks transfers to the subsequent non-diagramming assignment. If this is, in fact, the case we will have validated the hypothesis that argument diagrams generally may be used to diagnose students' ability to produce novel future arguments.

4.2   Feedback and Analysis

While the ability to manually grade student diagrams is beneficial it is does not completely address our central goal of making diagrams automatically comparable. To that end we plan to apply graph grammar induction algorithms to the task of automatically aligning and classifying students' argument diagrams. Graph grammars are a form of augmented context-free grammars that operate on graphs (Rekers and Schürr 1977). The grammars are defined by a set of production rules that match subgraphs of a given type, such as the the subgraph containing the citation and current study nodes along with the comparison arc (Figure 1, # 31, 36, & 37), and optionally map it to a different graph such as one that splits the comparison arc into two separate analogy and distinction arcs or that compresses it to a single "Supporting Feature" node that summarizes the text of the three contributions and is linked to node 23 via arc 32.

Manually-generated grammars were employed in LARGO to provide on-demand help (Pinkwart et al. 2009). Here the grammar productions were written to detect deviations from the desired argument model. When a given production applied an associated message was sent. Grammar induction algorithms such as SubdueGL (Jonyer 2002) take an elegant reduction approach to rule induction. Given a single graph or potentially a set of graphs G, the algorithm looks for a frequently occurring subgraph gi that can be collapsed to reduce the full set. It then generates a production rule mapping this subgraph to a smaller node or arc and then repeats with the, now reduced, graph or graph set. Thus in the case of our argument diagrams if most of the high-scoring diagrams include a citation and current study node connected by a comparison arc, then the *Supporting Feature* rule described above could be applied to the entire set to compress the diagrams. This approach has been applied successfully to the task of anomaly detection in large graphsets.

We plan to apply these algorithms in a similar manner to identify common error features. Given the graph grades we have collected, we plan to segment the set of graphs by grade and to identify frequently-occurring subgraphs that are present in particularly good or poor student diagrams. These common subgraphs can then be presented to domain experts to form the basis of manually-generated diagnostic rules to improve the help system. We plan to move beyond the binary approach taken by the SubdueGL algorithm to take advantage of the weighted data that the scores provide. In that way we can favor not just frequent subgraphs but those prone to occur, or not occur, in lower-scoring diagrams.

For example, based upon our initial semantic rules the use of the comparison arc #37 is correct and consistent with the instructions given to students. Here the student is not just presenting a supporting citation but highlighting the key similarities between it and the current study that make it relevant while at the same time pointing out the differences between them that make this study novel. Thus we expect that subgraphs of this type would be present in high-scoring diagrams. The lack of supporting citation nodes for the broad claim made in node #33, by contrast, is a basic argumentation error; we expect unsupported claims of this type to be found in poor diagrams. While these are a-priori rules that we have defined they illustrate the type of basic graph features that we will search for. As part of this search process we also plan to test the validity of a-priori rules of this type.

5 PRELIMINARY CONCLUSIONS

The ability to process and make persuasive arguments is a *core* skill. It is a necessary precursor to real-world problem solving in ill-defined domains from science to law. Moreover, it is a basic requirement for participation in an informed democracy. Argumentation skills however, are not always taught explicitly and students, even at the undergraduate level, often face difficulties such as getting "lost in the text" which limit their ability to benefit from authentic practice. For this reason educational researchers have investigated diagrammatic models of argument which provide a scaffolding effect for students' argumentation skills. While these tools have been successful at scaffolding individual argumentation and group discussion it has not yet been shown that they can effectively diagnose students' planned arguments or general argumentation skills. In this paper we laid out an ongoing series of studies to test the diagnostic utility of argument diagrams for human graders, and the potential for automatic diagnosis and feedback based upon grammar induction. The preliminary data that we have obtained has been promising.

Our overriding goal is to find a middle-ground approach to argument instruction. While formal models can be easily processed by an expert system they do not readily permit the real-world arguments that students need to practice. Free-text arguments, by contrast, allow student problem solvers to address real-world tasks but are not currently amenable to robust analysis and feedback. By having students use diagrammatic models of argument we can scaffold their argument production and make the resulting work amenable to human and machine analysis while at the same time permitting students to work on real-world arguments. In so doing we believe that we can exchange a relatively small, and ultimately beneficial, constraint on the users' interactions for a much larger opportunity for diagnosis and feedback.

REFERENCES:

Aldisert, Ruggero J. (1989) Logic for Lawyers: a Guide to Clear Legal Thinking. Clark Boardman Company Ltd.

van den Braak, S.W., van Oostendorp, H., Prakken, H. and Vreeswijk, G.A.W. (2006) A critical review of argument visualization tools: Do users become better reasoners? In *Workshop Notes of the ECAI-2006 Workshop on Computational Models of Natural Argument (CMNA VI)* eds. Grasso, F. and Kibble, R. and Reed, C. 67-75.

Chryssafidou, Evi, and Sharples, Mike., (2002) Computer-supported planning of essay argument structure. In *Proc. of the 5th Int. Conference of Argumentation.* Amsterdam

Easterday, Matthew W., Aleven, Vincent., Scheines, Richard., and Carver, Sharon M. (2009) Constructing Causal Diagrams to Learn Deliberation. In *International Journal of Artificial Intelligence in Education* 4(19):425-445.

Fernandes, R., and Simon, H. A. (1999) A Study of how Individuals Solve Complex and Ill-Structured Problems. In *Policy Sciences* 32:225-245

Gordon, Thomas F. (2011) Analyzing Open Source License Compatibility Issues with Carneades. In *Proc. of the 13th International Conference of AI and Law* ed. van Engers, T. 51-55 New York, New York, ACM.

Greene, Laurence. (2009) Writing in the Life Sciences. Oxford University Press: New York.

Jonyer, Istvan., Holder, Lawrence B. and Cook, Diane J. (2002) Concept Formation Using Graph Grammars In *Proc. of the KDD Workshop on Multi-Relational Data Mining.*

Llewellyn, Karl N. (1951) The Bramble Brush; On our Law and it's study. Oceana Publications Inc, Dobbs Ferry, New York.

Loll, F., & Pinkwart, N., (2011). Guiding the Process of Argumentation: The Effects of Ontology and Collaboration. In *Connecting Computer-Supported Collaborative Learning to Policy and Practice: CSCL2011 Conf. Proc. V. I.* eds. Spada, H., Stahl, G. Miyake, N. & Law N. pp.296-303. Int'l Soc. of the Learning Sciences

Lowrance, John D. (2007) Graphical Manipulation of Evidence in Structured Arguments. In *Law, Probability and Risk* 6(1-4): 225-240

Lynch, Collin., Ashley, Kevin D., Pinkwart, Niels., and Aleven, Vincent (2012) Adaptive Tutoring Technologies and Ill-Defined Domains. In *Adaptive Technologies for Training and Education* eds. Durlach, P. J. and Lesgold, A. M.. New York, Cambridge.

Lynch, Collin., Ashley, Kevin D., Pinkwart, Niels., and Aleven, Vincent (2009) Argument Diagramming and Diagnostic Reliability. In *Legal Knowledge and Information Systems: Proceedings of Jurix 2009.* ed. Guido Governatori. Amsterdam, IOS Press.

Pinkwart, Niels., Ashley, Kevin D., Lynch, Collin., and Aleven, Vincent (2009) Evaluating an Intelligent Tutoring System for Making Legal Arguments with Hypotheticals. In *International Journal of Artificial Intelligence in Education* 4(19):401-424.

Reed, Chris., Walton, Douglas. and Macagno, Fabrizio. (2007) Argument Diagramming in Logic, Law and Artificial Intelligence In *Knowledge Eng. Review* 22(1): 87-109.

Rekers, J. and Schürr, Andy. Defining and Parsing Visual Languages with Layered Graph Grammars In the *Journal of Visual Languages & Computing.* 8(1):27-55

Scheuer, O., McLaren, B. M., Loll, F., & Pinkwart, N. (in press). Automated Analysis and Feedback Techniques to Support Argumentation: A Survey. In *Educational Technologies for Teaching Argumentation Skills.* eds. Pinkwart, N. & McLaren, B. M. Bentham Science Publishers.

Scheuer, Oliver, Loll, Frank., Pinkwart, Niels and McLaren, Bruce. (2010) Computer-supported argumentation: A review of the state of the art. In *International Journal of Computer-Supported Collaborative Learning.* 5(1):43-102 New York , Springer.

Suthers, Daniel D. (in-press) Empirical studies of the value of conceptually explicit notations in collaborative learning. In *Knowledge Cartography* eds. A. Okada, S. Buckingham Shum and T. Sherborne. Cambridge, MA: MIT Press.

Voss, James F. (2006) Toulmin's Model and the Solving of Ill-Structured Problems In *Arguing on the Toulmin Model: New Essays in Argument Analysis and Evaluation* eds. Hitchcock, David and Verheij, Bart pp. 303-311 Berlin, Springer.

Voss, James F., Greene, Terry R., Post, Timothy A. and Penner, Barbara C. (1983) Problem Solving Skill in the Social Sciences. In the Journal of The Psychology of Learning and Motivation 17: 165 - 215 New York, New York. Academic Press.

CHAPTER 20

Modeling Student Behaviors in an Open-ended Learning Environment

Gautam Biswas, John S. Kinnebrew, and James R. Segedy

Vanderbilt University
Nashville, TN, USA
Gautam.Biswas@Vanderbilt.Edu

ABSTRACT

Present day Intelligent Tutoring Systems model student's knowledge and learning performance in detail, but the emphasis on knowledge and procedural performance does not lend itself to analyzing students' learning behaviors and strategies. We have adopted an integrated approach to help students learn domain content and metacognitive strategies in choice-rich learning environments. This paper presents an empirical approach imposed on a theory-driven model to extend performance-oriented student modeling by combining modeling students' learning performance and learning behaviors. We discuss the application of data mining techniques to build the content of such behavioral student models and demonstrate its effectiveness with case studies.

1   INTRODUCTION

Intelligent Tutoring Systems model students' knowledge and learning performance by monitoring and diagnosing their solution steps as they work on the system. Model-tracing tutors (Corbett and Anderson, 1992) track student performance at fine granularities by evaluating the correctness of every step in a students' solution. Student models that accurately capture students' knowledge and problem solving abilities (procedures) play an important role in adaptive and tailored instruction (VanLehn, 1988; Wenger, 1987; Woolf, 2009).

For lack of space, we do not review the different approaches to student modeling but refer the reader to seminal papers in the area (such as, Brown and Burton, 1978;

Brown and VanLehn, 1980; Brusilovsky and Millán, 2007; Conati, Gertner, and VanLehn, 2002; Goldstein, 1979; Johnson, 1986; Koedinger *et al.*, 1997; Reiser, Anderson, and Farrell, 1985; VanLehn and Niu, 2001). They describe a variety of applications for student modeling: (1) sequencing curricular units as students progress through a curriculum, (2) offering scaffolding and feedback (solicited or unsolicited) to help students overcome difficulties in their learning and problem solving, (3) tailoring explanations to the level of a student's knowledge, and (4) generating appropriate problems, *e.g.*, those within the students zone of proximal development (VanLehn, 1988).

Student modeling techniques and diagnosis of students' errors have become quite sophisticated, but the methods primarily focus on modeling knowledge and procedural performance. They do not easily lend themselves to analyzing students' general learning behaviors and strategies, aspects that are important to the transfer of learning (Schwartz, Bransford, and Sears, 2005) and preparation for future learning (Bransford and Schwartz, 1999). Preparation for future learning does not only emphasize declarative knowledge and procedural abilities for effective problem solving, but it also investigates the learner's readiness and capability to learn in new situations, especially situations that are unlikely to occur in traditional classroom settings.

The focus of our work has been on helping students simultaneously learn domain content and metacognitive strategies in open-ended, choice-rich learning environments (Leelawong and Biswas, 2008; Segedy, Kinnebrew, and Biswas, In Review). Our studies have shown that simultaneously tracking students' learning behaviors and performance provide better indications of preparation for future learning (Leelawong and Biswas, 2008; Schwartz *et al.*, 2009). We discuss an open-ended, choice-rich learning system, in which students learn science by constructing a causal model to teach a virtual agent.

This paper extends the traditional notion of knowledge- and performance-oriented student models to richer ones that also include models of students' learning behaviors and strategies. We discuss the application of sequence mining techniques to characterize the structure and content of such behavioral student models. We then demonstrate the effectiveness of our approach in identifying and analyzing students' learning behaviors in studies conducted in middle school science classrooms. We conclude the paper with a discussion of how these results inform the design of extended student models.

2   THE BETTY'S BRAIN LEARNING ENVIRONMENT

In Betty's Brain, shown in Figure 1, students learn by teaching a virtual agent. The system adopts a self-regulated learning (SRL) framework to help students develop cognitive and metacognitive learning strategies. As students explore hypermedia resources on a science topic, they construct a causal map to teach Betty, the virtual Teachable Agent (TA) (Leelawong and Biswas, 2008). Betty only knows what she has been taught by the student, but, once taught, she can use the map to answer questions like *"if deforestation increases, what effect does it have on polar sea ice?"* and explain her answers as a chain of causal relations (or links) from the map. The student can also ask their TA to take quizzes, which are a set of questions created and graded by a Mentor Agent named Mr. Davis. Betty's quiz performance is based on how closely her causal map (as created by the student) matches a (hidden) expert causal map.

Betty's quiz performance helps the students assess and reflect on their TA's, and, therefore, their own understanding of the topic under study. This assessment and subsequent reflection can help guide them as they continue their learning and teaching tasks. Our previous studies show that students are motivated to help Betty achieve high quiz scores by first learning/reviewing material related to the domain and then teaching this new or revised understanding to Betty (Biswas *et al.*, 2010; Segedy, Kinnebrew, and Biswas, In Review).

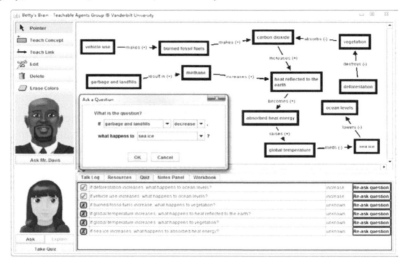

Figure 1 The Betty's Brain System

2.1 Cognitive and Metacognitive Model in Betty's Brain

The combined learning and teaching task in Betty's Brain is complex, open-ended, and choice-rich, so learners must employ a number of cognitive and metacognitive skills to achieve success. At the cognitive level, they need to identify and understand relevant information from the resources in the system, represent that information in the causal map format to teach their agent, and use questions and quizzes to explore Betty's understanding and assess her overall progress.

At the metacognitive level, they must decide when and how to: acquire information, build or modify the causal map, check Betty's progress, and revise their own understanding of both the science knowledge and the evolving causal map structure.

Figure 2 summarizes our model of an idealized student's cognitive and metacognitive activities for knowledge construction, monitoring, and seeking help when needed as they are involved in their learning and teaching tasks.

The students are scaffolded through dialogue and feedback provided by Mr. Davis. This feedback, explained in more detail elsewhere (Segedy, Kinnebrew, and Biswas, In Review), aims to help students progress in their learning, teaching, and monitoring tasks.

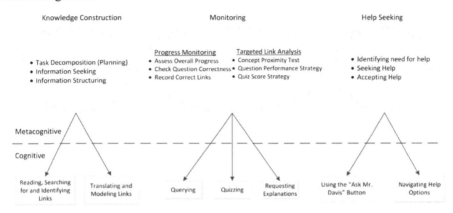

Figure 2 Idealized Student Behavior Model

3   STUDENT MODELING IN BETTY'S BRAIN

Betty's Brain tracks many details of students' learning interactions along with their teaching performance. This wealth of data provides opportunities to assess, model, and understand student learning behaviors and strategies. Realizing these opportunities requires effective methods for identifying important learning behavior patterns in the activity trace data.

3.1 How do we measure student's knowledge and performance?

In Betty's Brain, a student's work can be assessed by their performance on the learning task, an aggregate causal map score calculated as their correct minus incorrect causal links. This score can be computed repeatedly over the course of the students' interactions with the system by comparing the student's current map with the expert map (see Figure 3). We use these scores to track and study how students' learning and performance evolve throughout the period when they use the system (Kinnebrew, Loretz, and Biswas, In Review; Leelawong and Biswas, 2008).

However, the map score does not completely characterize students' learning states. To capture this type of information, the system is designed to track details of

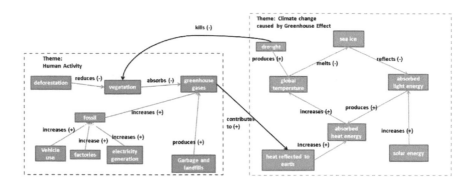

Figure 3 Example Climate Change Expert Causal Map

students' learning interactions with the system. By characterizing this data, we seek to gain an understanding of students' learning behaviors and strategies.

3.2 How do we characterize and study student behaviors?

In an earlier section, we discussed a cognitive and metacognitive model that represents and classifies the various learning behaviors. This model provides a framework for characterizing individual student's learning behaviors as they use the system. We start with the sequence of activities captured as log events. An example of a log event is an add link action, which is further characterized by the two concepts that are linked, as well as the direction and sign of the causal relation. In our work, we have characterized students actions into five primary categories: (1) *READ*: students access a page in the resources; (2) *LINK* or *CONCept Edit*: students edit the causal map, with actions further divided by whether they operate on a causal link or concept, and whether the action was an addition ("*ADD*"), a removal ("*REM*"), or a modification ("*CHG*"), e.g., *LINK-REM* or *CONC-ADD*; (3) *QUER*: students use a template to ask Betty a question, which she answers using a causal reasoning method; (4) *EXPL*: students probe Betty's reasoning by asking her to explain her answer to a question (she uses dialogue and animation on the causal map to demonstrate her causal reasoning and answer the question); and (5) *QUIZ*: students have Betty take a quiz, which is a set of questions chosen and graded by the Mentor agent.

This characterization is necessarily an abstraction where contextual details associated with the sequence of actions may be lost, e.g., whether the last *LINK-ADD* was related to the information that the student *READ* in a previous action. Adding many of these details may make the total number of action definitions unwieldy and hard to process and interpret. To maintain a balance between the number of distinct actions and retention of relevant context information, we augment action descriptions with several descriptive metrics. For example, we summarize the relatedness information among a small window of activities. This metric splits each categorized action into two distinct actions: (1) –*REL* if the current action is relevant to at least one of three recent actions; and (2) –*IRR* otherwise (Biswas *et al.*, 2010).

Similarly, to better understand students' knowledge acquisition (from the resources) in the context of their knowledge structuring (during map building), we apply two metrics to characterize each student *READ*-action: (1) *Read Time (SHRT/FULL)*, which is a determination of whether or not the student spent at least 30 seconds on the page. Given the length of resource pages, 30 seconds is typically sufficient for the students to read some of the material in detail (*FULL*) versus a brief look (less than 30 seconds) at the contents (*SHRT*), possibly to skim the material or to check whether the page was the one for which they were searching; and (2) *Read Repetition (FRST/REPT)*: whether the student has previously done a FULL read ($>= 30$ seconds) of the page (*REPT*), which suggests rereading of material on the page. Otherwise, they have not previously read this page in depth (*FRST*).

Using the extended action definitions, we apply sequence mining methods to characterize student behaviors in different contexts; *e.g.*, when a sequence of actions leads to an increase in their map scores versus a decrease. This forms the basis for defining patterns of actions that can be mapped back to the cognitive/metacognitive behavior model. To identify important activity patterns that distinguish learning behaviors between students, we employ a differential sequence mining technique (Kinnebrew, Loretz, and Biswas, In Review). This technique combines frequency measures and techniques from sequential pattern mining and episode mining to identify differentially frequent patterns that characterize behavioral differences between groups of students. The details of the sequence mining and related algorithms are presented elsewhere (Kinnebrew, Loretz, and Biswas, In Review).

4 CASE STUDIES: DERIVING LEARNING BEHAVIOR MODELS

We have collected a large amount of students' interaction trace data in recent studies that we conducted in middle Tennessee 5th through 8th grade science classrooms. Students were first introduced to the science topic (*e.g.*, global climate change, body thermoregulation) during regular classroom instruction, provided an overview of causal relations and building causal models, and given hands-on training with the system. Then the students spent 4-5 class periods teaching their agent by constructing a causal map. As they worked on the system, they received feedback on learning strategies from the Mentor agent.

For purposes of informing the behavioral aspects of the student model, this analysis focuses on two groups of students: (1) the *HI* group, whose members achieved map scores that were at least 80% of the maximum achievable score; and (2) the *LO* group, whose members achieved map scores below 40% of the maximum achievable score. This comparison helps us differentiate between productive and non-productive learning behaviors.

Results show that on average, when students in the *HI* group performed a link addition, deletion, or change, it improved the quality of Betty's causal map 60.3% (sd = 7.1%) of the time. The same metric for the low group was 42.5% (sd = 10.6%). An ANOVA showed that this difference was significant (p < .001, with an

effect size (Cohen's f) of 0.99). A similar difference was seen in the overall relevance of the students' actions: 53.3% (sd = 8.4%) to 40.6% (sd = 12.2%), level of
significance, $p < .001$; effect size, $f = 0.62$. Clearly, the *HI* group outperformed the
LO group, but the knowledge construction task was not easy for the *HI* group. They
made a number of errors but were more successful in correcting them than the *LO*
group.

Further, behavior pattern differences between the two groups indicate that the
HI-group performed their monitoring tasks using the quiz in a much more effective
way than the *LO*-group. For example, after the *QUIZ* action, the *HI*-group primarily
read relevant sections of the resource (*READ-REL*) to determine why Betty got a
quiz question wrong, or they used the quiz to remove an incorrect link they had recently added or changed (*LINK_REM-REL*). On the other hand, after the *QUIZ* action, the *LO*-group primarily read multiple pages, but the content of those pages
were not related to the quiz questions (sequence of *READ-IRR* actions).

Since reading was a primary component of the students' knowledge acquisition
and knowledge construction strategies, we probed deeper into their *READ* actions.
Both groups performed roughly equal numbers of read actions on pages they previously read in-depth (*REPT*) compared to ones they had not read in-depth (*FRST*).
On the average, the *LO*-group used more short (*SHRT*) reads (74%) than the *HI*-
group (69%), and the ratio of short to full page (*FULL*) reads was approximately 3:1
versus 2:1 for the *LO*- and *HI*-groups. Similarly, the ratio of irrelevant (*IRR*) to relevant (*REL*) reads was 3:1 (*LO*) versus 2:1 (*HI*).

Analysis of behavior patterns using differential sequence mining confirmed that
the *HI*-students were more likely to add a relevant (*REL*) link following a full-
length (*FULL*) re-read (*REPT*) of a page. The greater reliance on extended re-reads
before adding links suggests the *HI*-group employed a more careful approach to
identifying causal links in the resources, which may have helped increase their accuracy in teaching correct links, and also their ability to correct any incorrect links.
Further, the *HI*-group more frequently employed reading activities in a monitoring
context (*e.g.*, in conjunction with *QUIZ* actions). On the other hand, this analysis
showed that the *LO*-group was more frequently involved in short (*SHRT*) irrelevant
(*IRR*) reads.

Further details of the analyses can be found in (Kinnebrew, Loretz, and Biswas,
In Review). In the next section, we briefly illustrate how the data mining results are
used to update the student models.

5 BUILDING THE STUDENT BEHAVIOR MODEL STRUCTURE

We use the students' productive and unproductive behavior patterns derived by
our data mining analyses to build the structure of our idealized cognitive and metacognitive student behavior model that was illustrated in Figure 2. As an example,
we use the results discussed above to build the student modeling structure for progressive monitoring.

At the cognitive level, we define procedures to keep track of how frequently students use the query and quiz features, and what proportion of the queries and quizzes are relevant (*REL*) and irrelevant (*IRR*) to actions that the students performed immediately before the query or quiz action. At the metacognitive level, we have designed pattern detectors to determine which questions from the quiz the students follow up on, if any at all. For example, if the students' subsequent *READ* action (whether it is a *FRST* read of a page, or a *REPT* (re-read)) is linked to a question that had an incorrect answer (or an answer that was marked right, but for the wrong reasons), the student receives a check-mark for use of a good progress monitoring strategy. If the student reads a page that is considered irrelevant (*IRR*), we consider the student has not shown use of this progressive monitoring strategy.

We have developed preliminary mechanisms within the Betty's Brain system to systematically detect relevant behavior patterns as students work on the system. By collecting this information in an aggregated data structure, we can adopt frequency-based methods to develop levels of confidence on whether a student does or does not demonstrate the use of good strategies. Since the behavior patterns are ones that are derived from student activities, with time, we are most likely to capture a significant number of actual behavior patterns that can be mapped on to the idealized cognitive and metacognitive model. In the future, we will extend this further to devise Bayesian networks (e.g., VanLehn and Niu, 2001) that capture the structure of the cognitive and metacognitive models. We have performed some preliminary analyses in deriving aggregate student behavior models as hidden Markov model (HMM) structures (Biswas, et al, 2010). But this work will have to be extended to learning Bayes nets at the level of students' action and behavior patterns in future work.

Eventually, we will further extend the method presented in Kinnebrew, Loretz, and Biswas (In Review), where we characterize student behaviors for productive and unproductive segments of their work. During productive segments students' map scores increase progressively, and during unproductive segments, they remain flat or decrease. Using this information, along with information captured in the combined student behavior and performance models, we will devise systematic mechanisms to scaffold student learning and provide relevant feedback when students exhibit unproductive learning behaviors. Preliminary analysis of a classroom study employing this feedback has shown promising results (Segedy, Kinnebrew, and Biswas, In Review).

6 CONCLUSIONS

This paper discusses a method for combining student performance and behaviors to derive a more complete model of the student's learning states. A theoretical behavior modeling framework has been operationalized using exploratory data mining techniques. In future studies, we will extend these analyses using a combination of sequence mining and Bayesian learning techniques to design robust models to adapt and scaffold student learning in a variety of situations.

ACKNOWLEDGMENTS

This work has been supported by the National Science Foundation's Information and Intelligent Systems Award #0904387.

REFERENCES

Biswas, G., Jeong, H., Kinnebrew, J., Sulcer, B., and Roscoe, R. 2010. Measuring Self-regulated Learning Skills through Social Interactions in a Teachable Agent Environment. *Research and Practice in Technology-Enhanced Learning (RPTEL).* 5(2): 123-152.

Bransford, J. D. and Schwartz, D. L. 1999. Rethinking transfer: A simple proposal with multiple implications. In A. Iran-Nejad & P. D. Pearson (Eds.), *Review of Research in Education,* 24: 61-101.

Brown, J.S. and Burton, R.R. 1978. Diagnostic models for procedural bugs in basic mathematical skills. *Cognitive Science,* 2(2): 155-192.

Brown, J.S. and VanLehn, K. 1980. Repair theory: A generative theory of bugs in procedural skills. *Cognitive Science,* 4: 379-426.

Brusilovsky, P. and Millán, E. 2007. User models for adaptive hypermedia and adaptive educational systems. In: Brusilovsky, P., Kobsa, A., Neidl, W. (eds.) *The Adaptive Web: Methods and Strategies of Web Personalization.* LNCS, 4321: 3-53. Springer, Heidelberg.

Conati, C., Gertner, A., VanLehn, K. 2002. Using Bayesian Networks to Manage Uncertainty in Student Modeling. *User Modeling and User-Adapted Interaction* 12(4): 371-417.

Corbett, A. T. and Anderson, J. R. 1992. The LISP intelligent tutoring system: Research in skill acquisition. In J. Larkin, R. Chabay, C. Scheftic (Eds.), *Computer Assisted Instruction and Intelligent Tutoring Systems: Establishing Communication and Collaboration.* Hillsdale, NJ: Erlbaum.

Goldstein, I.P. 1979. The genetic graph: a representation for the evolution of procedural knowledge, 11(1): 51-77.

Johnson, W.L. 1986. Intention-Based Diagnosis of Novice Programming Errors. London: Pitman.

Kinnebrew, J.S., Loretz, K.M., and Biswas, G. In Review. A differential sequence mining method contextualized by student performance evolution to derive learning behavior patterns. *Journal of Educational Data Mining,*

Koedinger, K.R., Anderson, J.R., Hadley, W.H., Mark, M.A. 1997. Intelligent tutoring goes to school in the big city. *International Journal of Artificial Intelligence in Education* 8: 30-43.

Leelawong, K. and Biswas, G. 2008. Designing learning by teaching agents: The Betty's Brain System. *International Journal of Artificial Intelligence in Education,* 18(3): 181-208.

Reiser, B.J., Anderson, J.R., and Farrell, R.G. 1985. Dynamic Student Modeling in an Intelligent Tutor for Lisp Programming, *Proceedings International Joint Conference on Artificial Intelligence,* Los Angeles, CA, 8-14.

Segedy, J.R., Kinnebrew, J.S., and Biswas. G. In Review. Supporting cognitive and metacognitive skills in complex, open-ended learning environments. *Journal of Educational Psychology.*

Schwartz, D., Bransford, J., and Sears, D. 2005. Efficiency and innovation in transfer. In J. Mestre (Ed.), *Transfer of learning from a modern multidisciplinary perspective* Greenwich, CT: Information Age Publishing, 1-51.

Schwartz, D.L., Chase, C., Chin, D.B., *et al.* 2009. Interactive metacognition: Monitoring and regulating a teachable agent. In D. J. Hacker, J. Dunlosky, & A. C. Graesser (Eds.), *Handbook of metacognition in education* (pp. 340–358). New York, NY: Routledge.

VanLehn, K. 1988. Student modeling. In M. Poison & J. Richardson (Eds.), *Foundations of intelligent tutoring systems*. Hillsdale, NJ: Erlbaum.

VanLehn, K. and Niu, Z. 2001. Bayesian student modeling, user interfaces and feedback: A sensitivity analysis, *International Journal of Artificial Intelligence in Education*, 12, 154-184

Wenger, E. 1987. *Artificial Intelligence and Tutoring Systems*, Los Altos, CA: Morgan Kaufmann.

Woolf, B. 2009. Building intelligent interactive tutors: Student-centered strategies for revolutionizing e-learning. Amsterdam, The Netherlands: Elsevier.

CHAPTER 21

Detailed Modeling of Student Knowledge in a Simulation Context

Allen Munro and David Surmon

University of Southern California
Los Angeles, USA
munro@usc.edu

Alan Koenig, Markus Iseli, John Lee, William Bewley

University of California, Los Angeles
Los Angeles, USA

ABSTRACT

A detailed performance record and two types of models of tactical knowledge are produced when students conduct tactical planning and problem solving in the context of a tactical planning simulation, the *TAO Sandbox*. The detailed performance record includes every action taken by a user in an interactive simulation session. The detailed performance record can be utilized to play back the session in the simulation. This session record includes low-level analyses of session events that are produced by the Sandbox. Our team has developed methodologies for maintaining a model of user knowledge based on these observable performance events.

At the same time that it is recorded, each announcement of a performance event is reported to a collaborating software component, called the CRESST Assessment Application (CAA), which contains a detailed model of user knowledge at a higher level of representation. When a performance announcement is made to the CAA, it informs a Bayesian network that represents the student's exhibited competencies. The CAA, in turn, reports the new network values to a third software component, the CAA Performance Monitor, along with a brief textual message that it received from the Sandbox about the most recent measured performance. The performance monitor maintains student records that include complete histories of CAA estimations, together with the Sandbox-generated explanations. The performance monitor can present graphs of changing student models, with support for user-selected data filters to support the presentation of charts for different subsets of the Bayes net nodes.

Keywords: student model, Bayesian network, performance-based modeling, simulated task monitoring, adaptive training

1   THE TAO SANDBOX

The TAO Sandbox (Munro, Pizzini, and Bewley, 2009; Munro and Pizzini, 2011) is an evolving tool for building and delivering tactical planning problems for instruction that has been in experimental use at the Surface Warfare Officers School (SWOS) for several years. Using iRides Author (Munro, Surmon, and Pizzini, 2006; Munro 2007), we rapidly developed an initial approach to authoring scenarios and delivering tactical decision-making practice in the context of those scenarios. Feedback from the SWOS instructional staff guided the development of revised versions of this initial training and practice environment, which was initially called the ASW Sandbox, because it was designed to provide practice in planning surface tactics for Anti-Submarine Warfare (ASW). At the request of the instructors, we added features to support tactics planning practice for Air Defense and for Surface-Surface Warfare. Taken together, these warfare modes comprise the major aspects of tactics planning for Tactical Action Officers, and the new tool became known as the TAO Sandbox.

The Sandbox has two modes: Instructor Mode and Problem Mode. Instructor Mode is used for building scenarios and for conducting in-class demonstrations. Instructors create initial problem states largely by dragging ships, submarines, aircraft, missile bases, and air bases into position on selected maps. Problem mode is used for solving tactics problems in the context of scenarios authored by the instructors.

In Problem Mode, users can select problems that have previously been authored by instructors. Each problem has a map on which all action takes place. Typically the scenario has a mission briefing that explains the situation, including the commander's intent and intelligence on potential hostile units. Users solve the problem by running the scenario, ordinarily at some high multiple of real time, utilizing resources to try to avoid or to detect and attack hostile units, if required.

When using the Sandbox for tactics practice, users can pause scenarios, and they can speed up and slow down virtual time. They can deploy assets such as helicopters, sonobuoys of several types, the active and passive sonars of surface ships, and datums. The helicopters can be directed to utilize dipping sonars or to drop torpedoes. Several visualization tools are provided, including the Torpedo Danger Zone (TDZ), Advance Position, Limiting Lines of Approach, Air Defense Sectors, Cordon, Major Threat Axis, Vital Area, CIEA, and general purpose markers, including arrows and transparent shapes, which can be labeled.

When in Problem Mode, the Sandbox records all actions and independent events in a lightweight text format. Later, the Sandbox can be used to replay these recordings, so that instructor-led after-action-reviews can be conducted.

Users can choose to solve problems again to explore tactical alternatives. They can also replay automatic recordings of the actions that they took during a session. It is possible to interrupt a playback and continue the problem in a different way at any point. These features make it possible to use the Sandbox as a kind of "what-if" planning tool.

The instructors have used the Sandbox in a number of ways throughout the spiral development process. They have found five major ways to use this product to improve instruction and learning.

1.1   Demonstrations

One of Merrill's (2002) principles is that knowledge about procedures needs to be *demonstrated*. In fact, there is evidence that practice alone is less effective than a combination of viewing correct demonstrations plus practicing. Sweller and Cooper (1985) found that errors in an assessment were reduced by half if 12 math practice problems were replaced by 6 worked-out examples, each followed by a related practice problem.

The instructors apparently appreciate the importance of demonstrating solutions in the problem context. To illustrate a concept or a type of tactic, they create one or more scenarios that help to explicate the concept or that are suited to the use of the tactic. Then they demonstrate a solution to the problem in class. They have experimented with two ways of doing this. First, they have simply started the scenario and demonstrated the solution as the students watched, commenting on relevant features as the problem unfolds. The second approach is to perform the solution in advance with recording turned on. By playing back their recorded solution in class, they are able to attend more closely to the class, looking for questions or signs of confusion, rather than focusing on what must next be done with the mouse pointer. Instructors typically narrate the action. If a question arises, they can pause the playback in order to answer it. This second approach, using recorded demonstrations, has found favor with the instructors.

1.2   Practice and Remediation

Clark and Mayer (2008) recently reviewed the cognitive processing evidence for the importance of problem-solving practice, noting that "Generative cognitive processing occurs when a learner engages in deep processing in working memory in service of the learning goal.... For example an effective practice exercise can foster generative processing." (p. 6).

For a given class session, the instructor may prepare three problems, each of which requires some common knowledge and/or tactics, but each of which also has unique features that illustrate situations or tactical requirements that are not present in the others. The classroom is divided into three groups of students, and each group works on a different problem, discussing the issues and testing possible solutions. Ordinarily, students have enough time to attempt two or three different approaches. During this problem-solving time, the instructor is available to help groups that encounter thorny issues, providing an expert help system. In general, however, students take these problems as a challenge and do their best to find their own solutions.

As the students practice solving these surface problems, their actions are recorded. One to three students from each group then come to the head of the classroom and present their solution as the recording is played back. This gives the students who worked on different problems a chance to see this particular problem from another student's viewpoint. Finally, the instructor comments on the problem and the solution presented, pointing out issues that were not addressed and opportunities that were or were not exploited.

1.3 Student Assessment

Merrill's (2002) work and that of Clark, *et al* (2010) shows that assessment in conceptually realistic problem-based environments is more likely to predict post-training performance than do more conventional paper-and-pencil tests. Of course, schools such as SWOS have made a considerable investment in course materials, including test banks, and they cannot be expected to revolutionize every aspect of training overnight.

Previously developed assessment metrics (primarily short-essay questions) are utilized for assessing students. On occasion, however, a student who has demonstrated competence in the classroom suffers from test anxiety or some other problem that results in the production of marginal answers to a set of questions. To help in assessing such students, they are given a novel Sandbox problem and must describe the reasoning behind the actions that they take. This approach has been used to help students demonstrate their knowledge in an effective way.

1.4 Self-study

Some students have shown an interest in extending their tactics knowledge by performing additional problems on their own. If a question arises during self-study, the student can go to an instructor later and play back their attempt to solve the problem to the point at which the question arose.

1.5 What-if Exercises: Extending Knowledge

Later in the course, students are given the opportunity to take the Instructor version of the Sandbox with them. They can use this to author new tactical scenarios, either by editing an existing scenario from a growing library, or by building a new scenario from scratch. Any jpg file can be added as a map, and the Sandbox provides a simple method for assigning a scale to such new maps.

216

These five approaches to utilizing the TAO Sandbox, as of 2010, did not support *automated* assessment of student performances of practice problems. By adding detailed modeling of student knowledge in the context of surface warfare tactics, we have begun to support automated assessment and to experiment with instructional adaptations based on such assessments.

2   ARCHITECTURE TO SUPPORT MODEL-BASED ASSESS-MENT AND REPORTING IN A SIMULATION CONTEXT

There is a long history of utilizing student models in computer-based instruction and learning systems. In many cases, models are modified during training based on student answers to questions, student requests for help, or student choices of content. In the context of a free play simulation such as the TAO Sandbox, however, there is an opportunity to modify the estimation of student knowledge that the model reflects based on actions taken in the simulation. In effect, by watching a learner perform a task, we can estimate things about what the student knows and does not know.

A simple architecture for a system that assesses student knowledge based on actions and events in a simulation session is shown in the figure below. The simulation or game informs the assessment component about simulation events, such as user actions. At some point, the assessment component may generate an assessment event that justifies an instructional adaptation. The adaptive component selects the appropriate adaptation and directs other components to carry it out. In this figure the 'other' components are a pedagogy presenter (which could be a simple interface for displaying instructional text) and the simulator itself. Simulators can be prepared to offer services to adaptive components, services such as highlighting objects of interest, modifying a simulation scenario on the fly, or installing a new simulation state to start a different practice problem.

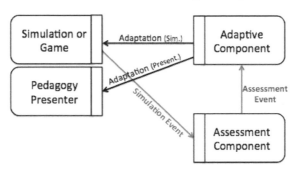

One way for an assessment component to manage its assessment of students is to maintain a detailed model of each student's knowledge.

3   TWO LEVELS OF MODELING IN THE TAO SANDBOX

Our approach tracks two types of student knowledge: low level knowledge that is directly evidenced through student actions and consequent simulation events, and higher level knowledge that is indirectly evidenced by the same student actions and simulation events. The simulator itself can gather evidence of the former type of knowledge. The higher level types of knowledge must be induced, based on relationships among the evidence data for the low level knowledge.

Distinctions between low level knowledge and higher level knowledge might be clarified by considering an example. The figure below represents a very small part of the knowledge domain, having to do with avoiding a dangerous submarine adversary, which requires learning about the adversary and making some estimates of its capabilities.

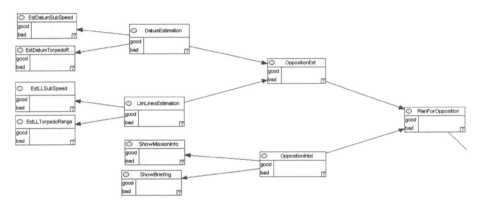

Specific evidences of learner knowledge are provided when the user interacts with the simulation. Using a slider that lets the user estimate the silent speed of a hostile submarine (which was earlier reported at an approximate position called a *datum*) is one example of evidence for low-level knowledge—represented by the EstDatumSubSpeed node at the upper left in the above figure. Depending on the accuracy of the estimation made, and on the user's estimation of the range of the hostile submarine's torpedo, an estimate can be made of the user's ability to carry out datum estimations (represented by the DatumEstimation node just to the right in the figure). This node, in turn, is one component that helps to determine the learner's overall ability for opposition estimation in undersea warfare. And such estimation ability itself feeds into a model value that represents the learner's ability to plan for undersea opponents. That value, in turn, contributes to an estimated value for other, higher-level student model values. The estimation of higher-level knowledge is further complicated by the fact that higher-level values can modify the lower-level values. If one type of evidence suggests that a learner is good at planning for the opposition, it is reasonable to guess that the learner might also know something about another contributor to that knowledge estimation.

We have implemented the estimations of lowest-level knowledge by instrumenting the TAO Sandbox to produce such values, as described in the next section. The higher-level knowledge estimates are maintained using a Bayesian Network, as described in the section titled "The Use of Bayesian Networks to Assess Student Performance," below.

4   INSTRUMENTING THE SIMULATION

Many types of actions in the TAO Sandbox now result in the simulation reporting an estimate of low-level knowledge. If the action is an appropriate one in the problem context, one or more specific estimates of user knowledge may be incremented. If the action is inappropriate in context, those same estimates of user knowledge will be decremented.

In addition to measuring the appropriateness of certain actions in context, the TAO Sandbox also 'scores' emergent events, which may not be the immediate results of user actions, but rather the consequence of earlier decisions. For example, if a hostile submarine fires a torpedo that damages the user's Mission Essential Unit (MEU), such as an aircraft carrier, that is a likely consequence of poor decisions made many (virtual) minutes or hours earlier in the session. But it is reported as an assessable event, nonetheless.

Most evaluation rules added to the TAO Sandbox are in some respects sensitive to particular problem contexts. It may be appropriate to use a ship's active sonar system when it is assigned to aggressively seek (and perhaps attack) a hidden submarine. In another problem context, where the task is to quietly slip past a threat while guarding an MEU, it would be a catastrophic mistake to utilize ship's active sonar, because that would pinpoint one's position for a hostile submarine. The goals of a given problem are laid out for the student in a mission briefing, the text of which cannot be 'read and analyzed' by the TAO Sandbox's low-level assessment system. Therefore, problem authors are provided with an interface that lets them use check boxes to specify aspects of the user's goal for a problem.

5   THE USE OF BAYESIAN NETWORKS TO ASSESS STUDENT PERFORMANCE

Bayesian Networks (BNs) are graphical models that represent probabilistic causal relationships between variables. They have many advantages such as being able to model prior knowledge, incomplete or missing data, clean or noisy observed data, and latent, uncertain, or unobserved variables. They do this by modeling abstract (hidden, latent) and complex constructs of higher-level knowledge together with the real-world manifestations of these constructs as observable measures and metrics. By instantiating a BN, we acquire the ability to assess higher order player abilities (such as situation awareness or decision making) directly by the capturing of lower level observables arising out of game/simulation-play. However, BNs *cannot* provide a definitive answer as to whether a given model is "correct," but they can calculate a *probability* of how well the model fits to some data.

Bayesian networks can be expanded to Dynamic Bayesian Networks (DBNs) to model time sequences of events, which have been used to model various warfare contexts. For example, Poropudas and Virtanen (2007) used a DBN to model an air combat simulation. They used DBNs to run what-if analyses for multiple events as they unfolded and used simulation data to populate the probability tables and dependencies. Iseli, Koenig, Lee, and Wainess (2010) used a DBN to model firefighting behavior in a damage control game and found that the DBN predicted performance as well as experts.

6   INSTRUCTIONAL ADAPTATION BASED ON CHANGES IN THE STUDENT MODEL

The BN is maintained by an application called the CRESST Assessment Application, or CAA. The CAA receives reports of low-level measurements of performance from the TAO Sandbox. It updates the Bayesian network based on this input. In addition, when certain value thresholds are exceeded for specific nodes in the network, it provides instructional adaptations. For example, when sufficient evidence has accumulated to suggest that the user is seeking out the available sources of information on the opponent, the CAA can direct the Sandbox to display an instructional message, such as the one shown in the figure below, "OK, you're getting the intelligence you need."

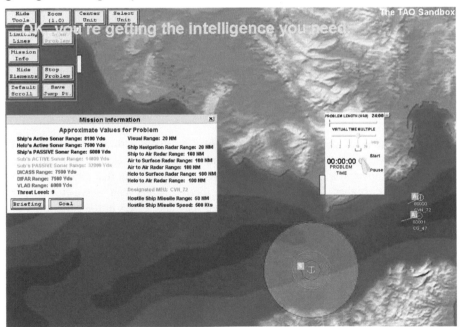

In addition to presenting instructional text, the CAA can direct the TAO Sandbox to modify the current problem, perhaps by changing the hostility of an opposing unit or by introducing an additional hostile or neutral unit into the problem. The CAA can also direct the Sandbox to present a different problem or to play a previously recorded problem session as an exemplar.

7   INTERACTIVE REPORTING OF STUDENT PROGRESS AS REFLECTED BY MODEL CHANGES

The CAA outputs every change in the Bayes net values to another application, the CAA Monitor. The CAA Monitor archives these values and presents a customizable charting interface that shows an individual student's progress in a selected subset of student model values.

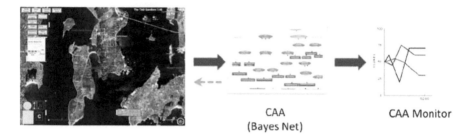

CAA
(Bayes Net)   CAA Monitor

The user of the CAA Monitor can define filters for viewing particular subsets of student model values. When the CAA Monitor is running with the CAA in a real time session, and the monitor has been attached to the current student's model, it will update the charted values in real time as the student carries out actions in a problem session. The CAA Monitor can also be used to inspect student student model data of students who are not currently using the TAO Sandbox, as well.

CONCLUSIONS

A simulation or game can be instrumented to make estimations of detailed student knowledge by evaluating actions and events. These estimations can then serve as the inputs to a Bayes net-based assessment component that models student knowledge at higher levels. Changes in these models can serve to trigger instructional interventions and adaptations that may take effect in the simulation or game or in a pedagogical interface.

REFERENCES

Clark, R. & Mayer, R. E. (2008). *E-learning and the science of instruction* (2nd ed). San Francisco: Jossey-Bass.
Clark, R. E., Yates, K., Early, S. & Moulton, K. (2010). An analysis of the failure of electronic media and discovery-based learning: Evidence for the performance benefits

of guided training methods In K. H. Silber, & R. Foshay, (Eds.), *Handbook of training and improving workplace performance, Volume I: Instructional design and training delivery*. Washington, DC: International Society for Performance Improvement.

Iseli, M. R., Koenig, A. D., Lee, J. J., & Wainess, R. (2010). Automatic assessment of complex taskperformance in games and simulations. (CRESST Report 775). Los Angeles, CA: University of California, National Center for Research on Evaluation, Standards, and Student Testing (CRESST).

Merrill, M. D. (2002). First principles of instruction. *Educational Technology Research and Development, 50* (3), 42-59.

Munro, A., Pizzini, Q. A., & Bewley, W., Learning Anti-submarine Warfare in the Context of a Game-Like Tactical Planner. In *Proceedings of the Interservice/Industry Training, Simulation & Education Conference (I/ITSEC)*, 2009. http://ntsa.metapress.com/ling.asp?id=u1q691546361167

Munro, A. & Pizzini, Q. (2011). *The TAO Sandbox Instructor Guide*. Center for Cognitive Technology, University of Southern California.

Munro, A., Surmon, D., and Pizzini, Q. A. (2006). Teaching procedural knowledge in distance learning environments. In Perez, R. and O'Neil, H. (Eds.) *Web based learning: Theory, research, and practice*. Inglewood, N.J.: Lawrence Erlbaum Associates.

Munro, A. (2007). Foundations for software support of instruction in game contexts. In O'Neil, H.F. and Perez, R. (Eds.) *Computer Games and Team and Individual Learning*. Amsterdam: Elsevier.

Poropudas, J., & Virtanen, K. (2007). Analyzing air combat simulation results with dynamic Bayesian networks. In Proceedings of the 2007 Winter Simulation Conference, pp.1370–1377. Washington, DC: Institute of Electrical and Electronics Engineers, Inc.

Sweller, J. & Cooper, G. A. (1985). The use of worked examples as a substitute for problem solving in learning algebra.

CHAPTER 22

Framework for Instructional Technology

Paula J. Durlach, Randall D. Spain

U. S. Army Research Institute
Orlando, FL, USA
Paula.Durlach@us.army.mil

ABSTRACT

There are various ways in which an adaptive learning environment could adapt—both in terms of the student data used to make instructional decisions and in the types of instructional decisions that are made. This paper will describe a framework for instructional technology (FIT), aimed at delineating four types of instructional decisions that a technology-based learning environment can make: corrective feedback, support, micro-sequencing, and macro-sequencing. Corrective feedback addresses ways that explicit feedback concerning errors could be given. Support addresses how the technology supports the student within a task with hints or prompts. Micro-sequencing addresses how a technology-based instructional environment determines what content or task to present next within a module. Macro-sequencing addresses how a technology-based instructional environment determines what module to present next. For each decision, FIT lays out a continuum of complexity, which roughly maps to levels of adaptation, and the corresponding sophistication of the student model required to support those decisions. Within any particular instructional system, the level can be different for corrective feedback, macro-sequencing, micro-sequencing, and support decisions.

Keywords: adaptation, remediation, feedback, interaction, learning, sequencing

1   INTRODUCTION

Vanderwaetere, Desmet, and Clarebout (2011) defined adaptive learning environments as those that accommodate different learning needs and abilities of learners by providing individualized instruction. There are various ways in which an adaptive learning environment could adapt to the learner—both in terms of the student data used to make instructional decisions and in the types of instructional decisions that are made. This paper describes a framework for instructional technology (FIT), aimed at delineating four types of instructional decisions that a technology-based learning environment can make: corrective feedback, support, micro-sequencing, and macro-sequencing. Corrective feedback addresses ways that explicit feedback concerning errors could be given. Support addresses how the technology supports the student within a task with hints or prompts. Micro-sequencing addresses how a technology-based instructional environment determines what content or task to present next within a module. Macro-sequencing addresses how a technology-based instructional environment determines what module to present next. For each decision, FIT lays out a continuum of complexity, which roughly maps to levels of adaptation, and the corresponding sophistication of the student model required to support those decisions. Within any particular instructional system, the level can be different for corrective feedback, macro-sequencing, micro-sequencing, and support decisions.

One of the intentions of FIT is to help bridge the divide between practitioners and researchers. FIT builds on top of an existing framework familiar to many instructional developers, but perhaps not so familiar to educational researchers: Levels of Interactive Multimedia Instruction (IMI). These levels are outlined in Table 1. Level I represents the lowest level of interactivity, in which the learner is a passive receiver of information. Learner level of interaction may be limited to setting the pace by which the learning content is displayed. At Level II, there is somewhat more interaction. The learner may have some control over the sequence of content and there is usually some form of assessment, normally in the form of recognition tests such as multiple-choice or matching. At Level III, the learner may have multiple ways of interacting with the learning environment and has more influence on its behavior. Finally, at Level IV IMI, the learner is acting in a real-time simulated environment. For some organizations, procurement of IMI is specified according to IMI level (e.g., TRADOC, 2010).

Table 1 Standard IMI Levels

Level I	Passive. The student acts solely as a receiver of information.
Level II	Limited participation. The student recalls information and responds to instructional cues.
Level III	Complex participation. The student applies information to scenarios and interacts with simulations.
Level IV	Real-time participation. The student engages in a life-like set of complex cues and responses.

Levels III and IV IMI are sometimes referred to as adaptive because the actions of the learner affect the behavior of the instructional environment. For instance, a Level III branching scenario exercise may cast students as a character in an unfolding story. At predetermined points, the student is asked to make a decision (from a list of options). Their choice determines the subsequent story path, illustrating implicitly the consequences of their decision. Likewise, in a Level IV real-time simulation, student actions affect what happens in the simulation. For example, in a driving simulator, stepping on the brake slows down the apparent motion of the vehicle in the virtual world. Thus, in Levels III and IV IMI, the system does alter its behavior in response to the student. In so doing, it may provide implicit feedback. Implicit feedback certainly can be instructionally useful, if it is noticed and interpreted by the student as resulting from their input. However, implicit feedback is not based on student needs, abilities, or other characteristics, and is therefore not the same as adaptation. We suggest that the distinction between interactivity and adaptation hinges on whether the reaction of the system is based solely on the consequences of a student decision in the simulated environment (interactivity), or whether it is based on an evaluation of that decision and an associated pedagogical choice made by the technology (adaptation). Thus, FIT treats interactivity and adaptation as separate aspects of instructional technology.

In technology-based adaptive learning environments, learner needs and characteristics can be represented in a student model. The information in the model is used to select instructional tactics most suitable to the student's current state. An example of such a decision is the selection of what content or activity to present next (VanLehn, 2006). To make a decision like this, an adaptive learning environment may use a computational model that uses a student's pattern of performance on prior exercises to select content specifically aimed at addressing their performance weaknesses. Alternatively, it could simply use a summary score on an assessment test. FIT attempts to illustrate how a particular type of instructional decision (e.g., feedback) can vary based on the richness of the information in the student model. Simple instructional decisions may be driven by simple student data (e.g., a summary score), whereas more sophisticated decisions likely require a more complex student model. Practitioners need to know when increasing complexity or adaptation in student models provides comparable payoffs in improved learning outcomes. FIT provides a model by which researchers and practitioners can systematically evaluate the impact of each instructional decision on learning outcomes. Specifically, the framework could be used to evaluate the available evidence (e.g., is there empirical evidence that suggests corrective feedback better than contextually adaptive feedback?) or to guide future research, specifically, to help answer: At what levels, for each instruction decision, is the effort required practically justified with learning gains?

2   CORRECTIVE FEEDBACK

Under corrective feedback, we address different ways in which an instructional system could provide explicit feedback at the item-level. By "item" we mean individual questions, decisions, or actions, or specific learning objectives. In the language of intelligent tutoring (VanLehn, 2011) this would be feedback at the step or sub-step level. FIT's levels of corrective feedback are listed in Table 2A.

Table 2 FIT Categories of Adaptive Instructional Decisions

A. Corrective Feedback Level

Level 0	No explicit item-level feedback – only summary score
Level I	Minimal feedback (item accuracy information)
Level II	Correct answer or explanation of correct answer
Level III	Error-sensitive feedback
Level IV	Contextually-adaptive feedback

B. Support Level

Level 0	No support
Level I	Fixed hints on request (problem determined); other fixed sources of information
Level II	Locally-adaptive hints, prompts, or pumps
Level III	Contextually-adaptive hints, prompts or pumps (True Scaffolding)
Level IV	Same as Level III, with interactive dialog

C. Micro-Sequencing Levels

Level 0	Recycling
Level I	Supplemental Remediation
Level II	Supplemental Remediation Levels
Level III	Adaptive Remediation
Level IV	Adaptive Content

D. Macro-Sequencing Levels

Level 0	No sequencing decisions, only one module or learning event.
Level I	Fixed, Student Choice, or Hybrid
Level II	Test-out
Level III	Role Adaptation
Level IV	Performance-Adaptive

It can be seen in Table 2A, that advancing from Level 0 to Level II, the feedback content provided to the student is increasingly detailed. Yet, for these three levels, the system only needs to know what the correct response is and whether the student's response matched it. At Level III, error-sensitive feedback, the system requires additional information – information about *how* the student erred, because error-sensitive feedback is tailored to the specific way in which an error was committed. Thus, a student erring by choosing the answer -2 to a request to solve the equation x=6-2x would receive different feedback from a student who erred by choosing 6. For a multiple-choice item, incorrect options need to be constructed taking into account the likely misconceptions that would lead to the specific erroneous choices. The associated error-sensitive feedback messages need to provide information aimed at repairing the specific error committed. For open-ended questions (the student can answer with any number, for example), certain likely erroneous answers need to be anticipated and appropriate error-sensitive feedback created.

We think of a student model as any data structure used to influence instructional decisions, where the data accumulates over multiple interactions, and represents the student with more fidelity than a single score. If we define a student model in this way, then error-sensitive feedback does not require a student model. It needs information about the student response for the current item and it uses information about how a hypothetical student might go wrong; but, it does not use data collected from prior student-system interactions, nor save any information about how the student erred to influence feedback on future problems. We think the term "local adaptation" is a good way to refer to this type of instructional move. Local adaptation requires knowledge of how the student responded at an item level, and enables the system to respond differently for different errors; but it does not save this information to determine how to give feedback for a subsequent student input.

This can be contrasted with contextually-adaptive feedback (Level IV). Contextually-adaptive feedback uses stored information about the student, in combination with local information, to make adaptive decisions. This stored information is typically data about past interactions with the system – the student model. With contextually-adaptive feedback, the student may be given different feedback for the same response, under different circumstances. For example, keeping track of prior student-system interactions potentially can allow the system to diagnose careless errors (e.g., Baker, Corbett, & Aleven, 2008). If the student model suggests that a student already has learned the information required to respond correctly, then an error may produce feedback like, "hmm, I thought you knew this one, please check your answer." Alternatively, contextually-based feedback might affect feedback timing: if the student is nearing mastery, then the feedback might be delayed until the end of a multi-step problem, whereas a more novice student might be given step-based feedback immediately.

3 SUPPORT

Levels of Support address how instructional supports such as hints and prompts are determined by a system while the student is interacting with the instructional materials, or applying knowledge to problem solving. The FIT support levels are presented in Table 2B. At Level 0, there is no support. At Level I, the student has access to information that is static. Access to it does not require any assessment of student actions or inputs. This could be a links to reference materials or a hint button, which students can select to get a hint. Hints can be delivered in various ways. They can be informational (distance = velocity x time), cuing (how are distance and time related?), or attentional (highlighting a particular part of the screen where relevant information is located).

A popular method of providing hints is to allow the student to request multiple hints, where the first hint is rather vague (what is the goal at the beginning of a meeting?), the next hint is more directive (try to build rapport), and so forth. The final hint may actually provide the answer (discuss the weather). Even though multiple levels of hints are available, FIT still considers this to be Level I, because the hints available to all students are the same, and do not take into account why this particular student may need a hint. This can be contrasted with Levels II and III, which are locally-adaptive, and contextually-adaptive, respectively. Locally-adaptive support adapts on the basis of the most recent student-system interaction. An example would be prompts that help a student self-correct, based on the type of error committed (similar to error-sensitive feedback, but without telling the correct answer). Contextually-adaptive support adapts on the basis of contextual or historical student data (a student model) and is equivalent to true scaffolding (i.e., it is faded as learner competency increases). Level IV adds collaborative dialog and problem solving using a natural language processing interface. The capability for natural language supports interactive tutorial dialogs, during which the tutor and the student interact to build student understanding (VanLehn, 2011).

4   MASTERY AND SEQUENCING

When learning materials or problem sets are chunked into conceptual modules instructional decisions can determine how to transition from one module to the next. If mastery criteria are used, students cannot progress to the next module unless mastery on the current one has been demonstrated. A strategy must be in place for selecting learning activities for students not yet at mastery. Likewise, a strategy must be in place for selecting the next module when mastery has been demonstrated. Thus, employment of mastery criteria involves two decisions: What to do when mastery is achieved, and what to do when mastery has not yet been achieved. FIT refers to these two decisions as Macro-sequencing and Micro-sequencing, respectively.

Micro-sequencing is essentially a remediation strategy. Table 2C illustrates FIT Levels of Micro-sequencing, and captures many common approaches to remediation. At Level 0, Recycling, students must meet a mastery criterion based on a single summary score to be given credit for a module. All students experience the same core content until mastery is achieved. Students who have not achieved the mastery criterion go through the core content again. Level I, Supplemental Remediation is the same, except that when remediation is required, specially prepared remedial content is provided. All students requiring remediation are presented with the same remedial content. Level II, Supplemental Remediation Levels is the same as Level I, except now different versions of the remedial content are available. Students may be assigned to different remedial content depending on their current level of mastery. For example, selection of version could be based on the student's distance from mastery or whether they have already gone through remediation for this module already.

Level III, Adaptive Remediation, is the first level of micro-sequencing that requires a student model of any complexity. Instead of keeping track of student mastery in terms of a single summary score, the model represents student mastery in terms of separable knowledge components. Knowledge components are associated (either one-to-one or many-to-one) to the learning objectives for the module. The student model keeps track of mastery for each knowledge component. Students unable to demonstrate proficiency on all knowledge components are given supplemental materials or problems targeted at their own specific areas of conceptual weakness. Remedial content is different for different students, depending on their particular pattern of deficiencies. Thus two students with the same summary score could be assigned different remedial content. VanLehn (2006) refers to this as macro-adaptation; however, others use that term differently (e.g., Park & Lee, 2004; Shute & Towle, 2003).

Level IV, Adaptive Content, is the most complex version of micro-sequencing. This technique can be used to sequence selection of problem-based learning tasks, where problem difficulty can be varied parametrically (e.g., the number and skill level of enemy forces in a first-person shooter serious game). In this type of situation, each student's progression path through the various challenge levels will be shaped by their own performance. Some students may have to complete exercises at each progressive level, while other students may be able to "jump" levels. If there are multiple dimensions that can be advanced, many different paths through the problem space may occur. This can be accomplished by altering the path of the learner through a library of pre-scripted scenarios, or by adapting the scenario parameters on-the-fly (i.e., without having to stop one scenario and begin a new one).

In general, each progressive level of micro-sequencing going from Level 0 to Level II requires increasingly more resources, because a greater amount of remedial content is required. Levels III and IV, require a student model and the instructional designer must be able to specify the association between performance assessment measures and the knowledge components as represented in the student model. Also requiring specification are the micro-sequencing rules by which the student model values are used to select the remedial content (Level III, Adaptive Remediation) or the path the student should take through the learning space (Level IV, Adaptive Content). This likely will require a more intense effort of up-front domain analysis (e.g., cognitive task analysis) and a more rigorous approach to assessment design, compared with less adaptive methods. Durlach and Ray (2011) concluded that there is clear evidence that using a mastery approach enhances learning outcomes; however, they failed to find any well-controlled research that would support inferences about the relative benefits of the different micro-sequencing levels. While some experiments varied micro-sequencing level across conditions, there were other confounded differences, which prevented attribution of learning benefits uniquely to this manipulation.

Macro-sequencing (or between module sequencing) addresses what to do when one module has been completed and selection of the next module must be decided. The FIT levels of macro-sequencing are listed in Table 2D. At Level 0, there is no decision, because there is only one module or learning activity. Level I covers instructional environments that require a fixed sequence of modules or leave it up to the student. In either case there is no adaptation. Level II, Test-out, refers to the policy of selecting the specific modules a student should complete, based on the results of a knowledge pretest. Typically the test items are associated with modules, so that above criterion performance on a subset of pretest questions allows the student to skip the associated module. Thus, each student may complete a personalized set of modules based on their pretest performance. Level III, Role Adaptation, also uses pre-learning information about the student, but bases decisions about the personalized set of modules scheduled for a student based on their motivation for learning. For example, a person brushing up on Spanish in preparation for a vacation may be given a different set of modules than a person brushing up on Spanish in preparation for helping in disaster relief.

Only Level IV, Performance-Adaptive, is adaptive based on an evaluation of student performance. At Level IV, modules cover the same topics in the same order for all students; however, there are multiple versions of each conceptual module. Higher performing students may be given a more advanced version, or be supported with less explanation of terminology, or fewer examples. Lower performing students may be given a more basic version, with a lot of supporting examples. A system by Tseng, Chu, Hwang, and Tsai (2008) fits this category. They had three versions of each module: easy, middle, and difficult. A student could be switched from one level to another between modules, with the difficulty level increasing for better performing students and decreasing for worse performing students. Test performance at the end of each subsequent module determined the level of the next module. Using this system to learn about mathematical sequences, students given Level IV Performance Adaptive macro-sequencing had better learning outcomes, compared with students given materials fixed at the middle level of difficulty.

5 CONCLUSIONS

The state of the evidence for recommending instructional decisions and levels of adaptation within FIT is varied. There is fairly convincing evidence that Level III Corrective Feedback (error-sensitive feedback) improves learning outcomes, compared with Levels 0 – II (Durlach & Ray, 2011). For support, micro-sequencing, and macro-sequencing there is a paucity of unambiguous evidence. Durlach and Ray (2011) found that much of the empirical research on the effect of different instructional decisions in technology-based learning environments confounded multiple factors across instructional conditions, making it difficult to attribute the relative contribution of each factor to observed benefits in learning outcomes. Nevertheless, they concluded that the evidence was suggestive for the benefits of mastery learning, error-sensitive feedback, metacognitive prompts, adaptive fading of worked examples, and adaptive spacing and repetition for drill and practice learning.

VanLehn (2011) presented the results of a meta-analysis on instructional effectiveness, not on the types of instructional decisions make, but rather on their granularity. In particular, he compared learning outcomes for answer-based, step-based, and substep-based instructional systems. Answer-based systems interact with the learner at the whole-problem level. For example, if a problem takes five steps or decisions for completion, the system provides feedback about the final answer, but not the intermediate steps. A step-based system interacts with the learner at the step level, providing coaching and/or feedback for each step (either immediately or in summary format after the entire problem has been attempted). Finally, a substep-based system interacts with the learner to help them (if required) arrive at the solution to each step, via an interactive dialog (typically supported by a natural language interface). From the meta-analysis, VanLehn (2011) concluded that answer-based systems were inferior to step-based and substep-based systems, which were equivalent in effectiveness. His interpretation of the results was that with answer-based systems students have difficulty identifying where they erred in the chain of steps to get to an answer, so they can not self-correct. In contrast, step-based and substep-based systems provide sufficient information through support and feedback that allows students to see where they have erred, to self-repair misconceptions, and to self-generate correct solutions for most problems. In addition, because students keep working on a solution instead of giving up, they encounter more learning opportunities. Mapping this interpretation into FIT suggests that the best return on investment when designing training technology should be to use error-sensitive feedback (Corrective Feedback Level III) and locally-adaptive hints, prompts, or pumps (Support Level II). Corrective feedback would provide students with diagnostic information to help them identify and repair their own misconceptions. Locally adaptive hints, prompts, and pumps would help students work through challenging problems. If a student were to get stuck, the system could give them prompts and hints to get them moving again. Such support encourages students to finish problems correctly rather than give up after several failed attempts. More importantly, it encourage students to do most of the reasoning themselves.

Unfortunately, VanLehn's (2011) interpretation about the benefit of step and sub-step based tutors does not offer much insight with regard to micro and macro-sequencing (any level might be just as good as any other). More research is needed in this area to determine if moving from lower to higher levels of remediation and sequencing provide better learning outcomes, or if learning outcomes plateau at a certain level.

In conclusion, FIT is a framework that describes the types of instructional decisions that an adaptive learning environment can make. For each instructional decision, FIT lays out a continuum of adaptation and the corresponding sophistication of the student model required to support those decisions. Some levels require simple student input that could be provided without a student model, others require complex student data that allow for contextually adaptive feedback, hints, and prompts, and performance adaptive remediation and sequencing. One aim of FIT is to help developers of training technology understand the range of possibilities, and select the most effective options in keeping with their resources when designing technology-based learning environments. We acknowledge that FIT does not cover the entire instructional design process; however, it does provide a reference akin to the Levels of IMI. FIT describes different possibilities for each factor (i.e., corrective feedback, support, micro-sequencing, and macro-sequencing). FIT does not address all the types of decisions a system could make. For example, it does not address decisions about managing student affect (D'Mello, Strain, Olney, & Graesser, in press). Yet it does address the instructional decisions most practitioners of IMI development face today. Additionally, they may require recommendations for the applicability of each FIT level to each level of IMI.

REFERENCES

Baker, R., Corbett, A. T., & Aleven, V. (2008). More accurate student modeling through contextual estimation of slip and guess probabilities in Bayesian knowledge tracing. In E. Aimeur, & B. Woolf (Eds.), *Proceedings of the 9th International Conference on Intelligent Tutoring Systems* (pp. 406-415). Berlin: Springer Verlag.

D'Mello, S. K., Strain, A. C., Olney, A., & Graesser, A. C. (in press). Affect, meta-affect, and affect regulation during complex learning. In R. Azevedo and V. Aleven (Eds.), *International Handbook of Metacognition and Learning Technologies*. Berlin: Springer Verlag.

Durlach, P. J. & Ray, J. M. (2011). *Designing adaptive instructional environments: Insights from empirical evidence*. ARI Technical Report 1297. U.S. Army Research Institute, Arlington, VA.

Park, O., & Lee, J. (2004). Adaptive instructional systems. In D.H. Jonassen (Ed.), *Handbook of Research on Educational Communications and Technology 2nd Edition* (pp. 651-684). Mahwah, NJ: Lawrence Erlbaum.

Shute, V. J. & Towle, B. (2003). Adaptive e-learning. *Educational Psychologist, 38*(2), 105-114.

TRADOC (2010). Distributed learning education and training products delivery order for new/redesign of dL IMI and blended distributed learning. Retrieved from http://www.atsc.army.mil/itsd/imi/BLDO/DLETPDOTemplates.asp.

Tseng, J. C. R., Chu, H., Hwang, G., & Tsai, C. (2008). Development of an adaptive learning system with two sources of personalization information. *Computers & Education, 51*, 776-786.

Vanderwaetere, M., Desmet, P., & Clarebout, G. (2011). The contribution of learner characteristics in the development of computer-based adaptive learning environments. *Computers in Human Behavior, 27*, 118-130.

VanLehn, K. (2006). The behavior of tutoring systems. *International Journal of Artificial Intelligence in Education, 16*, 227-265.

VanLehn, K. (2011). The relative effectiveness of human tutoring, intelligent tutoring systems and other tutoring systems. *Educational Psychologist, 46*(4), 197-221.

Section VI

Advances in Modeling for User Centered Design

CHAPTER 23

Assessment of Manikin Motions in IMMA

Erik Bertilsson[1,2], Ali Keyvani[2,3], Dan Högberg[1], Lars Hanson[2,4]

[1]Virtual Systems Research Centre
University of Skövde
Skövde, Sweden
erik.bertilsson@his.se

[2]Department of Product and Production Development
Chalmers University of Technology
Gothenburg, Sweden

[3] Innovatum AB
Trollhättan, Sweden

[4]Industrial Development
Scania CV
Södertälje, Sweden

ABSTRACT

When evaluating human-machine interaction in a virtual environment using Digital human modelling (DHM) it is important to ensure that the predicted motions lie within the range of behavioural diversity for different people within a population. This paper presents a study in which a comparison is made between motions predicted by the DHM tool IMMA (Intelligently Moving Manikin) and motions from real humans stored in a motion database. Results show similar motions but the predicted motions were in total statistically significantly different compared to the motions performed by real persons. The differences are most likely due to the balance function and joint constraints that the IMMA tool uses for predicting motions. Differences can also be due to other factors, aside of body size, such as age, gender or strength that affects the movement behaviour.

Keywords: Digital human modelling, Motion Prediction, Posture, Motion

1 INTRODUCTION

Digital human modelling (DHM) systems can be used to simulate production processes and analyse the human-machine interaction, particularly at early development stages (Chaffin, 2001). Consideration of anthropometric diversity, i.e. dimensional variation of human body measurements among targeted user group is one important aspect to consider when working with DHM tools (HFES 300 Committee, 2004). Another aspect is the modelling of human movements where the simulations need to represent human characteristics and behaviour. Methods for how to predict motions in DHM software can be classified into two groups (Pasciuto et al., 2011). The first group is data-based methods which base motion simulations on a database of captured motions and by doing so achieves motions of high credibility for specific tasks (Park, 2008). The other group, physics-based methods, bases their motions prediction on kinematic models of human body. These methods employ several inverse kinematic (IK) techniques while considering joint constraints such as range-of-motion (ROM), joint velocity and strength to solve and predict a motion. Using these methods makes it possible to predict motions for any given task (Abdel Malek and Arora, 2008). Additional hybrid methods do also exist using both data of captured motions and data on joint constraints to predict motions.

When using a DHM tool to generate simulations and evaluating motions the goal is to get a motion that answers well to the motion of a real person doing the same task. However, there are many factors that create uncertainty to whether a predicted motion is representative or not. These factors can be related to modelling of the manikin's biomechanical system, such as joint placement and joint constraints, but also related to behavioural diversity, which is more difficult to model and predict (Garneau and Parkinson, 2009). A DHM tool under development is the IMMA (Intelligently Moving Manikins) tool (Hanson et al., 2010). IMMA predicts motions based on IK techniques and considers joint constraints such as ROM, speed and strength, but also considers ergonomic conditions and additional functionality to avoid collision with surrounding objects. The goal is to be able to predict a motion that is realistic and lies within the range in behavioural diversity for any given task. Data for these joint constraints are under implementation where literature data is incorporated in the system and absent data is measured and defined.

As a part of the development process this paper describes the evaluation of the predicted motions of the IMMA tool. The different motions are compared and evaluated where focus is put on how well the motions predicted by the IMMA tool answers to the motions performed by real persons. The goal is to evaluate if motions predicted by the IMMA tool lie within the range of behavioural diversity for different people within a population, as well as highlight potential problems with the current model for predicting motion related behavioural diversity.

2 METHOD

The evaluation was done based on using motion data of tasks done by real subjects. The motion data was gathered from different sources including the HUMOSIM experiments (Reed et al., 2006). The HUMOSIM motions were first converted from joints global location format (.LOC) to hierarchical rotation matrices (Euler angles) and next, both types of data were stored into a unified database platform (Keyvani et al., 2011). The platform is capable of indexing motions both by joint values and joint positions, and it is also storing detailed information about the performed motions in form of annotation tables. One specific set of experiments was searched through the platform for further analysis. The type of task analysed consisted of lifting a 0.3 meter wide box, from a height of approximately 1.6 meter down to a height of approximately 0.9 meter. The motion database platform retrieved 23 motions performed by 18 different subjects for this specific task constrained by type of action, wrist joints starting position and wrist joints final position. The time length of the motions varies from 63 to 100 frames with a frame rate of 25 Hz.

Motions done by the human models, so called manikins, in the IMMA tool are currently controlled by locked end effectors connected to predefined paths. A path consists of a number of frames in where the position and direction of a coordinate system is defined in 3D space. The manikin is connected to the path via tool centre points (TCP) situated at chosen joint centre, e.g. both wrists. The connected joint centres then follow the path and all other joint positions are predicted using previously mentioned joint constraints and additional algorithms to balance forces and torques (Bohlin et al., 2011). Because motions are controlled by paths in the IMMA tool it was required to create paths for the box based on motion data. The motion data was analysed and a path for the box for each motion was defined as the point midway between the right and left wrist. During this analysis, four motions were discarded because they showed considerable differences when comparing the vertical position of the right and left wrist. This gave that further analysis was done on 19 motions performed in total by 16 subjects. The direction of the box was assumed to be constant, thus was the direction values set to zero. For each motion a manikin with the same bodily dimensions as the person performing the motion was created. For each manikin the biomechanical model was based on 12 anthropometric measurements (Table 1) and defined using the manikin creation process in the IMMA tool (Bertilsson et al., 2011). The manikins' wrists and feet were positioned at the same place in space as the humans' hands and feet according to the motion data. The manikins' wrists were connected to the box and one degree of freedom was left unlocked making it possible for radial abduction and adduction. In a similar manner the feet were locked in their position and one degree of freedom was left unlocked making it possible for flexion and extension of the metatarsophalangeal joints.

After the path, box and manikin had been positioned correctly a simulation was done for each corresponding measurement of a captured motion simulation. During each simulation the wrists of the manikins followed the path and all other joint

238

positions were predicted using joint constraints and additional algorithms to balance forces and torques. Motion data for each simulation was extracted from the IMMA tool and used for comparison with the motion data from the captured motions in the database. The motions were compared using data of joint angles and joint centre positions but also visually, evaluating the body postures at different time frames during the simulations. To get the same time length for each measurement only the last 60 frames were included in the evaluation. The time to perform each motion was considered to be approximately the same for all persons. Evaluation of motion data for joint angles and joint centre positions focused on the elbow and shoulder joints. A paired-difference t test was done with the hypothesis that there is a difference between the simulations created in IMMA and the HUMOSIM measurements. A two-sided test was chosen as this test is robust and works even if the normality assumption is violated (Agresti and Franklin, 2009). This test was done for both total mean values for each simulation and values for each frame comparing elbow joint angles and elbow and shoulder joint centre positions, a total of seven measurements for each simulation.

Table 1 Anthropometric measurements used when creating the virtual manikins in IMMA

Anthropometric measurement	Mean	Standard deviation
Body mass (weight) [kg]	71.0	13.0
Stature (body height) [mm]	1685.7	110.3
Tibial height [mm]	497.7	60.7
Sitting height (erect) [mm]	1563.9	49.7
Eye height, sitting [mm]	1457.2	50.0
Cervicale height, sitting [mm]	1329.3	42.4
Shoulder-elbow length [mm]	321.6	32.9
Shoulder (biacromial) breadth [mm]	377.7	28.2
Hand length [mm]	189.5	16.3
Hand breadth at metacarpals [mm]	81.8	6.9
Head length [mm]	194.0	9.0
Head breadth [mm]	148.3	7.8

3   RESULTS

When visually evaluating the motion from real subjects in the database compared to motion created in IMMA it is possible to see evident differences (Figure 1). Real persons work mainly with their arms and shoulders when performing the task and show very little movements in their spine and lower body. The IMMA manikins, on the other hand, tend to move spine, hip and ankles as well as arms and shoulders. These differences between real persons and IMMA simulations were similar for all manikins and simulations.

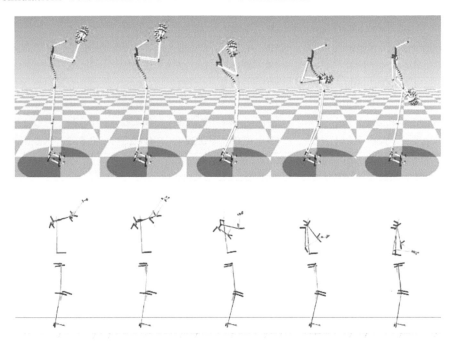

Figure 1 Visual comparison of the last 60 frames based on motion data in IMMA, seen on top, and HUMOSIM, seen below, for one person (subject 20 shown).

When only the elbow joints and shoulder joints are assessed the difference is not as evident and this is shown in the result from the paired-difference t test (Table 2). Nevertheless, there is still a difference for a 0.05 significance level for five of seven of the measured total means (Table 2). Evaluations were done for both left and right elbow and shoulder but due to space limitations in this paper, and that the results were similar, only the results of the right side evaluation are shown. The denotation X indicates horizontal distance in the coronal plane and the denotation Y indicates horizontal distance in the sagittal plane. The denotation Z indicates the vertical distance.

The paired-difference t test for each frame showed a difference between HUMOSIM and IMMA, similar to the results of the paired-difference t test for the total mean values (Figure 2, 3 and 4). A difference for a 0.05 significance level could be determined for all 60 frames when evaluating the distance difference in the Y-direction of the shoulder. For the other six measurements a statistically significant difference could not be seen for all 60 frames, only for parts of the motion (grey marked areas in Figure 2). For all seven measurements 62% of all frames could be determined to have statistically significant difference between HUMOSIM and IMMA data.

240

Table 2 Statistical comparison between HUMOSIM and IMMA based on the total mean for each person.

Difference ($x_{HUMOSIM}-x_{IMMA}$)	N	Mean	SD	Median	T-value	P-value
Right elbow angle [Degrees]	19	2,895	12,19	6,027	1,035	0,314
Right elbow X [m]	19	-0,023	0,030	-0,018	-3,323	**0,004
Right elbow Y [m]	19	-0,018	0,037	-0,020	-2,148	*0,046
Right elbow Z [m]	19	0,015	0,033	0,006	2,007	0,060
Right shoulder X [m]	19	-0,023	0,037	-0,018	-2,695	*0,015
Right shoulder Y [m]	19	-0,049	0,028	-0,050	-7,624	**0,000
Right shoulder Z [m]	19	-0,021	0,024	-0,023	-3,852	**0,001

**, Difference is significant at the 0.01 level (2-tailed).

*, Difference is significant at the 0.05 level (2-tailed).

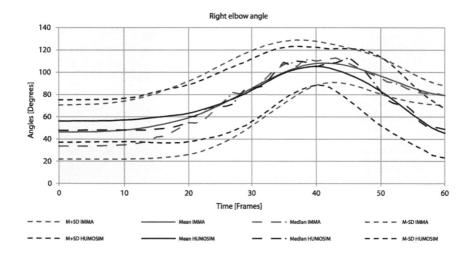

Figure 2 Statistical comparison of the right elbow joint angle between HUMOSIM and IMMA. Areas marked grey shows a significant difference for a 0.05 significance level.

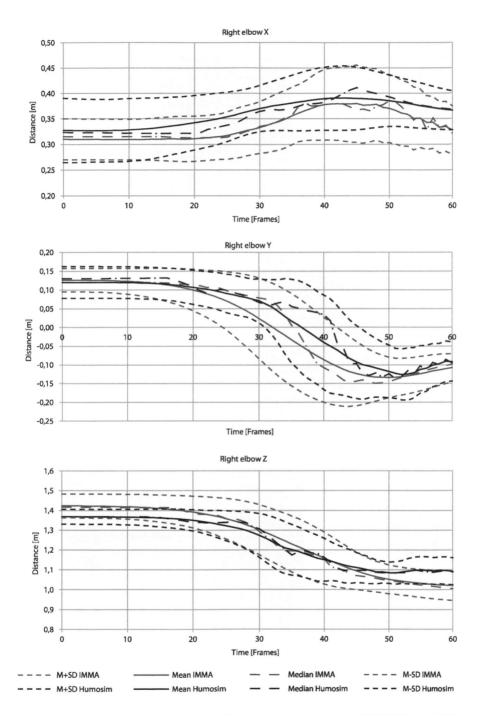

Figure 3 Statistical comparison of the right elbow joint position between HUMOSIM and IMMA. Areas marked grey shows a significant difference for a 0.05 significance level.

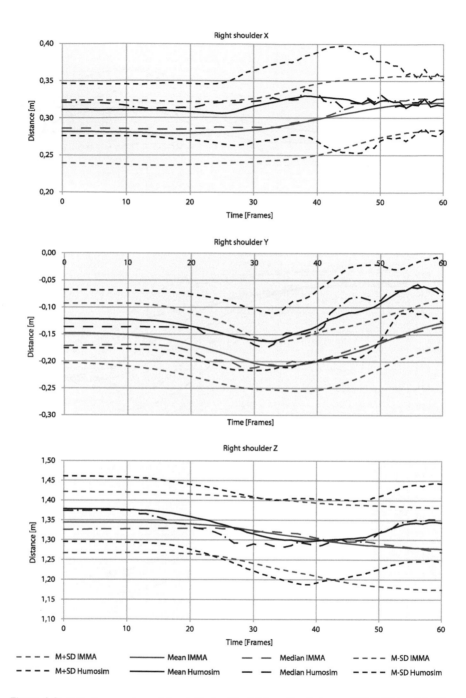

Figure 4 Statistical comparison of the right shoulder joint position between HUMOSIM and IMMA. Areas marked grey shows a significant difference for a 0.05 significance level.

4   DISCUSSION

The result from the study shows that the IMMA tool is able to create simulations with similar motion as a real person would do when the end-effector points are controlled by a defined path. However, it is evident that the manikins for the given task bend their spine, hip and ankle more than a real person would do. This is most likely due to the function for keeping the manikin in balance. This function bases its calculations on joint constraints and the weight of body parts. A better simulation would most likely be achieved if these joint constraints, weights and centre of mass definitions would be reviewed. The rotation of the box was considered to be zero for the path that the manikins followed in IMMA. This is not the actual case and adding the rotation of the box would probably also give a better simulation.

Though a statistically significant similarity was not possible to determine for more than two of seven total mean values, an average 38% of all frames showed a statistically significant similarity between HUMOSIM and IMMA data. This is especially evident when looking at the evaluation of the elbow joint angle where no significant difference could be determined for the major part of the motion where actual movement took place. In addition, the total mean showed no significant difference for the elbow joint angle. It is possible that using a method for evaluating the ergonomics by assessing the elbow joint angles would give the same result for both HUMOSIM and IMMA data. However, if more joints would be assessed at the same time the total result would not be the same.

Reasons for the significant difference between the HUMOSIM and IMMA data can be, as mentioned before, connected to the balance function and joint constraints that are not defined well enough. This difference can also to some degree be due to other factors, aside of body size, such as age, gender or strength that affects the movement behaviour. Further studies should therefore focus on reviewing the joint constraints as well as evaluating factors in addition to body size that can have an effect on the movement behaviour. The joint's velocity profile is not directly studied in this research. However, the statistical comparison diagrams for the elbow joint angle indirectly reflect the effect of velocity profile. If further research with the velocity profile in focus is performed, it is possible to better realize the underlying behaviour.

5   CONCLUSION

This paper describes a study in where simulations created in a DHM tool is compared to motions measured on real persons. The different data showed similar motions but the predicted motions were in total statistically significantly different compared to the motions performed by real persons. However, for some specific measurements a statistically significant level of similarity could be determined.

ACKNOWLEDGEMENTS

The motion data used in this study has been made available by the University of Michigan through its Center For Ergonomics (HUMOSIM). Access to this data is gratefully acknowledged. This work has also been made possible with the support from the Swedish Foundation for Strategic Research/ProViking and by the participating organizations. This support is gratefully acknowledged.

REFERENCES

Abdel-Malek, K. and J. Arora. 2008. Physics-based digital human modeling: predictive dynamics. In: Duffy,V.G. (Eds.), Handbook of Digital Human Modeling. CRC Press-Taylor & Francis Group, Boca Raton, Florida, USA, pp. 5.1-5.33.

Agresti, A. and C. Franklin. 2009. Statistics: The Art and Science of Learning from Data. Second edition, Pearson Education, Upper Saddle River, New Jersey.

Bertilsson, E., L. Hanson, D. Högberg, and I.M. Rhén. 2011. Creation of the IMMA manikin with consideration of anthropometric diversity. Proceedings of the 21st International Conference on Production Research, Stuttgart, Germany.

Bohlin, R., N. Delfs, L. Hanson, D. Högberg, and J.S. Carlson. 2011. Unified solution of manikin physics and positioning – Exterior root by introduction of extra parameters. Proceedings of the 1st International Symposium on Digital Human Modeling, Lyon, France.

Chaffin, D.B. 2001. Digital Human Modeling for Vehicle and Workplace Design. SAE International, Warrendale, PA.

Garneau, C.J. and M.B. Parkinson. 2009. Including Preference in Anthropometry-Driven Models for Design. Journal of Mechanical Design, ASME, Volume 131, Issue 10.

Hanson, L., D. Högberg, R. Bohlin, and J.S. Carlson. 2010. IMMA – Intelligently Moving Manikin – Project Status. Proceedings of 3rd Applied Human Factors and Ergonomics (AHFE) International Conference 2010, Karwowski, W., Salvendy, G. (Eds.), USA.

HFES 300 Committee. 2004. Guidelines for Using Anthropometric Data in Product Design. Human Factors and Ergonomics Society, Santa Monica, CA.

Keyvani, A., H. Johansson, M. Ericsson, D. Lämkull, and R. Örtengren. 2011. Schema for Motion Capture Data Management. In: Duffy, V. (ed.), Third International Conference on Digital Human Modeling, ICDHM 2011, Held as Part of HCI International 2011. Orlando, USA: Springer Berlin / Heidelberg.

Park, W. 2008. Data-based human motion simulation methods. In: Duffy,V.G. (Eds.), Handbook of Digital Human Modeling. CRC Press-Taylor & Francis Group, Boca Raton, Florida, USA, pp. 9.1-9.17.

Pasciuto, I., A. Valero, S. Ausejo, and J.T. Celigüeta. 2011. A hybrid dynamic motion prediction method with collision detection. Proceedings of the 1st International Symposium on Digital Human Modeling, Lyon, France.

Reed, M.P., J. Faraway, D.B. Chaffin and B.J. Martin. 2006. The HUMOSIM Ergonomics Framework: A New Approach to Digital Human Simulation for Ergonomic Analysis. Warrendale, Society of Automotive Engineers. SAE Technical Paper 2006-01-2365.

CHAPTER 24

The Use of Volumetric Projection in Digital Human Modelling Software for the Identification of Category N_3 Vehicle Blind Spots

Steve Summerskill, Russell Marshall, Sharon Cook

Loughborough Design School, Loughborough University, Loughborough, UK
s.j.summerskill2@lboro.ac.uk

ABSTRACT

The paper described the contribution of the Design Ergonomics Group (DEG) in the Loughborough Design School, UK, to a research project examining blind spots in Category N_3 vehicles. The project was commissioned by the UK Department for Transport (DfT) with the aim of understanding the nature of blind spots in driver's vision for Category N_3 vehicles. The project was instigated by the DfT due to a perception that a large number of accidents are caused by the blind spots associated with Category N_3 vehicles. The initial focus of the project was to understand the problem, with the processing of national accident data to establish if Category N_3 vehicle blind spots contribute to a significant proportion of accidents. The findings indicated that the number of side swipe accidents, and accidents with vulnerable road users, warranted further exploration. In order to establish the cause and nature of blind spots three Category N_3 vehicles were digitized using a FARO ARM scanning system, and imported into the SAMMIE Digital Human Modeling (DHM) system. To allow the exploration of blind spots a new feature was developed in the DHM system that allowed a representation of the three dimensional space visible to the driver of the Category N_3 vehicles through mirrors (indirect vision) and window apertures (direct vision). This allowed multiple mirror and window aperture projections to be created at the same time, allowing the identification of blind spots that exist. This technique identified a key blind spot that had the potential to be associated with the accidents that were identified in the accident data. This led to the definition of a proposed change to directive UN Regulation 46 which was presented at the 100th United Nations GRSG committee.

Keywords: DHM, Category N$_3$, blind spot, mirrors, HGV, volumetric

1   INTRODUCTION

The current design of large goods vehicles (Category N$_3$ vehicles) in Europe generally places the driver above the engine of the tractor unit which results in the driver's eye position typically being more than 2.5m above the ground plane. Figure 1 shows the eye height of a driver above the ground plane that was captured during the research described in this paper. The height of the driver above the ground combined with the obscuring effect of the vehicle structure makes it very difficult for the driver to directly see other vehicles, pedestrians, and cyclists that are in close proximity to the side and front of the tractor unit.

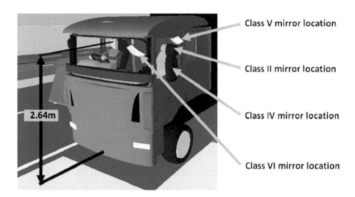

Figure 1 The Category N$_3$ vehicle mirrors that allow greater visibility of the areas directly adjacent to the vehicle

To account for these issues mirrors have been added to Category N$_3$ vehicles that increase the visibility of these areas as shown in Figure 1. Figure 2 shows the area of the ground plane that is required to be visible to the driver as specified by standard 2003/97/EC (2004) and UN Regulation 46. (2009). There have been a number of high profile accidents that involved collisions between Category N$_3$ vehicles, vulnerable road users and other vehicles. This led the UK government Department for Transport to commission the Loughborough Design School to determine if blind spots exist for the drivers of Category N$_3$ vehicles, and if these blind spots have the potential to contribute to accidents. In order to explore the issue a two phase research project was defined. The work packages of the research project aimed to define the current understanding of the issue through stake holder consultation. This involved interviews with a range of UK national bodies, charities and also interviews with the drivers of Category N_3 vehicles. In addition to this, accident data statistics were analysed to determine the proportion and severity of accidents that were caused by the drivers of Category N_3 vehicles being unable to see other road users. The results of these initial research activities highlighted the need to explore the potential blind spots in a manner that allowed the interaction between direct vision (looking through windows) and indirect vision (the use of mirrors) to be modelled, allowing potential blind spots to be identified. The results from these analyses allowed the identified blind spots to be quantified and compared to the accident data. The following paper describes the initial research activities that were performed and the subsequent analysis of three Category N_3 vehicles in order to identify blind spots in the combined direct and indirect vision of drivers.

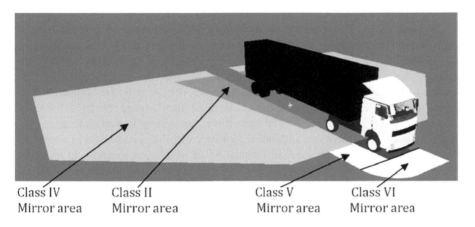

| Class IV | Class II | Class V | Class VI |
| Mirror area | Mirror area | Mirror area | Mirror area |

Figure 2 The ground plane areas that are required to be viewed by the driver through the vehicle mirrors as defined in 2003/97/EC for a left hand drive vehicle

2   THE ANALYSIS OF ACCIDENT STATISTICS

In order to examine the prevalence of accidents that occur with Category N_3 vehicles where blind spots are a potential contributing factor the UK STATS 19 accident database from 2008 was interrogated using a cluster analysis technique. The full methodological description and data tables from this cluster analysis can be seen in the main project report (Cook et al 2011). The results from the analysis identified key scenarios;

1. Articulated left-hand drive HGVs over 7.5 tonnes changing lane to the right and colliding with cars (25% all casualties, 14% of serious, 6% of fatal) (N_2 & N_3 vehicles)
2. HGVs over 7.5 tonnes changing lane to the left and colliding with cars (24% of all casualties, 14% of serious, 6% of fatal) (N_2 & N_3 vehicles)

3. HGVs (N_2 & N_3 vehicles) changing lane to the right and colliding with cars (11% of all casualties, 2% of serious)
4. Goods vehicles (all N Classes) turning left and colliding with vulnerable road users (5% of all casualties, 10% of serious, 19% of fatal)

In order to explore these scenarios in further detail, the UK 'On The Spot' accident database (Hill J.R. *et al* (2005) and Cuerden, R. *et al* (2008)) was used to identify specific accidents that relate to the identified scenarios. This allowed specific road junctions where accidents occurred to be simulated.

3 THE USE OF DIGITAL HUMAN MODELLING SOFTWARE TO IDENTIFYING BLIND SPOTS

Previous research has used a number of modelling techniques and real world tests to identify blind spots in driver's vision for a range of vehicle types (Ehlgen *et al* (2008), Tait, R. & Southall, D. (1999), Porter J.M & Stern M.C (1986)). In general, these attempts have focused upon the area of the ground plane that is visible to the driver, which is understandable, as this is how the standards that define the visible area to the driver are defined (see Figure 2). In defining a methodology for the analysis of blind spots it was considered important to allow the visualisation of how the direct and indirect vision capabilities combined in a manner that allowed a three dimensional understanding of blind spot location. In order to allow this the existing features of the SAMMIE DHM system were modified to allow the projection of multiple mirrors and multiple window apertures. The standard mirrors that exist on Category N_3 vehicles are shown in Figure 1, and the area of the ground plane that each mirror is designed to cover is shown in Figure 2.

The Class II mirror provides a view down the side of the vehicle in same manner as the wing mirrors on M_1 passenger cars. The Class IV mirror provides a wide angle view of the area adjacent to the side of the vehicle; the Class V mirror provides a wide angle view of the area directly adjacent to and under the passenger door of the vehicle, and the Class VI mirror provides a wide angle view of the area directly in front of the vehicle. The standards that define the area of coverage on the ground plane provided by these mirrors, also define the minimum radius of curvature of these mirrors. For example, the Class V mirror has a minimum radius of curvature of 300mm and the Class VI mirror has a minimum radius of curvature of 200mm. The SAMMIE DHM system was therefore required to allow the projection of four mirrors and multiple window apertures to provide a volumetric analysis of the space visible to the driver through direct and indirect vision. The projection technique is based upon ray tracing and is defined by the location of the eyes of the driver and the location and radius of curvature of the mirror. A ray is projected from the eye of the Digital Human Model and intersected with a vertex in the path that describes the edge of the mirror surface.

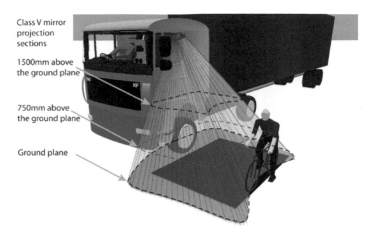

Figure 3 The projection of the Class V mirror to allow an analysis of the volume of space that is visible to the driver through the mirror

The resultant direction of the ray is defined by the radius of curvature of the mirror. The multiple rays that result from the multiple vertices that describe the edge of the mirror form a bounded volume of space, the interior of which can be seen by the driver. Figure 3 illustrates an example where the volume of space that is visible to the driver through the Class V mirror is shown. The importance of this volumetric approach is shown by the example in Figure 3. The black square on the floor in Figure 3 shows the area of the ground plane that is required to be visible in the standard 2003/97/EC for the Class V mirror. The Class V mirror on the vehicle has been adjusted so that is covers the four corners of this standard area. A Cyclist is shown to be within the specified area, and yet there is no intersection with the projection showing the volume that can be seen by the driver, making the cyclist invisible to the driver.

3.1 The data collection process to support the SAMMIE DHM analysis

In order to perform an accurate analysis of blind spots it was important for the DEG team to have a clear understanding of the postures adopted during the driving task by a range of driver sizes. In order to support this understanding twenty drivers with a stature range of 4th%ile UK male – 99th%ile UK male were photographed in their vehicles. In addition to this, each driver was interviewed to gain insights into potential blind spots and their implications for accidents. It was also considered important that a suitable sample of vehicles were selected for the analysis. In order to inform the selection of Category N_3 vehicles new vehicle registration data was analyzed. This allowed the top ten selling Category N_3 vehicles to be identified. The DAF XF 105, the Volvo 480 and SCANIA R420 were selected for analysis. The Volvo vehicle was left hand drive to allow an analysis of the situation for vehicles driving on UK roads (right hand drive layout) that originate from mainland Europe.

Figure 4 The variety of driver postures exhibited by drivers with a 4th%ile to 99th%ile stature range using UK anthropometric data (ADULTDATA 1998)

The DAF and SCANIA vehicles were right hand drive. Each of these vehicles was digitized in three dimensions using the FARO ARM data capture system. These data were then imported into the SAMMIE DHM system for analysis.

3.2   DHM system methodology

Each of three vehicles were setup in the SAMMIE DHM system to allow the driver adjustable mirrors (Class II and Class IV) to be adjusted using the center of rotation that was identified using the scanned data of the mirrors in four extreme adjusted positions (e.g. fully up fully right, and fully down fully left). The Class V and Class VI mirrors, which are not easily adjusted by the driver, were initially left in the positions that were found on the actual vehicles. Digital Human Models were then created using the anthropometric data gathered from the smallest (4th%ile stature UK male 1635mm) and largest (99th%ile stature UK male 1912mm) found in the interview phase of the project. The seat and steering wheel adjustability in the virtual vehicles were then adjusted to allow the joint angles identified in the interview phase to be recreated. In this way actual driving postures were recreated for a range of driver sizes. Using the projection technique shown in Figure 3, and the standard 2003/97/EC ground plane areas shown in Figure 2, the virtual mirrors were adjusted to ensure that the standards were being met. Following this the blind spots were analyzed by simultaneously projecting the mirrors and window apertures as shown in Figure 5. The identified blind spots were then tested by modeling scenarios described in section 2 in the DHM system. Scenarios 1-3 were associated with change lane maneuvers and so a three lane highway was modeled in the DHM system with a standard lane width of 3.5m. A car (width of 1.9m , length of 3.6m and height of 1.45m) was used as a visual target and used for the analysis of 'side swipe' scenarios as shown in Figure 5. Scenario 4, involves Category N_3 vehicles turning left at junctions in the UK (cars drive on the left it the UK), and colliding with cyclists. The number of variables associated with this scenario (junction size, orientation, and positioning of the cyclist and the Category N_3 vehicle) provided numerous combinations and potential outcomes. In order to provide structure for the analysis, a specific OTS case was identified that met the conditions of the scenario. The OTS case data includes detailed location information, allowing an accurate model of the junctions and vehicle positioning to be defined. The visual target for the analysis was a 50th%ile UK female adopting a cycling posture, whilst stationary at the junction.

3.3 The results from the initial identfication of blind spots

The results of the volumetric analysis highlighted a key blind spot that was consistent across all three of the Category N₃ vehicles that were analyzed, although the size of this blind spot did vary. This blind spot exists between the volume of space that is visible to the driver through direct vision (through the passenger window and windscreen) and the volumes of space that are visible through the Class II, IV and V mirrors. The identified blind spot was then tested in terms of the scenarios that were defined in the accident data analysis discussed in section 2.

3.3.1 Results from the analysis of scenarios 1, 2 & 3

The analysis of the 'side swipe' scenarios highlighted in the accident data showed that locations exist to the side of the Category N₃ vehicle that can completely obscure the target vehicle from direct vision, and only allow very small sections of the target vehicle to be seen in the Class V mirror. This is illustrated in Figure 5 which shows that the target vehicle cannot be seen in the Class IV mirror (left hand image) with the small portion of the target vehicle's door that can be seen in the Class V mirror highlighted with a white circle.

Figure 5 The target vehicle can be obscured from the vision of the driver of the Category N₃ vehicle using direct vision, with only a small portion of the target being visible at the bottom edge of the Class V mirror

A small portion of the front of the visual target vehicle was able to be seen in to the Class VI mirror. However, the distorted nature of the image of a part of a car at the edge of a Class 6 mirror with a radius of curvature of 200mm, combined with the interview data which illustrated that drivers only use this mirror when the vehicle is stationary, highlights the potential difficulty in relying upon the identification of a vehicle in this way. The results were consistent across all three vehicles that were analyzed. Further detail can be seen in the full project report (Cook et al 2011).

3.3.2   Results from the analysis of scenarios 4

The analysis of the specific OTS scenario that resulted in the injury of a cyclist was based upon the premise that the driver should be aware of the cyclist being in a location that is to the left of the Category N_3 vehicle before pulling away and starting the left hand turn. Each vehicle was placed into a suitable location at the junction, and it was determined that it is possible for the cyclist to be completely obscured from the drivers vision for all of the test vehicles. Figure 6 shows the vehicle junction that was modeled from the OTS data (top image) and the complete obscuration of the cyclist from the Category N_3 vehicle driver for all three tested vehicles.

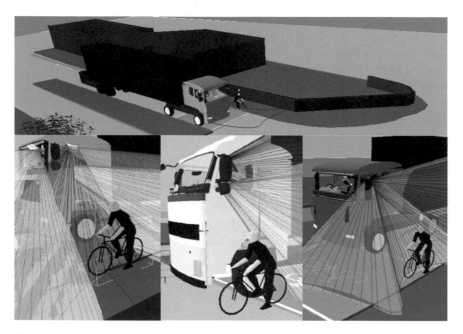

Figure 6 The vehicle junction that was modeled from the OTS data (top image) and the complete obscuration of the cyclist from the Category N3 vehicle driver for all three tested vehicles

3.3.3   Potential solutions to the issues identified in the results

The analysis of the four scenarios performed as part of the research highlighted the potential benefits of allowing the driver of Category N_3 vehicles to recognize the presence of other vehicles and vulnerable road users that are adjacent to the Category N_3 vehicle, and outside of the current 2m wide ground plane area that is specified in the standard 2003/97/EC. As part of the research performed potential solutions were analyzed. This included technologies such as radar and ultrasound based detection systems (see work package 5 in the project report (Cook 2011)). In addition to this mirror based solutions were also explored. An aftermarket mirror that has the potential to fill the blind spot between the volume of space visible through direct vision, and the volume of space visible through current Class V mirrors was identified. The SPAFAX VM5 mirror has the same radius of curvature as standard Class V mirrors, but has a larger surface area in the vertical orientation. This mirror was tested in both real world and DHM analyses and was shown to allow the identified blind spot to be eliminated. Figure 7 shows an example of the DHM analysis using the DAF XF 105 Category N_3 vehicle.

Figure 7 The virtual testing of the SPAFAX VM5 mirror. The left hand image shows the standard Class V mirror fitted to the vehicle, orientated to allow the floor area specified by standard 2003/97EC to be seen by the driver. The right hand image shows the volume of coverage by the SPAFAX VM5 mirror

4   SOLUTIONS

The vehicles that were analyzed as part of the research project all had the capability to achieve the visibility of the ground plane that is defined in the standard 2003/97/EC using direct and indirect driver vision. The blind spot that was identified to be common across all vehicles that were tested has the potential to be a contributory factor in side swipe accidents, and accidents with vulnerable road users. The results from the research project were used by the research team to specify an amendment to the definition of the standard 2003/97EC at the request of the UK Department for Transport. This amendment specified that the area of the ground plane that should be visible to the driver of Category N_3 vehicles should be increased laterally from 2m, to 4.5m. This was defined to allow the blind spot that was identified in the research project to be eliminated. The proposed amendment was presented to the United Nations GRSG committee at the 100th meeting in April of 2011.

5   CONCLUSIONS

The research project has highlighted the need to expand the volume of space that is visible to the drivers of Category N₃ vehicles. The blind spots that were identified in the research have been shown to have the potential to be a causal factor in the accident scenarios that were identified using accident statics data. The resultant amendment to the standard 2003/97/EC therefore has the potential to reduce accidents where vehicles and other vulnerable road users occupy the space that is adjacent to the cab of the Category N₃ vehicle, but further than 2m from the side of the vehicle.

REFERENCES

2003/97/EC. 2004. *Directive 2003/97/EC of the European Parliament and of the Council on the approximation of the laws of the Member States relating to the type-approval of devices for indirect vision and of vehicles equipped with these devices.* Official Journal of the European Union.

Cook, S., Summerskill, S., Marshall. R., et al., 2011. The development of improvements to drivers' direct and indirect vision from vehicles - phase 2. Report for Department for Transport DfT TTS Project Ref: S0906 / V8. Loughborough: Loughborough University.

Cuerden, R., Pittman M., Dodson, E. and Hill, J. (2008). The UK On The Spot Accident Data Collection Study – Phase II Report. Department for Transport Road Safety Research Report No. 73. http://www.dft.gov.uk/pgr/roadsafety/research/rsrr/theme5/

Ehlgen T., Pajdla T., Ammon D.: 'Eliminating blind spots forassisted driving', IEEE Trans. Intell. Trans. Syst., 2008, 9, (4),pp. 657–665 Larue, C. & Giguere, D. (1999),

Hill J.R. & Cuerden R.W. (2005). Development and Implementation of the UK On the Spot Data Collection Study – Phase 1. Department for Transport Road Safety Research Report No. 59. http://www.dft.gov.uk/pgr/roadsafety/research/rsrr/theme5/

Porter, J. M. and Stern, M. C. 1986, A technique for the comparative assessment of external visibility characteristics in road cvehicles, in A. G. Gale et al. (eds), Vision in Vehicles (Amsterdam: Elsevier), 313 – 322.

Sodhi, M. & Reimer, B., (2002). Glance analysis of driver eye movements to evaluate distraction. *Behavior Research Methods, Instruments, & Computers2002, 34 (4), 529-538*

Tait, R. & Southall, D. (1999). Drivers' field of view from large vehicles. Phase 3 Report. http://hdl.handle.net/2134/549

UN Regulation 46. 2009. *Uniform Provisions for Devices for Indirect Vision and of Motor Vehicles with Regard to the Installation of these Devices.* UN ECE Vehicle Regulations.

CHAPTER 25

The Use of DHM Based Volumetric View Assessments in the Evaluation of Car A-Pillar Obscuration

Russell Marshall, Steve Summerskill, Sharon Cook

Loughborough Design School,
Loughborough University, LE11 3TU, UK
R.Marshall@lboro.ac.uk

ABSTRACT

This paper concerns the development of a new volumetric vision assessment for Digital Human Modelling (DHM) and its application to transport research. The research was commissioned by the UK Government who identified an ongoing concern with the potential for car A-pillars (A-posts) to obscure driver's vision and be a contributory factor in road accidents. Road accident scenarios were identified from UK accident data, modelled and then evaluated within the DHM environment SAMMIE. A new assessment method was developed that creates a 3D projection of the volume of space visible through an aperture, or reflected through a mirror. Using these projections vision was assessed for the scenarios using three different category M_1 vehicles (cars). The assessments identified that A-pillar obscuration could have been a contributory factor in the accidents. The research also highlighted that if the driver was aware of the obscuration it could be relatively easily negated. Conclusions indicate that the volumetric vision assessment is a very useful tool and much more illustrative than the more traditional 2D ground plots of visibility. In addition, car A-pillars form a potential obscuration to vision and could lead to accidents especially when combined with factors such as inexperience, or tiredness.

Keywords: DHM, ergonomics, field of view assessment, a-pillar, SAMMIE

1   INTRODUCTION

This paper concerns the development of a new volumetric vision assessment for use within digital human modelling (DHM) systems and its application to research in the transport realm. The research was commissioned by the UK Department for Transport (DfT) who identified a perceived concern with the potential for category M_1 vehicle A-pillars (vehicles designed and constructed for the carriage of passengers and comprising no more than eight seats in addition to the driver's seat) to obscure driver's vision. Whilst the degree of obscuration is regulated there is a concern that it is still a potential contributory factor in road accidents.

This research addressed the UK market and right hand drive vehicles, however mirroring the situation in left hand drive vehicles makes the findings equally applicable for both configurations of vehicle and international markets.

To investigate these issues the approach taken was to recreate a number of scenarios identified from UK accident data in which A-Pillar obscuration may have been a contributory factor. The scenarios would be evaluated using multiple M_1 vehicles based upon UK/EU sales statistics and to specifically evaluate one vehicle with the split A-pillar configuration. The effect of different size drivers would also be evaluated.

To support these evaluations a volumetric field of view assessment method was developed that allows the volume of space visible through an aperture, or reflected through a mirror to be projected within the SAMMIE DHM environment. The method was designed to support multiple aperture / mirror projections in order to describe a full 360 degree field of view assessment.

2   A-PILLAR OBSCURATION

The obscuration to the driver's field of view caused by the A-pillar (the A-pillar is defined as the body structure of a vehicle between the front windscreen and the front side windows) of a vehicle is a long standing issue and the subject of significant research (Fosberry and Mills 1959; Pipkorn et al 2012). Research has been performed into attempting to understand both the magnitude of the problem and into solutions (Quigley et al 2001; Millington et al 2006).

Over the last 20 years developments in vehicle design have seen an increase in A-pillar size. This trend can be partly attributed to increased crash protection through the use of larger sections for the underlying body-in-white structure and the inclusion of airbag systems within the interior trim of the pillar. This increase in size has the potential to increase obscuration to the driver's field of view, increasing the size of any potential blind spot and increasing the risk of the driver being unable to see something important in the driving environment. The permitted obscuration from the A-pillar is governed by vehicle standards such as EEC regulation 77/649 (1977) that limits the angle of obscuration subtended from the driver's eye point. However, even with the regulation of this area it is recognised that A-pillars pose a potential risk and are likely to be a contributory factor in some road accidents.

3   VOLUMETRIC FIELD OF VIEW

Driver's field of view assessments in vehicles have been performed using a number of tools and methods in recent years (Tait and Southall, 1998, Millington et al 2006). A common approach to identify the field of view afforded to the driver is to use a 2D mapping approach in which the field of view through an aperture such as a window, or via a mirror, is projected onto the ground plane to plot the visible area (Figure 1). This approach is also employed in regulatory standards to describe the minimum field of view for vehicles of various types e.g. 2003/97/EC (2004) and UN Regulation 46 (2009).

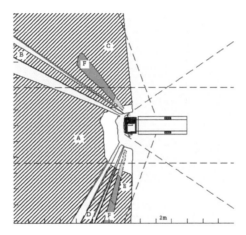

Figure 1 Example of a Direct field of view plot (Leyland Daf truck) (From Tait and Southall, 1998)

The definition of these areas has been performed using both real-world and virtual techniques. Real-world methods typically require direct observation by a driver and the positioning of target markers placed on the ground to identify the limits of view. Virtual methods typically involve the projection of the aperture or mirror boundary onto the ground plane. Comparisons of real-world and virtual assessments have shown a high degree of correlation of both approaches (Cook et al 2011). However, a fundamental issue in the use of these real-world or virtual assessments is the simplification of the field of view to a 2D area. In reality the field of view is a complex 3D volume, only part of which may intersect the ground plane. Taking a 2D approach has the advantage of being relatively easy to define in a standard and to subsequently evaluate which is important if vehicles are to be assessed in their compliance with such standards. However, a 2D approach is also a considerable compromise and distortion of the real-world environment. The use of the ground plane itself is questionable as most targets that a driver would wish to be aware of are not at this level, or rather the part of a target that you would wish to see is at a higher vertical level. For example you would want to see the torso, or head of a pedestrian, and not their feet, or the body of a car and not its tyres. In addition, the use of virtual techniques to create these projections raises issues such as the interpretation of the projections. The areas defined in Figure 1 can be interpreted that the driver would be able to see the area defined and suggest that something in that area would also be clearly visible. Yet this is not as clear as it may first appear as the volume of viewable space is not necessarily uniform in the vertical component. Thus a target positioned in the visible area on the ground plane may only stay visible for a small portion of its height, or a target positioned outside of the visible area may become visible part way up its height. Figure 2 shows the 3D volumetric projection of the field of view from the driver's side mirror of a 2010 model UK Volkswagen golf. A human is stood within the visible area as projected at the ground plane but the visible volume indicates that only their feet and part of their lower leg would be visible to the driver.

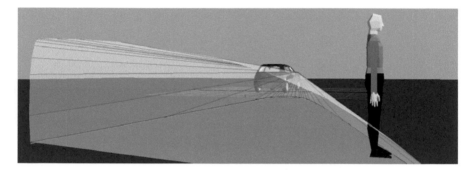

Figure 2   3D volumetric field of view projection through a car's wing (side) mirror and the ability to view a target within the visible 2D area projected on the ground plane.

To address these issues a 3D volumetric field of view projection tool was developed and implemented within the SAMMIE DHM system (Porter et al. 2004). The aim of the development was to provide a tool that could be used to project the viewable volume as afforded by direct vision through an aperture such as a window, or by indirect vision through a mirror. The viewable volume should be shown as a graphical object to be able to identify the interaction of this volume with objects in the virtual environment. In addition, multiple apertures / mirrors should be able to be projected simultaneously to support a full 360° evaluation of field of view around a vehicle.

The implementation of the projection tool involved projecting a ray from the driver's eye point through each vertex of the polygonal window or mirror object and then tracing that ray to a user definable cutoff. The tracing of the projected ray is a straightforward continuation of the eye-vertex vector for aperture projection. For mirrors the ray is reflected taking into account the curvature of the mirror using standard calculations. In both cases the projected rays were used to create polygons that ultimately create a boundary representing the visible volume of space. Finally, this volume is also used to create the 2D area projections at a user definable height to support assessments that can be directly compared with the standards but also to evaluate visibility at any height of horizontal plane.

4   SCENARIOS

The UK STATS 19 police accident database and the On-The-Spot (OTS) road accident database were interrogated for accidents in which A-pillar obscuration was a potential contributory factor.   These databases provide detail on UK road accidents to varying degrees of detail but together describe the scene, with photographs and a schematic of the road layout around the immediate vicinity.   In addition, the vehicle types involved are named together with a description of the event.   A review of the accident data shows that is it very difficult to differentiate between 'looked but failed to see' and 'failed to look' scenarios.   As such it is impossible to be certain that obscuration from the A-pillar was the cause of any accident, however it is possible to identify scenarios where A-pillar obscuration may have been a causal factor.

The scenario described here involves a car entering a roundabout with three exits.   The car was positioned in the right hand lane of the road approaching from the right of Figure 3.   The car stops at the roundabout to check the coast is clear and then proceeds onto the roundabout with the intention of taking the exit directly ahead (the left of Figure 3).   The car collides with a scooter rider already on the roundabout, passing across the path of the car intending to take the exit to the bottom of Figure 3.

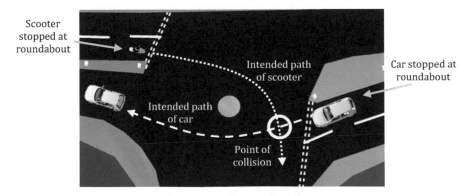

Figure 3 Roundabout Scenario in which the vehicle approaching the roundabout from the right collides with the scooter already on the roundabout travelling from the left to the exit at the bottom.

The rationale for selection of this scenario is based on a number of factors.   The first is that the car stops at the roundabout.   Presumably the driver then makes their observations and then proceeds to move onto the roundabout to take the exit opposite.   The driver statement indicates that the driver never saw the scooter, and that they "appeared out of nowhere".   The low speed of the accident indicates that the car did indeed stop or at least slow down to a point where they should have been able to observe the scooter already on the roundabout.   Thus it is possible that the driver did look but failed to see the scooter obscured by the driver's side A-pillar.   The accident occurred on a clear day, visibility was good and the roads were dry.

5 ASSESSMENTS

Assessments were performed using the scenario outlined above in order to establish if A-pillar obscuration could be a casual factor in accidents of this nature. Thus, the exact type of vehicle involved in the accident was not used in the evaluations. Instead three category M_1 vehicles were selected to provide a range of A-pillar configuration and to represent common vehicles on UK roads. There were: a 2010-on model Volkswgen Golf, a 2010-on model Hyundai i10, and a 2010-on model Volkswagen Touran. The Touran was selected due to its split A-pillar configuration that is compliant with A-pillar size and visibility regulations but has an increased effective area of A-pillar over standard configuration vehicles.

Figure 4 The three assessment vehicles. From left to right: Volkswagen Golf, Hyundai i10, Volkswagen Touran.

The assessment methodology began by modelling the vehicles and road layout of the accident scenario. Firstly the assessment vehicles were scanned using a FARO arm mounted contour scanner. The scanning process captured window apertures, mirror perimeters, exterior and interior panel outlines, and a range of internal seating and control details: seat cushion contours and adjustability, steering wheel contours and adjustability, pedal and gearstick locations etc. This resulted in a series of curves that were imported into Pro/ENGINEER and surfaced to create an accurate 3D model of the vehicle. Next, using the accident data the roundabout junction was modelled in the SAMMIE DHM system. Finally the vehicles were exported from Pro/E into the SAMMIE roundabout model.

The scenario was then replicated six times using all three vehicles with two extreme sizes of driver established by taking the smallest and largest human models capable of driving the vehicle (99%ile Dutch male (1941mm) for all vehicles and %5ile UK female (1534mm) for the Golf and i10, and 35%ile UK female (1615mm) for the Touran). The human models were positioned in the seat with the assistance of a virtual H-point manikin, seat compression data and representative real-world driving postures.

Figures 5 and 6 show the volumetric projection technique being used to project the field of view through the windscreen and the driver's side window for the Hyundai i10 and the 5%ile UK female driver. The analysis shows that there is a clear 'corridor' of obscuration from the A-pillar. At the point where the car is stationary the scooter entering the roundabout is completely obscured (Figure 5).

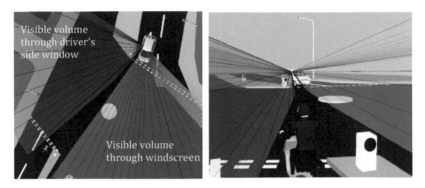

Figure 5 Showing the two volumes of visibility projected from the windscreen and driver's side window. This results in a 'corridor' of obscuration for the driver.

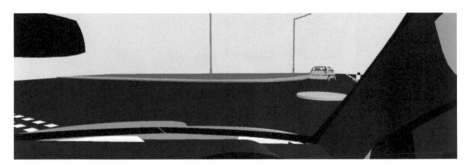

Figure 6 5%ile UK female driver's eye view of the roundabout from the Hyundai i10. The A-pillar fully obscures the scooter and its rider.

For all of the assessed M_1 vehicles the same blind spot caused by the A-pillar is apparent. This blind spot is greater for the smaller drivers of each vehicle as the angle subtended by the eye has to be greater to clear the A-pillar obscuration due to the eye point being closer to the pillar. It is possible that whilst the M_1 driver is performing their observations prior to setting off that the scooter could have travelled the length of this blind spot and be in the process of turning to their right, across the front of the M_1 vehicle. If the scooter rider was focused on their direction of travel and exit from the roundabout, it is possible that the M_1 vehicle could set off and cause the collision. This sequence of events would need very particular timing but Figure 7 below gives an indication of how this may come together to cause the accident.

Figure 7 The scooter travelling along the corridor of obscuration from the A-pillar of the Hyundai i10 and the 5%ile UK female driver.

The obvious obscuration caused by the A-pillar suggests that it could have contributed to this accident. However, this assumes the driver is unaware, or fails to manage the obscuration. The obscuration can effectively be eliminated by the driver tilting their head to 'look around' the A-pillar (Figure 8).

Figure 8 The scooter can be seen if the driver leans their head

All of the evaluations shown in this paper thus far have used a mean eye point and monocular vision. This provides a consistent generic approach, however, it does not take into account the binocular view afforded to a real person. Further evaluations using binocular vision revealed that a small visual target completely obscured by the A-pillar in monocular vision becomes partially visible in binocular vision. A larger target partially obscured becomes significantly less obscured with binocular vision. However, the pillar never fully disappears and the effect is markedly greater for eye points further from the A-pillar such as that for the 99%ile Dutch male. For smaller drivers, the A-pillar is a greater obscuration to vision and the effect of binocular vision is much reduced.

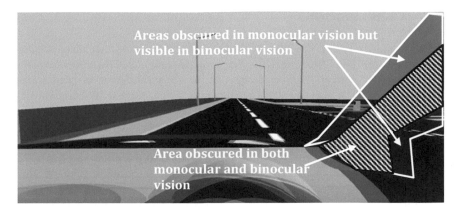

Figure 9 The impact of binocular vision on A-pillar obscuration

The evaluations performed in this research identified that A-pillars can create a blind spot for the drivers of M_1 vehicles. In addition these blind spots appear to be a contributory factor in the recreation of accident scenarios taken from accident data. However, the blind spots are variable in both size and position based on the design of the pillar, the position of the pillar and the eye point of the driver. Thus a smaller pillar combined with the driver being relatively close to the pillar may provide an equivalent blind spot to a larger pillar with the driver being relatively further away, as observed with the Volkswagen Touran and the Hyundai i10.

Blind spot awareness is typically part of driver training and so all drivers should be looking to manage blind spots in a particular vehicle. The evaluations performed showed that all blind spots identified that are attributed to A-pillar obscuration could effectively be eliminated by altering the posture and thus eye-point of the driver and essentially 'looking around' the pillar. However, other contributory factors such as tiredness, the driver being in a hurry, busy road conditions with lots of visual demand, other distractions such as passengers, etc. could all lead to the driver failing to check the blind spot. Thus, any solutions that could limit the effective blind spot would be helpful in maximising the driver's field of view.

Solutions are limited to this particular issue. Ideally pillars would be designed to be non-existent, transparent, or as slim as possible such as those proposed in the Volvo Safety Concept Car (Ross, 2003). However, crash testing requirements and standards for occupant safety place significant structural demands on the pillar structure and alternative configurations whilst showcased in concept vehicles have yet to appear in mainstream designs.

Whilst it is not possible or practical to recommend a target size of pillar, manufacturers should be made aware of the importance of this area to both primary and secondary safety. They should be encouraged to seek solutions in engineering and materials design with a view to reducing pillar size and should ensure that any treatments in this area such as darkened screen edges do not exacerbate the problem of A-pillar obscuration.

6   CONCLUSIONS

This research reports on the evaluation of category M_1 vehicle A-pillar obscuration. The UK Department for Transport commissioned an evaluation into the potential for current A-pillar designs to be a causal factor in road traffic accidents. Using UK accident data, accidents in which A-pillar obscuration may have been a factor were recreated in the SAMMIE DHM system together with three models of M_1 vehicles. Evaluations were performed using a new volumetric field of view evaluation tool and clear blind spots were identified that could easily obscure another vehicle. The ability to clearly visualise the blind spots in a 3D environment proved to be a powerful tool in assessing complex field of view scenarios. Whilst this work was not intended to perform in-depth research into A-pillar blind spot solutions, it is recommended that vehicle manufacturers investigate engineering and materials solutions to minimise A-pillar size.

REFERENCES

2003/97/EC. 2004. *Directive 2003/97/EC of the European Parliament and of the Council on the approximation of the laws of the Member States relating to the type-approval of devices for indirect vision and of vehicles equipped with these devices.* Official Journal of the European Union.

77/649/EEC. 1977. Council Directive on *the approximation of the laws of the Member States relating to the field of vision of motor vehicle drivers.* Official Journal of the European Communities.

Cook, S. E., S. Summerskill, R. Marshall, et al. 2011. *The development of improvements to drivers' direct and indirect vision from vehicles. Phase 2.* Loughborough University. Report for Department for Transport.

Fosberry, R.A.C. and B. C. Mills. 1959.   Measurement of Driver Visibility and its Application to a Visibility Standard. *Proceedings of the Institution of Mechanical Engineers: Automobile Division* January 1959 13: 50-81.

Millington, V.A., R. Cuerden, S. Hulse, et al. 2006. *Investigation into 'A' Pillar Obscuration – A Study to Quantify the Problem Using Real World Data.* PPR159 S0414/V8.

Pipkorn, B., J. Lundström and M. Ericsson. 2012. Improved car occupant safety by expandable A-pillars, *International Journal of Crashworthiness*, 17:1, 11-18

Porter, J. M., R. Marshall, M. Freer, et al. 2004. SAMMIE: A Computer Aided Ergonomics Design Tool. In *Working Postures and Movements, Tools for Evaluation and Engineering,* eds. Delleman, N, Haslegrave, C, Chaffin, D. pp. 454-470. CRC Press .

Quigley C. L., S.E. Cook and A. R. Tait. 2001. *Field of vision (A-pillar geometry) - a review of the needs of drivers.* Loughborough University.   Department of the Environment, Transport and the Regions (DETR) report - PPAD 9/33/39.

Ross, P. 2003. Top 10 techno-cool cars, *Spectrum*, IEEE , 40:2, 30- 35.

Tait A.R. and D. Southall. 1998. *Drivers' field of view from large vehicles – Phase 3 Report.* Loughborough University. Report for Department for Transport.

UN Regulation 46. 2009. *Uniform Provisions for Devices for Indirect Vision and of Motor Vehicles with Regard to the Installation of these Devices.* UN ECE Vehicle Regulations.

CHAPTER 26

Using Ergonomic Criteria to Adaptively Define Test Manikins for Design Problems

Peter Mårdberg[† a]*, Johan S. Carlson*[a]*, Robert Bohlin*[a]*, Lars Hanson*[b,c]*,
Dan Högberg*[d]

[a] Fraunhofer-Chalmers Research Centre for Industrial Mathematics, Gothenburg, Sweden
[b] Department of Product and Production Development, Chalmers University of Technology, Gothenburg, Sweden
[c] Industrial Development, Scania CV, Sweden
[d] Virtual Systems Research Centre, University of Skövde, Skövde, Sweden

[†]mardberg@fcc.chalmers.se

ABSTRACT

Digital manikins give a powerful aid in evaluating assembly station ergonomics. A proper verification of an assembly station can avoid costly changes that might occur later on due to injuries and discomfort among the workers.

However, how to select relevant test manikins for an evaluation is a non-trivial task, since the work of identifying which anthropometric variables that are critical is a tedious and time consuming process.

Usually only a few variables are selected as a base for building the test manikins. Even if there exist Digital Human Modeling (DHM) software which allow the user to evaluate batches of manikins, the designer still have to select the anthropometric variables of those batches. When several dimensions are considered, the designer have to either use a set of predefined manikins, or determine which anthropometric variables to test and generate manikins based on the confidence intervals of these variables.

When considering more complex assembly tasks, is it then true that these predetermined test manikins cover all the cases, or does there exist manikins that suffer from bad ergonomics even though all the test manikins turned out well?

In this paper, we propose a new algorithm for automatically building a set of test manikins. The set is iteratively constructed from the ergonomics results obtained by simulating the assembly operation. Different manikins perform the assembly operation and the ergonomics is evaluated. The anthropometric variables which affect the ergonomics are identified and used to iteratively build up the next manikin. In this way the test manikins are always selected throughout the whole set instead of only considering the boundary manikins, or assuming that the same set of predetermined manikins represents the entire set in every assembly operation.

The algorithm has been compared with a boundary method, and the results shows that the algorithm can find manikins with worse ergonomics than those tested by the boundary method.

Keywords: Digital Human Modeling, Sampling Algorithm, Response Surface Method

1   INTRODUCTION

Although many production systems in the manufacturing industry have been automated, there still exist a lot of manual assembly operations.

The assembly ergonomics is an important factor for reducing the cost and to ensure a sustainable and high quality production (Falck, et al., 2010). With the usage of virtual environments the ergonomics can be included and tested early in the development, long before physical environments are available.

However, to select relevant manikins for an accurate simulation is a non-trivial task since the anthropometric variables have to be identified in each assembly station. Even if there exist Digital Human Modeling (DHM) software which allows the user to evaluate batches of different test manikins, the anthropometric variables of those batches still have to be selected by the designer.

There exist two major strategies on how to select test manikins in the case of multiple dimensions: distributed or boundary cases (HFES 300 Committee, 2004; Bertilsson, et al., 2011a). In the latter the designer is specifying which anthropometric variables to use and the desired confidence intervals. The interval defines the boundaries of the variables, from which the set of manikins is generated. The distributed test manikins can either be selected randomly or they are predefined to represent the whole data set.

When consider more complex assembly tasks is it then true that the predetermined test manikins covers all the cases, or does there exist manikins that suffer from bad ergonomics even though all the test manikins turned out well?

In this paper we propose a new algorithm for selecting test manikins, called Adaptive Ergonomic Search (AES), where the entire assembly motion is considered, and the whole set of anthropometric data is used to build up representative test manikins. The algorithm is implemented in the Intelligent

Moving Manikins (IMMA) (Hanson, et al., 2011) software and tested on an assembly case from the automotive industry. The results are compared to the boundary method, and it is shown that there exist cases where the proposed algorithm finds manikins with worse ergonomics compared to the manikins selected by the boundary method.

2   RELATED WORK

The boundary manikins are built from a set of key dimensions (Bertilsson, et al., 2011b). One of the advantages of the boundary cases is shown in Bittner, where 17 representative test manikins are generated and achieve the same accommodation as a sample of 400 manikins (Bittner, 2000; HFES 300, 2004). However, the boundary cases only give answers to the ergonomics on the max and min of an interval. Furthermore, it is assumed that the ergonomics of the interior in between the intervals will always be less critical than the result from those on the boundaries. In fact, it is even assumed that there will not exist any other critical manikins on the boundary which might yield a worse result than those sampled.

The boundary method is also suffering from the fact that the critical variables have to be set by the user. Since manikins generated from regression equations with few dependent variables may have large errors in the estimated variables due to low correlation (You & Ruy, 2005), this can result in inaccurate manikins.

Distributed test manikins can either be selected randomly or they can be predefined to represent the whole data set.

An alternative method is presented in (Speyer, 1996) which provides a typology from which the user can build up appropriate set of test manikins.

However, there is no automatic solution on how to choose the distributed test manikins to ensure that the whole targeted population is accommodated (HFES 300 Committee, 2004).

3   SELECTING RELEVANT MANIKINS

In this section we present the kinematical manikin model, the comfort function, the anthropometric data set, the variable reduction method and our proposed AES algorithm.

3.1   The Manikin

The manikin model is built by links which model the rigid parts of the human skeleton. A link consists of a translation and a rotation, where the translation represents the length of a bone and the rotation represents the joint. Multiple links can be combined to model more complex parts of the body. For example, human joints which are allowed to rotate about more than one axis can be modeled by adding additional zero-length links (Bohlin, et al., 2011). See Figure 1.

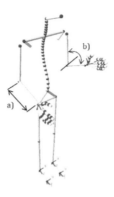

Figure 1: The manikin is built with links, which are composed of a translation a) and a rotation b).

The manikin has more than 160 degrees of freedom, which results in an infinitely number of ways to position the manikin in many cases. Although many solutions are eliminated due to task constraints and to human limitations there will still be redundant solutions. However, this gives us the opportunity to consider ergonomic aspects and to select the most comfortable solution.

Let $\theta = [\theta_1, \cdots, \theta_n]^T$ be the n dimensional joint angle vector which determines the position of the manikin. A position is called pose and we call a set of m poses a motion, i.e. $M = \{\theta(t): t \in [1, m]\}$, where each pose represent a time step t in the assembly motion.

For a pose to be realistic it has to fulfill some rules. The manikin needs to maintain the balance while holding the position. The balance has to consider both the body parts and objects being carried as well as exterior forces and moments in the environment.

The ergonomics in an assembly motion is defined as the sum of all evaluated poses in the motion. The comfort function c is defined to penalize high torques on the joints and poses where the joints are close to its limits. Let $p \in M$, and let $\hat{c}$ give the comfort for the entire motion of a manikin m

$$\hat{c}(M, m) = \sum_{p \in M} \left(\sum_{i=0}^{n} tan^2 \left(\frac{\theta_i \pi}{2\theta_{max,i}} \right) + \sum_{i=0}^{n} w_i (\alpha_i T_i)^2 \right),$$

where T_i is the torque, α_i is the rotational axis, and w_i a weight indicating the difference in the allowed stress of joint i.

3.2   The Anthroprometric Data Set

The link lengths of the manikin are generated from 56 anatomical variables according to the ISO 7250 standard using Swedish anthropometric data (Hanson, et al., 2009). However, due to missing data, only 39 variables are available for usage. The missing data are estimated by using regression equations on additional data sets (Bertilsson, et al., 2011a).

Principal Component Analysis (PCA) is used to reduce the dimensionality of the anatomical variables by removing unrelated data and creating a new independent set. The eigenvectors in the new set are built to capture the maximal variances of the underlying structure, and the eigenvalues λ are arranged according to the amount of variance that is captured by the corresponding eigenvector (Jolliffe, 2002).

The set of variables can be reduced from n to p dimensions by keeping enough eigenvectors to fulfill the desired reduction ratio τ which can be calculated using the eigenvalues

$$\frac{\sum_{i=1}^{p} \lambda_i}{\sum_{i=1}^{n} \lambda_i} \geq \tau.$$

The number of eigenvectors kept defines the dimensionality of the hyper ellipsoid, where the lengths of the principal axes are defined by scaling the normalized eigenvectors with

$$k = \sqrt{\chi_p^2 (1 - \rho) \lambda_i},$$

where p is the number of dimensions kept and ρ the sought accommodation level, see Figure 2.

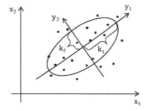

Figure 2: The confidence ellipsoid for two variables x_1 and x_2. The covered region of the ellipsoid is determined by weighting the principal axes y_1 and y_2 with k_1 and k_2 respectively.

In this case it was possible with the PCA approach to reduce the number of variables from 56 to 5 while keeping 75% of the total variation.

3.3 The Adaptive Ergonomic Search (AES) Algorithm

To identify which anthropometric variables that affect the ergonomics in an assembly motion the manikin is considered as a black box system. The inputs are the five reduced anthropometric variables and the comfort function is the system response. Let the correlation matrix Σ_X of the anthropometric data set X be decomposed into matrices of eigenvectors A and eigenvalues D as $\Sigma_X = ADA^T$. Let $\tilde{A}^T$ be the matrix of eigenvectors kept when reducing the dimensionality of Σ_X by PCA. An element $x \in X$ can be transformed onto the reduced set Y as

$$y = \tilde{A}^T \left(\frac{(x - \bar{x})}{\sqrt{\sigma}} \right),$$

where $\bar{x}$ is the mean vector and σ the vector of standard deviations of X.

By systematically perturbing the inputs y the interaction with the underlying system can be identified and used to build a response surface. Since the comfort function is differentiable in the interval of the joints it can be used to identify the sensitivity of influencing variables in y for a motion M as

$$\nabla \hat{c}(M, y^i).$$

The response surface can be traversed by iteratively making local estimations of the comfort as

$$y^{i+1} = \frac{\nabla \hat{c}(M, y^i)}{\|\nabla \hat{c}(M, y^i)\|} d + y^i$$

for a small step size d.

The manikins are sampled throughout the entire set to ensure that all regions where the comfort function reaches optima are found with high probability (Husslage, et al., 2011), see Figure 3 a).

The surface is traversed in two ways, either according to, or along the boundary condition, in the ascent direction of the gradient as shown in Figure 3 b) and c). Hence, if there exist an optimum on the hyper ellipsoid boundary, then the algorithm starts to traverse the boundary until it converges. When all sampled points have converged to an optimum, see Figure 3d), the set of test is manikins is returned to the user.

Figure 3: a) Samplings are selected throughout the set. b) The ergonomics are identified in each sample point is and used to traverse the set until optima is found. c) The algorithm follows the boundary until it converges to optima. d) The set of local optimum test manikins is returned to the user.

4   RESULTS

The algorithm is compared to the boundary method (Bertilsson, et al., 2011a) on an assembly operation. Four boundary and one centrum manikin were generated with regression equations by using stature and weight as dependent variables.

The methods were compared on a subsection of an assembly motion where a protective block is placed inside a car door. The assembly path is generated in Intelligent Path Solutions (Fraunhofer-Chalmers Research Centre , 2012) and three samplings along the path are shown in Figure 4. The path was made sparser by removing some intermediate points.

Figure 4: Three sample positions and orientations of the assembly path. Courtesy of Volvo Cars.

In the next step the IMMA manikin automatically calculates poses that follows the assembly path with as good comfort as possible. The manikins were not allowed to move their feet when they followed the path. However, before the manikins started to follow the path they were able to move their feet and place themselves in their most comfortable position. Five samples of the full path are shown in Figure 5.

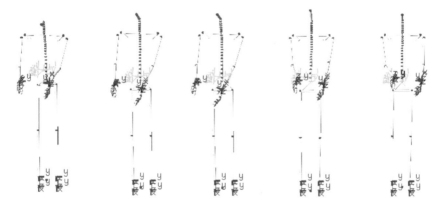

Figure 5: Five samples of the manikin when performing the assembly operation.

The algorithm was used with sample size of 15, and the results from both algorithms are shown in Table 1. An examination of the returned test manikins showed that they were interior manikins.

Figure 6 and Figure 8 show the ergonomic function per frame for the whole assembly motion for the manikins selected by the boundary method and by the proposed algorithm. Figure 7 and Figure 9 show a close up region of the lower left region in Figure 6 and Figure 8 respectively.

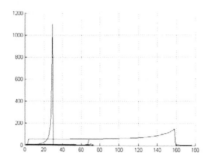

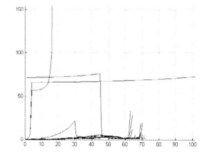

Figure 6: The comfort function (y-axis) per frame (x-axis) for the five boundary manikins.

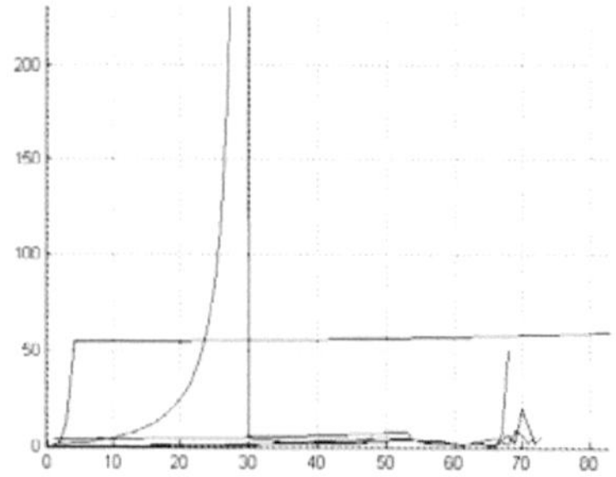

Figure 7: A close up on the lower left region in Figure 6.

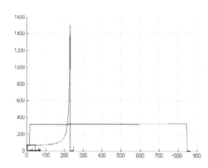

Figure 8: The comfort function (y-axis) per frame (x-axis) for the manikins that have been found by the AES algorithm.

Figure 9: A close up on the lower left region in Figure 8.

Table 1: The comfort of the compared manikins.

Boundary method		AES algorithm			
Manikin	$c()$	Manikin	$c()$	Manikin	$c()$
1	159.81	1	267063.71	9	114.85
2	10574.05	2	74.79	10	215.22
3	329.61	3	3329.27	11	153.43
4	144.01	4	96.86	12	100.52
5	22487.79	5	112.65	13	97.36
		6	112.59	14	215.22
		7	30146.24	15	153.43
		8	135.38		

5   CONCLUSION AND FUTURE WORK

A new automatic algorithm for selecting relevant test manikins based on surface response methods is proposed. In this work ergonomics is considered throughout the whole assembly motion, and the entire test set of anthropometric data is used to iteratively build up the set of test manikins. Furthermore, the algorithm is compared with a boundary method on an assembly test case. The results show that there exist manikins in the interior of the ellipsoid which have worse ergonomics than the manikins that were selected by the boundary method.

This shows some of the advantages of using the proposed algorithm. It eliminates the problem of identifying which anthropometric dimensions that are relevant to test, and it creates a set of test manikins which considers the fraction of the population that are relevant to test in each unique assembly situation. Moreover, the algorithm offers an adaptive evaluation since more ergonomic criteria are allowed to be included in the comfort function. The manikins are not allowed to move during the assembly operation since there are no penalties in the comfort function for moving on the floor. Thus, this lack of cost tends to make the manikin moving more on the floor instead of using the joints to follow the path. Therefore, a subsection of the path was chosen such that the manikin can follow it without moving the feet. These problems will be investigated during future research.

It is clear from Table 1 that there are manikins not covered by the boundary cases. From Figure 6 to Figure 9 it can be seen that the function has three characteristic shapes. Manikins that are performing the operation with no problem have an almost linear function, as shown in Figure 7 and Figure 9, while manikins where the function takes big steps indeed are suffering from the operation. The third shape represents manikins which only suffer from a part of the assembly operation.

There is no defined threshold to distinguish between good and bad comfort. Instead the comfort is compared between all the manikins, and we say that a manikin is suffering more from an assembly motion if there is large gap in the comfort between the compared manikins. Hence, a topic of future research is to develop a better definition of what is considered as comfortable motion and what is not in terms of the comfort function.

Another topic for future research could be to adjust the weights of the joints and to extend the comfort function by adding more ergonomic criteria. This although the comfort function is fairly simple and gives realistic motions in most cases.

ACKNOWLEDGMENTS

This work was carried out within The Swedish Foundation for strategic Research (SSF) ProViking II program and the Wingquist Laboratory VINN Excellence Centre, supported by the Swedish Governmental Agency for Innovation Systems (VINNOVA).

This work is a part of the Sustainable Production Initiative and the Production Area of Advance at Chalmers University of Technology.

REFERENCES

Bertilsson, E., Hanson, L., Högberg, D. & Rhén, I.-M., 2011a. *Creation of the IMMA manikin with consideration of anthropometric diversity.* Stuttgart, International Conference on Production Research.

Bertilsson, E., Högberg, D., Hanson, L. & Wondmagegne, Y., 2011b. *Multidimensional consideration of anthropometric diversity.* Lyon, First International Symposium on Digital Human Modeling.

Bittner, A., 2000. *A-Cadre: Advanced family of manikins for workstation design.* Santa Monica, CA, Proceedings of the IEA/HFES 2000 Congress. Human Factors and Ergonomics Society, pp. 774-778.

Bohlin, R. et al., 2011. *Unified solution of manikin physics and positioning - Exterior root by introduction of extra parameters.* Lyon, First International Symposium on Digital Human Modeling.

Falck, A.-C., Örtengren, R. & Högberg, D., 2010. The Impact of Poor Assembly Ergonomics on Product Quality: A Cost–Benefit Analysis in Car Manufacturing. 20(1), p. 24–41.

Fraunhofer-Chalmers Research Centre , 2012. *(IPS) Industrial Path Solutions.* [Online] Available at: www.industrialpathsolutions.com

Hanson, L., Högberg, D., Bohlin, R. & Carlson, J., 2011. *IMMA – Intelligently Moving Manikins – Project Status 2011.* Lyon, First International Symposium on Digital Human Modeling.

Hanson, L. et al., 2009. Swedish anthropometrics for product and workplace design. *Applied Ergonomics,* 40(4), pp. 797-806.

HFES 300 Committee, 2004. *Guidelines For Using Anthropometric Data In Product Design.* Santa Monica, CA: Human Factors and Ergonomics Society.

Husslage, B., Rennen, G., van Dam, E. & den Hertog, D., 2011. Space-filling Latin hypercube designs for computer experiments. *Optimization and Engineering,* 12(4), pp. 611-630.

Jolliffe, I., 2002. *Principal Component Analysis.* 2nd ed. New York: Springer-Verlag.

Speyer, H., 1996. *On the definition and generation of optimal test samples for design problems,* Kaiserslautern: Human Solutions GmbH.

You, H. & Ruy, T., 2005. Development of hierarichial estimation method for anthropometric variables. *International Journal of Industrial Ergonomics, 35,* pp. 331-343.

CHAPTER 27

Automatic Fitting of a 3D Character to a Computer Manikin

Stefan Gustafsson [†], Sebastian Tafuri, Niclas Delfs,

Robert Bohlin, Johan S. Carlson

Fraunhofer-Chalmers Research Centre for Industrial Mathematics,
Gothenburg, Sweden
[†]gustafsson@fcc.chalmers.se

ABSTRACT

Most advanced manikin software packages of today come with human meshes which deform as the manikin moves. The ability to import models made in external programs is usually limited or non-existent and the end-user is restricted to the manikins supplied by the program maker. This hinders e.g. the ability to perform ergonomic simulations of 3D models made from body scans of the actual factory workers. There are also large offerings of professionally created models of very high quality which may be used to diversify the simulated work force, more akin to the actual situation in many factories rather than using a one-size-fits-all manikin.

We use a method for automatically attaching a rigged mesh to a biomechanical skeleton, given some information about the bones as well as the binding pose. With this method ready-made characters exported from professional modeling programs can be imported into a manikin program with only minimal human effort.

Keywords: Digital Human Modeling, Computer Graphics, Rigging, Skinning

1 INTRODUCTION

The workers are the heart of many assembly factories and as such their long-term health is of prime importance. In planning new workstation layouts and assembly lines, virtual humans, also called *manikins*, may be used to evaluate the ergonomics. Having anthropometric diversity among the test manikins is important

(Bertilsson, et al., 2011a; Bertilsson, et al., 2011b) to be able to simulate a diverse work force. Traditionally, such evaluations are done by measuring humans working on physical mock-up stations, but computer simulations are becoming increasingly common. They have the advantage of being easier and cheaper to perform, and also support simulating scenarios for a variety of workers simultaneously. A realistic visual appearance of a computer manikin can be an important aid in ergonomic assessments of postures (Lämkull, et al., 2007), since most evaluation methods require human judgment and are not fully automated.

We therefore present a novel method for automatic fitting of a 3D character to a computer manikin.

Ensuring correct mesh deformations as the manikin moves requires a connection between the polygon mesh and an underlying skeleton, also called a *rig*. This rig is supplied by the 3D modeling program and is separate from the biomechanical skeleton of the manikin. Therefore, a mapping between these two must be made so that the rig's bones' transformations are updated correctly as the manikin's joints move. A typical rig used for animations may have around 50 bones, whereas the biomechanical skeleton of a manikin often have many more, perhaps modeling each vertebrae unlike the simplified spines found in most rigs.

If the naming and layout of the bones is consistent across different models, then this fitting can be re-used for other models as well, making it possible to run simulations on a diverse workforce.

2   RELATED WORK

A method to adapt a generic rig to a mesh is found in (Baran & Popovic, 2007), which can be useful to create complete rigged models from 3D scans and together with the method presented in this paper these models may then be used as manikins. To create a variety of body types the method presented here needs to be complemented with other methods. In (Kasap & Magnenat-Thalmann, 2009) the authors present a way to change the appearance of a model using the skin weights. A more sophisticated method is detailed in (Seo & Magnenat-Thalmann, 2003), where sizing parameters gained by statistical methods from a database can be used to alter the body type. A relatively simple way to allow for local scaling of the mesh at a joint is described in (Kavan, et al., 2008).

3   THEORY

We will first review some of the necessary background theory and definitions, including dual quaternions to express rotations, and skinning to calculate deformed vertex positions.

3.1   Quaternions and Dual Quaternions

Quaternions are an extension of the complex numbers commonly used to describe rotations. A quaternion q may be written as

$$q = (q_v, q_w) = iq_x + jq_y + kq_z + q_w = q_v + q_w \,,$$

where

$$i^2 = j^2 = k^2 = ijk = -1.$$

q_w is called the scalar part of q, q_v is the vector part, and i, j, k are the imaginary units. A *unit quaternion*, a quaternion of length one, gives a very compact representation of any rotation in 3D space, and the *shortest-path property* ensures that interpolating between two unit quaternions gives optimal results (Kuipers, 2002).

One limitation of quaternions is that they can only represent rotations whose axes pass through the origin. To remedy this, one can use *dual quaternions*, which are defined as quaternions whose elements are dual numbers. A dual quaternion $\hat{q}$ may be written as

$$\hat{q} = (\hat{q}_v, \hat{q}_w) = i\hat{q}_x + j\hat{q}_y + k\hat{q}_z + \hat{q}_w = \hat{q}_v + \hat{q}_w \,,$$

where $\hat{q}_w$ is called the scalar part (a dual number), $\hat{q}_v$ is the vector part (a dual vector), and i, j, k are regular quaternion imaginary units. A dual quaternion may alternatively be defined as

$$\hat{q} = q_0 + \varepsilon q_\varepsilon,$$

where q_0 and q_ε are regular quaternions and ε is the *dual unit* satisfying $\varepsilon^2 = 0$.

A unit dual quaternion $\hat{r} = q_0$ with dual part $q_\varepsilon = 0$ will represent a rotation just like a regular unit quaternion. However, a unit dual quaternion

$$\hat{t} = 1 + \frac{\varepsilon}{2}(it_0 + jt_1 + kt_2)$$

corresponds to a translation by vector (t_0, t_1, t_2). Composing $\hat{r}$ and $\hat{t}$ as $\hat{q} = \hat{t}q_0 = q_0 + \frac{\varepsilon}{2}(it_0 + jt_1 + kt_2)q_0$ yields a dual quaternion $\hat{q}$ which can express a rotation around an arbitrary axis (Kavan, et al., 2006).

3.2 Skinning

The process of deforming a mesh according the current pose of an attached rig is called *skinning*. The fundamental skinning equation which calculates a weighted position $f(v)$ for a vertex v influenced by n bones is (Akenine-Möller, et al., 2008):

$$f(v) = \sum_{i=1}^{n} w_i B_i(t) M_i^{-1} v \,,$$

where $B_i(t)$ is the transformation of bone i which changes during animation and M_i is the transform of bone i in the *bind pose*, i.e. the pose when the mesh was bound to the rig. The weights w_i are generally non-negative and sum to one, although certain effects, like for example muscle bulging, can be achieved by having weights that sum to more than one (Woodland, 2000). A thorough derivation of this equation is found in (Dalgaard Larsen, et al., 2001), where the authors also present some alternative viewpoints.

A transformation can be expressed in many ways, for example as a 4x4 matrix, a quaternion/translation pair, or a dual quaternion. By computing the total skinning transformation $S_i(t) = B_i(t)M_i^{-1}$ we can rewrite the skinning equation as

$$f(v) = \sum_{i=1}^{n} w_i S_i(t)v = \left(\sum_{i=1}^{n} w_i S_i(t)\right)v.$$

Note that the equation amounts to a blending of transformations. This explains why matrices are a poor choice, as the blended matrix is no longer necessarily a rigid transformation, but may contain scale and shear factors. The problem with using quaternion/translation pairs is that the model's vertices rotate around the origin of the body-space, and for skinning this is not what is desired. In (Kavan , et al., 2007) they show that dual quaternions are a better choice as they have all the required attributes of a transformation used for skinning: preservation of rigidity, coordinate-invariance, and shortest-path interpolation.

Implementing skinning using dual quaternions is relatively straightforward, and suffers only about a 1.5x cost compared to using matrices (Akenine-Möller, et al., 2008). The bone transformations B_i and M_i^{-1} can still be represented as matrices, and their product can be converted to a dual quaternion before blending. To convert a 4x4 rigid body transformation matrix S to a dual quaternion $\hat{q}$, first convert the rotation part of S to a quaternion q_0 using standard methods. Then, if (t_0, t_1, t_2) is the translation part of S, let $q_\varepsilon = \frac{1}{2}(it_0 + jt_1 + kt_2)q_0$. Setting $\hat{q} = q_0 + \varepsilon q_\varepsilon$ yields a dual quaternion $\hat{q}$ expressing the same rigid body transformation as S.

The dual quaternions can be uploaded as 2x4 matrices to the GPU where the skinning computations are done in the vertex shader. There the dual quaternions are blended using the vertex weights and the resulting dual quaternion is normalized and converted to a matrix, which is then used to transform the vertex' position and normal. Shader pseudocode can be found in (Kavan , et al., 2007).

For good results is also desirable to handle antipodality, which is the property of quaternions that q and $-q$ represent the same rotation. This is important to prevent skin flipping artifacts when joint angles become large, as the other rotation direction becomes shorter. A simple algorithm for this is given in (Kavan, et al., 2008), which is to pick an arbitrary pivot dual quaternion and possibly change signs of the others so that they all lie in the same hemisphere as the pivot element.

4   METHOD

Here follows a description of the test data and then the method for fitting the rigged mesh to the manikin.

4.1   Test Data

The developed method was tested within the Intelligently Moving Manikins (IMMA) (Hanson, et al., 2012) program. The IMMA manikin is comprised of a hierarchy of *links*, each having one degree of freedom and a number of characteristics such as length, min and max rotation angle, parent link, etcetera. One or more links make up a *joint*. This biomechanical skeleton is shown in Figure 1.

For a detailed description of the IMMA manikin, see (Bohlin, et al., 2011). The 3D models used for testing were taken from Poser Pro 2012 (Smith Micro Inc., 2012) and exported as COLLADA (Khronos Group, 2008) files.

4.2   Fitting the Model

The idea is to associate each bone in the rig with one of the manikin's joints, and then update the bones in the rig with the current joint transformations as the manikin moves. This will in turn deform the mesh using dual quaternion skinning as described earlier. Since the number of bones in the rig may differ from the number of joints in the manikin, a 1-to-1 mapping is generally not possible. Figure 1 shows one example of a rigged model alongside the IMMA manikin.

The bones in the rig may have a variety of names depending on the artist and program used. Therefore, a mapping between the rig's bones and the manikin's joints must be done. To avoid exposing internal details of the manikin, which also may change at any time, it is useful to have a robust intermediary format which is readily available. The H-Anim (Humanoid Animation Working Group, 2003) ISO/IEC standard provides such a useful intermediary.

H-Anim is designed to enable the exchange of humanoid characters between programs and defines a set of VRML/X3D nodes for animated models. It defines four different "level of articulation" (LOA) skeletons. The highest level, LOA 3, consists of the full hierarchy of 94 H-Anim joints. LOA 3 is likely to contain all relevant joints of the manikin, and probably many more, like joints for animating facial expressions. Equipped with a mapping file detailing which bone in the 3D model corresponds to which joint in H-Anim, the manikin program can then use this information to update the rig's bone transformations as the manikin moves, by using its own internal mapping of the manikin's joints to H-Anim joints.

In case the supplied mapping file contains bone/H-Anim joint mappings for joints not currently supported by the manikin, such bones need to be removed from the rig as they will not be updated during rendering. A simple way to do this is to loop though all vertices, and for each bone a vertex depends on that is not supported by the manikin replace that bone with its closest enabled parent bone using the same weight.

The bind pose gives rise to each M_i^{-1} of the skinning equation. Since we will be using the manikin's joint transformations to update each $B_i(t)$ as the manikin moves, ideally the inverse bind pose rotations should come from the manikin's joints. It is common to model in the familiar "T-pose" (see Figure 1), and the manikin should assume a pose as close as possible to this so that we can use the rotation part of the joint transformations to set the rotation parts of each M_i^{-1}. The translation parts may come from the 3D model's bones so that we get the exact world space positions for each bone in the rig. The more similar the poses are, the more natural the mesh deformation will be.

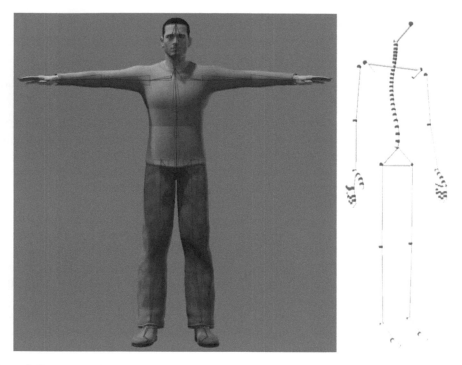

Figure 1: The 'James' figure from Poser Pro 2012 in the common T-pose (left). This figure has 53 bone segments, whereas the IMMA manikin's biomechanical skeleton consists of 72 bone segments (right).

5   RESULTS

Models were imported into IMMA and fitted with acceptable results. Figure 2 shows an imported model from Poser Pro 2012 that was fitted to a manikin of similar size and where care was taken to match the poses of the two skeleton as precisely as possible. If the two skeletons don't line up well, the mesh will be distorted, as can be seen in Figure 3.

Figure 4 shows the results of using the same mesh for manikins of different sizes. While the mesh will stretch to accommodate manikins of different sizes, this stretching will not give natural looking results if the manikins are much larger or smaller than the size of the original mesh.

Figure 5 illustrates what can happen when a bone is prolonged beyond its original length. The part of the mesh close to the elbow is not stretched as those vertices are not weighted to the wrist, but the vertices that are weighted to the wrist will follow the moving joint and lead to stretched triangles near that joint, which in turn can lead to poor mesh deformations.

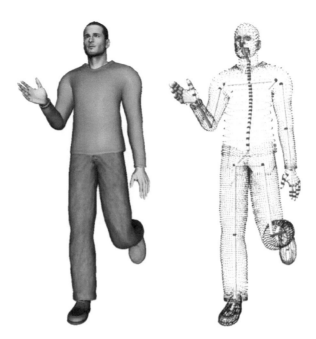

Figure 2: Poser Pro 2012's "James" model imported into IMMA and posed there. This is the result when carefully matching the poses, and when the height of the manikin closely matched that of the mesh. To the right is the same pose with the mesh rendered as points with the manikin's joints visible inside the mesh.

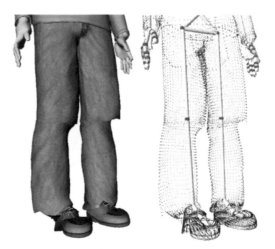

Figure 3: A close-up of the result when the two skeletons don't line up well during fitting. One problem here is that the manikin's legs were completely straight, whereas the legs in the 3D model were bent at an angle, resulting in a distorted mesh. Similar problems are visible for the wrists and ankles.

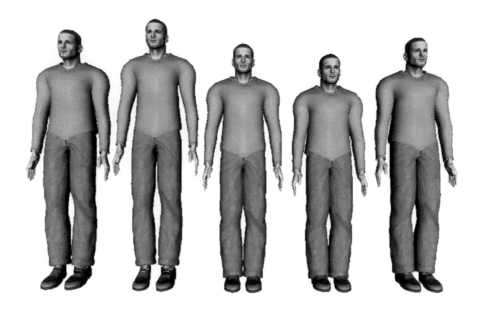

Figure 4: Five manikins of different sizes using the same base mesh.

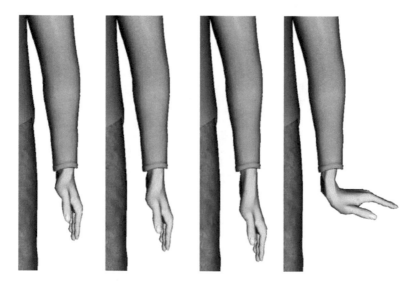

Figure 5: Prolonging the lower arm far beyond its original length leads to unnatural looking scaling and deformation as only the triangles with vertices weighted to the wrist are stretched towards that joint.

6 CONCLUSIONS AND FUTURE WORK

While the methodology described in this paper allows for the importing of external models with differing rigs, perfect mesh deformations would require an artist to model a mesh around the actual manikin skeleton. This will ensure that the joints of the manikin and the corresponding joints in the 3D model line up perfectly and the mesh deformations look natural as a result.

To ameliorate the scaling problems, one improvement would be to have a set of different sized meshes to select from. Another improvement would be to expand the H-Anim mapping table to also contain information about joint centers, such as whether the knee joint is located in the middle of the mesh or close to the edge. This information might be used to avoid artifacts such as the ones seen in Figure 3.

The method presented in this paper will lead to a rather simplistic scaling of meshes, and does not support changing the body type as the mesh is scaled, but for manikins of similar size as the original mesh it can give acceptable results.

ACKNOWLEDGMENTS

This work was carried out within The Swedish Foundation for strategic Research (SSF) ProViking II program and the Wingquist Laboratory VINN Excellence Centre, supported by the Swedish Governmental Agency for Innovation Systems (VINNOVA).

This work is a part of the Sustainable Production Initiative and the Production Area of Advance at Chalmers University of Technology.

REFERENCES

Akenine-Möller, T., Haines, E. & Hoffman, N., 2008. *Real-Time Rendering.* 3rd ed. Wellesley, MA: A.K. Peters Ltd..

Amstrup Andersen, K., 2007. *Spherical Blend Skinning on GPU,* Copenhagen, Denmark: University of Copenhagen.

Baran, I. & Popovic, J., 2007. Automatic rigging and animation of 3D characters. *ACM Trans. Graph.,* 26(3).

Bertilsson, E., Hanson, L., Högberg, D. & Rhén, I.-M., 2011a. *Creation of the IMMA manikin with consideration of anthropometric diversity.* Stuttgart, Germany, Proceedings of the 21st International Conference on Production Research (ICPR).

Bertilsson, E., Högberg, D., Hanson, L. & Wondmagegne, Y., 2011b. *Multidimensional consideration of anthropometric diversity.* Lyon, France, Proceedings of the 1st International Symposium on Digital Human Modeling.

Bohlin, R. et al., 2011. *Unified Solution of Manikin Physics and Positioning. Exterior Root by Introduction of Extra Parameters.* Lyon, France, First International Symposium on Digital Human Modeling.

Dalgaard Larsen, B., Steen Petersen, K. & Jakobsen, B., 2001. *Deformable skinning on bones,* Copenhagen, Denmark: Technical University of Denmark.

Hanson, L., Högberg, D., Bohlin, R. & Carlson, J. S., 2012. *IMMA - Intelligently Moving Manikins*. Lyon, France, Proceedings of DHM 2011, First International Symposium on Digital Human Modeling.

Humanoid Animation Working Group, 2003. *International Standard ISO/IEC FCD 19774:200x, Humanoid animation (H-Anim) specification.* [Online] Available at: http://h-anim.org/Specifications/H-Anim200x/ISO_IEC_FCD_19774/ [Accessed 28 02 2012].

Kasap, M. & Magnenat-Thalmann, N., 2009. *Sizing avatars from skin weights.* Kyoto, Japan, ACM.

Kavan , L., Collins, S., Zara, J. & O'Sullivan, C., 2007. *Skinning with Dual Quaternions.* Seattle, WA, ACM Press.

Kavan, L., Collins, S., O'Sullivan, C. & Zara, J., 2006. *Dual Quaternions for Rigid Transformation Blending,* Dublin, Ireland: Trinity College Dublin.

Kavan, L., Collins, S., Zara, J. & O'Sullivan, C., 2008. Geometric Skinning with Approximate Dual Quaternion Blending. *ACM Trans. Graph.,* 27(4), p. 105.

Khronos Group, 2008. *COLLADA - Digital Asset and FX Exchange Schema.* [Online] Available at: http://collada.org [Accessed 28 02 2102].

Kuipers, J. B., 2002. *Quaternions and Rotation Sequences: A Primer with Applications to Orbits, Aerospace and Virtual Reality.* Princeton, NJ: Princeton University Press.

Lämkull, D., Hanson, L. & Örtengren, R., 2007. The influence of virtual human model appearance on visual ergonomics posture evaluation. *Applied Ergonomics,* Volume 38, pp. 713-722.

Seo, H. & Magnenat-Thalmann, N., 2003. *An automatic modeling of human bodies from sizing parameters.* Monterey, CA, ACM.

Smith Micro Inc., 2012. *Poser Pro 2012.* [Online] Available at: http://poser.smithmicro.com/poserpro.html [Accessed 28 02 2012].

Woodland, R., 2000. Filling the Gaps--Advanced Animation Using Stitching and Skinning. In: M. DeLoura, ed. *Game Programming Gems.* Hingham, MA: Charles River Media, pp. 476-483.

CHAPTER 28

Comparison of Algorithms for Automatic Creation of Virtual Manikin Motions

Niclas Delfs[†], Robert Bohlin, Peter Mårdberg,

Stefan Gustafsson, Johan S. Carlson

Fraunhofer-Chalmers Research Centre for Industrial Mathematics,
Gothenburg, Sweden
[†]Delfs@fcc.chalmers.se

ABSTRACT

Although the degree of automation is increasing in manufacturing industries, many assembly operations are performed manually. To avoid injuries and to reach sustainable production of high quality, comfortable environments for the operators are vital. Poor station layouts, poor product designs or badly chosen assembly sequences are common sources leading to unfavorable poses and motions. To keep costs low, preventive actions should be taken early in a project, raising the need for feasibility and ergonomics studies in virtual environments long before physical prototypes are available.

Today, in the automotive industries, such studies are conducted to some extent. The full potential, however, is far from reached due to limited software support in terms of capability for realistic pose prediction, motion generation and collision avoidance. As a consequence, ergonomics studies are time consuming and are mostly done for static poses, not for full assembly motions. Furthermore, these ergonomic studies, even though performed by a small group of highly specialized simulation engineers, show low reproducibility within the group.

Effective simulation of manual assembly operations considering ergonomic load and clearance demands requires detailed modeling of human body kinematics and motions as well as a fast and stable inverse kinematics solver. The focus in this paper is to evaluate and compare the performance of different optimization algorithms. This is done by letting them find an optimum of the ergonomic function while kinematic constraints are fulfilled.

The manikin used in this work has 162 degrees of freedom and uses an exterior root. To describe operations and facilitate motion generation, the manikin is equipped with coordinate frames attached to end-effectors like hands and feet. The inverse kinematic problem is to find joint values such that the position and orientation of hands and feet matches certain target frames during an assembly motion. This inverse problem leads to an underdetermined system of equations since the number of joints exceeds the end-effectors' constraints. Due to this redundancy there exist a set of solutions, allowing us to pick a solution that maximizes a scalar valued comfort function.

Finding an optimum for the non-linear comfort function can be done with different algorithms, but the choice is not obvious. Therefore, in this paper we will implement, evaluate and compare the algorithms Interior Point and the so called Resolved Motion Rate. This will lead to a better understanding of their pros and cons, and give us the possibility to further improve the current way of generating motions for this type of digital human models.

The methods will be tested on a large number of random motions and on a set of challenging assembly operations taken from the automotive industry.

Keywords: Advanced Digital Human Modeling, Algorithms

1   INTRODUCTION

Although the degree of automation is increasing in manufacturing industries, many assembly operations are performed manually. To avoid injuries and to reach sustainable production of high quality, comfortable environments for the operators are vital, see (Falck, et al., 2010). Poor station layouts, poor product designs or badly chosen assembly sequences are common sources leading to unfavorable poses and motions. To keep costs low, preventive actions should be taken early in a project, raising the need for feasibility and ergonomics studies in virtual environments long before physical prototypes are available.

Today, in the automotive industries, such studies are conducted to some extent. The full potential, however, is far from reached due to limited software support in terms of capability for realistic pose prediction, motion generation and collision avoidance. As a consequence, ergonomics studies are time consuming and are mostly done for static poses, not for full assembly motions. Furthermore, these ergonomic studies, even though performed by a small group of highly specialized simulation engineers, show low reproducibility within the group.

Effective simulation of manual assembly operations considering ergonomic load and clearance demands requires detailed modeling of human body kinematics and motions, as well as a tight coupling to powerful algorithms for collision free path planning. The focus in this paper is to investigate the possibilities of stating a general optimization problem and then use a general local optimization algorithm. Hence, we will study the differences in pose predictions and motion generations between the common feasible direction method and the interior point method.

To describe operations and facilitate motion generation, it is common to equip the manikin with coordinate frames attached to end-effectors like hands and feet. The inverse kinematic problem is to find joint values such that the position and orientation of hands and feet matches certain target frames. This leads to an underdetermined system of equations since the number of joints exceeds the end-effectors' constraints. Due to this redundancy there exist a set of solutions, allowing us to consider ergonomics aspects, collision avoidance, and maximizing comfort when choosing one solution. This paper extends the work presented in (Bohlin, et al., 2011) and is a part of the IMMA (Intelligently Moving Manikins) project (Hanson, et al., 2011).

2 RELATED WORK

How to solve the inverse kinematic problem of a digital human have been studied extensively and different ways have been proposed. For efficiency reasons many uses a feasible direction method called Resolved Motion Rate which is a local optimization algorithm that only allows equality constraints, see (Baerlocher, 2001; Tevatia & Schaa, 2008; Chiaverin, 1997). Some need better than the local optimum found by resolved motion rate and therefore use general global optimization methods (Khwaja, 1998).

Creating a general optimization problem for the inverse kinematics and then comparing different local optimization routines has not been done before to the best knowledge of the authors.

3 MANIKIN MODEL

In this section we present the manikin model and the inverse kinematic problem which includes positioning, balance, contact forces, collision avoidance, and comfort function (Bohlin, et al., 2011).

3.1 KINEMATICS

The manikin model is a simple tree of rigid links connected by joints. Each link has a fixed reference frame and we describe its position relative to its parent link by a rigid transformation $T(\vartheta)$. Here ϑ is the value of the joint between the link and its parent. For simplicity, each joint has one degree of freedom, so a wrist for example, is composed by a series of joints and links.

To position the manikin in space, i.e. with respect to some global coordinate system, we introduce an exterior root as the origin and a chain of six additional links denoted exterior links – as opposed to the interior links representing the manikin itself. The six exterior links have three prismatic joints and three revolute joints respectively. Together, the exterior links mimic a rigid transformation that completely specifies the position of the lower lumbar. In turn, the lower lumbar represents an interior root, i.e. it is the ancestor of all interior links.

Note that the choice of the lower lumbar is not critical. In principal, any link could be the interior root, and the point is that the same root can be used though a complete simulation. No re-rooting or change of tree hierarchy will be needed.

Now, for a given value for each of the joints, collected in a joint vector $\theta = [\vartheta_1, \ldots, \vartheta_n]^T$, we can calculate all the relative transformations $T_1, \ldots, T_n$, traverse the tree beginning at the root and propagate the transformations to get the global position of each link. We say that the manikin is placed in a pose, and the mapping from a joint vector into a pose is called forward kinematics. Furthermore, a continuous mapping $\theta(t)$, where $t \in [0,1]$, is called a motion.

3.2   KINEMATICS CONSTRAINTS

In order to facilitate the generation of realistic poses that also fulfill some desired rules we add a number of constraints on the joint vector. These kinematic constraints can be defined by a vector valued function g such that

$$g(\theta) = 0 \qquad (3.1)$$

must be satisfied at any pose. Finding a solution to (3.1) is generally referred to as inverse kinematics. In the following subsection we describe in more detail two specific parts of g dealing with balance and contact forces, and positioning constraints.

3.3   BALANCE AND CONTACT FORCES

One important part of g ensures that the manikin is kept in balance. The weight of its links and objects being carried, as well as external forces and moments due to contact with the floor or other obstacles, must be considered. In general, external forces and moments due to contacts are unknown. For example, when standing with both feet on the floor it is not obvious how the contact forces are distributed between the feet. In what follows we let f denote the unknown forces and moments, so the kinematic constraint can be written

$$g(\theta, f) = 0$$

3.4   POSITIONING CONSTRAINT

Another common type of constraints restricts the position of certain links, either relative to other links or with respect to the global coordinate system. Typical examples of such constraints keep the feet on the floor, the hands at specific grasp positions and the eyes pointing towards a point between the hands. Positioning constraints are also expressed in the form (3.1).

3.5   COMFORT FUNCTION

When generating realistic manikin poses and motions, it is essential to quantify the ergonomic load. To do so, we introduce a scalar comfort function
$$h(\theta, f)$$
capturing as many ergonomic aspects as we desire. The purpose is to be able to compare different poses in order to find solutions that maximize the comfort.

The comfort function is a generic way to give preference to certain poses while avoiding others. Typically h considers joint limits, forces and moments on joints, magnitude of contact forces etcetera. Note that it is straightforward to propagate the external forces and moments and the accumulated link masses trough the manikin in order to calculate the load on each joint. The joint loads are key ingredients when evaluating poses from an ergonomic perspective (Westgaard R & Aarås, 1985).

It seems like fairly simple rules can be very useful here. Research shows that real humans tend to minimize the muscle strain, i.e. minimize the proportion of load compared to the maximum possible load (Rasmussen, et al., 2003), so by normalizing the load on each joint by the muscle strength good results can be achieved. A suitable comfort function may also depend on the motion speed and how many times the motion will be repeated.

It is fitting to let h include the collision avoidance, which is important for industrial simulations. It was shown in (Bohlin, et al., 2012) that a potential function which depends on distance makes the manikin able to avoid most collisions in a desirable way.

4   OPTIMIZATION

In this paper we use a unified treatment of balance, contact forces, position constraints, ergonomics and collision avoidance. Often in practice, the number of constraints is far less than the number of joints of the manikin. Due to this redundancy there exist many solutions, allowing us to consider ergonomics aspects and maximizing comfort when choosing one solution.

Let $x^T = [\theta^T, f^T]$ be the unknowns, i.e. the joint variables and the unknown forces and moments. Then we formulate the problem as follows:

$$\begin{cases} maximize & h(x) \\ while & g(x) = 0 \end{cases} \qquad 4.1$$

4.1   RESOLVED MOTION RATE (RMR)

Here we present a feasible direction solver called Resolved Motion Rate, which is the most common inverse kinematics solver for digital humans (Tevatia & Schaa, 2008). It solves the nonlinear optimization problem by iteratively linearizing the problem. Let x_0 be the current state. Then, for small Δx,

$$g(x_0 + \Delta x) \approx g(x_0) + J(x_0)\Delta x$$

Where $J(x_0) = \frac{\partial g}{\partial x}$ is the Jacobian at x_0. In order to satisfy $g(x_0 + \Delta x) = 0$ while increasing $h(x)$, we project $\nabla h(x_0)$ onto the null space of J. Here we use a Penrose pseudo inverse to project the gradient onto the null space and therefore get the step

$$\Delta x = -J^\dagger g(x_0) + \lambda(I - J^\dagger J)\nabla h(h_0)$$

where λ is a scalar defining the step length, calculated by doing a line search.

Using the pseudo inverse to project to the null space creates instabilities in the solver when J is close to a singularity. This instability can be reduced by perturbating the Jacobian matrix away from the singularity (Buss, 2009). Let $\gamma > 0$ be the damping factor, then the new step is

$$\Delta x = -J^T (JJ^T + \gamma^2 I)^{-1} g(x_0) + \lambda(I - J^\dagger J)\nabla h(h_0)$$

It can be problematic to find a γ that is good in all situations because it should be large close to a singularity to hinder oscillations but zero when J far away from a singularity to give fast convergence.

4.2   PRIMAL-DUAL INTERIOR POINT METHOD (IPM)

The term interior point method is used to describe any method solving a sequence of unconstrained minimization problem using a decreasing multiplier μ to find a local solution within the interior of the feasible region of the problem (Doyle, 2003). The problem (4.1) would then be

$$minimize \; -h(x) - \mu_k \sum_{i=1}^{m} ln\,(\mu_k - |g_i(x)|)$$

where μ_k is from a decreasing sequence $\mu_1 > \cdots > \mu_\infty > 0$. Classic methods for solving this sort of unconstrained problems converge slowly because of the ill-conditioning of the hessian close to the solution. The problem of slow convergence is mitigated by minimizing primal variables at the same time as the dual variables which are dependent on μ_k. This minimization is done for each μ_k and is done by the Newton method.

5   RESULTS

In order to demonstrate the difference between the RMR method and the IPM, we set up three cases where the first is to solve a set of random postures, the second and third are industrial cases. The RMR implementation is presented in (Bohlin, et al., 2011) whereas the IPM implementation is taken from the Ipopt library (Wächter & Biegler, 2006). We have used the same settings for scaling, m, radians, N, Nm for the quantities length, angles, force and moment respectively for all cases. The comfort function penalizes joint values near the limits unless otherwise stated. Furthermore, in the start pose, all joint angles are zero, see Fig. 1.

5.1   RANDOM POSES

To validate the solvers' applicability we randomize targets for the hands and let the solvers find the poses. The poses will be split between success and failure, where success is that the kinematic constraints are fulfilled. The target positions for the hands will be taken randomly in space as long as the hands does not cross each other and that they are not further apart than that a restricting box can fit around them; this box is used in industry as the accepted workspace.

We will use a restricting box from (Scania, 2008) which is

- Upper bound at shoulder.
- Lower bound at knuckles when the arms are relaxed at the sides.
- Left and right bounds are at elbows when both arms are extended at maximum to the sides.
- Back bound at the body's vertical line.
- Front bound at the knuckles when the arms are extended at maximum, forward.

The target rotations for hands should deviate less than a certain angle from a straight forward position. Furthermore, the manikin is free to move on the floor but heals and toes need to be on the floor. The tests are done with three different max-deviations. For each deviation level, 1000 problems are created and then solved by the solvers. Only the successful solutions are counted for the average iterations. The result can be seen in the tables 1 and 2 which show the success difference between RMR and IPM.

Figure 1. The starting pose of the manikin, all angles are set to zero.

Deviation	Successes	Average iterations
$\pi/8$	965	194
$\pi/2$	957	226
π	884	328

Deviation	Successes	Average iterations
$\pi/8$	994	345
$\pi/2$	989	349
π	981	326

Table 1. Success rate for resolved motion rate to the left and the interior point to the right. Hands are free to rotate in the sagittal plane (y-axis).

Deviation	Successes	Average iterations
$\pi/8$	643	512
$\pi/2$	559	585
π	352	766

Deviation	Successes	Average iterations
$\pi/8$	953	423
$\pi/2$	886	481
π	833	612

Table 2. Success rate for resolved motion rate to the left and the interior point to the right.

5.2 ASSEMBLY OPERATION

The second case is a real assembly situation from the automotive industry and the goal is to generate a feasible motion for the manikin assembling the tunnel bracket into the car, while the first is a simplification of this.

In both test cases we have added kinematic constraints to hands and feet. The hands should match the grasp positions on the tunnel bracket with five degrees of freedom locked – the palms are allowed to rotate around the surface normal. The feet are allowed to slide on the floor and rotate around the z-axis – the remaining three degrees of freedom are locked. The resulting pose predictions for the simplified assembly operation can be seen in Fig. 2 and Fig. 3.

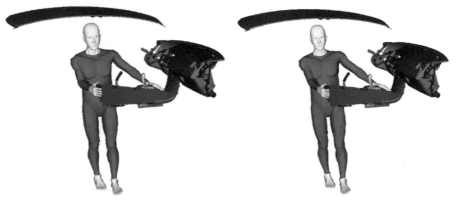

Figure 2. Using RMR for 120 iterations   Figure 3. Using IPM for 214 iterations

5.3 COLLISION FREE PATH

In the last case we add collision avoidance and path following to the tunnel bracket assembly. To be able to fulfill the need of convex second order derivatives in the IPM we decomposit the relevant parts of the car mesh with a series of convex meshes.

We let the manikin follow a path, that is guaranteed to be collision free, for the tunnel bracket and is calculated by industrial path solutions (Fraunhofer-Chalmers Research Centre, 2012). The solutions from the two solvers where similar and Fig. 4 shows the solutions. Furthermore, the number of iterations was increased from 1034 to 18830 when changing from RMR to IPM.

Figure 4. Snapshots of the solution for the tunnel bracket assembly with collision avoidance.

6   DISCUSSION

In order to use a path planner on a manikin the inverse kinematics for the manikin needs to be fast and reliable. A path planner may need many thousands of pose predictions to generate one collision free path. Usually a deterministic path planner only uses small steps from a know position to find its way.

From the results we see that when the initial guess is good and the deviation is small, the RMR solves the problems with much less iterations than the IPM and have almost as good success rate. But when the hand constraint does not allow rotation the RMR solves much fewer problems than the IPM and needs more iterations, which gets more obvious when deviation is accepted up to π radians.

Hence, it seems that RMR works well for problems with equality constraints and with good initial guesses, whereas when the initial guess is further away the IPM is to prefer. When doing pose predictions for short distances both methods find the same local minimum as seen in cases one and two but with a larger number of iterations needed for IPM. This is extremely apparent on case two where the IPM needs almost 19 times more iterations to calculate the motion. This seems to be because the IPM always needs to solve the relaxation variables even when the starting point is in optimum.

7   CONCLUSION

We used both the RMR and the IPM to solve the inverse kinematic system and compared the success rate and the number of iterations needed for solving the problems. It could be seen that the RMR solved almost all problems with good initial guesses and used fewer iterations. Hence, when creating an assembly motion the IPM needed almost 19 times as many iterations for the same solution. But with bad initial guesses RMR had both much lower success rate and needed more iterations than the IPM. We can therefore conclude that

- IPM is more robust and more general and therefore preferable for single pose predictions
- RMR efficient on easy problems and therefore preferable for path following

294

Both solvers were successfully tested in a challenging assembly operation taken from the automotive industry.

ACKNOWLEDGEMENT

This work was carried out within The Swedish Foundation for Strategic Research (SSF) ProViking II program and the Wingquist Laboratory VINN Excellence Centre, supported by the Swedish Governmental Agency for Innovation Systems (VINNOVA).

This work is part of the Sustainable Production Initiative and the Production Area of Advance at Chalmers University of Technology

REFERENCES

Baerlocher, P., 2001. *Inverse Kinematics Techniques For The Interactive Posture Control Of Articulated Figures.* Lausanne: EPFL.
Bohlin, R. et al., 2011. Unified Solution of Manikin Physics and Positioning. Exterior Root by Introduction of Extra Parameters. *First International Symposium on Digital Human Modeling.*
Bohlin, R. et al., 2012. *Automatic Creation of Virtual Manikin Motions Maximizing Comfort in Manual Assembly Processes,* s.l.: To Appear.
Buss, S. R., 2009. *Introduction to inverse kinematics with jacobian transpose, pseudoinverse and damped least squares methods,* San Diego: University of California.
Chiaverin, S., 1997. Singularity-Robust Tast-Priority Rdundancy Resolution For Real-Time Kinematic Control Of Robot Manipulators. *IEEE Transactions on Robotics and Automation.*
Doyle, M., 2003. *A Barrier Algorithm For Large Nonlinear Optimization Problems,* Stanford: Stanford University.
Falck, A. C., Örtengren, R. & Högberg, D., 2010. The Impact of Poor Assembly Ergonomics on Product Quality: A Cost–Benefit Analysis in Car Manufacturing. *Human Factors and Ergonomics in Manufacturing & Service Industries, vol. 20.*
Hanson, L., Högberg, D., Bohlin, R. & Carlson, J. S., 2011. IMMA–Intelligently Moving Manikins. *First International Symposium on Digital Human Modeling,* Issue June 2011.
Khwaja, A. A., 1998. Inverse Kinematics Of Arbitrary Robotic Manipulators Using Genetic Algorithms. *Advances in robotic kinematics: Analysis and Control.*
Rasmussen, J. et al., 2003. AnyBody – A Software System for Ergonomic Optimization. *World Congress on Structural and Multidisciplinary Optimization.*
Scania, 2008. *SES (Scania ErgonomiStandard),* Skövde: Scania.
Tevatia, G. & Schaa, S., 2008. Efficient Inverse Kinematics Methods Applied to Humanoid Robotics. *CLMC Technical Report.*
Westgaard R, H. & Aarås, A., 1985. The Effect of Improved Workplace Design on the Development of Work-related Musculo-skeletal Illnesses. *Applied Ergonomics, vol. 16.*
Wächter, A. & Biegler, L. T., 2006. On the Implementation of a Primal-Dual Interior Point Filter Line Search Algorithm for Large-Scale Nonlinear Programming. *Mathematical Programming 106,* pp. 25-57.

CHAPTER 29

An Algorithm for Shoe-Last Bottom Flattening

Xiao Ma[1], Yifan Zhang[1], Ming Zhang[2], Ameersing Luximon[1*]

[1]Institute of Textiles and Clothing
The Hong Kong Polytechnic University
Hung Hom, Hong Kong
[*]tcshyam@polyu.edu.hk

[2]Health Technology and Informatics
The Hong Kong Polytechnic University
Hung Hom, Hong Kong

ABSTRACT

The alignment of the foot and shoe-last is the first as well as the most important step in shoe-last fit evaluation. The traditional fit evaluation by comparing the lengths or widths of the foot and shoe-last, which are measured manually, may be costly and time consuming. It is a trend to evaluate the fit of a shoe-last by computer based on the 3D methods. In this way, efficient and accurate fitting result can be achieved through comparing several dimensions, even every point of the 3D shoe-last model with the foot model. The correct and precise alignment can ensure that the following fitting procedures and the final result are significant and effective. This paper presented an algorithm for shoe-last bottom flattening.

Keywords: shoe-last, alignment, bottom flattening

1   INTRODUCTION

Footwear fit is one of the most important considerations for customers (Savadkoohi & Amicis, 2009). For footwear products, manufacturers are facing quite a challenge to provide consumers with good fitting shoes (Ruperez et al., 2010). There are close relationships between shoe fit and foot problems. Clinical reports of foot problems such as blistering, chafing, black toes, bunions, pain, and tired feet are evidence of poor fitting shoes (Goonetilleke & Luximon, 1999). The chief resident of New York University Hospital Ran Schwarzkopf and his colleagues have done the study on the proportion of adults who are unaware of their own shoe size in 3 different New York City populations (Schwarzkopf, Perretta, Russell, & Sheskier, 2011). Their findings suggested that proper footwear sizing is lacking among a large proportion of our patients and that an adequate shoe size can be achieved with proper counseling. Improper fit can lead to pain, functional limitations, and falls in the elderly population (Menz & Lord, 1999). Properly fit footwear may not only provide the comfortable and satisfactory feeling to the customers, but also the right protection and support to the feet when it's in rest or motion. Rossi stated that many foot problems and diseases are known to be a result of poor fitting shoes (Rossi, 1988).

Goonetilleke and Luximon (2000) used principal component analysis to determine the foot axis so that shoe-lasts that match feet can be produced, resulting in a good fit. Some researchers have proposed methodologies to enable the design of shoe-last with different toe type, heel height and custom shoe-last using the existing shoe-last design standards and the sections from existing shoe-lasts, design tables and relationship equations which included comfort and fit aspects as well as design aspect (Luximon & Luximon, 2009).

Since there are great interests in manufacturing and design automation in footwear industry, the evaluation of footwear fit can be represented by comparing the 3D shoe-last model with the 3D foot model. This paper proposed an algorithm for shoe-last bottom flattening to realize the alignment between the shoe-last and foot model.

2   ALGORITHMS

A shoe-last is the representation of the inner shape of a shoe. So a foot that is wearing a shoe can be simulated by matching the 3D models of the foot and the shoe-last correctly. In order to control the matching on the directions of length and width, the bottom centerline and the tread point were selected as the landmarks. The tread point is the intersection point of the bottom centerline and the line through the lateral and medial metatarsophalangeal joints (MPJ). Therefore, the aim of the methodology is to find the bottom centerlines and the tread points of the foot and the shoe-last. Normally, the standard shoe-last is size 36. Its pternion point and bottom centerline are set on the special position in the coordinate system. Additionally, the point cloud data consists of sections which are parallel to one coordinate plane as well as perpendicular to another. In this case, the X axis is the bottom centerline; the pternion point is on the Z axis; at the same time, all sections are parallel to plane YOZ and perpendicular to plane XOY. So the bottom centerline of the shoe-last can be obtained easily. But to compare the difference between a foot and a shoe-last by a computer, the flattened bottom of the shoe-last should be obtained as it contained many dimensions of it. In this paper, a set of algorithms was proposed to get the centerline and the tread point of a foot, and flattened bottom of a shoe-last for the alignment, which is the first and primary proceeding in 2D or even 3D fit evaluating between a foot and a shoe-last.

2.1 Scanning and Orientation

A female foot, size 36, was chosen as the subject of this research. The personal information of the subject is shown in Table 1. Firstly, foot length and width were measured manually by using Witana's method (Witana, Xiong, Zhao, & Goonetilleke, 2006). To compare the manual measurements (MM), these two values were obtained again from the scanner software (Software Measurements, denoted as SM) after the foot was scanned by 3D Foot Laser Scanner II (Frontier Advanced Technology Ltd). Table 2 shows the results of the measurements.

Table 1 The subject's information

Characteristics	Information
Gender	Female
Age	25
Height	156
Weight	43.5

Table 2 The measurements of the foot

Dimensions	MM (mm)	SM (mm)	Average (mm)
Foot length	228.3	228.99	228.65
Foot width	88.1	88.74	88.42

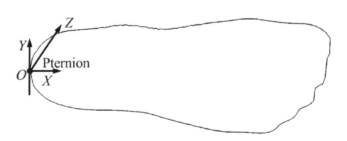

Figure 1   The orientation of the foot

Orientation is to define the coordinate system which will be used to the algorithms later. The foot length direction is defined as X axis, while the foot width direction is defined as Y axis, and Z axis went along the height direction from the foot bottom to the ankle (Figure 1). The point cloud data of the foot and shoe-last are both represented by N points, while $N = N_s \times n$ (N_s was the points number of every section, and n represents the number of sections, which was the ratio of the foot or shoe-last length L and the interval distance between sections D_i, $n = \frac{L}{D_i}$). All of these parameters were optional in the scanner software so that the data volume of every model can be controlled as required.

2.2   Extract the Bottom Contour

Once the coordinate system was established, the bottom contour of the foot or shoe-last should be extracted before detecting the centerline. The algorithm of the foot contour extraction is based on the projection on plane XOY. As sections of the 3D foot point cloud data are perpendicular to plane XOY, the XOY projection should consist of several straight lines with the same interval. Thus, the contour can be obtained when two extreme points of every line (points with minimum and maximum Y coordinate) were found.

The bottom contour of the shoe-last is different and not easy to be obtained comparing with the foot. First of all, the measuring of the foot for shoe-last design is based on the contour, not the print outline, for example, the definition of foot width is the distance between the minimum and maximum points on Y direction. Whereas, the design and measuring of a shoe-last is started from the bottom pattern that resembles the foot print. So, the useful bottom contour of a shoe-last is not the projection directly, but the shoe-last print actually. On the other hand, due to different heel heights of shoe-lasts, the projection will be distorted, which will interfere with the future processing based on the alignment, such as preliminary size evaluating based on the lengths of foot and shoe-last. The bottom contour should be extracted accurately for flattening later so that the real bottom pattern can be obtained correctly. Accordingly, an algorithm for extracting the bottom contour and its outline was proposed as follows:

(1)   Get section $S_i(x_i, y_i, z_i)$, $i = 1, 2, ..., n$. The section parallels to the plane YZ so as to all of its points have the same X value. Thus, the following processes will just be considered in 2D space YOZ.

(2)   Translate the origin point to the geometric center of the section $O_{ci}(y_{ci}, z_{ci})$. Where $y_{ci} = \frac{|y_{imax} - y_{imin}|}{2}$, $z_{ci} = \frac{|z_{imax} - z_{imin}|}{2}$, y_{imax}, y_{imin}, and z_{imax}, z_{imin} are the two most boundary points of S_i on the Y and Z direction respectively. The new coordinate system can be named $Y_c O_c Z_c$ as showing in Figure 2(a).

(3)   Calculate the discrete curvatures on every point in the third and fourth quadrant respectively, then the possible inflexion points set I_3 of the points in the third quadrant and set I_4 in the fourth quadrant can be found. The curvature is the amount by which a geometric object deviates from being flat, or straight in the case of a line. The greater the curvature on certain point of a curve, the bigger of the angle change near the point is. The discrete curvature K_i on point

s_i can be calculated by Formula (1).

$$K_i = \frac{H_i}{L_i} \tag{1}$$

Where $H_i = \frac{|k(y_i - y_{i-1}) - (z_i - z_{i-1})|}{\sqrt[2]{k^2+1}}$, is the distance from point s_i to line $s_{i-1}s_{i+1}$; $L_i = \sqrt[2]{(y_{i+1} - y_{i-1})^2 - (z_{i-1} - z_{i+1})^2}$, is the distance between points s_{i-1} and s_{i+1}; and k is the slope of line $s_{i-1}s_{i+1}$. Figure 2(b) shows the procedure of calculating the curvature on one point.

(4) The sets I_3 and I_4 contained N_c ($N_c < N_s$) points that have the greatest curvatures in all the section points were selected. These sets must contain the inflexion points of the third and fourth quadrant respectively.

(5) Transform the Cartesian coordinates of I_3 and I_4 to polar coordinates $P_3(\gamma, \theta)$ and $P_4(\gamma, \theta)$.

(6) Find the inflexion points p_3 and p_4 with the greatest γ of I_3 and I_4 respectively. Figure 2(c) shows an example. s_1, s_2 and s_3 were the possible inflexion points in I_3. Finally, s_1 was selected as the inflexion point in the third quadrant.

(7) Transform their polar coordinates back to Cartesian coordinates.

(8) Find the inflexion points of all the sections that constitute the contour of the shoe-last bottom. Then, all the bottom point can be found.

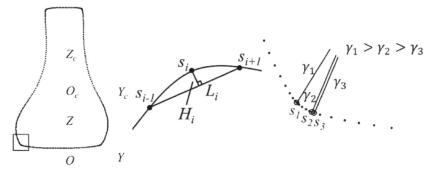

(a) A shoe-last section   (b) Calculating the curvature   (c) Finding the inflexion point
on point s_i

Figure 2   Extracting the bottom based on the inflexion points of the section

In this method, the inflexion point was not determined by selecting the point with the greatest curvature value directly. Actually, the shape near the corner, seeing the magnified part in Figure 2(c), was not very sharp. That's to say, the differences between the curvatures on the points near this area were not noticeable enough to judge that which the real inflexion point was. Thus, the possible inflexion points of every quadrant were selected at first. Then the inflexion points were determined by value γ in the polar space.

Since the heel height and toe spring, the bottom extracted was three dimensional. In order to compare it with the foot model, it needed to be flattened in a plane. The flattened bottom obtained by the proposed algorithm was the same as the bottom pattern that was the initial drawing of shoe-last design, which also contained most of the dimensions on both the length and width directions. The flattening was executed both on the length (X direction) and width direction (Y direction). On the width direction, as showing in Figure 3(a), the length between the flattened contour point $p_{fb}(x_{fb}, y_{fb}, z_{fb})$ and point O' on the bottom centerline was equal to the real length of the original bottom. So $y_{fb} = \sum l_s$ was determined by flattening on the width direction.

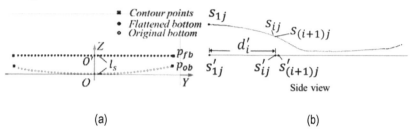

(a)   (b)

Figure 3   Flatten the bottom section on the width direction (a) and length direction (b)

On the other hand, as sections of the shoe-last were paralleled each other, the real length of the bottom was the sum of the distances between every two neighbor sections. Since the process on the length direction can only decide x_{fb}, the flattening algorithms on the length and width directions can be executed at the same time accordingly.

Hypothetically, $s_{i1}(x_{i1}, y_{i1}, z_{i1})$, $s_{i2}(x_{i2}, y_{i2}, z_{i2})$ and $s_{(i+1)1}(x_{(i+1)1}, y_{(i+1)1}, z_{(i+1)1})$, $s_{(i+1)2}(x_{(i+1)2}, y_{(i+1)2}, z_{(i+1)2})$ were the contour point pairs of section S_i and S_{i+1} respectively (Figure 3(b)). In the standard aligned shoe-last model, line $s_{i1}s_{i2}$ was parallel to line $s_{(i+1)1}s_{(i+1)2}$ as well as the Y axis. So, the distance $d_{i(i+1)}$ between line $s_{i1}s_{i2}$ and line $s_{(i+1)1}s_{(i+1)2}$ was $(x_{i1} - x_{(i+1)2})$. Then, the distance set $D = \{d'_i | i = 2, ..., n\}$ can be obtained, where d'_i was the distance between line $s_{i1}s_{i2}$ and line $s_{11}s_{12}$ that also was the rear point of the shoe-last bottom. The pair of points $s'_{ij}(x'_{ij}, y'_{ij}, z'_{ij}) | (i = 1,2, ..., n; j = 1,2)$ of the flattened bottom can be obtained from the original point $s_{ij}(x_{ij}, y_{ij}, z_{ij})$ by the Formula (2):

$$\begin{cases} x'_{ij} = x_{11} + d'_i \\ y'_{ij} = y_{ij} \\ z'_{ij} = 0 \end{cases} \tag{2}$$

The flattened bottom is also called bottom pattern, which is the first step in shoe-last design based on several dimensions of the foot. It's the reverse engineering of shoe-last design. Thus, this algorithm can be applied in shoe-last measuring and CAD, and so on.

2.3   Find the Bottom Centerline and Tread Point

A foot that is composed of bones, muscles, tissues, etc. like other organs of human body, is flexible and unsymmetrical. It is hard to find a centerline that can divide the foot into two parts equally, except near the heel area. The back part of the heel area is correspondingly regular and symmetric, which is from the pternion point to the heel point p_{heel} (about 18% of foot length based on the statistical data of Chinese)(Qiu, Wang, Fan, & Jin, 2006). The foot length (OA in Figure 4) was defined as the distance along the longitudinal axis from pternion to the tip of the longest toe (Xiong, Zhao, Jiang, & Dong, 2010). Although there is no appropriate approach to find this centerline directly, it still can be obtained through finding the heel centerline, which has an angle φ with the bottom centerline at the heel point p_{heel}. Firstly, the heel centerline was found using the algorithm proposed by Luximon (Luximon, 2001). Then, the contour points were rotated by angle $\pm|\varphi|$ based on point p_{heel}. The negative sign means rotating in a clockwise direction when it is a left foot, contrarily, the positive sign means counterclockwise for right foot. The X axis is the bottom centerline of the foot (Figure 4). The angle φ can be determined by finding the maximum length along X-axis for different angle $(\varphi = -n°, -n + 1, ..., 0, 1°, 2° ..., n°)$ and choosing the maximum one to determine φ.

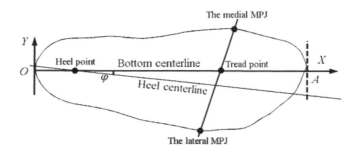

Figure 4   The example of finding the foot bottom centerline and the tread point

Once the bottom contour and bottom centerline have been found, the process of locating the tread point is correspondingly simple, since the algorithms of the foot and shoe-last are a little different.

As showing in Figure 4, the tread point of the foot is the intersection point of the bottom centerline and the line between the medial MPJ and the lateral MPJ. The bottom centerline of a standard shoe-last is just on the X axis as explained in advance, while its tread point is the point with the most minimum Z value.

2.4   Match the Foot and Shoe-last

Once matching the foot bottom contour and shoe-last bottom contour, some of their dimensions can be obtained, such as lengths, widths, etc. In this paper, the subject's foot length and width were 226.40mm and 90.74mm originally, while they were 228.80mm and 88.47mm respectively obtained by applying the algorithm, which were equivalent to the corresponding measurements of MM.

Figure 5 shows the matching of the foot bottom, the flattened shoe-last bottom and the shoe-last outline obtained by the projection on *XOY* plane.

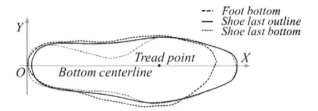

Figure 5   The matching of the foot bottom, the flattened shoe-last bottom and the shoe-last outline

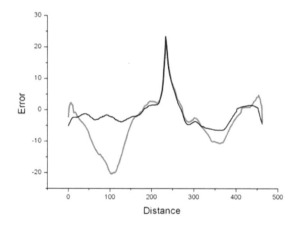

Figure 6   The dimensional errors (mm) between the foot bottom and the flattened shoe last bottom (gray); and the dimensional errors between the foot bottom and the shoe last outline (black). The x-axis is the distance (mm) from the foot contour point to the pternion point along the foot contour in the clockwise direction.

3   DISCUSSION

This paper discussed an algorithm for extracting and flattening the shoe-last bottom as well as a set of algorithms together for matching the foot and shoe-last. The matching of the foot bottom and the flattened shoe-last bottom is simulating the bottom pattern design of a master shoe-last or customized shoe-last. Thus, some basic comparisons and fit evaluation can be executed based on the matching. First of all, it can be used to preliminary fit evaluation by comparing 2D dimensions, such as the length, width, and etc. For example, Figure 6 shows the dimensional errors between the foot bottom contour and the shoe-last outline (a) or the flattened shoe-last bottom (b). These errors can predict the tightness (negative value) and the looseness (positive value) of the shoe made by the shoe-last (Goonetilleke, Luximon, & Tsui, 2000; Witana, Goonetilleke, & Leng, 2003).

Moreover, the alignment can be also applied to 3D fit evaluation. There are a lot of research on 3D fit evaluation or measurement of feet and shoe-lasts, and all of them were started from the alignment mostly. Thus, it is necessary to standardize its procedure to make sure that the alignment has highly simulated the situation that a foot was worn in the shoe made by the shoe-last. Otherwise, the results of the evaluation would be inaccurate or even invalid.

REFERENCES

Frontier Advanced Technology Ltd. Handbook of 3D Foot Laser Scanner II. Hong Kong, http://www.frontiertech.com.hk/english/products_FootLaserScanner.html .

Goonetilleke, R. S., & Luximon, A. (1999). Foot Flare and Foot Axis. *Human Factors,* 41(4), 596-607.

Goonetilleke, R. S., Luximon, A., & Tsui, K. L. (2000). *The Quality of Footwear Fit: What We Know, Don't Know and Should Know.* Paper presented at the Human Factors and Ergonomics Society Annual Meeting Proceedings, Proceedings 2, Multiple-Session Symposia.

Luximon, A. (2001). *Foot Shape Evaluation for Footwear Fitting.* Unpublished PhD Thesis, Hong Kong University of Science and Technology, Hong Kong.

Luximon, A., & Luximon, Y. (2009). Shoe-last Design Innovation for Better Shoe Fitting. *Computers in Industry,* 60(8), 621-628.

Menz, H. B., & Lord, S. R. (1999). Footwear and postural stability in older people. *J Am Podiatr Med Assoc,* 89(7), 346-357.

Qiu, L., Wang, Z., Fan, K., & Jin, Y. (2006). *The shoe last design and manufacture*: China Textile&Apparel Press.

Rossi, W. A. (1988). The Futile Search for the Perfect Shoe Fit. *Journal of Testing and Evaluation,* 16(4), 393-403.

Savadkoohi, B. T., & Amicis, R. (2009, December 10-12). *A CAD System for Evaluating Footwear Fit.* Paper presented at the Multimedia, Computer Graphics and Broadcasting, the 1st International Conference, MulGraB 2009, Held as Part of the Future Generation Information Technology Conference, FGIT 2009, Jeju Island, Korea.

Schwarzkopf, R., Perretta, D. J., Russell, T. A., & Sheskier, S. C. (2011). Foot and shoe size mismatch in three different New York City populations. *Journal of Foot & Ankle Surgery,* 50(4), 391-394.

Witana, C. P., Goonetilleke, R. S., & Leng, J. J. (2003). *A 2D Approach for Quantifying Footwear Fit* Paper presented at the The IEA 2003 XVth Triennial Congress, Seoul, Korea.

304

Witana, C. P., Xiong, S. P., Zhao, J. H., & Goonetilleke, R. S. (2006). Foot Measurements from Three-dimensional Scans: A Comparison and Evaluation of Different Methods. *International Journal of Industrial Ergonomics,* 36(9), 789-807.

Xiong, S. P., Zhao, J. H., Jiang, Z. H., & Dong, M. (2010). A computer-aided design system for foot-feature-based shoe last customization. *International Journal of Advanced Manufacturing Technology,* 46(1-4), 11-19.

Section VII

Validation for Human Interaction in Various Consumer, Ground Transport and Space Vehicle Applications

CHAPTER 30

Occupant Calibration and Validation Methods

Joseph A. Pellettiere and David M. Moorcroft

Federal Aviation Administration
Washington DC, USA

ABSTRACT

The Federal Aviation Administration has a number of standards and regulations that are designed to protect occupants in the event of a crash. Compliance with these regulations is described in the Code of Federal Regulations 14 CFR 25.562 for transport category aircraft, with similar regulations for other types of aircraft in parts 23, 27, and 29. Compliance with these regulations is typically met through physical testing; however AC 20-146 describes a method of compliance through the use of analytical methods. In order to successfully apply AC 20-146, a virtual Anthropometric Test Device (v-ATD) is needed for the aviation environment. This v-ATD must achieve a level of calibration that is specific to its intended use. Previous models had poor performance for some configurations, such as a typical lap belt only configuration. A rigorous methodology is needed to assess the level of correlation between a simulation and test data. The v-ATD calibration methods should be the first step and can be considered the corollary to the physical calibration methods that an ATD would be subjected to prior to certification testing. Specific performance objectives must be met in order for the v-ATD to support a simulation that meets the certification requirements and will affect the specific uses of the model. These performance objectives maintain flexibility at the system level to allow customization to the detailed design, but must provide some hard guidelines in order for the analysis to be useful

Keywords: Occupant Validation, Calibration, Modeling and Simulation

INTRODUCTION

The Federal Aviation Administration (FAA) has a number of standards and regulations that are designed to protect occupants in the event of a crash. As part of these regulations, dynamic testing and occupant injury assessment have been required for seats in newly certified aircraft since the adoption of Title 14 of the Code of Federal Regulations (CFR) Part 25.562, and similar regulations in Parts 23, 27, and 29 (USC Title 14). There are two tests that must be conducted. Test 1 is a primarily vertical impact test with a minimum impact velocity of 35 fps with peak acceleration of 14 G's and an impact angle of 30 degrees off vertical. Test 2 is a frontal test with a minimum impact velocity of 44 fps with peak acceleration of 16 G's and an impact angle of 10 degrees yaw. Both tests have limits on the rise time with associated injury metrics that must be met before a seat is certified for use in aviation. Other aircraft categories have similar requirements. The injury metrics include limits on lumbar and leg loads, Head Injury Criterion (HIC), shoulder strap loads when used, and requirements that belts remain in place. For complete details, please see the applicable regulations. While compliance with these regulations is typically met through dynamic testing, the rule as stated does allow for computer modeling and simulation to be conducted in support of a dynamic test program.

§§ 25.562(b) states that "Each seat type design approved for crew or passenger occupancy during takeoff and landing must successfully complete dynamics tests or be demonstrated by rational analysis based on dynamic tests of a similar type seat…" It is the demonstration by rational analysis that led to the development of AC 20-146 and the ability to use M&S in certification programs. AC 20-146 (FAA, 2003) identifies three main groups to which the use of M&S may be applicable, these include: applicants including the seat as part of the aircraft type design, seat manufacturers, and seat installers. Two possible uses for M&S in certification programs include: the establishment of the critical seat installation or configuration which will be tested, potentially reducing the number of required certification tests or to demonstrate compliance of changes made to an existing certified seat system. M&S is not limited to just the certification of the seat. The uses of M&S in aircraft seat design and evaluation are numerous and begin in the early phases of any development program. Computer aided engineering tools are common and lead to the direct development of M&S of prototype designs. The use of M&S here allows tradeoffs to be conducted, evaluation of injury risks, investigation of potential failure areas, and the selection of successful design parameters. M&S in these early phases will help develop the test plans to successfully validate a system.

Any computational occupant model used as part of an analysis must undergo a proper validation (Pellettiere et al, 2011). To support AC 20-146, this validation must be supported by dynamic testing. The steps necessary for conducting this validation can be separated into several process steps (Pellettiere and Knox, 2011) with each step assigned a role and responsibility (Pellettiere and Knox, 2004). This step wise process leads to a model with different levels of validation during development. The Analytical Tools Readiness Level (ATRL) describes these levels (Harris and Shepard, 2009) (Table 1). The ATRL was developed based upon the Department of Defense Technology Readiness Level for research and development where a system becomes ready for advanced development and field testing once it has reached level 6. An ATRL of 6 would be a reliable model that is calibrated and supported with testing. An ATRL of 8 could be the future application of M&S for certification programs to reduce the number of necessary tests.

Table 1 Analytical tools readiness levels

Level	ATRL
1	Back of envelope analysis with no verification or validation – no correlation data
2	Analysis correlates roughly with simple tests
3	Analysis provides some guidance but prediction track record is poor (coarse qualitative trade studies possible)
4	Predicts trends in a controlled environment, but not reliable in a relevant environment
5	Predicts trends reliable enough to discriminate between designs (good qualitative trades possible)
6	Trends are very reliable, but absolute values must still be calibrated with testing (quantitative trades possible after calibration)
7	Process predicts absolute value, no development testing is required (quantitative trades possible)
8	Process predicts absolute value, certification/qualification authorities accept analytical results with reduced testing
9	Certification/qualification authorities accept analytical results without testing

VERIFICATION AND VALIDATION

The use of computer models in seat certification requires the creation of accurate models and the assessment of that accuracy. "Verification and validation (V&V) are the processes by which evidence is generated, and credibility is thereby established, that the computer models have adequate accuracy and level of detail for their intended use. (ASME, 2006)" Verification is "the process of determining that a computational model accurately represents the underlying mathematical model and its solution. (ASME, 2006)" This includes code verification, which seeks to determine if the equations are properly implemented into the code and software quality assurance, and calculation verification, which evaluates the numerical error from discretization. Validation is "the process of determining the degree to which a model is an accurate representation of the real world from the perspective of the intended uses of the model (ASME, 2006)." It is during validation that the results of the simulation are compared to the test data by use of validation metrics.

Validation Metrics

Validation metrics provide the quantitative comparison of data to determine the level of agreement between simulation and experiment. This is a mathematical measure of the mismatch between model predications and experimental results. For aircraft seat models, it is typical to do a deterministic comparison, i.e. compare a single test to a single model prediction. When making a comparison two features are of high interest: error on the peak and waveform error (shape error). Error on the peak is the most intuitive, as many injury criteria are based on a peak value (such as lumbar load) and is simple to calculate using a simple difference or a relative error. The shape error provides a more general evaluation of whether the model predicts the same physical phenomenon as the test. Of the available waveform metrics, the Sprague and Geers Comprehensive Error was chosen as the most appropriate for aircraft seat models (Sprague and Geers, 2004, Moorcroft, 2007). This error compares each data point throughout the waveform and has a magnitude and phase component. Given two curves of equal length and frequency, measured m(t) and computed c(t), the following time integrals are defined:

$$I_{mm} = (t_2 - t_1)^{-1} \int_{t_1}^{t_2} m^2(t)dt \quad I_{cc} = (t_2 - t_1)^{-1} \int_{t_1}^{t_2} c^2(t)dt \quad I_{mc} = (t_2 - t_1)^{-1} \int_{t_1}^{t_2} m(t) \cdot c(t)dt$$

The magnitude component is defined as:

$$M_{SG} = \sqrt{I_{cc}/I_{mm}} - 1$$

The phase component is defined as:

$$P_{SG} = \frac{1}{\pi}\cos^{-1}(I_{mc}/\sqrt{I_{mm}I_{cc}})$$

The comprehensive error is defined as:

$$C_{SG} = \sqrt{M_{SG}^2 + P_{SG}^2}$$

v-ATD requirements

§§ 25.562 call out the use of the Title 49 Part 572 Subpart B, referred to as the Hybrid II, or its equivalent. The only ATD that has been determined to be equivalent is the FAA Hybrid III (Van Gowdy et al., 1999). Each physical ATD used for testing has a series of calibration requirements that are necessary to ensure that the ATD is performing as designed and that any measurements collected can be deemed as reliable. A Virtual ATD (v-ATD) must have similar requirements so that its use and data predictions can be consistent. If this were not the case, then it would be difficult to determine if any data differences were the result of the v-ATD, the software, or the seat system performance. This virtual calibration is also necessary since there are criteria that must be met for a system to be accepted. The calibration can be broken into four parts: mass and geometry evaluation, sub-assembly evaluation, pelvic shape evaluation, and dynamic response evaluation.

Mass and Geometry Evaluation

Design and drawing packages exist for ATDs. Details are provided for both the geometry and the mass of each component as well as for the as built ATD. These details also include the locations and range of motion for each joint. The v-ATD should match these design specifications within a tolerance of +- 0.1 in. To demonstrate compliance it is acceptable to provide a table of joint and segment information for both the v-ATD and the physical ATD listing the joint locations on the physical ATD and v-ATD denoting the dimensional differences.

Sub-assembly Evaluation

The ATD is built up of several different components that generally have a limited life. There are requirements for periodic assessment to ensure that the sub-assembly components are still within calibration. There are tests that must be conducted on the head, neck, thorax, lumbar spine, abdomen, pelvis, and limbs.

These tests would be different for each ATD type. The specific tests and acceptance criteria are cited in the Part 572 regulations. It should be noted that when each ATD is built, it is originally tested and calibrated to these standards. The v-ATD should also be subject to these same calibration tests and the results should demonstrate compliance to the same tolerances as the physical ATDs.

Pelvic Shape Evaluation

While the mass, geometry, and sub-assembly evaluations do provide basic calibration of the v-ATD with the design goals of the physical ATD, it is still necessary to conduct an analysis of the pelvic shape. The shape of the pelvis affects how the ATD interacts with the seating surface and can affect the lumbar loads. Various parameters affect this evaluation including the material properties and physical discretization. A strict procedure should be followed that subjects both the physical ATD and the v-ATD to the same conditions. The difference in the location of the H-points of the ATDs can be compared and should differ no more than 0.2 in and the difference in the orientation of the pelvis should be less than 2 degrees.

Dynamic Response Evaluation

The first three evaluations relate to how the v-ATD is constructed and is straightforward in that they follow already accepted guidelines for the physical ATD. However, these are only at the component level and the response under dynamic conditions of v-ATD must be evaluated since it will be used to assess the response seat systems. To aid in this effort, a set of dynamic impact scenarios was developed. For each scenario, a physical test was conducted three times. Specific geometry was chosen to be representative of aircraft seats, however, the seats were made rigid so as only to elicit the ATD response. An additional change was that Teflon sheets provided a consistent level of friction on the seat pan and floor. The four scenarios chosen for a forward facing dummy are as follows:

1. 2-point belt without a toe stop. Input acceleration pulse is 16 G, 44 ft/s defined in Part 25.562 for the horizontal test condition.
2. 60 degree pitch test with a 2-point belt. Input acceleration pulse is 19 G, 31 ft/s defined in Part 23.562 for the combined horizontal-vertical test condition.
3. 3-point belt. Adjust shoulder belt to produce 1.25 in of initial slack. Input acceleration pulse is 21 G, 42 ft/s defined in Part 23.562 for the horizontal test condition. The geometry of the 3-Point restraint system shall be such that the shoulder belt to lap belt attachment point is 4 in to the right of the ATD centerline.
4. 4-point belt. Adjust shoulder belt to produce 1.25 in of initial slack. Input acceleration pulse is 21 G, 42 ft/s defined in Part 23.562 for the horizontal test condition.

All four scenarios were developed to be representative of typical aircraft configurations that would be tested for various seating systems. For each scenario there are a number of parameters that should be measured during the tests. It should be noted that these parameters may differ slightly from those that would be collected during typical certification tests (Table 2). Scenario 1 is typical of a configuration for transport category aircraft where the restraint

312

system is of a 2-point type. Here, the key is the amount of dummy forward flexion as it loads into the restraint system. Scenario 2 is a combination transport category and general aviation category configuration. The seat orientation is the same for both, while the 2-point configuration is typical for transport aircraft. The chosen pulse is the more severe general aviation pulse. In the combined test, there is little forward flexion, so the influence of the shoulder belt should be minimal on the seat and ATD lumbar loading. Scenario 3 is typical of a general aviation category aircraft with the 3-point restraint that induces torso twist while scenario 4 is similar except with a 4-point restraint that gives chest loading and increased neck flexion that are missing from the component level tests and to check for submarining motion. Again, here the primary concern is ATD forward motion and restraint system loading.

Table 2 Dynamic Calibration data set - forward facing ATD

Channel Description	Scenario 1	Scenario 2	Scenario 3	Scenario 4
Sled Ax	X	X	X	X
Upper Neck Fx,Fz, My *			X	X
Upper Neck Fy , Mx*			X	
Chest Ax (CFC 180)			X	X
Lumbar Fz, My		X		
Right/Left Lap Belt Load	X		X	X
Right Shoulder Belt Load				X
Left Shoulder Belt Load			X	X
Seat Pan Fx, Fz, My	X	X	X	X
Head CG X, Z Position	X	X	X	X
H-point X Position	X		X	X
H-point Z Position	X	X		
Knee X, Z Position	X			X
Ankle X, Z Position	X			
Shoulder X, Z Position			X	X
Opposite Shoulder X, Z Position			X	
Head Angle	X			X
Pelvis Angle	X	X		X

* FAA Hybrid III only, blank cells intentionally left blank

A simulated v-ATD was evaluated for scenario 2 as an example. A couple of points to first consider, the seat geometry, restraint anchor locations, and initial dummy position of the model should match the test conditions within a pre-defined error bounds. This information is available from the test data set. For all four scenarios, the input acceleration is required. Ensuring that the sled acceleration between the test and simulation match should be an automatic process it is requested to report those errors as a check on the validity of the simulation process. To conduct the full evaluation, all the channels from Table 2 would need to be investigated, but for brevity only selected channels will be reported here. These include: lumbar load, pelvic angle, H-Point Z position, and head cg X position.

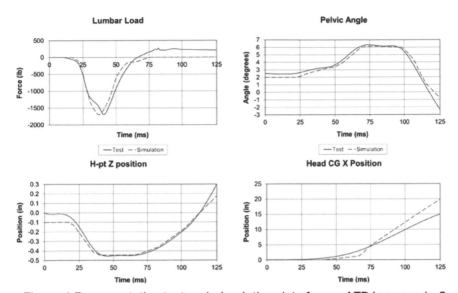

Figure 1 Representative test and simulation data for a v-ATD in scenario 2

Lumbar load was chosen as that is a primary variable for passing this test scenario. The pelvic angle was chosen as its initial orientation and motion can affect the lumbar load. The H-point motion was chosen as that parameter represents the lower torso motion compression of the cushion, if installed; in this case here it also represents a parameter that has a relatively small value making calibration difficult. The head cg X position was chosen as it is a point further on the body and in a different direction than the others, giving an indication of the overall body motion and to how the ATD head is performing, affecting predictions of a head strike.

Upon visual inspection, all the channels would appear to be reasonable (Figure 1) with the correlation referred to as good, well, acceptable, etc. Notice there is nothing in those descriptors to determine when a parameter is on the edge of acceptability. In fact, if you look at the position of the H-point in the Z direction, it starts with a 0.10 in difference yet the peak position match has a much lower difference. Calculating the shape and magnitude errors for each of the channels provides details into the differences between the test and the simulation (Table 3).

Table 3 Errors for simulation calibration

Channel	Magnitude	Shape
Lumbar Fz	0.57%	8.8%
Head CG X	32.2%	25.2%
H-point Z	0.13 in	4.5%
Pelvic Angle	0.19°	4.4%

While a visual inspection of the data provides little insight to the true behavior of the model, an investigation into the differences in the error metrics is more meaningful. The four parameters chosen for example demonstrate four different behaviors. The lumbar load has a peak that with an error of less than 1%, but the shape has a much larger error of almost 10%. If an analyst was only investigating the peak, then this might be acceptable. The H-Point Z position is in almost the opposite situation, the magnitude error is 0.13 inches or almost 25% of the peak, but the shape error is much smaller at 4.5%. The pelvis angle has both errors less than

5%. Finally, the head CG X position has both errors in excess of 25% even though this is not readily apparent. This type of analysis underscores that different types of error metrics should be used and that the acceptance values need to be situational dependent. In this scenario, the lumbar Fz load is a primary measure that is in the loading direction. This is an injury criteria number that must be met to use the simulation for certification purposes. This contrasts with the head position which may not even need to be tracked for this scenario. The purpose of this scenario is to investigate lumbar and seat loading. For this purpose, the head cg position provides little insight into the system and the behavior of the v-ATD. One of the other scenarios would be better suited to investigate the response of those parameters.

A v-ATD that demonstrates compliance with all four scenarios for all parameters of interest within the error bounds would be considered compliant. A compliant v-ATD would be suitable in a variety of simulations of seat systems. Since the four scenarios span different orientations, restraints, and energy levels, it may not be possible for a v-ATD to meet all the requirements. In this case, the v-ATD would be conditionally compliant. Depending on the intended use of the M&S, conditional compliance may be sufficient. Some examples of conditionally compliant v-ATDs and their uses are listed below:

A v-ATD that meets all performance requirements in dynamic test scenarios 1, 3, and 4, but cannot meet the performance requirements for scenario 2, would only be approved for simulations matching the restraint and load application direction of scenarios 1, 3, and 4.

A v-ATD with acceptable correlation for a significant portion of the head path in scenario 1 could be approved on the condition that the model can only be used in installations where the head path is prevented from exceeding the correlated area by external factors such as structural monuments.

A v-ATD that does not meet the shoulder belt loads for scenarios 3 and 4 could be allowed with the proper application of engineering judgment. For example, if the v-ATD greatly over-predicts the belt load, it would not be appropriate for a simulation focused on determining head path since the extra belt load would most likely shorten the head trajectory. Conversely, if the v-ATD significantly under-predicts the shoulder belt load, the v-ATD may be used in a head path simulation, where it would likely produce a conservative result.

System requirements

The previous section discussed requirements for the v-ATD. The goal of the v-ATD is for it to be used simulations of actual seat systems to support development and certification programs. A brief discussion of what this would entail is provided. In 2003, the FAA released Advisory Circular 20-146 to provide general guidance for the use of computer modeling and analysis within seat certification. This includes general guidance on how to validate the computer models and under what conditions the models may be used in support of certification.

The validation of the system level model needs to be more customized and use good engineering judgment. The reason for this customization is that the intended use of the model will dictate the validation plan, in a similar fashion as to which parameters were important for validating the specific scenarios for the v-ATD. The goal here is to validate against those parameters that are important to the specific analysis. The results of the system model may be used to demonstrate compliance to 2X.562. This determination will be made based upon the validation of the system model. It is important to note that in order to validate the model, a dynamic test to generate the validation data is still required. There are several parameters of interest that should be validated. There may be others that the analyst should also investigate depending on the specific model. These parameters include: occupant trajectory, structural response, internal loads, structural deformation, restraint system, Head Injury Criteria, spine loads, and femur loads. Some of these parameters were calibrated as part of the v-ATD development process. AC 20-146 gives general guidance that peak responses should be within 10 percent of the measured values in order to validate the system model. Some adjustment may be necessary for some of these parameters depending on the level of v-ATD calibration. For example, if the seat pan loads are only calibrated to an acceptable level of 25% as part of the v-ATD calibration, then it would be unreasonable to expect them to correlate any better than this during the system level validation.

DISCUSSION

There were four parts to determine a v-ATD calibration: mass and geometry evaluation, sub-assembly evaluation, pelvic shape evaluation, and dynamic response evaluation. The mass and geometry evaluation is a check on model discretization to ensure the same size and weight of the v-ATD matches the ATD. The sub-assembly evaluation determines that the components of the v-ATD perform the same as the physical ATD. These first two evaluations are straight forward with acceptance values already developed; the v-ATD should meet the same specification as the physical ATD. The pelvic shape evaluation begins to investigate slight differences that exist between ATDs. The final evaluation, the dynamic evaluation, evaluates the entire v-ATD, and has a much higher level of complexity. For the dynamic evaluation, the calibration parameters must be chosen to be representative of the conditions being simulated.

For aviation seating simulations, four scenarios were developed. Each parameter that is selected for evaluation is dependent upon the scenario to be simulated. For instance, in scenario 2, the lumbar load is an important parameter. In addition to just selecting which parameters to investigate, acceptability levels should also be chosen. The Sprague and Geers error metrics were demonstrated to be able to discern differences in responses that might otherwise seem acceptable. As such, these error metrics would be suitable for determining test and simulation correlation and can be successfully applied to the task of v-ATD calibration.

The missing process is the judicious selection of acceptable error levels for each scenario. While sounding simplistic, this in itself can be a difficult task. While it might be easy to just set a level of 5% or 10%, the question arises, is that the correct choice? For instance, what is the level of test variation that can be expected for each parameter? If the testing cannot be reproduced within the chosen error level, then a model developed will have difficulty in achieving this as well.

While these error metrics were determined suitable for the v-ATD calibration, they can also be extended for use at the system level. A system level model will contain the v-ATD, among the other components. Suitable parameters and acceptability levels would need to be chosen and the Sprague and Geers error metrics could be applied to determine the validity of the model.

Another consideration is what about situations where a v-ATD cannot be calibrated for all four scenarios. The scenarios were developed to elicit different types of responses that would be expected of system level simulations. A choice here is to consider such a v-ATD as conditionally compliant with appropriate use limitations placed thereupon. This would be considered acceptable if an analyst was investigating effects on a particular test condition, then a v-ATD that is conditionally compliant for that scenario can be chosen. It may then be possible to use several conditionally compliant v-ATDs to simulate the different scenarios.

DISCLAIMER

The findings and conclusions in this paper are the opinions of the authors and should not be construed to represent any agency determination or policy.

REFERENCES

US Code of Federal Regulations, Title 14 Parts 23.562, 25.562, 27.562, 29.562, Washington DC, US Government Printing Office.

FAA Advisory Circular 20-146, "Methodology for Dynamic Seat Certification by Analysis for Use in Parts 23, 25, 27, and 29 Airplanes and Rotorcraft," May 19, 2003.

Van Gowdy, DeWeese R, Beebe MS, Wade B, Duncan J, Kelly R, and Blaker JL, "A Lumbar Spine Modification to the Hybrid III ATD for Aircraft Seat Tests," SAE Paper No. 1999-01-1609.

Pellettiere J and Knox T., "Occupant Model Validation," in Advances in Applied Digital Human Modeling, ed. Duffy VG, CRC Press 2011.

Pellettiere J, Crocco J, and Jackson K, "Rotorcraft Full Spectrum Crashworthiness and Occupant Injury Requirements," 67th AHS Annual Forum, Phoenix, AZ, 2011.

Pellettiere J and Knox T, "Verification, Validation, and Accreditation (VV&A) of Human Models," SAE Paper No. 2003-01-2204.

Harris W and Shepard C, "Strategic Research Planning and Management," 65th AHS Annual Forum, Grapevine, TX, 2009.

ASME V&V10- 2006, "Guide for Verification and Validation in Computational Solid Mechanics," 2006

Sprague MA and Geers TL, "A Spectral-Element Method for Modeling Cavitation in Transient Fluid-Structure Interaction," Int J Numer Meth Eng. 60 (15), 2467-2499. 2004.

Moorcroft D, "Selection of Validation Metrics for Aviation Seat Models" The Fifth Triennial International Fire & Cabin Safety Research Conference, Oct, 2007.

CHAPTER 31

Model for Predicting the Performance of Planetary Suit Hip Bearing Designs

Matthew S. Cowley[1], Sarah Margerum[1], Lauren Harvill[1], Sudhakar Rajulu[2]

[1]Lockheed Martin
1300 Hercules
Houston, TX 77258
matthew.s.cowley@lmco.com

[2]NASA Johnson Space Center
2101 NASA Parkway
Houston, TX 77058
sudhakar.rajulu-1@nasa.gov

ABSTRACT

Designing a space suit is very complex and often requires difficult trade-offs between performance, cost, mass, and system complexity. During the development period of the suit numerous design iterations need to occur before the hardware meets human performance requirements. Using computer models early in the design phase of hardware development is advantageous, by allowing virtual prototyping to take place. A virtual design environment allows designers to think creatively, exhaust design possibilities, and study design impacts on suit and human performance.

A model of the rigid components of the Mark III Technology Demonstrator Suit (planetary-type space suit) and a human manikin were created and tested in a virtual environment. The performance of the Mark III hip bearing model was first developed and evaluated virtually by comparing the differences in mobility performance between the nominal bearing configurations and modified bearing configurations. Suited human performance was then simulated with the model and compared to actual suited human performance data using the same bearing configurations.

The Mark III hip bearing model was able to visually represent complex bearing

318

rotations and the theoretical volumetric ranges of motion in three dimensions. The model was also able to predict suited human hip flexion and abduction maximums to within 10% of the actual suited human subject data, except for one modified bearing condition in hip flexion that was off by 24%. Differences between the model predictions and the human subject performance data were attributed to the lack of joint moment limits in the model, human subject fitting issues, and the limited suit experience of some of the subjects. The results demonstrate that modeling space suit rigid segments is a feasible design tool for evaluating and optimizing suited human performance.

Keywords: space suit, design, modeling, performance

1 INTRODUCTION

Designing a space suit can be very difficult with the numerous and complex requirements, which include the constraint that a crewmember be able to move without being overly encumbered by the suit. To verify that new suit designs meet these requirements, prototypes must eventually be built and tested with human subjects. However as is common with design, numerous iterations will occur before the hardware is finalized and designing a suit to uphold human performance requirements often negatively affects the suit's complexity. These factors often lead to quickly escalating development costs as multiple prototypes are built and tested.

The current planetary-type space suits use complex and multi-component joints with rigid segments and bearings. These rigid segments allow the suit joint to move while maintaining a constant air volume (and therefore air pressure) inside of the suit. Because of the difficulty in modeling the flexible cloth components in the suit as well as the compound motions needed by the multi-bearing joints, computer models have rarely been used in the development phase of suit hardware. With increases in the amount of quantitative data of suited human performance and in modeling technologies, modeling complex space suit systems has become much more feasible. Using hardware design models early in the design phase of suit development would be very advantageous, and allow virtual, concurrent engineering to take place (Cutkosky, Engelmore, Fikes, 1993).

The purpose of this study was limited to specifically evaluating a hip joint model of the Mark III Technology Demonstrator Suit (Mark III) for use as a design tool. This suit was chosen because the hip joint is made of rigid sections that can be modeled with current technology. The results of this test will pave the way for further model development and tests to evaluate all rigid sections of the Mark III suit and will eventually be enhanced with the flexible cloth sections.

The study commenced by examining general changes in mobility performance due to hip bearing modifications (e.g., individual bearings were fixed). Human performance data from the Mark III were examined with the nominal and modified bearing configurations. The Mark III hip bearing model was then created, tested, and validated with the suited human performance data.

2   METHODS

2.1   Suited Human Subject Performance Testing

Four male subjects participated as test subjects for this evaluation. The test subjects were selected for similar hip and leg anthropometry in an effort to reduce human variation in the data. Table 1 shows how the subject's anthropometry compared to the rest of the astronaut population. Suited human subjects performed isolated motions of maximal hip flexion and hip abduction while being recorded with motion capture equipment. There was some difficulty with matching the motion of the suit to the pure definition of the hip angles as the mechanics of the suit forced the subjects into mixed flexion/abduction while "flexing" and mixed abduction/external rotation when "abducting" as the leg follows the hip join bearings' arcs of movement. For simplicity's sake, these compound motions were defined as flexion and abduction.

Table 1. Distribution of the Subjects' Average Among the Astronaut Population

Anthropometry	Stature	Hip Breadth	Thigh Circumference
Average (cm)	179.4	35.8	63.5
Std. Dev.	5.2	1.1	3.4
Population Percentile	52.6	67.4	74.5

Retro-reflective markers were placed on key landmarks of the suit (see Figure 1). Data was processed with a custom-made inverse-kinematic model and processed with a fourth order, zero-lag, high-pass Butterworth filter with a cut off frequency of 6 Hz. Definitions, reference frames, and reference planes commonly used were prescribed by the International Society of Biomechanics (Wu and Cavanagh, 1995). The motion capture data was processed using standard motion capture analysis techniques. Joint angles computed for the hip joint used a flexion, rotation, adduction Euler angle sequence.

320

Figure 1. Motion capture of a subject in hip flexion (left), and the resulting 3D motion data (right).

2.2   Suit Model Development

A 3D solid model of the suit and a representative human manikin were created in SolidWorks (Dassault Systèmes, Vélizy-Villacoublay, France) (see Figure 2). The rigid sections of the suit models were reverse-engineered from the Mark III brief and hip bearings. The measurements taken were accurate to within ± 1 mm. The model includes the upper hip bearing (UHB), mid-hip bearing (MHB), lower hip bearing (LHB), and the rolling convolute joint (RCJ) at the mid-thigh.

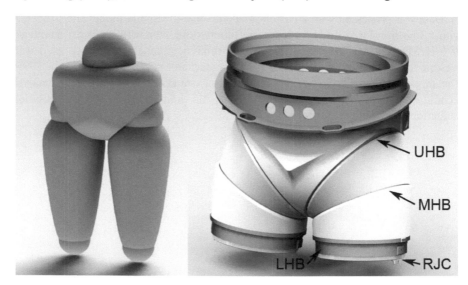

Figure 2. Illustration of a 3D solid model of a subject's pelvis and thigh (left), and the Mark III Technology Demonstrator Suit brief and hip bearing model (right).

The partial human manikin was scalable with respect to specific anthropometric measures for the pelvis and thigh (see Figure 2). Only the hip breadth, thigh circumference, and femur length were used as scaling dimensions for this study. Subject measurements were taken either with an anthropometer or tape measure or extracted from 3D scans.

The SolidWorks platform allows for the creation of 3D assemblies of multiple solid parts that have some degrees of freedom constrained to each other through geometry, while remaining fully articulating in the other degrees of freedom. SolidWorks also allows for exact measurements to be made between points, lines, planes, or surfaces while in any configuration.

The manikin was aligned so that the sagittal plane of the pelvis was co-planar with the sagittal plane of the brief. The x-axis of the hip socket is fixed coincidentally with the UHB centroid. Keeping the hip in a fixed location between conditions was necessary as the true location of the hip would be dependent on a myriad of human-related variables and was the best option outside of allowing the hip to free-float within the brief. Co-locating the hip socket axis and the UHB centroid also provided the largest possible ROM for all of the test conditions in a fixed hip state.

Maximum flexion was determined for each configuration by determining when the knee joint center of the manikin was at its most anterior position (see Figure 3). This was done because of the non-planar and circular paths of the bearings. The most anterior position measurement allowed for a consistent measurement to be taken between conditions and still achieved a large flexion angle. Once the suit was taken to the end range, joint angles and bearing rotations were measured within the SolidWorks software.

Figure 3. Illustration of the Mark III model with a subject manikin in the maximum flexion (left) and abduction (right) positions for the nominal condition.

Maximum abduction was determined for each configuration by putting the manikin knee joint center at its most superior and lateral position (see Figure 3). This orientation of the leg was also chosen because of the non-planar and circular paths of the bearings and the inability to move in true abduction. The most superior and lateral position measurement allowed for a consistent measurement to be taken between conditions, achieve the largest abduction angle, and provide a simple method of comparing the bearing model to the human subject data.

3   RESULTS AND DISCUSSION

3.1   Suited Human Performance and Suit Model Data

Figures 4 and 5 illustrate the differences between the model data and the human subject data for the hip joint angles. The charts show the mean and span of the maximum angle ranges across all four subjects for flexion and abduction against the bearing model's estimated maximum angles. The hip joint angles were measured from the thigh relative to the pelvis. The angles are reported as positive if in flexion or abduction. The suit model predicted an approximate 60% reduction in maximum flexion and abduction with the UHB locked and an approximate 25% reduction with the LHB locked, but only a negligible change with the MHB locked in comparison to the unlocked state.

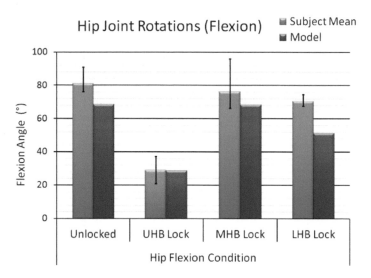

Figure 4. A comparison of the maximum hip flexion mean (blue) with an error bar of the highest maximum and lowest maximum across subjects as compared to the model (red).

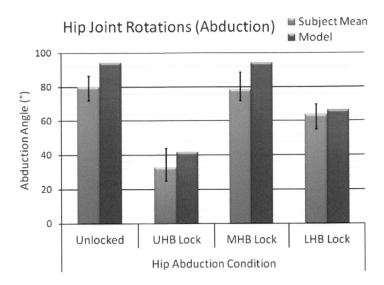

Figure 5. A comparison of the maximum hip abduction mean (blue) with an error bar of the highest maximum and lowest maximum across subjects as compared to the model (red).

Although not all of the model predictions fit within the range of the peak angles of the human tests, the performance trends are kept. The error between the model prediction and the subject data range was < 10%, except for the LHB locked condition in flexion (Figure 4), which had a 24% error.

There were some physical differences between the 3D solid model of the joint bearings and the human test subject environment. For this study the manikin model was locked into a specific orientation/location within the brief and hip flexion and abduction definitions were standardized between conditions, whereas the human subjects were allowed to move freely within the brief and choose for themselves where their maximum flexion and abduction angles were. Similarly, there was a fair amount of slop in the suit's fit around the knee and lower thigh, allowing for greater motion, as seen in the LHB locked condition. Other factors included the lack of dynamic elements in the model, such as mass, gravity, resistance, suit-human contacts, or joint moment restrictions.

3.2   Suit Model Visualization

Mobility visualizations were created to show the generally unused regions of motion that the hip bearing of the Mark III allows for. Figure 6 is an illustration of the hip mobility in the Mark III compared to unsuited hip motion. Each shape represents the possible locations of the knee joint in 3D, the blue disk-like area is the maximum allowable motion of the Mark III in the nominal condition, and the green is unsuited hip motion for common tasks (walking, kneeling, climbing, etc.). Functional hip motion is taken from published walking data (Gage, DeLuca, and Renshaw, 1995) and previous NASA functional mobility tests (England, Benson,

324

and Rajulu, 2010). The general area of unsuited hip motion not accounted for by the Mark III hip joint is shown as the green volume not overlapped by the blue. Approximately 60% of the unsuited hip motion area would have been unreachable by someone in a Mark III suit trying to move in the exact same way. The inverse of the previous statement is also represented visually where only a fraction of the allowable mobility of the Mark III would be used to perform common tasks without compensating with other joint motion (i.e., waist and leg rotation).

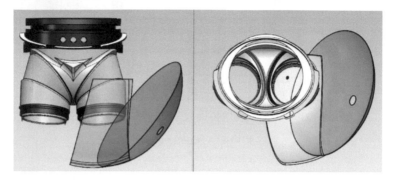

Figure 6. Mark III hip mobility (blue shape) vs. estimated human functional hip use (green shape).

The mobility of each bearing configuration can also be visualized and quantified into an area as was shown in Figure 6 to compare configuration changes and the impact each bearing has on the overall mobility. For example, Figure 7 illustrates the model's capability to represent all possible locations of the knee joint center in 3D space (blue disk-like area) when the UHB was locked. For this condition the available range of motion was greatly reduced and was limited to a ring shape. The motion volume was created by superimposing the allowed motion for each of the bearing segments. Comparing the UHB locked condition to the nominal condition revealed a reduction of 96% in total movement area of the knee joint center.

Figure 7. Visualization of the allowable mobility of the knee joint center with the UHB locked.

4   CONCLUSIONS

The bearing model results demonstrate that modeling the Mark III brief and hip bearing sections for use as a performance design tool was feasible and useful. Although not all of the model predictions fit within the range of the peak angles of the human tests, they are representative to a reasonable degree considering the testing limitations. This was especially true for the conditions that limited the performance, such as with the UHB and LHB locked conditions. It is reasonable to state that such models of the hard segments and bearings of the suit are feasible and useful in studying and analyzing the behavioral characteristics of these joint bearings. Models cannot replace human-in-the-loop performance and verification tests, but may offer a huge benefit in conjunction with human testing to improve efficiency within the human-centered design process and for making sure that later stage designs adhere to requirements. The benefits of a suit joint model offers an added advantage to traditional suit design by allowing designers to visualize performance changes, instead of theorizing outcomes and building prototypes to test those theories out. It also gives designers the ability to think outside the box and exhaust design possibilities without additional resources, thereby getting to the optimal bearing designs much quicker.

This model will form the basis for the evolution of a full suit and human dynamics model, capable of calculating workloads, efficiencies, and injury predictions. This model and analysis are initial steps in the development of a more complex virtual design tool that will aid suit developers in creating the next generation space suits with increased efficiency, reliability, and performance. Advances in software capabilities and computer processing power have enabled us to efficiently create dynamic bearing models that could be used to predict suit-human contact forces and internal suit resistances. A biomechanical model of this magnitude could eventually lead to predict injury potentials for planned tasks, suit architectures, hardware designs, or contingency situations.

ACKNOWLEDGMENTS

This project was funded by the National Aeronautics and Space Administration.

REFERENCES

Cutkosky, M.R., R.S. Engelmore, R.E. Fikes, et al. 1993. PACT: an experiment in integrating concurrent engineering systems. *IEEE Computer*, 26: 28-37

England, S.A., E.A. Benson, and S.L. Rajulu. 2010. Functional Mobility Testing: Quantification of Functionally Utilized Mobility among Unsuited and Suited Subjects. Houston, TX: National Aeronautics and Space Administration, NASA/TP-2010-216122.

Gage, J.R., P.A. DeLuca, and T.S. Renshaw. 1995. Gait analysis: principle and applications with emphasis on its use in cerebral palsy. *J Bone Joint Surg Am*, 45: 1607-1623.

Wu, G. and P.R. Cavanagh. 1995. ISB recommendations for standardization in the reporting of kinematic data. *J Biomech*, 28: 1257-1261.

CHAPTER 32

Integration of Strength Models with Optimization-Based Posture Prediction

T. Marler, L. Frey-Law, A. Mathai, K. Spurrier, K. Avin, E. Cole

The University of Iowa
Virtual Soldier Research (VSR) Program

ABSTRACT

Although much work has been completed involving the study of human strength from the perspective of individual muscles and with regards to static analysis of joints, strength models have not yet been fully integrated within a whole-body predictive digital human model. This work presents an approach to modeling joint-based strength, and to incorporating it as a constraint within optimization-based posture prediction, such that variations in strength automatically affect human performance. Initial results have been successful, and based on subjective validation, the predicted postures are realistic.

Keywords: Digital human modeling, strength modeling, posture prediction

INTRODUCTION

As the use of digital human models (DHMs) increases and involves more applications, it becomes critical that such models take into consideration variations and limitations in strength as well as anthropometry. With regards to product and process design, a primary benefit of *DHMs* is the ability to evaluate quickly and inexpensively different scenarios with changes in problem parameters that define the DHM as well as the virtual environment. Strength characteristics constitute a key parameter for providing realistic results. In addition, strength limits ultimately provide a cornerstone for injury prevention.

Especially within the manufacturing arena, data-based ergonomic tools have been developed to help assess the risks associated with strenuous tasks. While such tools provide well accepted sources for design and process standards, they are limited in the types of scenarios to which they apply. Furthermore, traditional DHMs are limited in their predictive capabilities and versatility. Currently, there exist some DHM tools for evaluating static strength limitations a posteriori. They allow the user essentially to freeze an avatar in a specified posture and analyze the percent of a population capable of performing the specified static task. However, there are no tools for predicting posture and human performance while concurrently taking into consideration strength limitations. Consequently, this paper presents new work with predictive strength models in the context of the direct optimization-based posture-prediction approach developed at the University of Iowa's Virtual Soldier Research Program. This work is based on Santos, a virtual human who uses optimization to predict whole-body postures while considering external forces and without requiring pre-recorded data.

The musculoskeletal system is inherently redundant, with multiple muscles acting synergistically about each joint to produce the joint torque necessary to perform a task. While some approaches focus on individual muscle models (Marler *et al* 2008), normative data on joint strength is only available at the joint level as noninvasive direct muscle force measurements are not possible in humans. Thus, all models of strength must be validated against assessments of net joint torque. For individual muscle models, assumptions regarding load sharing between muscles are necessary. For some computations, such as joint compression and shear estimates, these individual muscle models are valuable. However to assess task difficulty through relative contraction intensity, estimated joint torques need only be standardized by maximum joint strength. Thus, we modeled peak static strength (i.e., maximum voluntary contractions, MVCs) at the joint level as a function of joint angle.

With posture prediction, the required joint torques needed to accomplish a task are determined along with the corresponding joint angles. By plotting the predicted joint torques as a function of joint position, we compare the strength requirements for a particular task to a normative database of the strength capability for each joint. This provides us with a measure of task difficulty, represented as a percent of maximum strength (considering position effects on peak strength capabilities). As a post processing tool, the model provides output indicating at which time, if any, during a simulation joint torques are likely to exceed strength capabilities, which are influenced by fatigue.

In addition, while leveraging this optimization based approach to posture prediction, we have developed constraints for modeling strength, which can be altered on the fly and integrated in the predictive process. Data-based representations of maximum allowable joint torque provide strength limits, and are based on on experiments involving isometric and isokinetic contractions at several major joints of the body. This data is averaged to provide curves (and thus mathematical constraints) for maximum-torque versus joint angle for the ankle, knee, hip, torso, shoulder, elbow, wrist, and neck for men and women. Although the raw data represents subjects with approximately 50th percentile strength, the curves can be shifted on a continuous scale to represent any percentile, based on the means and standard deviations of the experimental data. These curves are fully integrated as an optimization constraint and allow one to study how strength limitations affect predicted performance. Furthermore, for a given predicted posture and set of joint torques, one can determine the percent of the population capable of completing such a task.

SANTOS MODEL

Santos (Marler *et al*, 2008) is a DHM built on a biomechanically accurate musculoskeletal model with approximately 109 predicted degrees of freedom (DOFs. Santos's anthropometry can be altered on the fly based on existing or user-specific standards. This human structure, originally developed at VSR, is shown in Figure 1. There is one joint angle for each DOF, and the relationship between the joint angles and the position of points on the series of links (or on the actual avatar) is defined using the Denavit-Hartenberg (DH)-notation (Denavit and Hartenberg, 1955). This structure is becoming a common foundation for additional work in the field of predictive human modeling and has been successfully used with various research efforts (Ma *et al*, 2009; Howard *et al*, 2010).

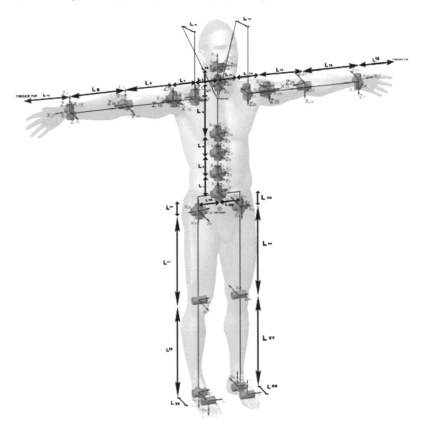

Figure 1: Computational skeletal model

The core functionality for any virtual human is the ability to model human posture and motion realistically. In this respect, Santos is *predictive*. Posture and motion are not based on prerecorded motion or data that must be updated when there are alterations in the virtual environment. Rather, Santos predicts how people strike various poses and how they move. Such predictive capabilities afford the model considerable autonomy and stem in large part, from an optimization-based approach to human modeling. This approach is extremely fast, allowing Santos to operate and react in real time in most cases. The most unique aspect of our approach is the ability to handle external forces and complex dynamics problems. With this approach, the joint angles (one for each degree-of-freedom) essentially provide design variables that are determined through optimization. The objective function(s) is one or more human performance measures, such as energy, discomfort, joint displacement, etc. A detailed discussion of the underlying method for predicting posture while considering external loads is provided by Liu *et al* (2009) and Marler *et al* (2011), and an overview is provided here.

The fundamental formulation is given as follows:

Find: $\mathbf{q} \in R^{DOF}$

To minimize: $f(\mathbf{q})$

Subject to: $\text{Distance} = \left\| \mathbf{x}(\mathbf{q})^{\text{end-effector}} - \mathbf{x}^{\text{target point}} \right\| \leq \varepsilon$; $q_i^L \leq q_i \leq q_i^U$; $i = 1, 2, \text{K}, DOF$

q is a vector of joint angles, x is the position of an end-effector or point on the avatar, ε is a small positive number that approximates zero, and DOF is the total number of degrees of freedom. $f(q)$ can be one of many performance measures (Marler, 2005; Marler *et al*, 2009). The primary constraint, called the *distance* constraint, requires the end-effector(s) to contact a specified target point(s). q_i^U represents the upper limit, and q_i^L represents the lower limit. These limits are derived from anthropometric data. In addition to these basic constraints, many other constraints can be used as boundary conditions to represent the virtual environment.

Including conditions of static equilibrium allows one to consider joint torques that are constrained using the strength model. The basic structure presented above is extended conceptually for such use as follows:

Given:
1) End-effectors and associated target points/lines/planes
2) External loads
3) Position and orientation of ground

Find: Joint angles

To minimize: Joint torques and/or joint displacement

Subject to:
1) Distance constraints involving given target points/lines/planes
2) Joint limits
3) Torque limits
4) All contact points remain on the contact plane
5) Balance
6) Static equilibrium, including body mass, external forces, and ground reaction forces

The torque limits (one upper and lower limit for each DOF) can be fixed based on data from the literature. However, with this work, analytical expressions are used to relate maximum joint torque to joint angle.

STRENGTH MODEL

Of the many capabilities being developed for Santos, one of the most critical is a realistic model of strength. Thus, in this section, an overview of our approach to modeling strength is presented, followed by details regarding implementation, determination of percent capable, and an example of the strength model for a single joint.

In general, data-based representations of maximum allowable joint torque provide models of strength limits and are easily implemented as toque limits in the formulation above. Although much effort has been spent on modeling individual muscle strength over the past half century, the ability to represent the complete muscular system in real time is not yet feasible or easily validated. Consequently, we have developed a new method for mathematically representing strength, based on the net muscle torque produced about a joint, such as elbow flexion. It has long been recognized that muscle moment generating capability depends on 1) contraction velocity (Hill, 1938); 2) muscle length tension relationship (Lieber, 1992); and 3) muscle moment arm. Although each of these factors has been well studied in isolation, only recently have their interactions been considered. We have developed an empirical method for modeling strength while incorporating these factors as a curve (for use with posture prediction) that relates joint torque to joint position, or as a three-dimensional (3D) surface (when used with motion prediction) of joint torque relative to joint position and movement velocity (Frey Law *et al*, 2009), as shown in Figure 2.

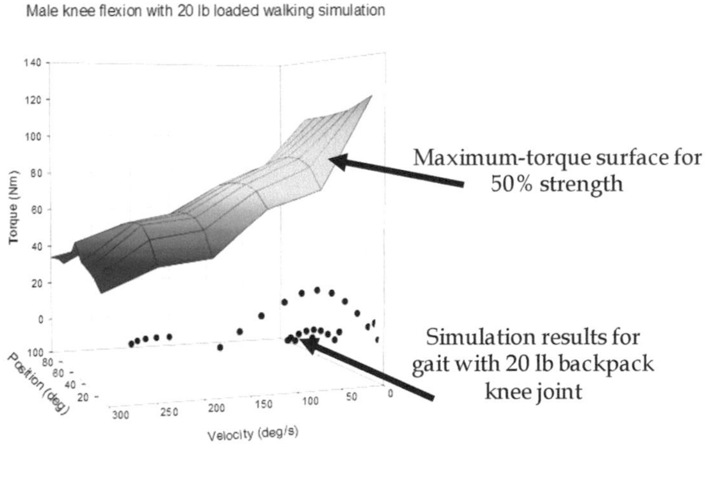

Figure 2: Example strength surface

Using isokinetic (constant velocity) and isometric (constant length) strength testing through a normal range of motion, we measure accurate peak torques for a given joint position and velocity, accounting for internal limb effects. In the context of dynamic motion prediction (Xiang *et al*, 2009), this data is then mapped as a 3D surface of peak torque relative to angular velocity and joint position. These surfaces provide a reference to interpret Santos's predicted joint torque data required to complete a task as a percentage of maximum. When one considers joint-angle velocities of zero (i.e. isometric torque-angle relationship), the surfaces essentially reduce to curves that can be used as differentiable constraints in the posture-prediction formulation.

Strength Constraints

As described above, static strength can be modeled as a function of joint angle (i.e. equation representing peak torque vs angle) and using population percentiles (i.e. to represent normal variation in strength capability between individuals) for men and women. Curves are fit to experimentally-derived mean isometric and isokinetic torque data for subjects with approximate 50th percentile strength (see Frey Law et al, in press for more information on experimental data collection). After considering several possible equation families, including polynomial, Chebyshev, and cosine series, we chose to model peak torque as a function of both joint angle and contraction velocity using logisitic equations (Tablecurve 3D, Systat Inc, Chicago, IL). Because we are interested specifically with static strength for this application, we used these best fit equations with velocity set to zero. The corresponding torque angle relationship takes the following form:

$$\text{mean peak torque} = a + b \text{ angle} + c \text{ angle}^2 + d \text{ angle}^3 \qquad (1)$$

This relatively simple polynomial provides reasonable fits for the experimental data. Given the polynomial nature of this curve, it is easily differentiable and thus easily integrated in a gradient-based optimization formulation. Furthermore, implementation simply entails communication of the relevant model parameters (equation coefficients), which can be scaled with changes in percentile strength.

In order to represent mean peak torque for different percentile-strength characteristics, the coefficients are scaled. The coefficients in (1) are essentially mean values, and along with the coefficients the covariance (CV) is determined with each curve fit. Therefore, coefficients for any percentile strength can be determined as follows:

$$coefficient_{\%ile} = z_{\%ile}\left(CV\right)\left(coefficient_{mean}\right) + coefficient_{mean} \qquad (2)$$

Percent Capable

The above approach allows one to predict postures that automatically adhere to strength limitations. Nonetheless, given a specific task or posture, it is often necessary to determine the percent (of a population) capable of performing that task. In addition, given that many existing DHM applications only include percent-capable analysis, this capability has been incorporated within the Santos model for verification purposes.

Rating the percent capable relies on the inherent variations in strength observed between individuals. Because strength follows a normal or Gaussian distribution (Frey Law et al, in press) we can use standardized z-scores to transform mean and standard deviation (SD) data to population percentiles. We model torque as a function of joint angle, and similarly estimate SD using a constant coefficient of variation (SD/mean). The covariance (CV) varies from values of 0.298 to 0.480, averaging 0.350 across 15 joint torque directions (unpublished observations). This allows for a simplified means to estimate torque relative to population percentiles, as follows:

$$\tau_{\%ile} = \tau_{mean} + z_{\%ile} \times (CV * \tau_{mean}) \tag{3}$$

$$CV \text{ mean} = \frac{\sum_{k=1}^{n} \left(\frac{\tau_{sd}}{\tau_{mean}}\right)}{n} \tag{4}$$

To calculate percent capable, we solve (3) for the z-score based on the torque required to complete the task (using posture prediction) relative to the modeled mean and SD (or CV) torque values for each joint:

$$Z\%ile = \frac{\tau - \tau_{mean}}{CV * \tau_{mean}} \tag{5}$$

The resulting z-score is then compared to its corresponding percentile using standard z-score tables in order to determine the percentile. For example, a z-score of 0.0 represents the median or 50th percentile, -0.675 and +0.675 correspond to the 25th and 75th percentiles, respectively. Essentially, percent capable is determined as 100% - specified percentile. That is, if a task requires the strength of someone at the 75%ile for strength, then only 25% of the population will be capable of performing that task (and 75% will not be strong enough).

Example Model

An example of the static strength model relative to other reported experimental data for knee extension in men is shown in Figure 3. The solid and dotted lines represent the 5th, 50th, and 95th percentile strength models developed for Santos. The experimental data points used to generate this model (as part of the 3D model), and 4 additional studies (Stobbe, 1982; Kooistra *et al*, 2006; O'Brien *et al*, 2009; and Thorstensson *et al*, 1976) are also shown. Note that some studies demonstrate higher peak torques than our static strength model, but their cohorts often included only healthy college males and/or athletes. Note the highly curvilinear relationship between peak torque and knee joint angle for knee extension. Clearly, small offsets in joint angle can relatively large changes in joint torque, as could the normal variation in human strength observed between different test populations. However, overall these data demonstrate similar torque-angle relationships for knee extension and good agreement overall between torque magnitudes.

334

A curve like those shown in Figure 3 is developed for each major muscle group in the body, with the exception of the neck. These relationships are then mapped to the appropriate joint and DOF within the Santos model shown in Figure 1. This mapping process involves inherent approximations, given that the approximate, albeit extensive nature of Santos's computational skeletal structure. Nonetheless, the proposed process provides successful and useful results.

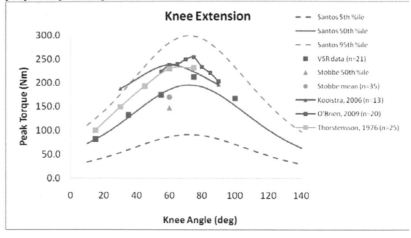

Figure 3: Santos strength curves for male knee extension

RESULTS

As a basic test case, posture prediction was used to predict human performance while holding a barbell weighing 130N. Santos's strength was varied from 10th percentile (relatively weak) to 90th percentile (relatively strong). The objective function was primarily composed of a model for effort, which models the tendency to remain in one's initial posture, with a small component representing joint displacement, which models the tendency to minimize joint articulation. The results are shown in Figure 4.

Santos tries to hold up the barbell, but as his strength decreases, this becomes less feasible. In general, the curve relating maximum joint torque to joint angle, for shoulder flexion is relatively flat until higher joint angles are experienced. As the percentile strength is reduced, this curve is scaled down (the maximum joint torque is reduced for all angles), and in order to accommodate reduced strength a smaller moment arm (between the barbell and the shoulder) is assumed. This in turn reduces the torque that is applied on the shoulder by the barbell load.

CONCLUSIONS

This paper has leveraged a proven approach for predicting postures with external forces, and unique method for modeling strength of critical muscle groups in order to simulate the effects of changes in strength characteristics on predicted performance. Previously, relationships between maximum joint torque, joint angle, and joint velocity had been used for post processing, to evaluate the feasible of completing various tasks a posteriori. This new method, however, allows one to integrate strength characteristics within a predictive model. With respect to the underlying strength model, initial results have been shown to coincide closely with exiting data in the literature. In addition, predicted postures are realistic.

A key aspect of any DHM is the ability to integrate with new capabilities and with pre-existing models. The human system is far too complex to be represented with a single effort. With this direction in mind, the proposed work represents a multi-scale model empowered in large part by the underlying optimization-based approach. The system level human model is coupled with a computationally tractable whole-body muscle model. Such coupling is also being pursued with high fidelity joint models, internal models, and clothing models; and represents an exciting direction for future work. Ongoing efforts also involve using the imbedded strength models to predict the maximum loads a DHM can apply in the context of various assembly and manufacturing tasks. In addition, methods for combining strength curves and surfaces for out-of-plane joint articulation are being explored.

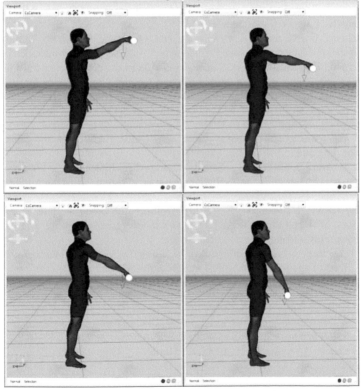

Figure 4: Predicted posture without strength constraints, with 90th% strength, with 50th% strength, and with 10th% strength

REFERENCES

Denavit, J., and Hartenberg, R. S. (1955), A Kinematic Notation for Lower-pair Mechanisms Based on Matrices. *Journal of Applied Mechanics*, 77: 215-221.

Frey Law, L. A. and Avin, K. G. (2010), Endurance Time is Joint-Specific: A Modeling and Meta-Analysis Investigation. *Ergonomics*.

Frey Law, L. A., Xia, T., and Laake, A. (2009), Modeling Human Physical Capability: Joint Strength and Range of Motion. Handbook of Digital Human Modeling. V. G. Duffy (Ed). Boca Raton, FL, CRC Press.

Hill, V. A., (1938), The heat of shortening and the dynamic constants of muscle. Procedings of the Royal Society of London B. 126, 136-195.

Howard, B., Yang, J., and Gragg, J. (2010), Toward a New Digital Pregnant Woman Model and Kinematic Posture Prediction. *3rd International Conference on Applied Human Factors and Ergonomics*, July, Miami, FL.

Kooistra, R. D., Blaauboer, M. E., Born, J. R., de Ruiter, C. J., and de Haan, A. (2006), Knee extensor muscle oxygen consumption in relation to muscle activation. *Eur J Applied Physiol*, 98: 535-545.

Lieber, R. L. (1992), Skeletal Muscle Structure and Function. Implications for Rehabilitation and Sports Medicine. Williams & Wilkins, Baltimore, Maryland.

Liu, Q., Marler, T., Yang, J., Kim, J., Harrison, C. (2009), Posture Prediction with External Loads – A Pilot Study. *SAE 2009 World Congress*, April, Detroit, MI, Society of Automotive Engineers, Warrendale, PA.

Ma, L., Zhang, W., Chablat, D., Bennis, F., and Guillaume, F. (2009), Multi-objective Optimization Method for Posture Prediction and Analysis with Consideration of Fatigue Effect and Its Application Case. *Computers and Industrial Engineering*, 57, 1235-1245

Marler, R. T. (2005), A Study of Multi-objective Optimization Methods for Engineering Applications. Doctoral Dissertation, University of Iowa, Iowa City, IA.

Marler, T., Arora, J., Beck, S., Lu, J., Mathai, A., Patrick, A., Swan, C. (2008), "Computational Approaches in DHM," in *Handbook of Digital Human Modeling for Human Factors and Ergonomics*, Vincent G. Duffy, Ed., Taylor and Francis Press, London, England.

Marler, R. T., Arora, J. S., Yang, J., Kim, H. –J., and Abdel-Malek, K. (2009), Use of Multi-objective Optimization for Digital Human Posture Prediction. *Engineering Optimization*, 41 (10), 295-943.

Marler, T., Knake, L., Johnson, R. (2011), Optimization-Based Posture Prediction for Analysis of Box-Lifting Tasks. *3rd International Conference on Digital Human Modeling*, July, Orlando, FL.

O'Brien, T. D., Reeves, N. D., Baltzopoulos, V., Jones, D. A., and C. Maganaris, N. (2009) The effects of agonist and antagonist muscle activation on the knee extension moment–angle relationship in adults and children. *Eur J Applied Physiol*, 106: 849-856.

Stobbe, T.J. (1982), The Development of a Practical Strength Testing Program for Industry. Doctoral Dissertation, The University of Michigan.

Thorstensson, A., Grimby, G., and Karlsson, J. (1976) Force-velocity relations and fiber composition in human knee extensor muscles. *J Applied Physiol*, 40: 12-16.

Xiang, Y., Chung, H. -J., Kim, J. H., Bhatt, R., Marler, T., Rahmatalla, S., Yang, J., Arora, J. S., Abdel-Malek, K. (2009), "Predictive Dynamics: An Optimization-Based Novel Approach for Human Motion Simulation," *Structural and Multidisciplinary Optimization*, 41 (3), 465-479.

CHAPTER 33

Effects of Changes in Waist of Last on the High-heeled In-shoe Plantar Pressure Distribution

Jin Zhou[1, 2], Weiwen Zhang[1], Petr Hlaváček[2], Bo Xu[1], Luming Yang[1], Wuyong Chen[1]*

1National Engineering Laboratory for Clean Technology of Leather Manufacture,
Sichuan University
Chengdu, P.R. China
Rufuszhou.scu@gmail
2 Faculty of Technology, Tomas Bata University
Zlin, Czech Republic
hlavacek@ft.utb.cz

ABSTRACT

In order to clear which kind of arch design of the last is optimal for high heeled shoes with 20mm, 50mm and 80mm heel height, ten healthy female subjects participated in this study. Their 3D model of the right foot in the normal weight bearing and in standing on the elevated heel was scanned and averaged foot arch was calculated. Three last designs with different arch shape were provided: control last in which the averaged foot arch was applied; the higher and lower last were constructed with 2.5mm higher or lower in the position of cuboids than the control last. The insole region was divided into hind, mid and fore foot and peak pressure (PP), contact area (CA) and pressure-time integrals (PTI) were selected for evaluation. Paired-t test with the significant level a=0.05 was selected for difference comparison between the control and lower last, control and higher last. The result shows that for the low heel-height shoes such as 20mm, the higher last received a lowest PP and PTI under the forefoot; however for the mid and high heel-height shoes for instance 50 or 80mm, the control last was shown to be the optimal one with lower value on both PP and PTI variables for the forefoot area. The optimal last design would be applied on the current high heel-heighted shoes manufacture.

Keywords: high heel-heighted shoe, plantar pressure, last design, foot arch

1   THE OPTIMAL LAST DESIGN FOR HIGH HEELED SHOES BASED ON THE CHANGE OF ARCH SHAPE

1.1   INTRODUCTION

Foot arch is a magic construction, for it not merely acts as the force transmitter, but also it functions as force absorption when walking with foot flat. But when the foot heel elevated, the length of medial and lateral longitudinal arch was shortened(Kouchi and Tsutsumi, 2000). The shorter fascia of arch on the one hand makes the foot prevail to be hurt during walking; on the other hand, it makes the mid foot unable to perform as the force absorption component and increases the pressure value under forefoot.

There were many studies concerning on how the high-heeled shoe affected the kinematic and kinetic characteristics during walking (Stefanyshyn et al., 2000, Kouchi and Tsutsumi, 2000, Speksnijder et al., 2005, Cho and Choi, 2005). Their results found that high-heeled shoe did have the side effects on the musculoskeletal system and lead to the redistribution of centre of body, then result the unnatural plantar pressure distribution. Another study about the influence of the shank curve on the plantar pressure distribution of 3 inches high-heeled shoes was carried out (Cong et al., 2008), where the shank is located on the arch position of shoe to support arch during walking. Although this study imitated the change of arch shape on last by inducing three kind of shank with different length and thickness, this imitation could not represent the real changes on the last, because the design of insert cannot meet the shape of sole of last and new problem will occur when these two unmatched part combine together.

Our study aimed to design the control last by following the arch data of the foot, and then made comparison between this last and other two type lasts where these two lasts with their arch heights being 2.5mm lower or 2.5mm higher than the control one. Finally, the question of which kind of arch design was optimal in alleviating the forefoot pressure concentration for high heeled shoes with 20mm, 50mm and 80mm heel height was answered. It was hypothesized that for each heel height the control last would achieve a better pressure distribution under the whole sole region and the control last was considered to be the optimal design for heel-heighted shoes.

1.2   METHODS

Subjects

Ten healthy female university students aged between 23-25 years old and with a foot length within 230±5 mm were recruited. Their basic information such as age, height and body weight was also recorded. Subjects with deformities such as pes planus, pes cavus, and hallux valgus or with foot skin diseases were excluded in this study. Purposes and methods of this study were explained to the participants, and their oral approvals were received before measurement start.

Foot measurement and Last design

Only right foot in upright barefoot condition and standing on the heel with heel height of 20mm, 50mm and 80mm was measured by the three dimensional laser scanning system (INFOOT USB: Standard type, I-Ware Laboratory Co.,Ltd, Japan). The main positions of right foot were marked and indentified by the software of system (INFOOT-digital measurement interface 2.9, I-Ware Laboratory Co.,Ltd, Japan) (De Mits et al., 2011). Prior to the measurement, calibration was made by following the manual of scanning system.

The sagittal profile along with foot longitudinal axis (the foot longitudinal axis is the line passing through the end of heel to the centre of second toe, based on this axis, the foot is divided into medial and lateral part) was extracted from the marked foot model in 3D software (Powershape, Delcam Co.,Ltd, UK) (Figure 1a). The arch curve was defined as the part of profile between the first metatarsophalangeal joint to the position of the heel contact (red line in Figure 1a). Then the arch curve of sole was equally cut by 10 perpendicular lines and produced 10 joints (Kouchi and Tsutsumi, 2000) and the coordinates of each joint were recorded (Figure 1b). At last all coordinates in one joint from 10 subjects were averaged and an average foot arch section was constructed.

The three last designs were provided (Figure 2): control last was the one by applying the average foot arch as the arch curve of last; while higher last was 2.5mm higher than the control last in the position of cuboids; similarly, lower last was 2.5mm lower in that position. The 2.5mm diminished in height of cuboids implied the 3.5 mm decreasing in the instep girth. The difference of 3.5mm is the degree of sizing system which means half size larger or smaller and this difference in the last industry is considered to be significant enough to cause the difference in the comfort perception. Therefore, this change in the shape of arch curve is enough to demonstrate the differences in the plantar pressure. Besides, the modification happened on the position of cuboids, for that this position is usually considered as the symbol of the midfoot when measuring the length or width of the midfoot.

Except the differences on the midfoot, the other part of last was kept the same for the same heel height, and all of the last used in this study were hand-made by the same last engineer.

In-shoe plantar pressure measurement

The Novel Pedar-X insole pressure system (Pedar-X, Novel GmbHgmbh, Germany) was applied in the data collection. At least three time successful data of right foot was required for each subject. The sole region of foot was divided by Novel multimask (Novel Multimask software, Novel GmbHgmbh, Germany) into 3 main parts: heel, midfoot, forefoot (Figure 3). The plantar pressure variables of peak pressure (PP) (kPa), contact area (cm2) (CA) and the pressure-time integrals (kPa*s) (PTI) for each region were calculated and for further analysis.

Statistical analysis

First of all, the three times plantar pressure data of right foot were averaged within the individual and then among all ten subjects. All the data was shown in the form of Mean (SD). Paired-t test was applied in order to compare the difference between the control and lower last and control and higher last on the pressure variables under each region of sole for different heel height condition. All the analysis was performed by the SPSS software with significant level a=0.05 (v.16.0; SPSS Inc, Chicago, USA).

1.3   RESULTS

The mean age of ten healthy female participants is 24(0.2) years old, mean height is 1.60 (0.04) m, mean body weight is 49.8(4.4) Kg, mean foot length is 230(2.4) mm and mean width is 87.7 (2.4) mm.

CONTACT AREA

The more arch height increasing (from lower last to control last, and then to higher last), the more midfoot contact is found in each heel height (Table 1). The control last recorded the significant larger CA under the hind foot than that of lower last with the 20mm heel height (p=0.01<0.05). In the heel height of 50mm, control last indicated a significant smaller CA (p=0.02<0.05) under the forefoot than that of lower last. The value of CA under mid foot (p=0.00<0.01) and fore foot (p=0.04<0.05) of control last has significantly increased by 18% and 2%, when comparing with the lower last in heel height of 80mm.

PEAK PRESSURE

With the increasing CA under mid foot, the PP of forefoot is decreasing (Table 1). In the 20mm heel height, the significant difference of PP was found between control last and higher last under forefoot region (p=0.03<0.05). The PP of control last under hind foot area significantly reduced in the 50mm heel height than that of lower last (p=0.00<0.05). While in contrasting with the higher last, 11% PP arising for control last under hind foot was found (p=0.01<0.05), in the condition of 80mm heel height.

PRESSURE-TIME INTEGRAL

Unlike variable of PP, the trend of PTI with arch height elevating fluctuates a lot (Table 1). In the heel height of 20mm, the PTI of control last under forefoot increased significantly in comparison with higher last (p=0.01<0.05). Similarly, in the heel height of 80mm, variable of PTI of control last under the hind foot (p=0.02<0.05) was significantly larger than that of the higher last. There were no significant differences existing in the condition of 50mm heel height.

1.4 DISCUSSION

Because the 10 female subjects were selected by matching the age, height, body mass and foot length as well as only the right feet were selected for data analysis, the variability of these factors would not affect the results of pressure values.

To ameliorate the comfort perception of high-heeled shoes was to make the last more close to the foot(Hawes et al., 1994). The method of inserting with insole, heel cup or arch support was to diminish the spaces between the shoe and foot. Hong et al (Hong et al., 2005) reported that the total contact insert was proved to be that it reduced the peak pressure in the medial forefoot and the the effectiveness of the this insert was greater in the higher heels than in the lower heels and in flat heels. While, heel cup insert for high-heeled shoes effectively reduced the heel pressure and impact force and arch support insert reduced the medial forefoot pressure (Yung-Hui and Wei-Hsien, 2005) . In the study of shank curve, Cong et al. (Cong et al., 2008) described that PTI in forefoot region decreased as the depth of last increased, while it increased in midfoot and hind foot regions. Besides, this research also concluded that higher depth of last generated the less peak pressure in the forefoot region in mid stance phase.

The results of our study were similar with the above ones; however, there were some differences existing. In term of similarity, as the increasing of arch height, the more midfoot contact was observed, and then the PP under the forefoot decreased; in term of distinction, in the 50 and 80mm heel height, the PTI was better distributed under the forefoot when wearing the shoes with control last, not the higher last which showed more mid foot contact; while in the lower heel height, higher last is the better choice in attenuating the PTI value. Hence, our hypothesis was partially manifested, for that wearing with heel height of 20mm the best last should be design with the more arch support, such as the higher last, because this type of arch height attenuated the peak pressure and pressure-time integral under the forefoot area; but if wearing the shoe with higher heel height, such as the height of 50 and 80mm, the optimal type was the control last which utilized the data of foot. Nevertheless the control last functioned inferior in reducing the forefoot peak pressure than higher last, it achieved the both variables of peak pressure and pressure-time integral reduced in the forefoot.

There were some limitations during this study, such as the criterion of degree of arch changing. Only 2.5mm degree was applied, although differences were enough in influencing the contact area under midfoot, but this change was not enough to trigger more explicit differences between the comparisons during this study, that's also the reason why only a few significant differences were found, therefore further studies would focus more degrees changes, such as 5mm or even 7.5mm.

2 CONCLUSIONS

The main results were concluded as below: for the low heel-height shoes such as 20mm, the best last design was the higher last, in which the arch height was 2.5mm higher than that of height of feet; while for the mid and high heel-height shoes for instance 50 or 80mm, the optimal last design was the control last, where the arch shape was the same as the arch shape of the foot.

ACKNOWLEDGEMENTS

This study was supported by the project of "new leathers and furs with microbiological resistance for medical use" (2009DFA42850) and by the items from Agency of Science and Technology of Sichuan Province for financial support (item No. 2009HH0004).

Table 1 Result of in-shoe plantar pressure measurement of the three types of last under the heel-heighted shoe with 20mm, 50mm and 80mm heel height

variables	region-Last	cases	20mm heel height mean	P value	50mm heel height	P value	80mm heel height	P value
CA	Forefoot-higher last	10	40.30(0.95)	0.27a	40.27(1.92)	0.76a	39.22(1.59	0.70a
	Forefoot-control last	10	39.72(1.59)	0.58b	†40.21(2.14)	0.02b	†39.53(2.94)	0.04b
	Forefoot-lower last	10	40.04(1.40)		40.99(1.60)		38.42(3.64)	
	midfoot-higher last	10	33.08(3.88)	0.57a	27.69(4.48)	0.97a	20.62(4.36)	0.18a
	midfoot-control last	10	32.88(3.94)	0.73b	27.72(4.86)	0.07b	†21.33(4.00)	0.00b
	midfoot-lower last	10	32.57(5.34)		26.50(4.53)		18.06(4.93)	
	Hindfoot-higher last	10	26.95(0.39)	0.40a	26.23(0.82)	0.26a	25.88(1.38)	0.94a
	Hindfoot-control last	10	†26.81(0.67)	0.01b	26.37(0.82)	0.60b	25.85(1.05)	0.66b
	Hindfoot-lower last	10	26.12(1.01)		26.20(1.08)		25.67(1.73)	
PP		10	*242.38(59.64)	0.03a	239.00(45.05)	0.32a	274.31(102.16)	0.92a
		10	260.94(68.03)	0.46b	251.88(69.63)	0.38b	275.13(94.91)	0.41b
		10	253.31(90.49)		264.06(76.83)		297.88(70.74)	
		10	164.00(32.59)	0.57a	162.13(47.11)	0.67a	177.50(101.54)	0.51a
		10	166.50(33.71)	0.23b	157.00(54.73)	0.82b	184.75(89.07)	0.81b
		10	177.06(26.30)		154.50(56.30)		190.06(88.50)	
		10	182.38(30.75)	0.28a	127.56(23.42)	0.99a	*103.69(15.09)	0.01a
		10	174.94(24.11)	0.08b	†127.5(16.88)	0.00b	116.25(19.94)	0.09b

	10	163.31(23.97)		145.69(24.99)		123.06(23.50)	
PTI	10	*70.05(16.52)	0.01a	91.19(12.87)	0.30a	126.46(48.81)	0.07a
	10	80.09(22.72)	0.77b	85.78(21.10)	0.29b	109.94(28.25)	0.12b
	10	81.10(28.63)		90.17(23.57)		123.10(24.16)	
	10	61.85(11.62)	0.55a	72.48(16.21)	0.25a	83.57(47.17)	0.66a
	10	63.0912.95)	0.21b	69.64(18.54)	0.39b	80.64(30.91)	0.97b
	10	67.5712.31)		66.89(18.08)		81.02(38.53)	
	10	65.15(14.27)	0.18a	57.34(10.49)	0.20a	*51.21(8.11)	0.02a
	10	59.69(12.81)	0.81b	59.76(10.32)	0.21b	58.36(10.19)	0.13b
	10	58.72(15.62)		62.79(13.02)		62.31(14.43)	

*: indicate the differences between higher last and control last are significantly and the value is < 0.05.

†: indicate the differences between lower last and control last are significantly and the value is < 0.05.

'a' implies the comparison between higher last and control last;

'b' implies the comparison between lower last and control last.

'CA' is short for the 'contact area'; 'MF' is short for the 'maximum force'; 'PP' is short for the 'peak pressure';

'CT' is short for the 'contact time'; "PTI/" are short for the 'pressure/force-time integrals'

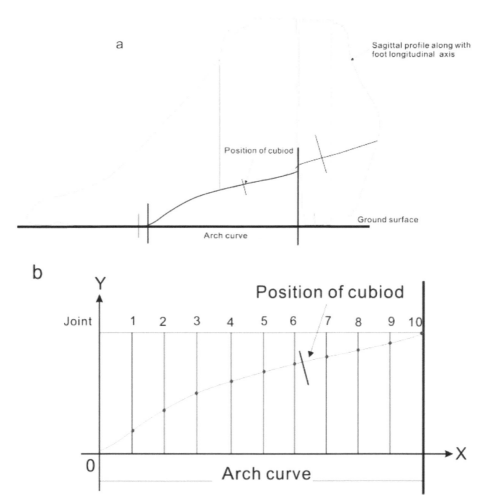

Figure 1 analysis of foot 3D model. (a) The sagittal profile along with foot longitudinal axis with hind foot elevated by the heel with height of 20, 50 and 80mm. The red section is the arch shape of foot which is the section of profile between the first metatarsophalangeal joint to the position of the heel contact; (b) The arch section was equally divided into 10 parts, each part has a pair of coordinate, and then all coordinates in one joint from 10 subjects were averaged to construct an average foot arch section.

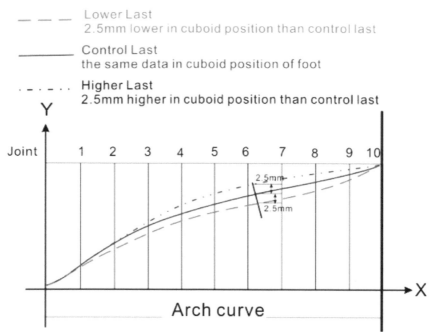

Figure 2 the last design with the modification on arch shapes. The red line was the control last in which the data was from the average foot arch. The other two lines indicated that the lower last and higher last which was 2.5mm lower or higher than the control last in the position of cuboids.

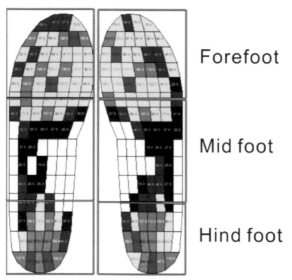

Figure 3 sole regions classification. The sole region was divided into forefoot, midfoot and hindfoot.

REFERENCES

W. H. Cho and H. K.Choi. 2005. The mechanism of postural balance control of high-heeled women. In. *Advances in Fracture and Strength, Pts 1- 4,* eds. Y. J. Kim, and H. D. Bae, (eds.), Asian Pacific Conference on Fracture and Strength. Jeju Isl, SOUTH KOREA.

Cong, Y., Luxinmon, Y. and Zhang, M. 2008. "Effect of Shank Curve of High-heeled Shoe on the Plantar Pressure Distribution." Paper presented at 7th Asian-Pacific Conference on Medical and Biological Engineering, Apcmbe, 2008.

Demits, S., Coorevits, P., De Clercq, D. et al. 2011. Reliability and Validity of the INFOOT Three-dimensional Foot Digitizer for Patients with Rheumatoid Arthritis. *Journal of the American Podiatric Medical Association* 101: 198-207.

Hawes, M. R., Sovak, D., Miyashita, M. et al. 1994. Ethnic-differences in forefoot shape and the determination of shoe comfort. *Ergonomics* 37: 187-196.

Hong, W. H., Lww, Y. H., Chen, H. C. et al. 2005. Influence of heel height and shoe insert on comfort perception and biomechanical performance of young female adults during walking. *Foot & Ankle International* 26:1042-1048.

Kouchi, M. and Tsutsumi, E. 2000. 3D foot shape and shoe heel height. *Anthropological Science* 108: 331-343.

Speksnijder, C., Vdmunckhof, R., MooneN, S. et al. 2005. The higher the heel the higher the forefoot-pressure in ten healthy women. *The Foot* 15:17-21.

Stefanyshyn, D. J., Nigg, B. M., Fisher, V. et al. 2000. The influence of high heeled shoes on kinematics, kinetics, and muscle EMG of normal female gait. *Journal of Applied Biomechanics* 16: 309-319.

Yung, Hui, L. and Weihsien, H. 2005. Effects of shoe inserts and heel height on foot pressure, impact force, and perceived comfort during walking. *Applied Ergonomics* 36: 355-62.

CHAPTER 34

Fingertips Deformation under Static Forces: Analysis and Experiments

Esteban Peña-Pitarch[1], Neus Ticó-Falguera[2], Adrià Vinyes-Casasayas[1] and Damián Martinez-Carmona[1]

[1] Escola Politècnica Superior d'Enginyeria de Manresa (EPSEM-UPC)
Av. Bases de Manresa, 61-73
(08242) Manresa (Spain)
Corresponding author: esteban.pena@upc.edu
[2] Althaia Xarxa Assistencial de Manresa, F.P.

ABSTRACT

Introduction: When designers, physicians, and engineers are working in a new design or calculation forces exerted by fingertips they need to consider that each person's fingertip is different. Taking into account the material properties for each person the Poisson's ratio, Young's modulus, and density are different. These properties will be influential in the fingertip force.

Objectives: Development of a three-dimensional human fingertip model to simulate with finite element methods a behavior of soft finger when it is under a force. In a soft finger, Poisson's coefficient can be different according to person and age. Other factors like temperature have influence on this coefficient. Young's modulus can be used for the dermis, epidermis, soft tissue, grease, and tendons. Once we have collected all of these dates we create a three-dimensional model.

Methods: With a caliper and a precision balance Tefal-EASY we measure three times the force and deformation of fingertips, with 0 N no deformation to 7 N maximum deformation in z-axis (axis normal to balance and fingertip) using the same conditions for each experiment. Once we have done experiments with a person we introduce constrains in a finite element program to compare deformation in z-axis and force with the results and experiments.

Keywords: Fingertip deformation, force, Young's modulus, Poisson's ratio.

INTRODUCTION

When designers, physicians, and engineers are working in a new design or calculation forces exerted by fingertips they need to consider that each person's fingertip is different. Taking into account the material properties for each person the Poisson's coefficient, Young's modulus, and density are different. These properties will be influential in the fingertip force. To calculate fingertip deformation is not new, Johnson [1] describes the stresses and deformations when the surfaces of two bodies are brought into contact and Wriggers [2] shows the equations to calculate fingertip deformation for different types of fingertip in a transversal plain. [3] shows the process developed to calculate deformation for a fingertip. Authors create a 3D model from images from Computer Tomography (CT) for a fingertip and submit with a compression stress with a surface and they use commercial software of Finite Elements (FE). In our case we use a similar method to calculate a fingertip deformation but using a virtual model fingertip drawn with a General Public License (GPL) program. However both process were a little different, the results and validation for two methods had similar results. Serina et al [4] explain how a soft part of fingertip works when it is under continuous forces in the time. The objective of this study was to look for a solution for hands with some pathologies in tendons and nerves of hands with long time stress over the fingertip. Same authors in [5] show the behavior of fingertip when they are subjected to continuous loads over time. Tests are performed to determine the physical properties of fingertip under various forces and angles. The goal that motivates this study is to find solutions to diseases suffered in the tendons and nerves of the hand due to prolonged efforts on the surface of the finger. Wu et al [6] examine the effects of variable time dynamic loads applied on the surface of the finger. That is, the influence effort applied during a given time (especially when pressing with your finger on a surface) in the tensions suffered by muscles and tendons of the human finger.

To make the calculations and put the conditions of the material -in this case the human finger-, we do not take into account all the small arteries or muscles that are inside the finger because when calculating deformations they do not suffer finger influence. Therefore, the finger will be distributed in three different materials or parts, which should look good and set its properties so that the calculations are reliable. These parts will be the nail, soft tissue and bone, see Figure 1, where its properties are those of the distal phalanx, to be in the area of application of force in most cases.

The nail, whose role is to protect and cover the terminal part of the fingers, is located at the upper end of the finger. Being a rigid part, its characteristics affect the way that the finger will undergo deformation when performing strength, which will be smaller. The phalanges of the fingers are bony structures, and are an important part, because without them, the finger movement it would be virtually impossible. The distal phalanx is located inside the finger, surrounded by soft tissue. As the nail is a rigid part that will also undergo deformation the finger looks small. The physical properties of the phalanges are similar to those of the nail. The soft tissue is the most visible part of the finger, surrounds the distal phalanx and protects a portion of the nail.

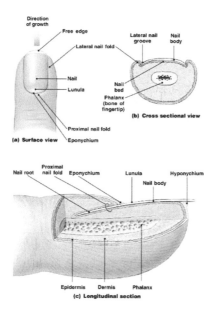

Figure 1. Composition and distribution of the interiors of the finger

Next table shows the properties obtained form [3].

	Nail	Phalanges	Soft tissue
Young's modulus	$1.7*10^4$ MPa	170 MPa	$65*10^{-3}$ MPa
Poisson's ratio	0.3	0.3	0.45
Density	1.8 g/cm^3	1.8 g/cm^3	1 g/cm^3
Specific weight	0.01764 N/cm^3	0.01764 N/cm^3	0.0098 N/cm^3

Table 1. Properties of materials of fingertip considered

MATERIAL AND METHODS

In this section we explain the material and methodology used in the experimental process. We use a caliper and a precision balance. We do the experiment with one subject and do three times the same experiment. The experiment is shown in Figure 2.

With a caliper and a precision balance Tefal-EASY we measure three times a force and a deformation of fingertips, with 0 N no deformation to 7 N maximum deformation in z-axis (axis normal to balance and fingertip) using the same conditions for each experiment. To obtain experimental dates we proceed to:

1. Measure by a caliper the size of the finger used in the experiment before being subjected to any force.
2. Activate the scale and to apply a desired finger pressure on the surface of the scale until the display shows the appropriate value.
3. Get the dimensions of finger subject to load using the caliper (Figure 2). Repeat each measure to ensure the correct data.
4. Repeat the second and third step for each of the load to obtain the deformations.
5. Enter in a table the results obtained.

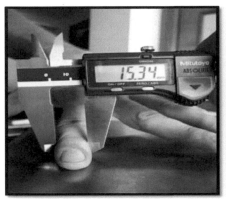 

Figure 2. Obtaining experimental dates.

Table 2 shows all the values obtained, where Z1 are the values obtained in z-axis direction for the first trial, Z2 is the second trial and Z3 is the third trial. Column average is the average of the three trials and column deformation is the rest of measure of the fingertip without any deformation minus the average in each row. Note that when the subject applies a force of 6 N and a force of 7 N the deformation is very small. For this reason when we talk about results we only consider until 6 N, a bigger force than this value does not have deformation compared with a 6 N force.

	Z1	Z2	Z3	Average	Deformation
0 N	15.08	15.08	15.08	15.08	-
1 N	14.14	14.20	14.15	14.16	0.92
2 N	13.57	13.50	13.49	13.52	1.39
3 N	13.22	13.185	13.20	13.201	1.857
4 N	12.84	12.80	12.88	12.84	2.36
5 N	12.09	12.12	12.10	12.103	2.867
6 N	11.84	11.85	11.86	11.85	3.33
7 N	11.83	11.84	11.83	11.833	3.353

Table 2. Deformation for each force measure on transversal section of fingertip

RESULTS

Once we have the finger model, we use commercial software to calculate the finger deformation from the finite element method (FEM), for various reasons: The program chosen for the calculations is the RamSeries.

This program is a software product for simulating and structural design in various fields of engineering, based on the finite element method, which includes the company Compassis.

The company Compassis offered to work with Custom GiD software, which allows working with complex geometries, meshes generated in the models, and, once the facts calculations are obtained, to analyze the results using various graphic techniques. First we introduce the restrictions; we fix the top part of the finger. It sets the top of the finger and the nail as the force applied under the finger, and thus prevents the movement of the finger. It also restricts the finger inside, the significant role that would make the phalanges of the finger. Once the fixed constraints are done, we restrict all other surfaces as elastic constraints. Once done, the constraints are already defined. The next step is to specify the material properties, a layer is chosen as the isotropic material. The last step is to apply the necessary force to the finger. We applied different forces to check the deformation suffered by each finger. We analyzed forces from 1 N to 6 N. We calculated up to 6 N, bigger force of the finger is not deformed as most have been tested in the experimental model. In Table 2 it is shown the experiment with a force of 6 N and 7 N, and we do not appreciate significant differences in displacements for z-axis.

An estimated area pressure was obtained in the experiment and the N divided by it. Once the force is already applied, proceed to the meshing of the model to be calculated using the finite element method. The tests results will be based on the deformation suffered by the finger. The following photos are images of a set of colors depending on the deformation of each area (at the bottom of the pictures) where there's a legend to interpret the different degrees of deformation suffered by each zone according to the color you have.

Figure 3. Fingertip deformation with 1 N force applied

DISCUSSION

When carrying out calculations with the finite element program, as explained above, the Young's modulus and Poisson's ratio can influence the results. A comparison was made with different values for Young's modulus to see which one is most similar to the experiment carried out and to do calculations with this value. In the case of Young's modulus four different values were calculated, and that comes closest to the results of the experiment was 65 kPa, as it is shown in Figure 4.

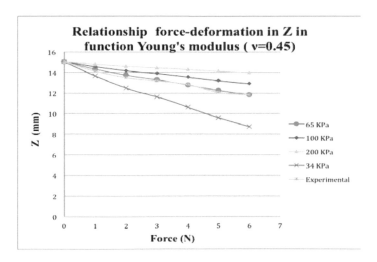

Figure 4. Graphic with different Young's modulus based on the Poisson ratio 0.45

Poisson's ratio has also compared the results to the different values that could exist. It has been observed that the Poisson's ratio has a little influence on the

deformation of the surface of the finger. As shown in the Figure 5 the differ
the values obtained with the different coefficients is minimal. Since the difl
between the values is minimal, we used for calculating the coefficient of 0.45

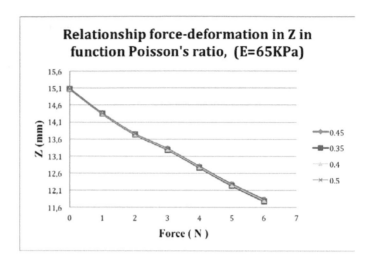

Figure 5. Graphic with different Poisson's ratio based on the Young's modulus 65 kP

Table 3 shown the experimental results in one column and the other c
shows the results of the finite element calculation.

FORCE (N)	EXPERIMENTAL	RAMSERIES (FEM)
1	0.92	0.72001
2	1.39	1.3258
3	1.857	1.7644
4	2.36	2.284
5	2.867	2.8006
6	3.33	3.2508

Table 3. Fingertip deformation for each force in mm. Experimental and FEM

The results obtained by the calculation show a similar pattern wi
experimental data. The displacements produced may not be exactly equal fo
methods, since the experimental method can have a margin of error having to
oneself to be difficult to maintain the same power for a few seconds while me
displacement. Therefore, the 3 replicates were performed for each measur
However, it is somewhat difficult to make accurate displacement. Therefore
differences obtained between both methods can be considered valid.

Figure 6. 1 N force applied in the experiment and simulation.

Figure 7. 6 N force applied in the experiment and simulation

CONCLUSIONS

When comparing the experimental values and those obtained by the simulation we can ensure that the model created by the finite element calculation is a model that ensures reliable and accurate approximation to the actual behavior of the tissues that make up the human finger when efforts are under compression. This model may be used in future studies to obtain accurate estimates of tissue deformations for any desired strength, being especially useful in cases where we want to establish a force determined to carry out some applications, such as activating a switch or detecting a minimum pressure of the finger on a surface with the help of sensors.

ACKNOWLEDGEMENT

This work was partially supported for Spanish government by the project DPI2010-15446.

REFERENCES

[1] Johnson, K.J., "Contact Mechanics", ISBN-0-521-34796, Cambridge University Press, 1999.

[2] Wriggers, P., "Computional Contact Mechanics", Second edition, ISBN-10 3-540-32608-1, Springer, 2002.

[3] Shimawaki, S., Sakai, N., "Quasi-static Deformation Analysis of a Human Finger using a Three-dimensional Finite Element Model Constructed from CT Images", *Journal of Environment and Engineering,* Vol. 2, N1 (2007), pp.56-63.

356

[4] Serina, Elaine R., Mockensturm, Eric, Mote Jr., C.D., Rempel, David, "A Structural Model of the Forced Compression of the Fingertip Pulp", *Journal of Biomechanics*, Vol. 31 (1998), pp.639-646.

[5] Serina, Elaine R., Mote Jr., C.D., Rempel, David, "Force response of the fingertip pulp to repeated compression-effects of loading rate, loading angle and anthropometry", *Journal of Biomechanics*, Vol. 30, No.10 (1997), pp.1035-1040.

[6] Wu, John Z., An, Kai-Nan, Cultip, Robert G., Krajnak, Kristine, Welcome, Daniel, Dong, Ren G., "Analysis of Musculoskeletal Loading in an Index Finger During Tapping", *Journal of Biomechanics*, Vol. 41 (2008), pp.668-676.

CHAPTER 35

The Arch Analysis with 3D Foot Model Under Different Weight-loading

Yu-Cheng Lin

Overseas Chinese University
Taichung City, Taiwan
yclin@ocu.edu.tw

ABSTRACT

The design of foot arch is a critical issue for manufacturing comfort and satisfied shoes. Traditionally, linear dimensions, just like arch height, are applied to describe the shape of arch. However, it is inappropriate to describe 3D geometric features with 1D data. Furthermore, the 3D change of arch is not so clear. The main purpose of this study was to discuss the change of arch during different weight-loading with foot 3D scanned data. A 3D Scanner was used to collect the three-dimensional foot models by scanning subject's foot and then construct relevant 3D feet models. The volunteers who aged from 18 to 25 were undergraduate and graduate students. Three levels of loading were considered, 0%, 50% and 100% of body weight. With the 3D scanned data, the size and trend of change on geometric features of arch during different loading levels will be obtained. The data of length curvature, circumference, area, angle and volume were collected from the 3D scans and were analyzed to discover the change and variation of arch during different weight- loading. According to the result, the features and trend of variation for arch at different loading weight were more obvious.

Keywords: arch, 3D scan, weight-loading

1   INTRODUCTION

With the social development and pursuit of quality life, the fitness and comfort of footwear become more and more important. The diversity of shoe's design and functions is also emphasized due to the variety of work and activity. The feet have to support full body weight and relevant external loading, and the arches play an important role on motion and shock-absorbing. A detailed knowledge of foot shape and foot structure is particularly important for those involved in the design and construction of shoes, as a fundamental prerequisite for a shoe to be comfortable is that the internal shape of the shoe closely approximates the shape of the foot (Hawes et al. 1994[b]). Despite the importance of foot shape to shoe comfort, the most omprehensive studies investigating foot morphology have been conducted on adults, particularly military personnel (Parham et al. 1992, Hawes and Sovak 1994[a]) or by footwear manufacturers when designing shoe lasts (Kouchi 1995, Kusumoto et al. 1996).

One of the more important and highly variable structural char acteristics of the human foot is its medial longitudinal arch (Cavanagh and Rodgers, 1987; Shiang et al., 1998), which provides necessary shock absorption for the foot during activity (Shiang et al., 1998; Williams and McClay, 2000). Variations in arches and severe gait problems are "treated" with orthotic devices (Williams and McClay, 2000). When orthotic devices are prescribed for use in running, they have positive effects in only approximately 70% to 80% of runners (Gross el al., 1991; Saxena and Haddad, 1998). The design, development, and fabrication of orthoses are critical to their effectiveness. Orthoses that effectively support the longitudinal arches of the foot have been found to significantly decrease strain in the plantar aponeurosis (Kogler, 1996).

Traditional method to analyze the arch is to measure the linear dimensions and curvature/circumference of arch directly. With the introduction of 3D scanning method, related arch data are taken from the 3D arch digital models instead of direct measurements. The optical 3D body scanning system is one of the major measuring instruments now (Tsai, 2000). Adopting optical measurement method can obtain body surface data clouds quickly and exactly within seconds and then construct the 3D digital model. Although the model may be fragmented because of shaded surface and poor reflection quality, the applications of 3D model are more extensive than traditional anthropometric methods. In order to avoid the fragmented scanned images, the standing posture was suggested (Daanen, 1998).

Traditionally, feet are classified as being high, normal, or low arched. A high-arched foot is supposed to be at increased risk for injury to the bony structures on the lateral side of the foot (oversupinated), whereas a low-arched foot can be at greater risk for soft-tissue damage on the medial side of the foot (Howard and Briggs, 2006). A proper arch design is the basis to manufacture modern footwear and insoles. Also, despite having many common anatomical characteristics, the shape and biomechanics of the foot differ greatly between individuals. Existing methods of foot type classification are typically based on the measurement of morphological parameters of the foot, mostly in the standing weight-loading position, or during movement. Methods of foot type classification based on foot morphology could be put into one of the following categories: visual non-quantitative inspection, anthropometric values, footprint parameters and radiographic evaluation (Razeghi, 2002).

It is important to have a relatively easy and reliable way to classify the foot arch. Thus, footprint parameter methods are usually adopted on most studies due to the simplicity and non-invasion measurement. An imprint of the foot, provided by a simple ink pad or recently developed ophisticated pressure transducers, has been utilized to classify feet into groups. In either case, the core assumption is that any changes in the shape and orientation of structural components of the foot, collected while the subject is standing or moving, would be reflected in the imprint. Measurement of width or area of contact on the imprint is suggested to provide a simple and objective means of foot classification. Some indices and angles have also been introduced to evaluate the static or dynamic position of the foot while it comes into contact with the supporting surface. The commonly current methods to classify foot type are arch index, modified arch index, arch angle, footprint index, arch-length index, truncated arch index and Brucken index (Razeghi, 2002).

3D measurement method substitutes for footprint parameter methods and direct measurement since the process to complete one scan is speedy and all data for foot type classification is able to calculate from the 3D data clouds. Typically, the 3D point data are then used to calculate heights, lengths, widths and angles (Liu et al, 1999; Luximon, 2003). Point cloud, mesh or surfaces of objects can be modeled even though the surface characteristics may not be perfect (Yahara, 2005). However, a 3D digital model should provide more 2D and 3D information, not only 1D data. Furthermore, the deformation of arch within different weight-loading and the change of relevant features are also important on the design of footwear and application.

The purpose of this study was to analysis the deformation of foot under different weight-loading from the 3D foot models scanned by 3D foot scanner and to derive the correlative critical measurements. The levels of weight- loading were considered as 0%, 50% and 100% of subject's body weight themselves. The subjects were asked to control the weight-levels in a small range. Thus, the shape deformation of arch was taken from the 3D foot models.

2   METHODOLOGY

2.1   Portable 3D body scanner and softwares

The portable 3D body scanner manufactured by LT-tech co. was employed to retrieve the children body surface data. A set of scanner contain three scan modules (Figure 1). The precision is ± 1.0 mm and the resolution is 2.5 mm. About eighty thousand 3D data points were taken within 5 seconds. The scanned images were edited and modified by the relevant software and anthropometric dimensions were retrieved from the 3D digital images. The scanning process was conducted in a darkroom to avoid the influence of light.

Figure 1. The scanning modules of the portable 3D body scanner.

The software, Beauty 3D, was utilized to collect the 3D data clouds and relative color images, and to merge all data points into single 3D models with noise reduction function. Another image process software, called Anthro 3D, was adopted to reduce the noise, and calculate the data desired, like linear distances, curvatures, circumferences, area, cross-section area, surface area, volume, angles, etc.

2.3   Experimental process and sample

The levels of weight-loading were considered as 0%, 50% and 100% of subject's body weight themselves. The subjects were asked to stand on of black support platform and then put the measured foot into the scanning area slightly. The heights of black support platform and scanning area are the same. A weight scale was set under the scanning area in order to assist subjects to control weight-loading. All subjects had to control their weight-loading in a small range. The 0% body weight-loading was considered as the measured foot was put on the scanning area lightly and center of body gravity was on the other foot. The 50% body weight-loading was considered when the body weight was put on two legs equally so the reading of weight scale was almost 50% of weight. When the subject put his full weight on the measured foot only, the 100% body weight-loading was obtained.

Five male and five female were invited to attend a pilot experiment in order to make sure the range of weight-loading control data. Each foot was measured six times. Table 1 lists the result of pilot experiment. Thus, all subjects have to control their weight-loading in the range of mean plus/minus one standard deviation.

Table 1 the test result of weight-loading control

	100% weight-loading	50% weight-loading	0% weight-loading
Left foot	98.70% ±1.66%	49.69% ±2.41%	6.04% ±1.89%
Right foot	98.78% ±1.47%	49.52% ±2.81%	7.30% ±1.74%

Thirty volunteers aged from 18 to 25, including 15 male and 15 female, were recruited to attend the former experiment. Several landmarks were set on the anatomic and critical points before scanning process. Each volunteer was asked to scan his both feet at the conditions of 0%, 50% and 100% body weight separately. Three typical data were derived from the 3D foot models, i.e. arch height of prominent navicular bone (pre-marked), arch height of soft tissue and angle of medial longitudinal arch (prominent navicular bone). The definitions of derived dimensions are shown as Table 2.

Table 2 the definitions of dimensions

Dimension	Definition
arch height of prominent navicular bone	The height of prominent navicular bone (highest point of the medial longitudinal arch) in the sagittal plane to floor.
arch height of soft tissue	The height of the highest point along the soft tissue margin of the medial plantar curvature may to floor.
angle of medial longitudinal arch	The angle between the line from prominent navicular bone to metatarsal head with the line from prominent navicular bone to calcaneus.

3   RESULT

The measurement results of derived data are shown in Table 3. There are significant difference to level of weight-loading and gender for all three dimensions, but no significant different between right and left foot. With the increase of weight-loading, both arch heights decreases. The same trend is obtained with observing data of angle of medical arch.

Table 3 Result of measurements (mean±standard deviation).

			0%	50%	100%
arch height of prominent navicular bone	Male	Right foot	4.8±0.5cm	4.6±0.5cm	4.4±0.5cm
		Left foot	5.1±0.9cm	4.6±0.8cm	4.4±0.7cm
	Female	Right foot	4.1±0.4cm	3.7±0.4cm	3.5±0.3cm
		Left foot	4.1±0.5cm	3.6±0.4cm	3.3±0.4cm
arch height of soft tissue	Male	Right foot	2.3±0.4cm	2.0±0.3cm	1.8±0.3cm
		Left foot	2.4±0.5cm	1.9±0.4cm	1.7±0.3cm
	Female	Right foot	1.8±0.5cm	1.4±0.4cm	1.3±0.3cm
		Left foot	1.9±0.3cm	1.6±0.2cm	1.3±0.2cm
angle of medial longitudinal arch	Male	Right foot	127.1±5.3°	129.0±5.5°	131.3±5.4°
		Left foot	123.4±8.9°	128.6±7.7°	130.9±7.4°
	Female	Right foot	127.6±5.9°	132.5±4.9°	135.3±4.8°
		Left foot	125.2±6.9°	130.1±6.4°	133.4±6.3°

Table 4 summaries the result of Duncan analysis. The Duncan analysis result of two kinds of arch heights shows the three levels of weight-loading are all significantly different. However, there is no significant difference between 100% and 50% of weight-loading on angle of medial longitudinal arch. Thus, the change of angle of medial longitudinal arch becomes smaller as the weight-loading increase, so the quantity of difference between 100% and 50% weight-loading is small enough to becoming insignificant.

Two kinds of arch heights of male are all larger than those of female, but the angle of medial longitudinal arch of male is smaller than that of female. Thus, males have higher arches than females. That is, males tend to be high arched.

Table 4.Summation of the Duncan analysis

Dimension	Weight-loading	Gender
arch height of prominent navicular bone	100% < 50% < 0%	Male > female
arch height of soft tissue	100% < 50% < 0%	Male > female
angle of medial longitudinal arch	100% = 50% > 0%	Male < female

4   CONCLUSION

In order to realize the deformation under different weight-loading, this study conducted an experiment to obtain the 3D foot scans and analyzed the data taken from the 3D foot scanned model. Arch height of prominent navicular bone, arch height of soft tissue and angle of medial longitudinal arch are calculated. The result indicated there are weight-loading and gender differences. There is no significant between 100% and 50% weight-loading for angle of medial longitudinal arch, but the difference between 50% and 0% is significant. Thus, the change of angle of medial longitudinal arch becomes smaller with the increase of weight. Also, males tend to being high arched than females.

A further study should be done to obtain other information that be only derived from 3D scanned models, like arch volume or cross-section of prominent navicular bone. Thus, the 3D shape deformation of arch could be clearer and benefits for design and construction of shoes.

REFERENCES

Cavanagh, P. R. and M. M. Rodgers. 1987. The arch index: a useful measure from footprint. *Journal of Biomechanics* 20: 547.

Daanen, H .A. M., and G. Jeroen. 1998. Whole body scanners. *Displays* 19: 111-120.

Gross, M. L., L. B. Davlin and P. M. Evanski. 1991. Effectiveness of orthotic shoe inserts in the long-distance runner. *The American Journal of Sports and Medicine* 19: 409-417.

Hawes, M.R. and D. Sovak 1994[a]. Quantitative morphology of the human foot in a North American population. *Ergonomics* 37: 1213–1226.

Hawes, M.R., D. Sovak, M. Miyashita, S. J. Kang, Y. Yoshihuku and S. Tanaka. 1994[b]. Ethnic differences in foot shape and the determination of shoe comfort. *Ergonomics* 37: 187–196.

Howard, J. S. and D. Briggs. 2006. The arch-height-index measurement system: a new method of foot classification. *Athletic Therapy Today* 11: 56-66.

Kogler, G. F., S. E. Solomonidis and J. P. Paul. 1996. Biomechanics of longitudinal arch support mechanisms in foot orthoses and their effect on plantar aponeurosis strain. *Clinical Biomechanics* 11: 243-254.

Kouchi, M. 1995. Analysis of foot shape variation on the medial axis of foot outline. *Ergonomics* 38: 1911–1920.

Kusumoto, A., T. Suzuki, C. Kumakua and K. Ashizawa. 1996. A comparative study of foot morphology between Filipino and Japanese women. *Annals of human biology* 23: 373–385.

Liu, W. et al., 1999. Accuracy and reliability of a technique for quantifying foot shape, dimensions and structural characteristics. *Ergonomics* 42: 346-358.

Luximon, A. et al. 2003. Foot landmarking for footwear customization, *Ergonomics* 46: 364-383.

Parham, K., C. Gordon, and C. Bensel. 1992. Anthropometry of the foot and lower leg of U.S. Army soldiers: Fort Jackson, S.C. Natick, MA: U.S. Army Natick Research, Development and Engineering Center.

Razeghi, M. and M. E. Batt. 2002. Foot type classification: a critical review of current methods. *Gait and Posture* 15: 282–291

Saxena, A. and J. Haddad. 1998. The effect of foot orthoses on patellofemoral pain syndrome. *The Lower Extremity* 5: 95-107.

Shiang, T. Y., S. H. Lee and S. J. Lee. 1998. Evaluating different footprint parameters as a predictor of arch height. *IEEE Engineering in Medical and Biology Magazine* 17: 62-67.

Tsai, C. C. 2000. Apply 3D anthropometric data to establish sizing system for elementary and high school students. *Master thesis*, Tsing Hua University, Taiwan.

Williams, D. S. and I. S. McClay. 2000. Measurements used to characterize the foot and the medial longitudinal arch. *Physical Therapy* 80: 864-873.

Yahara, H. et al. 2005. Estimation of anatomical landmark position from model of 3-dimensional foot by the FFD method. *Systems and Computers in Japan* 36(6): 1-13.

CHAPTER 36

Multi-Scale Human Modeling for Injury Prevention

Sultan Sultan, Tim Marler

The University of Iowa
Virtual Soldier Research (VSR) Program

ABSTRACT

In considering the human system, multi-scale modeling is critical for providing accurate modeling and simulation results. The human system is too complex to be modeled on all levels of fidelity by a single tool or a single group of researchers. Thus, a platform is needed that lends itself to multi-scale modeling, whereby different models with varying degrees of fidelity can be integrated. This is especially important for local injury prevention models that represent specific human joints or soft tissue. These more focused models can be useful independently, but often their benefits are fully realized only in the context of a complete system-level human model that essentially connects the local model to a virtual environment. Incorporating high-fidelity local models within a larger-scale digital human model (DHM) provides more accurate input to the local models and provides increased ease of use. Although multi-scale modeling is an active area of research, little work has been completed that integrates high-fidelity biomechanical models with complete system-level DHMs. This work focuses on integrating Santos, a joint-based predictive DHM, with an approximate static muscle model and high-fidelity finite element analysis (FEA) knee model. The system-level human model (Santos) and the high-fidelity FEA model have been successfully coupled. Predicted torque values and results from the FEA model coincide nicely with results in the literature. This integrated system allows one to study the effects of various postures on the knee joint and to modify tasks in order to reduce the likelihood of injury. Consequently, the newly developed multi-scale model promises to provide an extremely useful tool for injury prevention.

Keywords: Digital human modeling, multi-scale modeling, FEA modeling

INTRODUCTION

As one attempts to model more complex systems, depending on and growing a single model becomes less tractable. This is especially true of a system as complex as the human body. As the applications for digital human modeling (DHM) extend beyond the ergonomics of reaching tasks, the details of gait, or the medical issues surrounding a single human organ or component, coupling multiple models becomes necessary for answering critical questions, especially in the case of injury prevention. In fact some authors contend that almost all problems involve multiple scales (Weinan and Bjorn Engquist, 2003b).

Although the field of DHM has gained significant momentum and although many human-component models have surfaced, there are deficiencies with the current state of the art and thus many opportunities for advancement. Most current human models geared towards simulating whole-body activities are inherently limited with respect to fidelity; they generally represent system-level models that determine performance with respect to joint angles or Cartesian position of key landmarks. Currently, there is no single human modeling software platform that allows for easy expansion, growth, and integration. Component models are typically independent and uncoupled from other human systems, and they tend to be limited in scope addressing only a single section of the body.

Especially when considering the humans system, multi-scale modeling is critical for providing accurate and useful modeling-and-simulation results. The human system is far too complex to be accurately modeled on all levels of fidelity by a single tool or even a single group of researchers. Although multi-scale models are becoming more prevalent in the field of Biology, linking the molecular scale to the whole-organism scale (Martins et al, 2010), multi-scale modeling has not yet surfaced in the DHM field. To be sure, multi-scale modeling has been used with some human components. For instance, researchers propose multi-scale models for increasing the accuracy of bone models (Cacciagrano et al, 2010).

Recently, a number of multi-scale numerical methods, such as residual free bubbles (Sangalli, 2003), the variational multi-scale method (Hughes et al, 1998), the multi-scale finite element analysis (FEA) method (Hou and Wu, 1997), and the heterogeneous multi-scale method (HMM) (Weinan and Bjorn, 2003) have been introduced. Multi-scale modeling can lead to higher accuracy in various applications such as finite element analysis (Masud and Bergman, 2005), homogenization (Li et al, 2004), and biomechanics (Tang et al, 2007).

Despite the multitude of work with multi-scale modeling in general, little work has been completed within the context of a scalable and open DHM. Thus, this work focuses on integrating predictive joint-based system level DHMs with high-fidelity joint models. Santos, developed at The University of Iowa's Virtual Soldier Research (VSR) Program, provides the system level human model, and is linked with an FEA model of the human knee. The intent is not necessarily to develop a unique knee model, but to demonstrate the feasibility of linking complete human models with different levels of fidelity, and thus laying the groundwork for an injury prevention tool. Using joint torques predicted by Santos, as well as approximate shear and compression forces due to muscle contraction, high fidelity

FEA models are used to analyze displacement and stress responses within skeletal components and soft tissue.

METHOD

1. High-fidelity Model

MRI data was used as the basis for the solid models of the knee joints. The raw MRI data was processed using Osirix software for segmentation. After the segmentation process, the point data in VTK format was used to model three-dimensional geometry with Geomagic studio software. The 3D model in IGES format was then exported to Abaqus software for meshing and analysis. All ligaments and menisci were represented with very basic models in Abaqus 9.11, and the components were then assembled to form a complete knee joint model.

All joint components were meshed using the tetrahedron element. Constraints were used to tie all ligaments to the joint as shown in Figure 1. Contact problems were developed for three pair of components. The first pair includes the contact between femur cartilage and patella cartilage, the second between femur cartilage and tibia cartilage, and the third between femur cartilage and menisci. The coefficient of friction used for the contact problems was 0.01.

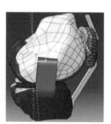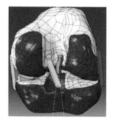

Figure 1 FEA knee model

Values for Young's modulus and Poisson's ratio were taken from the literature (Kubicek and Florian, 2009;Moglo and Shirazi-Adl, 2005; Chantarpanich et al, 2009) and are shown in Table 1. Zero-displacement boundary conditions were applied to the upper face of the femur. Three compression loading conditions were were applied to the lower face of the tibia. For the purpose of FEA model validation, results for contact pressure from the present study are compared with those in the literature, as shown in Table 2.

Table 1 Material properties

Descriptions	E (MPa)	ν
Femur and Tibia Bones	17600-18400	0.3
Cartilage and Menisci	12	0.45
Ligaments	300	0.45

Table 2 Comparison of max contact pressure (CPRESS) in MPa with literature

Compression Force (N)	890	1000	1500
Present study	2.234	2.515	3.230
Martine et al. (2009)	2.25	-	-
Shirazi et al. (2008)	2.89	2.96	3.42
Inaba et al. (1990)	1.4	1.6	2.3
Ahmed et al. (1983)	2.75	-	-
Fuku et al. (1980)	-	3	4
Walker et al (1975)	-	3.2	2.92

The results are generally in agreement with the existing literature. For each loading condition, the simulated compression force falls within the range of results in the literature and coincides with the average of literature-base results. Of course, every FEA model is unique. Nonetheless, these results suggest the proposed model is acceptable.

2 System-level Model

While the above-described FEA model provides the high fidelity joint model, the system-level model governing the biomechanical performance of the overall body is provided by Santos (Abdel-Malek et al, 2004). Santos is built on a biomechanically accurate musculoskeletal model with approximately 109 predicted degrees of freedom (DOFs). There is one joint angle for each DOF, and the relationship between the joint angles and the position of points on the series of links (or on the actual avatar) is defined using the Denavit-Hartenberg (DH)-notation (Denavit and Hartenberg, 1955).

The core functionality for any virtual human is the ability to model human posture and motion realistically. In this respect, Santos is *predictive*. Posture and motion are not based on prerecorded motion or data that must be updated when there are alterations in the virtual environment. Rather, Santos actually predicts how people strike various poses and how they move. The underlying optimization-based approach is extremely fast, allowing Santos to operate in real time. The most unique aspect of our approach is the ability to handle complex dynamics problems. With this approach, the joint angles essentially provide design variables that are determined through optimization. The objective function(s) is one or more human performance measures, such as energy, discomfort, joint displacement, etc. A detailed discussion of the underlying method for predicting posture while considering external loads is provided by Liu et al (2009) and Marler et al (2011).

Joint angles serve as the design variables, which are incorporated in various objective functions and constraints, the fundamental formulation for which is given as follows:

Find: $\mathbf{q} \in R^{DOF}$

To minimize: $f(\mathbf{q})$

Subject to:

    1) Distance constraints involving given target points/lines/planes

2) Joint limits
3) Torque limits
4) All contact points remain on the contact plane
5) Balance
6) Static equilibrium, including body mass, external forces, and ground reaction forces

q is a vector of joint angles, and DOF is the total number of degrees of freedom. $f(q)$ can be one of many performance measures (Marler, 2005), but in this case represents a function of joint torques. The primary constraint, called the *distance constraint*, requires the end-effector(s) to contact a specified target point(s). In addition, the joint angles are constrained to remain between limits derived from anthropometric data. In addition to these basic constraints, many other constraints can be used as boundary conditions to represent the virtual environment. Including conditions of static equilibrium allows one to consider joint torques that are provided as input to the FEA Model. The Santos model predicts not only joint angles and joint activation torques, but ground reaction forces (GRF) as well.

3 Muscle Model

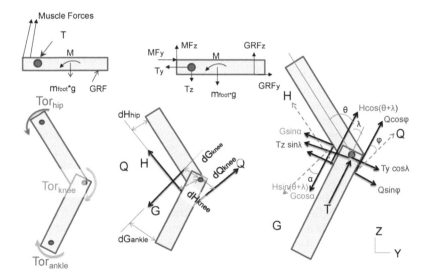

Figure 2 Static analysis of the ankle joint

Although Santos predicts activation torques for each DOF, these torques stem from dynamic motion only; they do not consider active muscles. Therefore, muscle forces must be determined. As an initial feasibility study, the muscle forces relevant to the knee are approximated using static analysis.

There are three main muscles acting on the knee joint: hamstring muscle, quadricep muscle and gastrocnemius muscle. For the ankle joint, there are two main muscles: the gastrocnemius and soleus muscles. The free body diagram for this analysis is shown in Figure 2, and is based loosely on the work by DeVita and Hortobagyi (2001).

MFy and **MF**z represent the y and z components of muscle forces acting on the ankle joint, respectively. They are analytically computed as 25% and 10% of the body weight (BW), respectively (DeVita and Hortobagyi, 2001; Lin et al, 2011). mfoot is the mass of the foot. Given the values of **MFy**, **MFz**, and values for **GRFy** and **GRFz** determined by the Santos model, equations of static equilibrium are derived for the ankle and used to determined **Ty**, **Tz**, and then **T**.

Using the value for **T**, the force of the knee muscles can be determined with joint torques from predicted postures. The contribution of the gastrocnemius muscle is 31.9% of the ankle joint torque, and the contribution of the hamstrings is 63% (DeVita and Hortobagyi, 2001) of the hip joint torque. The quadriceps muscles force is computed from the knee joint torque. These calculations are shown as follows:

$$\mathbf{H} = 0.63 * \left(\frac{\mathbf{Tor}_{hip}}{\mathbf{dH}_{hip}}\right)$$

$$\mathbf{G} = 0.319 * \left(\frac{\mathbf{Tor}_{ankle}}{\mathbf{dG}_{ankle}}\right)$$

$$\mathbf{Q} = \frac{\mathbf{Tor}_{knee} - \mathbf{G} * \mathbf{dG}_{knee} + \mathbf{H} * \mathbf{dH}_{knee}}{\mathbf{dQ}_{knee}}$$

where H is the hamstrings force, G is the gastrocnemius force, Q is the quadriceps force. Tor$_{hip}$, Tor$_{knee}$, and Tor$_{ankle}$ are torque values computed by Santos for the hip, knee, and ankle joint respectively. dH$_{hip}$ (0.042 m) and dH$_{knee}$ (0.047 m) are the moment arms for the hamstrings force with respect to hip joint and knee joint, respectively. dG$_{knee}$ (0.018 m) and dG$_{ankle}$ (0.051m) are moment arms for the gastrocnemius force with respect to knee joint and ankle joint, respectively. dQ$_{knee}$ (0.035m) is the moment arm for the quadriceps force with respect to the knee joint. The moment arm values are mean values throughout the stance phase (DeVita and Hortobagyi, 2001; Nisell, 1985). The angle φ changes based on θ and λ, and can be calculated based on the approach by Nisell (1985). Both the compression and shear forces at eth knee, F$_{shear}$ and F$_{comp}$, are then computed as follows (DeVita and Hortobagyi, 2001):

$$\mathbf{F}_{shear} = -\mathbf{G} * \sin\alpha - \mathbf{H} * \sin(\theta + \lambda) + \mathbf{Q} * \sin\varphi - \mathbf{T}_z * \sin\lambda + \mathbf{T}_y * \cos\lambda$$

$$\mathbf{F}_{comp} = \mathbf{T} - \mathbf{G} * \cos\alpha + \mathbf{Q} * \cos\varphi + \mathbf{H} * \cos(\theta + \lambda)$$

where α is the angle of gastrocnemius force with the tibia (α = 3 deg), and θ and λ are the angles of the femur and tibia with vertical axis, respectively (see Figure 2).

Having computed the muscle forces, and knowing the ground reaction forces from predicted postures, both compression and shear forces are applied to FEA model.

4 Multi-scale Integration

The automated process for integrating the components of the multi-scale model proceeds as follows. A user works within the Santos software environment. As one conducts analyses using posture prediction, one selects an option for high-fidelity analysis. Then, Abaqus FEA software is automatically spawned in the background, and Santos's current mesh is automatically exported to Abaqus representing whatever posture was being studied. The pre-existing FEA model is scaled/morphed and articulated to fit the avatar mesh. Concurrently, joint torques are exported to the FEA model and automatically imposed as boundary conditions. In addition, joint torques and GRF are exported to the static analysis module, which then exports muscle forces, also imposed on the FEA model as boundary conditions. The FEA model is then run. Currently, a finite set of pre-set joint angles is used for the FEA analysis, and capabilities for FEA morphing and scalarization are under investigation.

RESULTS

Using the model described above, the knee was analyzed with a series of articulation angles. The right knee is bent behind the avatar, with both feet remaining on a level surface. Table 3 shows the results under both compression and shear forces due to the body weight for three knee postures (0, 30, and 90 degrees). The body mass of Santos was taken as 75 kg. In general, as the articulation angle increases, so does the stress and strain.

Table 3 Results of FEA due to shear and compression of 75 kg body mass

Angle of Tibia with Femur (deg)	Max Contact Pressure (MPa)	Max Stress (MPa)	Max Strain %
0	2.718	6.441	2.771
30	3.571	32.682	2.112
90	8.022	60.091	3.722

Table 4 illustrates results for two cases, both involving coupling between eth Santos model and the FEA model. One case uses an articulation angle of 0 degrees (standing), and the other considers 60 degrees. These results are for the right knee joint under loading of 75 kg body mass.

Changing the angle of the tibia with femur changes the value of the ground reaction force in both legs and thus changes the value of compression and shear forces. Essentially, weight is shifted towards the left leg. As expected, with an increases articulation angle, compression forces decrease and shear forces increase. Note that both the maximum stresses and maximum strain increases in the case of a 60-degrees articulation angle, and this due to muscle contraction (increased force) surrounding the knee. Interesting, although the values of the stresses and displacements change, the relative distribution remains unchanged with knee articulation. The highest stress occurs at the ligaments, where injuries tend to be most common.

Parameters dictating the likelihood of knee injury typically fall into one of the following categories: 1) compressive load and external moments, 2) contact depth and area, 3) load-extension relationship, and 4) wear rate. This work does not address the latter but provides progress towards virtually determining metrics for eth first three parameters. Item 1) is related to the predicted joint torques and the compressive load determined form static analysis. Item 2) is related to strain determined form the FEA model. Work is ongoing to integrate the proposed work with item 3). Certainly, as loads on the knee increase, the risk of injury increases. However, developing multi-scale models with high fidelity components promise additional insight into the nature of potential injures.

CONCLUSION

This paper has presented initial work with a multi-scale finite element model as an essential part of the predictive model for an injury prevention strategy. An optimization-based whole-body predictive DHM provides the macroscale model for predicting posture while considering external forces. A validated FEA model for the human knee provides an example of a microscale model for determining the contact pressure, stress, and strain fields. Results for the FEA coincide well with the existing literature, and the feasibility of coupling these two models has been clearly demonstrated. Although multi-scale modeling is certainly an active field of research, coupling various models of different scales in the context of a predictive whole-body DHM, open doors for exciting research as well as powerful new tools. It provides the ability to concurrently use predictive real-time human models, and extract detailed information regarding stress and strain of joints and of soft tissue, both of which can be tied to the propensity for injury.

When addressing the issue of linking different models, we categorize the coupling as lose of tight. Lose coupling suggests that data is not passed between models seamlessly; it is exchanged as a separate operation (i.e. writing and then reading text files). Tight coupling entails automatically exchanging data in real time. Although Abaqus is automatically launched from the Santos software environment, there is some necessary user interaction. This process, nonetheless provide a novel DHM tool, but there are many opportunities for improved coupling. Currently, the FEM of the knee joint is designed for four angles of the tibia with the femur (0, 30, 60, and 90 degrees). Thus intermediate angles must be evaluated by interpolating between existing results. Work is ongoing to enable the FEA model to parametrically morph to whatever articulation angle is provided by Santos. This will allow for the evaluation of infinitely many postures and scenarios.

Future work also includes coupling Santos with a model for automatically predicting the forces resulting from muscles. Posture prediction with external forces yields torques that result from mechanical static analysis, but it does not inherently consider muscle forces. Existing software can provide muscle activation, given motion. Thus, we propose using not only posture prediction, but dynamic motion prediction developed at VSR. Because current muscle models do not provide the direction of eth muscle forces (just the magnitude of muscle activation), a complete muscle-wrapping model is being implemented that automatically predicts muscle elongation and orientation with changes in posture or motion. Integrating Santos with muscle models (internal to the Santos software or external) provides another exciting level of multi-scale modeling.

Table 4 Results for articulation angles of 0 degrees (standing) and 60 degrees

Description	Case 1: 0 deg (standing) (angle of tibia with femur)			Case 2: 60 deg (angle of tibia with femur)		
Santos: -Santos model						
-Knee model						
-Predicted torque (N.m)	**Hip** 1.061	**Knee** -7.700	**Ankle** 33.502	**Hip** 8.355	**Knee** 2.660	**Ankle** 100.860
-GRF vertical (N)	**R-foot** 386.906		**L-foot** 385.767	**R-foot** 256.325		**L-foot** 515.917
Static analysis (R-knee): -Compression Force (N) -Shear Force (N)	408.792 82.002			280.200 485.100		
FEA Results (R-knee): -Max CPRESS (MPa) -Max Stresses (MPa) -Max Strain (%)	2.718 6.441 2.771			4.491 15.561 4.582		

REFERENCES

Abdel-Malek, K., Yang, J., Kim, J., Marler, R. T., Beck, S., and Nebel, K. 2004, Santos: A Virtual Human Environment for Human Factors Assessment. *24th Army Science Conference*, November, , FL, Assistant Secretary of the Army, (Research, Development and Acquisition), Department of the Army, Washington, DC.

Cacciagrano, D., Corradini, F., Merelli, E., and Tesei, L. 2010, Multiscale bone remodeling with spatial P systems. *EPTCS 40*, 70-84.

Chantarpanich N., Nanakorn P., Chemchujit B. and Sitthiseripratip K. 2009, A Finite Element Study of Stress Distributions in Normal and Osteoarthritic Knee Joint, J Med Assoc Thai vol. 92 Suppl. 6.

Denavit, J., and Hartenberg, R. S. 1955, A Kinematic Notation for Lower-pair Mechanisms Based on Matrices. *Journal of Applied Mechanics*, 77: 215-221

DeVita P. and Hortobagyi T. 2001, Functional Knee Brace Alters Predicted Knee Muscle and Joint Forces in People With ACL Reconstruction During Walking. Journal of Applied Biomechanics,17.

Hou, T.Y., and Wu, X.H. 1997, A multiscale finite element method for elliptic problems in composite materials and porous media. *Journal of Computational Physics 134*: 169-189.

Hou, T.Y., Wu, X-H., and Cai, Z. 1999, Convergence of a multiscale finite element method for elliptic problems with rapidly oscillating coefficients. *Mathematics of Computation 68*(227): 913-943.

Hughes, T., Feijoo, G., Mazzei, L., and Quincy, J. 1998, The variational multiscale method - a paradigm for computational mechanics. *Comput. Methods Appl. Mech. Engrg, 166*: 3-24.

Li, S., Gupta, A., Liu, X., and Mahyari, M. 2004, Variational eigenstrain multiscale finite element method. *Comput. Methods Appl. Mech. Engrg., 193*: 1803-1824.

Lin, Yi-Chung, Kim, Hyung Joo and Pandy,M. G. 2011, A computationally efficient method for assessing muscle function during human locomotion. *Int. J. Numer. Meth. Biomed. Engng , 27*.

Liu, Q., Marler, T., Yang, J., Kim, J., Harrison, C. 2009, Posture Prediction with External Loads – A Pilot Study. *SAE 2009 World Congress*, April, Detroit, MI, Society of Automotive Engineers, Warrendale, PA.

Marler, R. T. 2005, A Study of Multi-objective Optimization Methods for Engineering Applications. Ph.D. Dissertation, University of Iowa, Iowa City, IA.

Marler, T., Knake, L., Johnson, R. 2011, Optimization-Based Posture Prediction for Analysis of Box-Lifting Tasks. *3rd International Conference on Digital Human Modeling*, July, Orlando, FL.

Martins, M.L., Ferreira Jr., S.C., and Vilela, M.J. 2010, Multiscale models for biological systems. Current Opinion in *Colloid & Interface Science 15*: 18-23.

Masud, A., and Bergman, L. A. 2005, Application of multi-scale finite element methods to the solution of the Fokker–Planck equation. *Comput. Methods Appl. Mech. Engrg., 194*: 1513–1526.

Moglo, K. E., Shirazi-Adl, A. 2005, Cruciate coupling and screw-home mechanism in passiveknee joint during extension–flexion. Journal of Biomechanics 38 ,1075–1083.

Nisell, R. 1985, A study of joint and muscle load with clinical applications. ACTA Orthopaedica Scandinavica Supplementum No. 216, Vol. 56.

Sangalli, G. 2003, Capturing small scales in elliptic problems using a residual-free bubbles finite element method. *Multiscale Model. Simul 1*: 485-503 (electronic).

Tang, C.Y., Guo, Y.Q., Tsui, C.P., and Gao, B. 2007, Multi-scale finite element analysis on biomechanical response of functionally graded dental implant / mandible system. *Journal of the Serbian Society for Computational Mechanics 1*,(1).

Weinan E, and Bjorn Engquist, 2003, The heterogeneous multi-scale methods. *Comm. Math. Sci. 1*(1): 87-133.

Weinan E, and Bjorn Engquist, 2003b, Multiscale modeling and computation. Notices of the AMS, 50 (9).

CHAPTER 37

Comparative Ergonomic Evaluation of Spacesuit and Space Vehicle Design

Scott England[1], Elizabeth Benson[1], Matthew Cowley[2], Lauren Harvill[2], Christopher Blackledge[1], Esau Perez[2], Sudhakar Rajulu[3]

[1]MEI Technologies Inc.
2525 Bay Area Blvd. Suite 300
Houston, TX 77058, USA

[2]Lockheed Martin
1300 Hercules
Houston, TX 77058, USA

[3]NASA Johnson Space Center
2101 NASA Parkway
Houston, TX 77058, USA

ABSTRACT

With the advent of the latest human spaceflight objectives, a series of prototype architectures have been developed for a new launch and reentry spacesuit that would fit with the new mission goals. Four prototype suits were evaluated to compare their performance and enable selection of the preferred suit components and designs. A consolidated approach to testing was taken: concurrently collecting suit mobility data, seat-suit-vehicle interface clearances, and qualitative assessments of suit performance within the volume of a Multi-Purpose Crew Vehicle mockup.

It was necessary to maintain high fidelity in a mockup and use advanced motion-capture technologies to achieve the objectives of the study. These seemingly mutually exclusive goals were accommodated with the construction of an optically transparent and fully adjustable frame mockup. The construction of the mockup was such that it could be dimensionally validated rapidly with the motion-capture system. This paper describes the method used to create a space vehicle mockup compatible with use of an optical motion-capture system, the consolidated approach for evaluating spacesuits in action, and a way to use the complex data set resulting from a limited number of test subjects to generate hardware requirements for an entire population.

Kinematics, hardware clearance, anthropometry (suited and unsuited), and subjective feedback data were recorded on 15 unsuited and 5 suited subjects. The selection of unsuited subjects was chiefly based on their anthropometry in an attempt to find subjects who fell within predefined criteria for medium male, large male, and small female subjects. The suited subjects were selected as a subset of the unsuited medium male subjects and were tested in both unpressurized and pressurized conditions. The prototype spacesuits were each fabricated in a single size to accommodate an approximately average-sized male, so select findings from the suit testing were systematically extrapolated to the extremes of the population to anticipate likely problem areas. This extrapolation was achieved by first comparing suited subjects' performance with their unsuited performance, and then applying the results to the entire range of the population.

The use of a transparent space vehicle mockup enabled the collection of large amounts of data during human-in-the-loop testing. Mobility data revealed that most of the tested spacesuits had sufficient ranges of motion for the selected tasks to be performed successfully. A suited subject's inability to perform a task most often stemmed from a combination of poor field of view in a seated position and poor dexterity of the pressurized gloves, or from suit/vehicle interface issues. Seat ingress and egress testing showed that problems with anthropometric accommodation did not exclusively occur with the largest or smallest subjects, but also with specific combinations of measurements that led to narrower seat ingress/egress clearance.

Keywords: Spacesuits, Anthropometry, Ergonomics, Biomechanics, Human System Integration, NASA, Motion Capture, Population Analysis

1   INTRODUCTION

The next-generation space vehicle being designed at the National Aeronautics and Space Administration (NASA) is required to accommodate a specific range of crewmember anthropometry while enabling suited operations at a variety of pressures and permitting all safety hardware to be used in all planned contingencies. The Human-System Integration Requirements (CxP 70024) specify these various human factors constraints including critical anthropometric, mobility, and strength requirements that must be accommodated by any spaceflight hardware such as spacesuits and space vehicles. These conflicting design objectives for vehicles and human systems necessitate a consolidated approach to testing and quantitative hardware evaluation, bringing multiple groups together to investigate integration issues. Historically, however, hardware testing has focused on qualitative evaluation of a single major hardware system at a time. One such test, labeled Functional Mobility Testing (England 2010), was conducted to determine the mobility requirements for the new generation of spacesuits. This testing became the cornerstone for Multi-Purpose Crew Vehicle (MPCV) suited mobility requirements, despite relatively immature operational concepts and lack of a high-fidelity test environment. As vehicle, operations, and suit concepts matured over several years, a consolidated test was envisioned to evaluate the integrated performance of the resulting spacesuits and the latest vehicle design.

The primary goals of this study were to quantitatively evaluate the performance of a series of prototype spacesuit architectures in the completion of a simulated mission to the International Space Station (ISS) and to estimate this performance for populations not currently accommodated by the prototype spacesuits. To accurately simulate performance of the tasks, a high-fidelity mockup of the Orion MPCV was needed. However, quantitative analysis of spacesuit performance required extensive visual access for the motion-capture cameras. These conflicting needs were resolved with the construction of an optically transparent and fully adjustable vehicle frame mockup.

2   METHOD

To accurately represent a mission-similar environment in which to functionally evaluate the prototype spacesuits, a high-fidelity mockup of the MPCV was required. The primary challenge was that the internal volume of the MPCV is too small to be used for a motion-capture study volume, and existing vehicle mockups were fully enclosed. To resolve these problems, NASA's Anthropometry and Biomechanics Facility (ABF) personnel designed a fully adjustable mockup of the vehicle's critical work areas with as little solid structure as possible. The adjustability of the mockup was critical as the MPCV's design was still evolving. Commercially available bird netting wrapped around the structure created a sense of the internal crew areas while permitting the use of an optical motion-capture system (Vicon, Oxford, UK). The ensuing mockup frame (Figure 1) was constructed chiefly out of extruded aluminum beams (80/20 Inc., Columbia City, Indiana).

Figure 1: Modeled drawing of reconfigurable mockup

A Vicon capture volume was created that encompassed the mockup such that subjects could be tracked while translating from the vehicle hatch to either recumbent seat shown in Figure 1. The Vicon system was also used to validate the dimensions of the mockup against a computer-aided design (CAD) drawing of the relevant design iteration of the MPCV by quickly enabling the calculation of distances between key points.

Fifteen unsuited and five suited subjects participated in this study. Selection of unsuited test subjects was chiefly based on anthropometry, in an attempt to find subjects who fit within defined categories for a medium male, large male, and small female. Suited test subjects were selected for their ability to adequately fit into multiple prototype suits. They also were required to complete the test in the unsuited state.

Test subjects were fully instrumented with a set of retroreflective markers positioned to enable the calculation of all major joint angles. After the markers were in place, subjects completed an array of functional tasks representative of major tasks performed in a mission to the ISS as kinematic data was recorded in Vicon at 100 Hz (Figure 2). In addition to kinematic data, hardware clearance, suited anthropometry, and subjective feedback were recorded at key times throughout the test. Relevant operational tasks were performed in both seat positions and with suited subjects at each of three pressure states: unpressurized, vent pressure, and nominal pressure. Suited anthropometry was recorded for critical dimensions at each pressure. Hardware clearance was recorded for hardware interference issues relevant to suited ingress and egress of the recumbent seats (Figure 2).

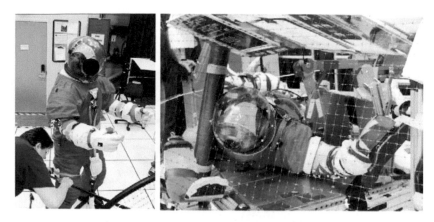

Figure 2: Left – Suited anthropometric data being collected in a suit at vent pressure. Right – Kinematic data, hardware clearance, and subjective feedback being recorded as a suited subject attempted pressurized seat ingress

Four prototype spacesuit concepts were evaluated in this study including the Pathfinder 1, Pathfinder 2, Demonstrator Suit, and Zipper Entry ILC suit. These suits had multiple designs for helmets, mobility components, and sizing adjustments, and had various pressurization strategies. Typically when a new suit design concept is fabricated, a single prototype is constructed for initial evaluation. These prototypes are generally fabricated in a single size to accommodate an approximately average-sized male. Because the suits accommodated only a narrow band of the potential population, findings from the suit testing were systematically extrapolated to the extremes of the required anthropometry for crewmembers through a variety of means.

3   RESULTS

3.1   Mockup for Motion Capture

The optically transparent mockup was a success in enabling the use of motion-capture technology while subjects performed high-fidelity tasks. Figure 3 shows an unsuited test subject reaching for the Displays and Controls (D&C) panel in Vicon and in video. Calculation of joint angles requires the reflective marker sets that comprise body segments on either side of a joint to be fully visible. Figure 3 illustrates that the upper body of the test subject was fully captured by the Vicon cameras, enabling calculation of all key joint angles for this task. Ranges of motion (ROM) were calculated for each joint by determining the maximum joint mobility necessary for each task performed.

Figure 3: Large, unsuited test subject touching the Display and Controls Panel in Vicon (left) and on video (right)

The fidelity of the mockup was such that, for any critical dimension, the reconfigurable mockup was never more than 1 inch from the dimensions in the official CAD file and was often substantially less. This could rapidly be verified by placing Vicon markers on key landmarks of the MPCV mockup and taking a short data capture (Figure 4). Once landmarks of the physical mockup were recorded in Vicon, they could be exported into a spreadsheet where distances between markers were calculated and compared to the intended design of the vehicle.

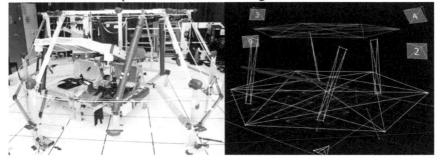

Figure 4: Key landmarks from the reconfigurable mockup reconstructed in Vicon

3.2 Spacesuit Evaluation

The use of a transparent space vehicle mockup permitted large quantities of both quantitative and qualitative data to be collected with human-in-the-loop testing. Failures to complete a task were generally attributed to problems with suit-vehicle integration, poor dexterity and tactility with pressurized gloves, or field-of-view issues when the subject was seated, rather than insufficient mobility from the new spacesuits. Kinematic trajectories were processed into major joint angles for every task by every subject in every suited condition. The extremes of these joint angles were then compared across tasks to yield subject total ranges of motion. This standardization of the test across suits yielded a technique to comparatively evaluate each suit's strengths and weaknesses. The results indicated the tested spacesuits had

380

sufficient ranges of motion for nearly all the selected tasks to be performed successfully. As a final step, the combination of all suited mobility results were then analyzed within suit condition and across subjects to update suit mobility requirements (Figure 5).

Joint Movement		Unpressurized	Vent Pressure	Nominal Pressure	Figure
SHOULDER		Range of Motion (°)			Figure
Flexion	Old Requirement	120	120	100	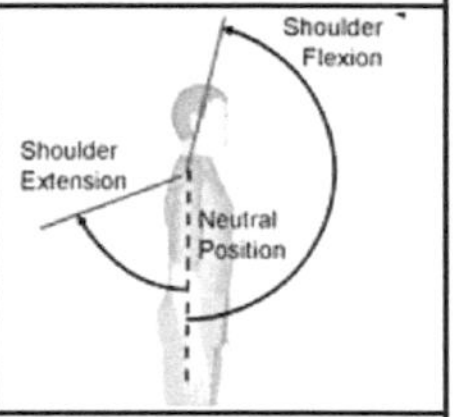
	New Requirement	130	130	120	
	Δ ROM	10	10	20	
Extension	Old Requirement	55	55	10	
	New Requirement	50	45	25	
	Δ ROM	-5	-10	15	
Adduction	Old Requirement	20	20	0	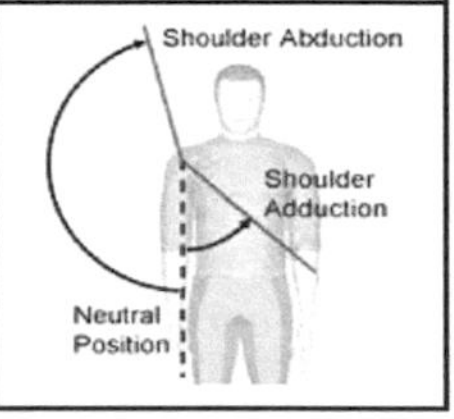
	New Requirement	40	30	20	
	Δ ROM	20	10	20	
Abduction	Old Requirement	100	100	90	
	New Requirement	100	100	70	
	Δ ROM	0	0	-20	

Figure 5: Subset of pre- and post-test mobility requirements for primary shoulder motions

3.3   Population Analysis

Observational data and feedback from suited test subjects indicated several challenges based on subject anthropometry existed for nominal operations of the suits. The recumbent seats, particularly during ingress and egress, provided the greatest possibility for problems to arise based on bulk of the suit and anthropometry of the test subjects. Subjects often encountered multiple unsuccessful suit-specific techniques for ingress and egress before finding an approach that worked for them. These techniques included sliding into the seat facing down, facing up, squeezing the helmet between the seat and D&C console then lying down, hugging the strut, and more (Figure 6). The smaller the test subject, the closer the seat pan must be adjusted toward the D&C panel to maintain proper eye alignment, the eyepoint, with the controls; such an adjustment would reduce the available ingress window. Even though the small, unsuited test subjects involved in testing were able to ingress the seat without severe difficulty, population analysis using suited anthropometry indicates that this will not be the case for small, suited crewmembers and suited crewmembers of other anthropometric characteristics are at risk as well.

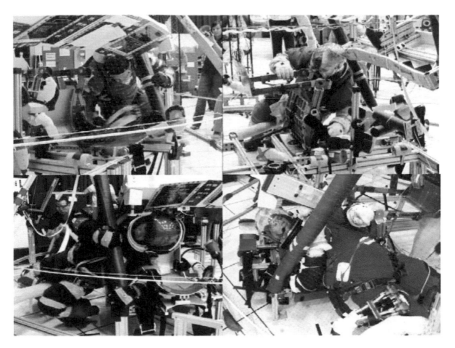

Figure 6: Multiple techniques were used for ingressing and egressing the seat including facing down and outward (top left), down and inward (top right), sliding in sideways (bottom left) and sliding in on one's back (bottom right).

Analytically, failures to ingress or egress the seats were determined by using population analysis to estimate and compare key unsuited and suited anthropometric dimensions to the restrictive points of seat and vehicle hardware surrounding the ingress window. Nine key landmarks outlining the ingress/egress path were recorded with the Faro Arm (Faro Technologies, Inc., Lake Mary, FL) for each seat setting. The anthropometric measurement of seated eye height determined the available ingress/egress area, since the seat was adjusted based upon subject anthropometry to ensure a consistent eyepoint in relation to the D&C panel. The analysis was multivariate and involved seated eye height, bideltoid breadth, chest depth, and thigh clearance, with the latter measurements critical to observed seat ingress/egress strategies (Figure 6). When the subject dimensions were greater than the hardware clearance dimension necessary for a specific mode of ingress, that ingress technique failed for that subject within the population. For example, Figure 7 illustrates a population analysis for unsuited individuals who are expected to pass or fail a particular ingress strategy based on the combination of eye height and bideltoid breadth. When no successful mode of ingress could be calculated for a subject across all ingress strategies, it can be inferred that subjects of those dimensions would face significant challenges with ingressing the seat.

Similar population-based analyses were performed evaluating other potential issues identified during the suited testing, including interference with a strut placed near suited subject's shoulders and reaching an umbilical restraint device located near the subjects' feet. These population analyses revealed the impact of both the suit and vehicle on subject accommodation and the results are corroborated with the observed issues in the performed human-in-the-loop test.

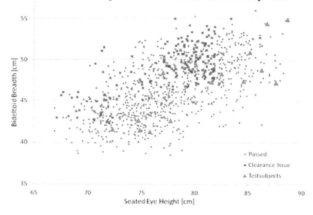

Figure 7: Example of population analysis performed on the unsuited data to estimate the impact of the ingress/egress volume on the population.

4 DISCUSSION

The construction and use of a high-fidelity mockup compatible with advanced motion-capture technology was quite successful. The mockup was validated to be within acceptable tolerances of other high-fidelity mockups of the Orion MPCV. Additionally, the mockup enabled suited ROM to be quantified for all suits in all conditions as subjects completed the critical functional tasks. Performance of the four tested suits were compared with respect to the specific architecture of each suit. That is, the unique mobility characteristics of each suit were presented to stakeholders to facilitate future design decisions based on the capabilities of existing suits. The mobility requirements update, as presented in Figure 5, reflects various needs and capabilities of the new prototype suits. For example, the requirements were updated to increase shoulder flexion and adduction, which reflects a more mature series of operational concepts in this round of tests and the need for greater arm mobility in a mission that now is geared toward microgravity intravehicular activities while the suit is unpressurized or at a low vent pressure. Previous mobility requirements included more significant operations for planetary extravehicular activity (EVA) operations including fall recoveries, geological exploration, and habitat fabrication, all of which require much more significant lower-body mobility.

Variation in subject anthropometry among unsuited and suited subjects did not exhibit any serious design accommodation issues during the test; however, population analysis of suited anthropometry and performance suggests that suited subject accommodation issues may exist when prototype spacesuits are developed for crewmembers of other sizes. Conditions may exist where specific subject anthropometry, limited mobility, and additional suit bulk combine to exceed hardware clearances or cause failures in performance of tasks. Care must be taken during crew selection and hardware verification to avoid creating excessive rates of failure for nominal mission operations.

5 CONCLUSION

Concurrent evaluation of prototype spacesuits in a vehicle mockup for multiple test subject anthropometries is a task that is difficult, yet necessary to provide meaningful insight to hardware designers about spacesuit and space vehicle requirements verification. The creation of an optically transparent, fully adjustable vehicle mockup was challenging yet successful in practice. It enabled quantitative analysis of the spacesuit prototypes while allowing inspection by a variety of stakeholders in real time. This ability to observe the subjects in real time was secondary to the three-dimensional kinematic data in initial test priority, yet it ended up being very useful for breaking down task completion beyond what is normally visible in a fully enclosed vehicle mockup. The objectives of the test were successfully completed, but it must be acknowledged that this test had several key limitations including all operations being performed at full gravity and a single test subject completing tasks in an open mockup where at least two astronauts would be present in actuality. Despite those acknowledged limitations, this series of consolidated experiments provided vast improvements in the knowledge base for the data collectors and for stakeholders in the involved hardware systems.

ACKNOWLEDGEMENTS

The authors thank NASA's EVA Systems group for providing funding for this study. We appreciate Oceaneering International Inc., David Clark Company Inc., and ILC Dover Inc. for providing spacesuits and relevant suit hardware. Additionally we thank Lockheed Martin's Orion Human Engineering group for their assistance in validating the construction of the reconfigurable Orion mockup.

REFERENCES

England, S.A., Benson, E.A., and Rajulu, S.L. Functional Mobility Testing: Quantification of Functionally Utilized Mobility among Unsuited and Suited Subjects (NASA/TP-2010-216122). NASA Johnson Space Center, Houston, TX, May 2010
Human-Systems Integration Requirements (HSIR) Revision D (CxP 70024). NASA Johnson Space Center, Houston, TX, December 2009

CHAPTER 38

Evaluation of the Enlarging Method with Haptic Using Multi-touch Interface

Kohta Makabe*, Masato Sakurai*, Sakae Yamamoto*

*Tokyo University of Science, Japan

ABSTRACT

Touch interfaces are becoming more and more common.In the touch interfaces, the operation can be divided into two types "tactile" and "haptic". This purpose is to examine the feature of operation by multi-touch, and operation of Haptic. The work of expanding a picture was done using three kinds of methods. As a result, it was suggested that it is not fit for the operation by multi-touch tuning finely. Furthermore, it turned out that operation of Haptic gives a subject a good impression compared with operation of Tactile.

Keywords: touch panel, haptic, pinch, multi-touch interface, subjective evaluation

1   INTRODUCTION

Touch interfaces have become popular to be used the ticket vending machines in public transportation, automated teller machines (ATMs), and so on. Recently, it is used the high-end mobile phone (smartphone) and personal digital assistant (PDA) as user interface. In the touch interfaces, the operation can be divided into two types "tactile" and "haptic". Tactile means only touching an object. Haptic means tracing an object. Most of smartphone are introduced in touch interfaces which can operate with haptic as well as tactile. In addition, the multi-touch interfaces begin to be distributed in smartphone and PDA, which recognize the more than two points with touch operation when performing it. Multi-touch interfaces are that it has an intuitive and easy to use interface[1].

According to the previous study[2], pinch is subjectively-assessed and objectively-rated when it is compared to the traditional operation with tactile on touch panel. It is considered that the continual change pinch has led to good evaluation. However, it is still remaining a question what characteristics affect in the haptic operation with two fingers to evaluate high.

The purpose of this paper is to reveal whether the method with haptic using multi-touch interface is better when compared to other operations with the continual changes and to examine the weaknesses of the method with haptic using multi-touch interface.

2 EXPERIMENT

2.1 Enlarging Methods Used in This Experiment

Three types of enlarging methods were used for the task in this experiment. All the methods could continually change the size of the object in the task. In this experiment, they were called pinch, slider, and button, respectively. Figure 1 shows each enlarging method in this experiment. Pinch was the enlarging method that puts two fingers on the touch panel and expands its distance as shown this figure. The size of the object was continually changed by the distance of two fingers. Pinch had both of characteristics of multi touch and haptic. Slider was the enlarging method that puts a finger on touch panel and slides it on the bar as shown this figure. The size of the object was continually changed by the position of the finger on the bar. Slider had both of the characteristics of single touch and haptic. Button was the enlarging method that touches the button on the touch panel and holds down it as shown this figure. The size of the object was continually changed by the time to hold down the button. Button had both of the characteristics of single touch and tactile.

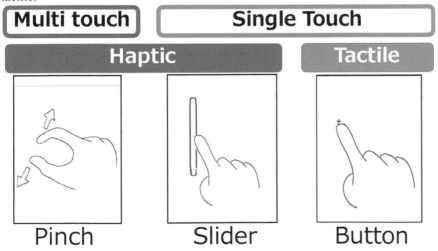

Figure 1 Enlarging methods used in this experiment.

2.2 Task

The task was to enlarge an object of rectangle on the touch panel using the above mentioned methods. Then the subjects were asked to drag it to the goal position. Figure 2 shows the screen of the touch panel in the task. Figure 3 shows the details of the task. At first, to start the task, each subject pressed the button of TIMER to measure the time for the task. And the subject was asked to enlarge the object to the same size of the goal with one of three methods instructed by the experimenter. The color of the object changed to yellow from red if it was the same size of the goal. After that, the subject dragged it to the goal position and its color changed to blue from yellow if it was the same position of the goal. To finish the task, the subject pressed the button of TIMER. For each enlarging method, the task was performed.

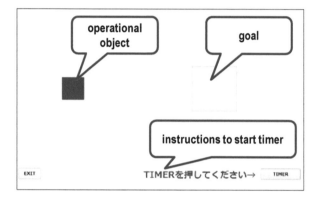

Figure 2 Screen of the touch panel in this experiment

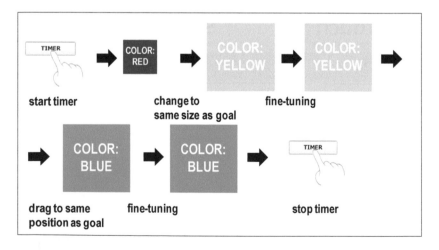

Figure 3 Details of the task.

2.3 Procedure

Figure 4 shows the procedure in this experiment. Before the task, each subject was asked to perform the calibration in the touch panel so that the touch panel could be correctly operated. The subject performed the task with one of three methods instructed by the experimenter. The method was randomly chosen for each subject. For each method, the task was performed five times. Then the subject was asked to subjectively evaluate about the operation with its method. The total in the task was 15 times through three methods.

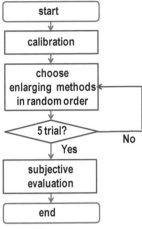

Figure 4 Procedure.

2.4 Measurements

In each task, three kinds of the operation time were measured. They were called "task completion time (completion time)", "first time that changed to the same size (rough-tune time)", "the time taken to fix the size to fine-tune (fine-tune time)" Completion time means the time for finishing the task (i.e. the time for pressing the timer button the second to finish the task). Rough-tune time is defined that time which the first change to the size you want from the start of the task (i.e. the time for yellow colored the first). Fine-tune is defined that time which is fixed to the size you want to fine-tune (i.e. the time for finishing the size change before dragging it).

The subjective evaluation carried out using a seven-rank (1: do not think at all, to 7: think very much) about the following five items; easy to operate, being able immediately to the size you want, intuitive movement, continual change, and sufficiently sensitive. And the questionnaire using 20 pair of adjectives carried out, too.

2.5 Apparatus

The task was run on a LENOVO Think Pad X201 Tablet. The camcorder (SONY Handycam HDR-HC3) was used in order to record the movement of the subject's hand during the task.

2.5 Subjects

Ten students participated in this experiment. The ages ranged from 21 to 24. All subjects were right-handed and the task was performed in the right upper limb. Eight of ten had smartphone holders.

3   RESULTS AND DISCUSSIONS

3.1   Rough-tune, Fine-tune, and Completion Times

Figure 5, 6, and 7 show the average results of the rough-tune, the fine-tune, and the completion times with a function as the enlarging methods in this experiment, respectively. The horizontal and vertical axes indicate each enlarging method and time of sec. Each bar represents the average results for all the subjects including the inter-subject deviation.

In the rough-tune time as shown in Figure 5, Pinch and Button are quicker than Slider. On the other hand, in the fine-tune time as shown in Figure 6, Pinch is slower than others. In the completion time as shown in Figure 7, Slider is the fastest in all the methods and Pinch is the slowest.

Firstly, think why the task accomplished by the pinch time is slow. The cause could be attributed to smallness of the object to change the size, failure due to parallax, length of time to fine-tune time.

In case of pinch, it is confirmed that their hands are trembling when fine-tuning for the operation by two fingers from the images recorded by the camcorder. Thus, it is relatively difficult and takes a time for performing fine-tune in the task compared with that of others. In cases of the methods of operation using only one finger, it is relatively easy for performing fine-tune since fine-tune times are quicker. It is considered it is useful for using only one finger when requiring fine-tuning.

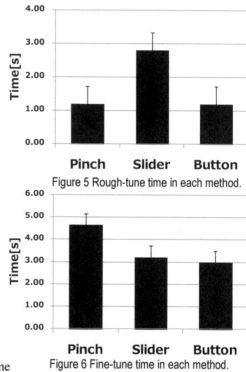

Figure 5 Rough-tune time in each method.

Figure 6 Fine-tune time in each method.

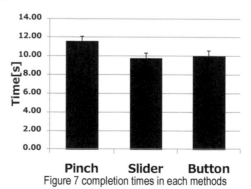

Figure 7 completion times in each methods

3.2   Subjective Evaluation

Table 1 and 2 show and the questionnaire of 20-pair adjectives for all the methods, respectively and the average results of the subjective evaluation in five items.

Table 1 Questionnaire using 20 pairs of adjectives.

		Pinch	Slider	Button
uncomfortable	comfortable	3.29	5.86	5.29
old	new	6.14	5.14	3.14
unrefined	refind	5.43	5.14	4.14
complected	sinple	4.43	6.29	6.57
boared	interest	5.57	4.71	3.57
vomity	pleasant	4.14	5.00	5.00
unprepossessing	prepossessing	4.43	5.00	4.86
slow	fast	4.29	5.71	5.14
dark	sunny	4.57	4.29	4.29
hard	soft	5.43	4.71	4.00
not easy	easy	5.14	6.14	6.57
jumpy	fluidity	5.43	6.00	4.71
not percision	precision	3.00	4.86	5.14
heavy	light	5.00	5.57	4.86
not free	free	5.00	4.86	4.29
not quick	quick	5.14	5.14	5.29
sober	dramatic	4.14	3.71	2.57
blunt	exquisite	4.71	4.29	4.43
cranky	firm	2.57	5.14	5.86
uncompanionable	companionable	5.14	5.86	5.86

Table 2 Subject evaluation in five items.

	Pinch	Slider	Button
easy to operate	3.43	5.86	6.14
being able immediately to the size you want	2.57	5.57	6.14
intuitive movement	5.86	6.29	4.57
continual change	6.14	5.14	4.71
sufficiently sensitive	3.57	5.57	6.00

When it looks at on the whole, evaluation of a slider is high compared with other two methods. Each method is looked at in detail.

Evaluation of pinch shows "easy to operate" and "being able to immediately" is very low compared with other two methods. It is thought that this reason is because the time of fine tuning was taken very much in operation by pinch. It is suggested also from this result that the pinch is not fit for fine tuning. Furthermore, the column of the pinch of an adjective pair shows attaching the evaluation "uncomfortable", "not precision", and "uncompanionable." After all, since it is hard to do fine tuning, the subjects think that this method was felt such. And it can be said that evaluation of "uncomfortable" was attached. However, pinch has obtained high evaluation by the "interest", "new", "free."

The column of button of an adjective pair shows attaching the evaluation "old", "boared", "hard", "sobar." This enlarging method is generally used from ancient times. Therefore, especially a subject can also say that he thought that it was pretty simple operation, without feeling newness. Moreover, magnification is being fixed, and since change is also mechanical, a subject is considered to have thought that this operation was hard. Since the evaluation very high in "simple", "easy" has been obtained, it may be for everybody.

When the column of the slider was looked at, it was called this and there was no bad item not much. That is, this experiment is considered to be operation in which the place with the pinch and a sufficient button is extracted. Since it is method of letting 1 finger sliding, it can be said that evaluation with a high item of "fluidity" has been obtained. Moreover, as compared with the button, the slider is improving the fault in a button. That is, by operation containing the element of Haptic, a subject is considered to give a better impression to operation.

4   CONCLUSION

It was suggested as a result of this experiment that the operation using a multi-touch was not useful for tuning finely. In operation of 2 fingers, it is mentioned as one of the reasons that a fingertip will tremble. In operation of the one point finger, there is no such problem and it can be said that it is easy to tune finely. It thinks also from subjectivity evaluation or a dialog with a subject. Moreover, by operation containing the element of haptic, a subject is considered to give a better impression to operation. It is thought that the evaluation "new", "interest", "soft", "fluidity", specifically increases. There is a possibility that it is the high method of evaluation since the operation using haptic can control magnifying power. Since the speed of expansion and reduction is controllable for subject itself by letting a finger slide as mentioned above, it is thought that it is operated more by itself. It is necessary to verify also about magnification in the future.

REFERENCES

[1] Furuichi, M.: The State of the Art in Touch Panel Based Human Interface Studies, *Information Processing Society of Japan*, Vol.50, No.4, pp.327-333, 2009 (in Japanese).
[2] Kohta,M., Masato,S. Sakae,Y.: Evaluation of the Enlarging Method with Haptic, *Journal of Ergonomics in Occupational Safety and Health*, Vol.13, Supplement, pp.77-80, 2011 (in Japanese).

CHAPTER 39

Reconstruction of Skin Surface Models for Individual Subjects

Yui Endo, Natsuki Miyata, Mitsunori Tada,
Makiko Kouchi and Masaaki Mochimaru

National Institute of Advanced Industrial Science and Technology
Tokyo, Japan
y.endo@aist.go.jp

ABSTRACT

In this paper, we describe a novel approach for the reconstruction of the skin surface model for an individual subject. In our method, we reconstructs the three-dimensional mesh model of the skin surface for an arbitrary individual subject, that is homologous for a template model and whose anthropometric dimensions equal to the ones of the subject. We use the optimization method of minimizing the energy cost function with respect to the mesh faces transformation, so that the output mesh keeps the local detail shape of the template model in the fitting process and satisfies the dimensional constraints for the subject.

Keywords: digital human, digital hand, homologous modeling

1   INTRODUCTION

Reconstruction of the three-dimensional skin surface models for individual subjects is the first important issue for the various types of ergonomic analysis for the human in the virtual environment, based on anatomy, anthropometry, kinematics, dynamics, and so on. Especially these models reconstructed as "homologous" with a template model can be useful to statistically analyze the dimensions between anatomical landmarks on the human body or to construct the skin surface deformation models based on the skeletal joint rotations. Here, the "homologous model" for the human skin surface is defined as the meshes that have the same topological structure in vertices, edges and faces as the template mesh model. Also, each vertex of these meshes corresponding to the anatomical landmark point on the skin surface must have the same index as the one of the template. We call this template mesh model the "generic model".

At the beginning, we must decide the inputs of this reconstruction process that can adequately represent the body shape for the target subject.

So far, some studies have been proposed, which uses the highly dense sample mesh (or point cloud) of a subject as the input (Allen 2003, Anguelov 2005, Seo 2003). The meshes are obtained from a three-dimensional laser scanner and positions of the anatomical landmarks are automatically estimated or manually specified from the meshes. In the research of Allen et al. (Allen 2003), the homologous model is generated by iteratively fitting the generic model to the sample mesh, using the landmarks as references. In this iteration, the generic model is refined as a subdivision surface and its vertices are fitted to the surface of the sample mesh by solving the optimization problem so as to minimize the energy-function. Sumner et al. (Sumner 2004) uses the correspondence algorithm for the deformation of the generic model into the sample mesh, which solves an optimization problem so that the deformed generic model smoothly keeps the detail of the local shape of the mesh surface. These approaches may be effective for simple poses of the subject, where the scanner can obtain the whole region of the subject's whole body (or some target parts of the subject) without dirty polygonal faces. For some complex poses like the grasp posture of the subject's hand, it may be difficult (or take a great deal of time and effort) to reconstruct the sample mesh from the scanner.

On the other hand, some studies only refer to positions of the anatomical landmarks on the subject's body in the generation process of the homologous model (Sumner 2004, Botsch 2008). These landmarks are obtained by the same way as the above-mentioned or by using the motion capture system. In most cases of these studies, the objective is to interactively deform initial pose of a character model into a desired pose. As their methods for surface deformation, some sort of global variational optimization problem is often used, based on the solution of differential equations. The problem is that there is no guarantee that the homologous models resulted from their methods always have desired anthropometric dimensions, because commonly we cannot allocate the landmarks on the subject's body enough to explicitly specify all such dimensions.

Therefore, we use anthropometric dimensions of the subject's body, in addition to anatomical landmark positions as the inputs. These dimensions are manually measured. In this paper, we describe the novel method to reconstruct the skin surface models for individual subjects that is homologous to the generic model. Our reconstruction approach solves the optimization problem for the minimization of the mesh deformation costs, under the condition that the landmark positions of the generic model is fitted to the ones of the subject and its anthropometry dimensions are respectively equal to the user-measured values (as shown in Figure 1). We show the details of our method in the next section.

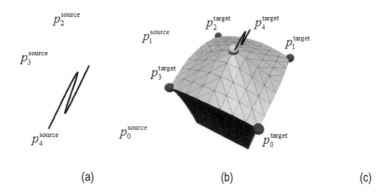

(a)   (b)   (c)

Figure 1 Concept of our optimization-based mesh deformation method. The inputs of the method are the source mesh (a) and the landmark points p_i^{source} and p_i^{target}. Our method obtained the deformed mesh (b) by fitting the landmark point p_i^{source} to the relative point p_i^{target} ($i \in \{0, ..., 4\}$). On the other hand, deformed mesh (c) is obtained by fitting the landmark point p_i^{source} to the relative point p_i^{target} ($i \in \{2, ..., 4\}$) so that the distance between p_0^{target} and p_1^{target} is a quarter of the side length of the cube before the deformation. As shown in (c), in our method, user-specified anthropometric dimensions are represented as the constraints set of the distance between two landmark points.

2   RECONSTRUCTION OF THE SKIN SURFACE MODEL

2.1   Overview

Figure 2 shows the overview of our method for the skin surface reconstruction. At the beginning, we need to prepare 3 types of the models as the input: the generic model, the landmark point set for the generic model and the target, and anthropometric dimension set for the target. The generic model M_{gen}^{gen} is a triangular mesh with 2-manifold and used as a template model for the output mesh models obtained from our reconstruction method. Thus these outputs are all homologous to the generic model. Where, superscript indices next to the symbolic model name represent the types of the topological structure G of the mesh and the subscript indices represent the types of the relative shape S which is based on. The landmark point set L_S includes the landmark points on the relative shape and is defined by the user. For each landmark point $p \in L_S$, a reference vertex $v_{(p)} \in M_S$ is specified, so the landmark position $\mathbf{p}$ of the point p is identical to the position $\mathbf{v}$ of the vertex v. Anthropometric dimension set D_S is also user-defined constraints used in the deformation process for the low-resolution generic mesh as will be shown later. Each anthropometric dimension $\delta = \{v_{(\delta 0)}, v_{(\delta 1)}, d_\delta\} \in D_S$ is represented as the two vertices $v_{(\delta 0)}, v_{(\delta 1)} \in M_S$ and the distance d_δ between them. By using these input models, three steps of our method are performed in sequence: 1) simplification of the generic model, 2) optimization for constructing the deformed low-resolution generic model, and 3) deformation transfer to the high-resolution generic model.

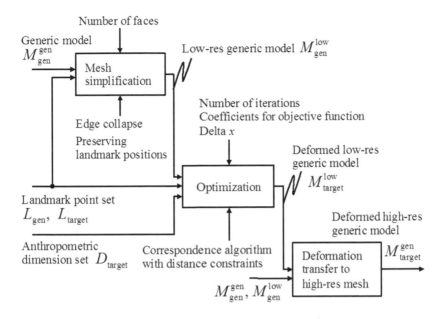

Figure 2 Overview of our method.

2.2 Simplification of the generic model

In order to represent the realistic whole body of human, at least several tens of thousands of triangular faces are needed as the generic model. However, for such a dense mesh, it may take several days to perform the optimization based on the correspondence algorithm described in the following section. On the other hand, the number of specified landmark points may be less than one hundred, so it is considered that this deformation affects the low-frequency shape information of the mesh. There, we simplify the detail of the generic model by reducing the number of faces. We use edge collapse algorithm based on the quadric error metric (Garland 1997), the classic and simple algorithm for the mesh simplification.

After performing the above simplification, we obtain the low-resolution generic model M_{gen}^{low}. During this process, the reference vertices for the landmark points may be moved or removed. So, for each landmark point $p \in L_{gen}$, we newly find the nearest vertex $v_{nearest(p)} \in M_{gen}^{low}$, move it to the position of the p, and redefine it as the reference vertex for the p.

2.3 Optimization for deformed low-res. generic model

The next step of our reconstruction method is to deform the obtained low-resolution generic model M_{gen}^{low} so that each landmark point $p_{gen} \in L_{gen}$ on the M_{gen}^{low} is fitted to the relative landmark point $p_{target} \in L_{target}$ and each constraint $\delta \in D_{target}$ of the distance d_{δ} between specified two vertices $v_{(\delta,0)}, v_{(\delta,1)} \in M_{target}^{low}$ is satisfied.

For this deformation, we solve the following optimization problem

$$\min \quad E(\mathbf{v}_1^{\text{low}},...,\mathbf{v}_{N_v^{\text{low}}}^{\text{low}}) = w_S E_S(\mathbf{v}_1^{\text{low}},...,\mathbf{v}_{N_v^{\text{low}}}^{\text{low}}) + w_I E_I(\mathbf{v}_1^{\text{low}},...,\mathbf{v}_{N_v^{\text{low}}}^{\text{low}})$$

$$\text{subject to} \quad \mathbf{v}_{\text{nearest}(p_i)}^{\text{low}} = \mathbf{v}_{\text{nearest}(q_i)}^{\text{target}} \quad (i=1,...,N_p) \tag{1}$$

$$\left\| \mathbf{v}_{(\delta_j,0)}^{\text{low}} - \mathbf{v}_{(\delta_j,0)}^{\text{low}} \right\| = d_{\delta_j} \quad (j=1,...,N_\delta)$$

where the variables $\mathbf{v}_i^{\text{low}}$ $(i=1,...,N_v^{\text{low}})$ is the position of the vertex $v_i^{\text{low}} \in M_{\text{gen}}^{\text{low}}$, p_j and q_j are respectively the landmark point in the set L_{gen} and the set L_{target}, δ_k is the distance constraint in the set D_{target}, w_S and w_I are the user-defined weights. E_S represents the "deformation smoothness" and E_I represents the "deformation identity", which are both defined as the "correspondence" between the source and target mesh (Sumner 2004). By solving this optimization problem, we obtain the deformed low-resolution generic model $M_{\text{target}}^{\text{low}}$.

2.4　Deformation transfer to the high-res. generic model

The final step of our reconstruction method is to transfer the deformation for the low-resolution generic model to the original high-resolution generic model. The new position $\tilde{\mathbf{v}}_i^{\text{gen}}$ of the vertex $\tilde{v}_i^{\text{gen}} \in M_{\text{target}}^{\text{gen}}$ $(i=1,...,N_v^{\text{gen}})$ is calculated from the position $\mathbf{v}_i^{\text{gen}}$ of the vertex $v_i^{\text{gen}} \in M_{\text{gen}}^{\text{gen}}$ as follows:

$$\tilde{\mathbf{v}}_i^{\text{gen}} = R_{\text{target}}^{\text{low}}(m)\left(R_{\text{gen}}^{\text{low}}(m)\right)^{-1}\mathbf{v}_i^{\text{gen}}$$

$$R_S^G(m) = \left[\frac{\mathbf{v}_{(m,1)}^{G|S} - \mathbf{v}_{(m,0)}^{G|S}}{l_x^{G|S}} \quad \mathbf{n}_m^{G|S} \quad \frac{\left(\mathbf{v}_{(m,1)}^{G|S} - \mathbf{v}_{(m,0)}^{G|S}\right) \times \mathbf{n}_m^{G|S}}{l_z^{G|S}} \quad \mathbf{v}_{(m,0)}^{G|S} \right]$$

$$l_x^{G|S} = \begin{cases} \left\| \mathbf{v}_{(m,1)}^{G|S} - \mathbf{v}_{(m,0)}^{G|S} \right\| & (G=\text{low}, S=\text{gen}) \\ 1 & (G=\text{low}, S=\text{target}) \end{cases} \tag{2}$$

$$l_z^{G|S} = \begin{cases} \left\| \left(\mathbf{v}_{(m,1)}^{G|S} - \mathbf{v}_{(m,0)}^{G|S}\right) \times \mathbf{n}_m^{G|S} \right\| & (G=\text{low}, S=\text{gen}) \\ 1 & (G=\text{low}, S=\text{target}) \end{cases}$$

where m is the index of the triangular face in the $M_{\text{gen}}^{\text{low}}$ which is the nearest to the v_i^{gen}, $\mathbf{v}_{(m,n)}^{G|S}$ $(n=0,1,2)$ is the position of the vertex $v_{(m,n)}^{G|S} \in M_S^G$ of the face $f_m^{G|S} \in M_S^G$, and $\mathbf{n}_m^{G|S}$ is the normal of the $f_m^{G|S}$. Now, we finally obtain the deformed high-resolution generic mesh $M_{\text{target}}^{\text{gen}}$. The concept of this process is shown in Figure 3.

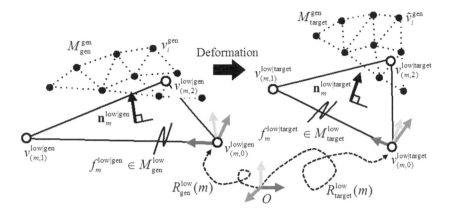

Figure 3 Concept of our deformation transfer process for the high-resolution generic model.

3   RESULTS

We applied our reconstruction method to the whole body of a subject and the right hand of another two subjects. The results for the whole body are shown in Figure 4. The subject is female adult. We manually specified 64 landmark points and did not use anthropometric dimension set as the constraints. We also obtained the dense sample mesh model for the subject by using the laser scan so as to check the error distribution between the sample model as the ground truth and generated generic model. Here, we define the "reconstruction error" as the distance between each vertex on the generic model and the nearest face on the sample model. The information of the reconstruction error for all results is shown in Table 1. Although we tried to reconstruct the model for the female body from the generic model for the male body and did not use any constraints for the anthropometric dimensions, we could obtain the result with high accuracy across the whole area of the body surface. The total processing time is about 50 minutes, where the number of the triangle vertices and faces of the initial (high-resolution) generic model are 13143 and 26248 and these of the low-resolution generic model are 668 and 1329, respectively.

The other results of our method are for the right hand of two subjects, as shown in Figure 5. Two subjects are both male adult. We manually specified 22 landmark points and 22 constraints for the anthropometric dimensions. We also obtained the dense sample mesh model for the plaster model of the subjects by using the CT scan so as to use them as the ground truth. Although the posture of each sample model is clearly different from the generic model, the reconstruction error at each finger region is still small. The total processing time for both is about 100 minutes, where the number of the triangle vertices and faces of the initial (high-resolution) generic model are 42617 and 85098 and these of the low-resolution generic model are 1026 and 1984, respectively.

In our optimization process for all trials, we set the number of iterations to 400,

398

set the weight w_S and w_I to 1.0 and 0.001 respectively, and set Δx for solving differential equations to 10^{-5} of the bounding box for the generic mesh.

4   CONCLUSIONS

We proposed the novel reconstruction method for the homologous mesh models of individual subjects by optimization-based deformation approach, where the inputs were manually specified landmark points and anthropometric dimension set. By applying this method to the sample models for the individual body and hands, we confirmed the validity of our method.

REFERENCES

Allen B., B. Curless and Z. Popović. 2003. "The space of human body shapes: reconstruction and parameterization from range scans." ACM Transactions on Graphics (ACM SIGGRAPH 2003), vol. 22, no. 3: 587-594.
Anguelov D., P. Srinivasan, D. Koller, S. Thrun, J. Rodgers and J. Davis. 2005. "SCAPE: Shape Completion and Animation People." ACM Transactions on Graphics, vol. 24, no. 3: 408-416.
Botsch M. and O. Sorkine. 2008. "On Linear Variational Surface Deformation Methods." IEEE Transactions on Visualization and Computer Graphics, vol. 14, no. 1: 213-230.
Garland M. and P. S. Heckbert. 1997. "Surface Simplification using Quadric Error Metrics." Proceedings of SIGGRAPH '97 Conference: 209-216.
Seo H., N. M. Thalmann. 2003. "An automatic modeling of human bodies from sizing parameters." Proceedings of the 2003 Symposium on Interactive 3D Graphics: 19-26.
Sumner R. W. and J. Popović. 2004. "Deformation Transfer for Triangle Meshes." ACM Transactions on Graphics (ACM SIGGRAPH 2004), vol. 23, no. 3: 399-405.

Table 1 Error for the results [mm]

Model	Mean error	Max. error	Min. error	Variance
whole body	4.25	38.12	0.03	12.42
hand (1)	1.63	15.13	0.004	2.03
hand (2)	2.61	13.73	0.001	4.32

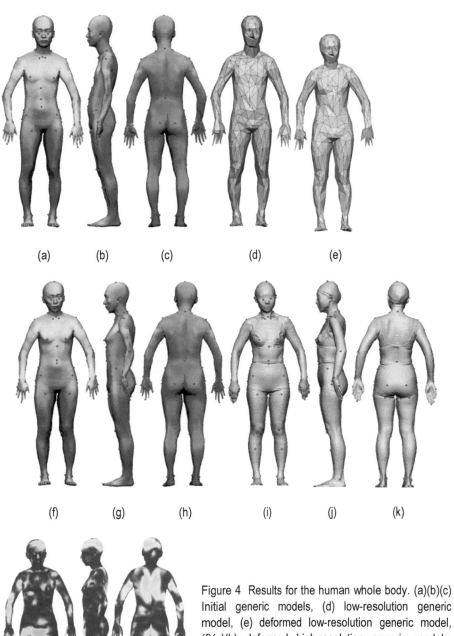

(a)   (b)   (c)   (d)   (e)

(f)   (g)   (h)   (i)   (j)   (k)

(l)   (m)   (n)

Figure 4 Results for the human whole body. (a)(b)(c) Initial generic models, (d) low-resolution generic model, (e) deformed low-resolution generic model, (f)(g)(h) deformed high-resolution generic models, (i)(j)(k) sample models and (l)(m)(n) error maps between sample model and the reconstructed model. The sample mesh was not referred to in our method but used for the validation. In this case, we did not use anthropometry distance set as the constraints.

400

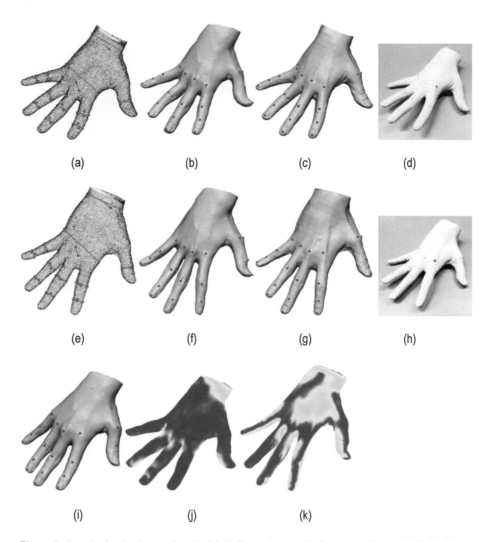

Figure 5 Results for the human hands. (a)-(f),(i) are the results for one subject and (e)-(h),(k) are the ones for an another subject. Both or them are reconstructed from the same initial genetic model (i). (a)(e) Anthropometry distance set, (b)(f) deformed high-resolution generic models, (c)(g) sample models, (d)(h) plaster models scanned by CT scanner for creating the sample meshes (c)(g), (i) initial generic model and (j)(k) error maps between the sample model and the reconstructed model. Also in these cases, the sample mesh was not referred to but used for the validation.

CHAPTER 40

Effects of High Heel Shape on Gait Dynamics and Comfort Perception

Makiko Kouchi, Masaaki Mochimaru

Digital Human Research Center, AIST
Tokyo, Japan
m-kouchi@aist.go.jp, m-mochimaru@aist.go.jp

ABSTRACT

Effects of the shape of high-heels on the gait and comfort perception were investigated using three experiment shoes that were different in the antero-posterior position of the top lift (base of the heel). Distance from the rearmost point of the heel to the rearmost point of the top lift was 15, 25, 35 [mm] in shoes S-15, S-25, and S-35, respectively. Walking motion and ground reaction force (GRF) were measured for the both feet of eight females. Significant differences were observed in impact peak and active peaks of vertical GRF measured for the heel and forefoot separately as well as the amount of plantarflexion of the ankle joint after heel contact. Shoe S-35 was least preferred because it felt unstable. These differences can be explained by the geometry of the heel, ankle joint, and forefoot and the plantarflexed position of the foot.

Keywords: shoe comfort, footwear biomechanics, gait

1   BACK GROUND

Previous studies on high-heeled gait revealed that comfort rating decreased with an increase in heel height (Lee and Hong, 2005). However, effects of heel design on comfort and stability remain unclear. The purpose of this study is to analyze the relationship between the position of the base of the heel, gait, ground reaction force (GRF), and perception of comfort for high-heeled shoes.

2   SUBJECTS AND METHODS

Three pairs of shoes with a heel height of 65 mm were prepared. Figure 1 shows the three shoes. The only difference between the three shoes was in the position of the top lift (base of the heel). Specifically, distance from the rearmost point of the back of the heel to the rearmost point of the top lift was 15, 25, and 35 mm for shoe S-15, shoe S-25, and shoe S-35, respectively. Open-toe pumps with ankle strap and adjustable foot circumference size were selected to minimize the effect of the fitting of the shoes on the evaluation of comfort.

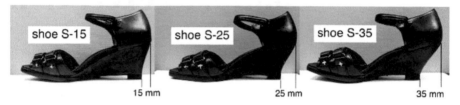

Figure 1 Three shoes used in the gait experiment.

Eight females (age 28-57 years) participated in the experiment in which walking motion (Vicon, 200 Hz) using a DIFF maker set (The Clinical Gait Analysis Forum Japan, 1999) and ground reaction force (GRF) (AMTI, 1000 Hz) were measured. GRF was measured for the heel and the forefoot separately. Walking speed was selected by each participant, and controlled with a metronome for all shoe conditions. The order of the shoes was randomly assigned for each participant and seven trials were measured for each condition. The following variables were calculated for the stance phase for both legs: maximum and minimum of the hip joint angle; angle at heel contact (HC), maximum during 0-50% of the stance, and minimum during 50-100% of the stance for the knee joint angle; angle at HC, minimum during 0-30% of the stance, and maximum during 50-100% of the stance for the ankle joint angle. Since the marker on the MP joint was on the shoe rather than on the foot, ankle joint angle was described as the difference from the angle when standing on both feet. For GRF, the following variables were calculated for both legs: impact peak and active peak of heel vertical GRF; active peak of the forefoot vertical GRF.

Mean value of the seven trials was used for the representative value of a subject/condition, and between-condition differences were tested using paired t-test. The level of significance was chosen as $p < 0.05$.

2   RESULTS

No significant between-condition differences were observed for the hip joint and knee joint variables. Shoe S-35 had significantly smaller amount of ankle plantarflexion from HC to max. Vertical GRF variables of the heel were largest in shoe S-35 and smallest in shoe S-15, while vertical GRF variables of the forefoot were smallest in shoe S-35 and largest in shoe S-15. Figure 2 shows example vertical GRF measurements of one subject.

Shoe S-25 was most preferred, and shoe S-35was the least preferred. Average rank of preference was 1.9, 1.6, and 2.5 for shoe S-15, S-25, and S-35, respectively. Shoe S-35 was not preferred because it was unstable. S-15 was preferred because it was stable, but was not preferred because it felt heavy and unnatural.

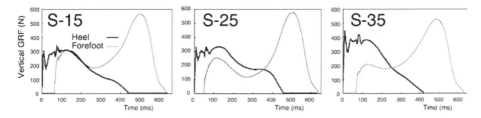

Figure 2. Examples of measured vertical ground reaction force of a subject in the three shoe conditions.

2   DISCUSSION

Reasons of the larger heel vertical GRF variables of the heel and smaller vertical GRF variables of the forefoot in shoe S-35 may be explained by the geometry of the top lift, the ankle joint (AJ), and for forefoot. Forces on the heel and the forefoot are balanced in terms of the moment around AJ. When the top lift position is more anterior and closer to the ankle joint, the moment arm of the heel is smaller, thus the force on the heel inevitably increases.

Smaller amount of plantarflextion of AJ contributes smaller negative power in early stance phase, which means absorption of the impact on the heel by AJ is smaller. This may also contribute to the larger impact peak of the heel.

Stronger sensation of the heel wobbling in shoe S-35 may be due to larger force on the heel as well as the unfavorable position of the AJ for constriction of triceps muscle of the calf to stabilize AJ.

The present study showed that the heel shape has significant effects on the sensation of comfort as well as the dynamics of the gait. There could be other design factors of the heel that have significant influence on the comfort. The present study suggests that experiment shoes should be selected carefully when evaluating effects of heel height.

3   CONCLUSIONS

Position of the top lift has significant effects on the shoe comfort and dynamics of the gait even when the heel height is the same. When effects of high-heels on the sensation of comfort are examined, shoes used in experiments should be selected carefully.

ACKNOWLEDGMENTS

This study was a collaboration study with Senshukai Co. Ltd.

REFERENCES

Hong WH, Lee YH, Chen HC, Pei YC & Wu CY 2005: Influence of heel height and shoe insert on comfort perception and biomechanical performance of young female adults during walking. Foot & Ankle International, 26:1042-1048.
The Clinical Gait Analysis Forum of Japan, 1992: DIFF Data Interface File Format (DIFF) User's Manual. English version. http://www.ne.jp/asahi/gait/analysis/DIFF-USA.pdf, http://www.ne.jp/asahi/gait/analysis/DIFF-JAPAN.pdf (in Japanese)

Section VIII

Cognitive and Social Aspects: Modeling, Monitoring, Decision and Response

Affective LED Lighting Color Schemes Based on Product Type

Jiyoung Park[1], Jongtae Rhee[1,2]

1 u-SCM center
2 Department of Industrial systems Engineering
Dongguk University
82-1, Pil-dong, Jung-gu, Seoul, South Korea
Jiyoungpark.dgu@gmail.com

ABSTRACT

This paper is to assess the effect of affective LED lighting color schemes for goods displayed in the retail stores on customer emotions based on products type using the survey. The satisfaction on the products or services can be increased in terms of customer emotions. The data are empirically collected from one of the largest retail stores in Korea. In addition, a proposed research model is built with the customer demographic, information of affective LED lighting color schemes and goods, and customer emotions. We did questionnaire survey with 80 University students to fine out LED lighting color schemes using RGB color codes. Survey results showed that the colors thought most appropriate for each product group were the representative colors most commonly associated with each product group. Based on these results, it is judged that associations had in memory of existing products have some effect on what colors are preferred or thought of as most appropriate. By identifying the relationships among factors of light, goods, and customer emotions, the results can provide determinants on the successful development of intelligently affective LED lighting color schemes. Consequently, it is ultimately expected to facilitate the customer relationship management and to increase sales of retail stores.

Keywords: affective LED lighting color schemes; customer emotions

1. INTRODUCTION

Although the success of a product in the market is strongly dependent on the aesthetic impression, the influence of emotions has been underestimated for a long while [1][2]. As shown the consequence of recent research, the concern of customer's emotion and affect is significant to the marketing strategy [3]. Not surprisingly, the light is one of the most critical factors to move personal emotion, and many researchers have believed that the customer's decision comes under the influence of the condition of lighting systems in market space [4]. From this academic point of view, affective LED lighting color schemes are defined as the systems that can recognize user emotions precisely and that can provide adequate lights according to the recognized emotion.

These lighting systems mentioned above are normally constructed for both emphasis of attractiveness of product and appealing environmental design under marketing conditions. In conjunction with psychological state and rhythm, various variables, such as temperature and brightness of light colors, can make a person comfortable and a product attractive [4]. As a result, the expected income can be improved by well-constructed affective lighting systems.

Despite this interest in affective LED lighting color systems, there is little research on finding relationships among lighting, products, and customer emotions, simultaneously. In previous research, evidence has suggested that the significant and direct effect of either light or product on customer emotions is demonstrated, but the interaction with three factors is less considered. Thus, this study aims to assess the effect of affective LED lighting color systems for goods on customer emotions with the Kansei engineering approach. The Kansei engineering is a useful approach to translating customer emotions to definite design factors [5][6]. Also, data can be analyzed through the systematic and formal procedure and the statistical evidence of customer satisfaction can be derived. For the case study, the goods in the retail store are especially targeted in this study and the results will provide the determinants and outcomes for development of intelligently affective LED lighting color schemes. It is expected to facilitate the customer relationship management and to increase the sales of retail stores in the marketing context.

The remainder of this paper is as follows. Section 2 briefly reviewed the LED lighting for customer's emotion. Then, Section 3 shows the research methodologies including the survey for architecture of affective LED lighting color scheme. Finally, this paper ends with conclusions and proposes future research in Section 4.

2. LED LIGHTING FOR CUSTOMER'S EMOTION

Color is an important factor that affects emotions through human senses within space. Color psychologist Louis Cheskin said that "colors consciously or unconsciously cause emotions in people either through intuition or symbols, and 90% of human behavior is caused by emotions, the remaining 10% by reason." As such, colors have a direct effect on human emotions caused by sensory responses, prompt time cognition judgments, and, integrated with social or physiological needs, cause motives. Emotions, as related to colors, refer to the capability to receive external stimuli, senses or cognition, or the evoking of various sentiments, therefore are used synonymously with sensitivity, which refers to the heightened psychological experience caused by senses or cognition, which in turn are the result of reason or intellect[7]. Therefore colors and emotions relate to the psychological meaning or communicative function embodied by colors. Here, emotions are called color emotions. Color emotions refer to all types of emotions that are felt by humans, and their study identifies the relationships that exist among visual colors and the emotional images of human psychology.

Applying, designing and using various color lights as emotional lighting according to the physiological and psychological effects of different colors is a necessary technology for the comfort and relaxation-seeking modern mind, and responds to the demand of the times, an emotion-centered era. Choreographing and designing emotional lighting with different colors and various functions can be called the designing of light, and is a scientific approach to 'well-being' design. Therefore, it is necessary that emotional lighting be re-examined in a scientific manner, to deduce various ways in which human emotions may be affected. This will shed new light on the significance of emotional lighting as a cutting-edge engineering tool of the emotional design age. Designing color emotional lighting for all spaces with lighting, including domestic spaces, business spaces and leisure spaces will make a realistic contribution to improving the quality of life.

Color lighting has various emotional effects on humans in specific spaces, and causes various changes in human behavior as well. The lighting design of living space, especially the method of lighting used, has an important effect on human perception of space. Also, as lighting causes changes in the optical information gained by a person, a bright living space heightens people's moods, whereas low luminosity, while relaxing, can also cause depression. Also, high saturation improves the mood, while low saturation induces a peaceful and stable atmosphere.

Gilbert Brighouse studied muscular activity in response to color lighting with a sample of hundreds of college students. In the experiment, muscular response was 12% faster than normal under red light, whereas under green light a relative delay was found. Kurt Goldstein asserted that in human life, each color has a different role, and that each has its own special importance[8]. He claimed that these combined allow humans to function normally. Color stimuli were given to human bodily organs and their behavior was observed. As a result, following discoveries of abnormalities in the cerebellum, it was claimed that 'colors have a greater effect on neurosis and psychosis patients.' Johann W. Goethe confirmed that light wavelengths have various effects on the mind and body, not only on the autonomous nervous system but the pituitary as well[9]. This signifies that colored light affects body function controls, including growth and sleep, temperature control, sexual urges, speed of metabolism and appetite. Based on such research, basic research on cognitive functions or physiological signal measurements was performed using various color filters with white halogen lamps. Results showed that warm colors had a relatively higher level of activation of the parasympathetic nervous system compared to the sympathetic nervous system, whereas with cold colors, activation of the sympathetic nervous system was relatively higher than the parasympathetic. However, results from an experiment wherein colors were presented in a color room, physiological response measurements were proportional to preference for colors.

With minimum light source efficiency requirements for energy savings in lighting, high-efficiency, long-life LED lighting is fast becoming the lighting of the future. An LED is an element with the properties of a semiconductor diode, with different electrical properties from existing lighting. Light emission of the diode responds to a one-way direct current. Response speed is very fast, and, with vastly improved efficiency of late, these devices are fast replacing existing lights.

As such, much research is being performed on LEDs. After the discovery of light emission by semiconductors in response to voltage, General Electronics first used a red LED in 1962, followed by the development of blue LEDs by Dr. Suzy Nakamura of Japan's Nichia Chemicals in 1993. In 1997, Nichia applied yellow fluorescent material to a blue LED to create a white LED. With the development of red, green, blue and white LEDs, various colors of light became possible. The development of white LEDs, in particular, signaled that LED lighting could be developed as lamps to replace existing lighting methods in electronic appliance displays.

Predicting luminance and color temperature is very important in the planning stages of indoor lighting. Dutch scientist Kruithof researched that luminance and color temperature are close correlated, and affect the emotions humans feel in response to lighting.

Kruithof's curve, which shows the range of luminance of various color temperatures that may be used for indoor lighting, shows that at low color temperatures of 2500K, luminance in the range of 50~100lx is most comfortable, and at high color temperatures of 6500K, at least 500lx luminance is required for persons to feel most comfortable.

However, Nakamura identified that how lighting feels can change according to what activity is done, even in the same lighting environment, proving that Kruithof's curve is not always useful. Based on this, Naoyuki Oi performed more detailed research into the regions of luminance and color temperature preferred by people according to indoor activity.

Oi researched the regions of luminance and color temperature preferred by individuals for indoor activities in the categories of gatherings, studying, relaxing, retiring, cooking, and dining. A schematic representation of the results of Oi's study, minus overlapping regions, showed that lighting evenly adequate for all four representative categories of indoor activities requires sufficiently high luminance at high color temperatures, and, at low color temperatures, must have as wide a range of luminance as possible, including areas overlapping with the uncomfortable zone of Kruithof's curve.

3. RESEARCH METHODOLOGY

3.1 Architecture of affective LED lighting color system

The prototype of affective lighting system architecture is structured to change the illumination and evaluate the customer emotions as shown in Figure 1. The system architecture comprises four parts: legacy systems, control servers, lighting systems, and data mining systems. First, legacy systems are used to manage the basic information on sales and products; based on this information, the control server appropriately changes and manages the color of lighting for enhancing customer satisfaction. Lighting systems receive the command from the control server and finally, the data mining is applied to increase sales based on the relationships between lighting factor and sales information.

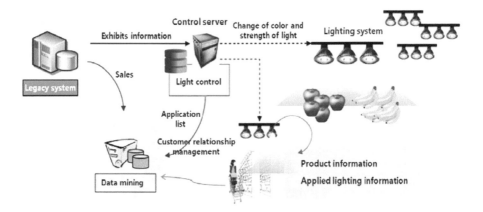

Figure 1. Schematic architecture of affective LED lighting color systems

Among others, the ultimate goal of this paper is particularly to build practical foundation for the relationship among lights, goods, and customer emotions for effectively developing the control server. As a hub of affective lighting systems, the control server plays a critical role in designing the intelligent lighting systems from the perspective of customer emotions. However, the evidence of the relationships has not been clearly suggested so far and the lack of studies should be pointed out. To this end, we adopt the Kansei engineering approach to systematically analyzing the customer emotions. As a concrete method, this Kansei engineering approach provides more effective results and insights for managers and developers of affective LED lighting color systems.

3.2 Survey data collection and analysis

The data were collected from one of the largest retail stores of a grocery retail chain in Korea which provide agriculture product, fruits, and marine products. As pointed out before, our study focuses on the determinants of affective lighting

412

systems for customer emotions and its outcomes like customer satisfaction. To test effect of affective lighting systems on customer emotions, several variables were investigated.

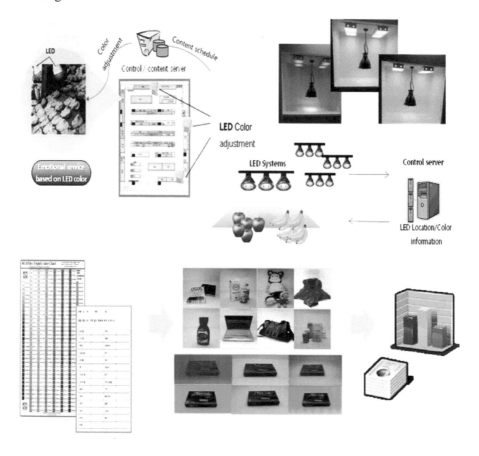

Figure 2. Process of affective LED lighting color schemes

We asked the division of product type. For example, agricultural, fishery, fresh meat, processed food, clothing, shoes, etc. about 24 kinds of products. We demand explanations to find the thinking of satisfaction color using the RGB code color sheet. Figure 2 shows this process. We did questionnaire survey with 80 University students. There are average age 22.3, Male 44, Female 36 during November 14~25, 2011. We analysed without blank data.

Survey results showed that the colors thought most appropriate for each product group were the representative colors most commonly associated with each product group. Based on these results, it is judged that associations had in memory of existing products have some effect on what colors are preferred or thought of as most appropriate.

Table 1. List of Finding color for product type

Product type	average RGB	Product type	average RGB
agricultural		kitchen, living utensils	
fishery		interior	
fresh meat		toy	
frozen and refrigerated		vehicle	
dairy		stationery	
bread		office supplies	
processed food		home electronic	
instant dish		clothes	
drinking		sundries	
coffee		sporting	
cookie		hygiene items	
layette		beauty	

4. CONCLUSIONS AND FUTURE RESEARCH

As human emotions are considered as the significant factor for marketing, many firms have highlighted the way to attract an interest of customers. The retail stores are recently concerned with the customer emotions to increase their sales in the marketing context. As one of the main issues, lighting systems are arguably emphasized to improve the product image, that is, the light makes products more attractive.

This study deals with the effect of affective lighting systems on customer emotions with respect to goods displayed in retail stores. In more detail, we design a research model based on the customer demographic, factors of lights and goods, and customer emotions. Moreover, the relationships with customer satisfaction should be identified. In response, the Kansei engineering approach is applied to systematically analyze the human emotions. It is expected that determinants and outcomes of lighting systems on customer emotions are clearly addressed for developers and managers in companies. The light on the products can be critical in the retail stores to attract the customer attention and enhance their performance. Due to the time limitation, we have described only research methodology, but in future, this paper will show its fruitful results through the Kansei engineering approach. Also, the whole systems for the intelligent LED lighting color systems are being developed. Furthermore, the research model is extended to examine the intention to purchase with customer emotions and satisfaction by the structural equation modeling. The structural equation modeling can provide the detailed and quantitative information about the relationships among the factors.

REFERENCES

H.M. Khalid and M.G. Helander, "A framework for affective customer needs in product design", Theoretical Issues in Ergonomics Science, vol. 5, no. 1, pp.27-42, 2004.

M.G. Helander and H.M. Khalid, Affective and pleasurable design, In: G. Salvendy (Ed.), Chapter 21 of Handbook on Human Factors and Ergonomics, Wiley, New York, pp.543-572, 2006.

H.M. Khalid, "Embracing diversity in user needs for affective design", Applied Ergonomics, vol. 37, no. 4, pp.409-418, 2006.

B. Son, Y. Park, H.S. Yang, and H. Kim, "A service platform design for affective lighting system based on user emotions", WSEAS Transactions on Information Science and Applications, vol. 6, no. 7, pp.1176-1185, 2009.

M. Nagamachi, "A study of emotion-technology", Japanese Journal of Ergonomics, vol. 10, no. 4, pp.121-130, 1974.

K. Yamamoto, "Kansei engineering", The art of Automobile Development at Mazda Special Symposium at Michigan University, 1986.

M. Kim, Illustrations to express emotional images, Ewha womans University, Master's thesis, pp. 14-15, 2001.

Goldstein, Kurt, Some Experimental Observations Concernin the Influence of Color on the Function of the Organism, Occupational Therapy and Rehabilitaion, 1942.

Suzy chiazzari, The Complete Book of Color, Paperback, pp.16-19, 1999

M. Nagamachi, Kansei Engineering, Kaibundo Publisher, Tokyo, 1993.

K. Miyazaki, Y. Matsubara, and M. Nagamachi, "A modeling of design recognition in Kansei Engineering", Japanese Journal of Ergonomics, vol. 29, pp.196-197, 1993.

K. Nakada, "Kansei engineering research on the design of construction machinery", International Journal of Industrial Ergonomics, vol. 19, pp.129-146, 1997.

C. Tanoue, K. Kshizaka, and M. Nagamachi, "Kansei Engineering: a study on perception of vehicle interior image", International Journal of Industrial Ergonomics, vol. 19, pp.115-128, 1997.

M. Nagamachi, "Kansei engineering as a powerful consumer-oriented technology for product development", Applied Ergonomics, vol. 33, pp.289-294, 2002.

S. Schutte and J. Eklund, "Design of rocker switches for work-vehicles – an application of Kansei engineering", Applied Ergonomics, vol. 36, pp.557-567, 2005.

K. Choi and C. Jun, "A systematic approach to the Kansei factors of tactile sense regarding the surface roughness", Applied Ergonomics, vol. 38, pp.53-63, 2007.

C.E. Osgood, The nature and measurement of meaning, in C.E. Osgood and J.G. Snider (Ed.), Semantic differential technique-a source book, Chicago: Aldine publishing company, pp.3-41, 1969.

S. Schutte, J. Eklund, J. Axelsson, and M. Nagamachi, "Concepts, methods and tools in Kansei engineering", Theoretical Issues in Ergonomics Science, vol. 5, no. 3, pp.214-231, 2004.

X. Gao and J. Xin, "Investigation of human's emotional responses on colors", Color Research and Application, vol. 31, no. 5, pp. 411-417, 2006.

H. Suk and H. Irtel, "Emotional response to simple color stimuli", Kansei Engineering, vol. 7, no. 2, pp.181-188, 2008.

CHAPTER 42

Human Activity and Social Simulation

Yvon Haradji[1] Germain Poizat[2] & François Sempé[3]

[1]EDF Recherche et Développement
Clamart, France
yvon.haradji@edf.fr
[2]Université de Bourgogne
Dijon, FRANCE
germain.poizat@u-bourgogne.fr
François Sempé AE
Paris, France
francoissempe@yahoo.fr

ABSTRACT

The purpose of this text is to clarify the role of understanding human activity within the context of a social simulation issue. We will examine this role on the basis of our experience with the SMACH[1] project (multi-agent simulation of human behaviour), the aim of which is to anticipate the behaviour of EDF's (Electricité de France) customers regarding energy consumption. We consider that the performance of a social simulation platform requires a structural relationship with the understanding of human activity in order to: model behaviours, validate the simulation and provide a framework for using the simulation platform.

Keywords: human activity, modelling, multi-agent, habitat, energy consumption.

---

[1] Simulation Multi-Agents des Comportements Humains

INTRODUCTION

The energy issue is becoming central to our debates on society and legislation is evolving as a consequence. The European Union has stated that we must divide greenhouse gases emissions by 4 by 2050. It is in order to meet this challenge to control our energy efficiency and to reduce overall consumption that we have undertaken to develop tools which will simulate the effectiveness of new technico-organisational systems. The SMACH platform, created in partnership with University Paris 6, the IRD[2] (Institute for Developmental Research) and ECCLEER (European Center Laboratories for Energy Efficiency Research), is part of this dynamic and focuses on two energy-related issues. The first concerns how human behaviour evolves in accordance with energy tariff offer, whilst the second articulates human behaviour with thermal behaviour of the building into simulations, in order to make them more effective (Bahaj and James, 2007; Kashif, Le, and Dugdale et al., 2011).

In this article, our aim is to show that in order to be successful, a social simulation platform needs to have a structural relationship with human activity in order to: define the behaviour model, validate the model and design the situation in which the simulation platform will be used.

1   SIMULATION IN ORDER TO ANTICIPATE HUMAN BEHAVIOUR

The aim of social simulation can be to question social theory (simulation of models) or to reproduce human interactions (Manzo, 2007; Sanders, 2007; Phan and Varenne, 2010). Our work is directed towards the latter: we define a set of constraints which help to imitate certain aspects of human performance. In this case simulation is a decision-making aid, the purpose of which is to anticipate human behaviour. The simulation platform becomes the equivalent of a virtual laboratory (Phan and Amblard, 2007; Sanders, 2007) and the decision is developed by varying the exogenous (the price, for example) and endogenous factors relating to human activity in a domestic setting (family make-up, individual preferences, characteristics of the habitat, etc.).

The rapid development of Multi-Agent Systems (MAS) (Heath, Hill, and Ciarallo, 2009) and more especially their use in social simulation (Manzo, 2007; Sanders, 2007; Phan and Varenne, 2010; Amblard and Dugdale, 2011) can be explained by the compatibility between certain properties of human activity and certain properties assigned to the agents. First of all, the human actor determines what is pertinent for him/her and progressively constructs his/her coupling with the environment (Varela, Thompson, and Rosch, 1991); MAS are autonomous in the

---

[2] Institut de Recherche pour le Développement

context of their interactions with the environment (Drogoul and Ferber, 1994). Furthermore, the collective activity of human actors cannot be described as the sum of individual activities (Hutchins, 1995); with MAS, interactions at a local level produce emerging collective behaviour at a global level (Drogoul and Ferber, 1994; Amblard, Ferrand, and Hill, 2001). Together, human activity and multi-agent approaches lead to the construction of situated behaviour and to dynamics between individual and collective entities.

The SMACH platform requires one to define a model corresponding to the pertinent characteristics to be used to simulate a family's activity. Once this reduction has been made, it is possible to simulate a family's organisation (for example, how they organise themselves before leaving for work in the morning) and to produce multiple simulations. Each simulation produces an energy-consumption curve resulting from the appliances used. It is then possible to compare the different simulations and determine which are the least energy-efficient, the most energy-efficient, etc.

The aim of the simulation we use in our work is to reproduce certain aspects of human behaviour and to provide a decision-making aid which predicts the energy consumption produced by varying human behaviour.

2 MODELLING HUMAN BEHAVIOUR

The performance of a social simulation platform depends first and foremost on its capacity to reflect the phenomenon being simulated. In this chapter we set out the structural elements of human activity that we have chosen to develop a minimal model of human behaviour which will produce realistic simulations.

Our works on energy management (Haué, 2004), on the modelling of human activities within a domestic setting (Poizat, Fréjus, and Haradji, 2009) and on identification of the contextual dimensions of individual and collective activity (Salembier, Dugdale, and Fréjus, et al., 2009; Guibourdenche, Vacherand-Revel, and Grosjean, et al., 2011) have allowed us to grasp the complexity of such domestic human activities. Using this knowledge, in this chapter we set out the three orientations leading to our multi-agent model of human behaviour. We thus explain the choices we have made in order to reflect an individual and collective dynamic within the simulation and, finally, we show that simplification in the description of the phenomenon is pertinent with regard to our aim of helping in the decision-making process.

2.1 Individual-centred modelling

Regarding the individual aspects of domestic activity, we accepted the possibility of a dynamic and progressive construction of decisions which leads to coherent and different actions as a situation gradually unfolds. In order to simulate human actors with their freedom of action and decision making, we opted for an individual-centred agent model (Sanders, 2007; Phan and Amblard, 2007; Gil-Quijano and Sabouret, 2010).

419

This orientation helps in the construction of a digital experience specific to the point of view of each agent. In our model, an agent corresponds to an individual person, and a family corresponds to a group of agents. The dynamic of each agent is determined by the combination of three aspects: a) *the dynamic coupling of the agent with the situation*. Each agent has a subjective view of the situation, i.e. a personal perception of the situation (for example, the mother-agent who is making breakfast in the kitchen does not know if it child-agent has got out of bed yet). As an agent gradually performs actions and moves around, it acquires knowledge about the situation and it possibilities for action: b) *Action priority based on a set of possibilities*. Every moment, based on the history of it previous actions, each agent calculates priorities for a set of possible actions (which are different for each agent) and personal preferences (one agent might prefer to read whilst another might want to play on the computer): c) *an agent's state linked to a quest for comfort*: This state concerns an individual's sensation of hot or cold. For example, doing the housework will increase the agent's sensation of warmth.

This modelling shows us that agents have their own individual interaction dynamic within the situation. At each moment, their universe is defined by their individual perception of the situation, their comfort level and their preferences regarding possible actions. So as with any real situation, autonomous behaviour results, i.e. each agent's specific, realistic and dynamic coupling with the situation.

2.2 Modelling to lead the emergence of family organisation

Human activity in a work or family environment cannot be resumed as a series of individual actions. Whether it involves helping the children (to get dressed, to do their homework, etc.) or sharing domestic tasks (doing the washing, cooking meals, etc.) a large part of human activity results from collective activity with its aspects of cooperation, sharing knowledge, negotiation and adjustments within interactions. The multi-agent model that we describe below aims to generate this type of collective behaviour.

In order to reflect this aspect of domestic activity, we used the following logic: a) *Cooperation* was essentially developed for when an action is shared (the agents do the housework together for example) and for mutual aid (getting the agent-child ready to leave home in the morning, for example); b) cooperation between agents is built from *information-gathering* in the environment (seeing that an agent-child is dressed) and from *acts of language* (requests for information, requests for action, response, etc.); c) *cultural aspects* characterise the family group in its lifestyle (the agents prioritise comfort, saving money, etc.) but also in its sociability (the agents wish to eat together); d) *the levels of responsibility* differentiate the agents involved in the action (between children and adults, for example).

Implementation of individual dynamics and of interactions between agents leads to the emergence of a simulated family organisation. From one simulation to another we observe a variety of behaviours along with regularities and variation in this family organisation. This general behaviour accurately reflects human activity in a real-life situation: as they perform actions, families gradually develop a form of general organisation, whereas locally there is constant variation in individual and collective activity.

Here, the organisation of the family of agents clearly appears as an emerging level of local interactions. Agents have a local view of their actions, a partial perception of a collective whole that they determine and which is return affects them. The emergence of the family organisation can thus be identified and analysed by an observer of the system, whilst at the same time having the concrete result of the conjunction of the two mechanisms, one relating to the autonomy of the agents and one relating to their collective dynamic.

2.3   Modelling channelled by its goal of helping decision making

Human activity is a complex phenomenon made up of numerous dimensions (cognitive, physical, psychological, cultural, social, etc.). In order to resolve our issue, we considered that the pertinent reduction consisted in focusing on the dynamic of individual and collective interactions within a domestic setting. This remains a very broad framework and our bias was to define the minimal properties which constituted our model and which produced a simulation that imitated the reality of human activity.

However, we also decided not to represent all of the aspects of human activity that our minimal model might have represented. This was the case for the simultaneous involvement of human actors to multiple concerns (Theureau, 2003; Poizat, Fréjus, and Haradji, 2009). We did not believe the growing complexity of the model (management of parallel actions, increasing number of possibilities, etc.) to be pertinent to our objective of a decision-making aid. Indeed, it is enough to reduce the activity to a sequential dynamic (it being possible to interrupt and restart sequences). For example, in order to represent a person who is simultaneously doing the ironing and watching television, we define an agent who starts by watching the television, stops in order to do the ironing, and then watches television again when the ironing is finished. The actions of the agents are sequential, but the objects which consume energy (the television and the iron) function simultaneously. The model does not take account of one important dimension of the activity, but the simulation nevertheless represents realistic consumption.

Our model produces human behaviour. We nevertheless restricted the power of expression by accepting not to reproduce certain aspects of human activity: realism can be ignored if it is not necessary to the objective of a decision-making aid.

3.   VALIDATING THROUGH PARTICIPATORY SIMULATION

The effectiveness of a social simulation platform is only proven when the results of the behaviour model have been validated. Once again, an understanding of human activity is a determining factor: a level of validation must be constructed by comparing simulated results with the natural phenomenon.

The SMACH simulation platform offers two types of simulation: automatic simulation (presented earlier) and participatory simulation. The latter allows human actors to interact on a virtual scene via their avatars and to thus take part in the simulation of a given life moment. The involvement of human actors increases the level of realism of the simulation and makes it possible to target a level of validation for the simulation's results.

This is the type of experiment we performed, and we used this method to validate our human behaviour model. In this chapter we set out the two conditions which make this level of validation possible: the creation of an individual-centred participatory simulation environment and the performance of a realistic experiment involving human actors.

3.1   Designing an activity-centred participatory simulation

Participatory simulation must make it possible to place human actors within a logic of action. It is important for us to put these actors into a situation of experience which is sufficient to allow them to live known or realistic life moments. It is therefore vital to construct an environment which approaches the virtual scene from the point of view of each actor.

We selected three principles which would allow us to achieve this level of realism. The first principle is based on *involvement.* It is not necessary to achieve a high degree of realism in order for actors to be able to project themselves into the situation. It is by using suggestion that we can enable users to immerse themselves in the simulation. We therefore targeted a scene with significant elements which would allow the participants to call upon previous experience (Horcik and Durand, 2011). The second principle is designed to encourage *action within the simulation.* To achieve this, the human actors have at their disposal a subjective view of the scene, real or supposed knowledge of what the other participants are doing or have done, opportunities to communicate through ready-made messages, and different possibilities of action. The final principle is designed to make *each person's level of comfort "perceptible".* Each human actor is represented by an avatar which changes colour in accordance with its level of comfort (hot, cold, neutral).

The interaction proposed via the participatory simulation is based on a logic of individual-centred design. In this way, human actors and automatic agents each have a subjective view of the scene and the same possibilities for action.

3.2   Validating the behaviour model

The conditions under which a participatory simulation takes place are extremely important (Brax, Amblard, and Becu, et al., 2010) in order to create sufficiently realistic scenes which will allow the behaviour model to be validated. To this end, we developed a flat-sharing scenario involving 5 agents. Three of these were controlled by human actors, with the remaining characters being automatic agents. The human actors were unaware of the roles played by the other human actors. Nor did they know what characters were played by the automatic agents. The simulation took place over a simulated period of 4 hours (a weekday evening) and the experiment was in three phases, representing three successive evenings (with each phase lasting approximately 15 minutes).

We found two levels of validation to be possible within this experiment. *Validation through observation of realistic behaviour.* The human actors succeeded in creating or recreating realistic life moments. For example, a human actor, himself sharing a flat in real life, told us he was reproducing his behaviour at home. We were also able to observe that on different days of the simulation a numerical experience of flat-sharing was constructed. For example, one person did not want to do the washing up because she had done it on the two previous days. *Validation through realistic interaction between automated agents and human actors.* The automated agents were not identified for the first two days but were partially identified on the final day. Demonstrations given during a congress confirmed this result: automated agents are rarely identified.

Participatory simulation leads to a dimension of validation. The creation of action sequences similar to those in real life, and the non-identification of automated agents in the scene help to validate the realism of the human behaviour model. We will conclude this chapter by saying that the level of validation is pertinent but still not sufficient. In the future, we would like to complete this level of validation with a comparison between simulation results and results from a real situation.

4   DESIGNING THE SITUATION IN WHICH THE SIMULATION PLATFORM IS TO BE USED

The use to be made of a social simulation platform can have a very major influence on the results. In this regard we consider the work of research and development engineers and the possibilities available to them to achieve a quality simulation. To this end we make a distinction between two aspects of the aid that we can offer: the first aims to understand and facilitate their reasoning and the second to define the methodological foundations of a study

Sempé and Gil-Quijano (2010) highlight the need for a reasoning aid for simulating complex phenomena. Indeed, how can one make the link between a model's static entry data and final behaviour which has been dynamically constructed by a set of multiple interactions? In the context of a social simulation, research and development engineers thus face a paradox: they must describe a non-determinist model and at the same time anticipate and manage productions emerging from the phenomena being simulated. Just like human activity in a real situation, simulation does not allow itself to be closed off in a prediction (Theureau and Pinsky, 1984).

Within the framework of SMACH, and in order to get beyond this paradox, Sempé and Gil-Quijano (2010) offer to help research and development engineers by using an iterative loop approach. The first phase involves a static description of the model. Tools are here used to define the parameters and, via a loop, to assess their pertinence. For example, research and development engineers define tasks one by one and will then be able to use a tool which allows them to assess or modify the pertinence of the logical links between tasks. The second phase concerns dynamic improvements to the behaviour model. Research and development engineers have access to an environment which enables them to examine the contexts at the origin of an agent's or a collective organisation's action. This contextual analysis will then allow them to understand unsatisfactory behaviour and to modify it by acting directly on the model (by making the agent perform a different action, for example). This initial work on a modelling-aid environment clearly demonstrates the need to help research and development engineers in the gradual construction of their simulations.

But in order to offer even more help, from a methodological standpoint it is vital to provide a framework for using the simulation platform. In order to construct their sets of data (persons, preferred profiles, distribution of tasks, etc.), research and development engineers require previous knowledge of behaviour in domestic settings. It would seem necessary for these data to be identified as pertinent by empirical studies in a real situation. More generally, we consider that the scope of social simulation can no longer be called into doubt when it becomes possible to link simulation with empirical studies. For example, regarding the evolution in European city populations, Sanders (2007) calibrated the simulation by using data from the 50 previous years in order to then be able to run different forward-looking scenarios relating to the next 50 years. As far as we are concerned, the simulation could be linked to field knowledge in order to generalise results beyond real observed situations, or, with regard to forward-looking studies, one might even link participatory simulation with the real baseline situation.

Our works are not sufficiently advanced, but on the basis of our preliminary results, we consider that a social simulation problematic should integrate the issue of the situation in which the simulation platform is used. The quality of the simulation will then depend on the capacity of research and development engineers to place their studies within a methodological framework which combines simulation data with activity data from real situations.

CONCLUSION

The interlinking of an understanding of human activity with the design of the SMACH platform allowed us to obtain conclusive results: it is possible to simulate human behaviour in a domestic setting, to obtain a certain level of validation and to highlight evolutions in consumption. These initial results have led us to continue our work with a view to going beyond certain limits. Indeed, our simulations relate to one family over a simulated period of time of approximately 4 to 5 hours. We would like to move forward and find answers to the following question: how can we progress from a short simulation of a single family to longer simulations which integrate a variety of populations and environments?

We will approach this issue of "scaling up" by combining three directions of

424

research. The first orientation relates to the possibility of performing lengthy simulations, i.e. of examining the question of how human activity is organised over different time scales. The second orientation relates to the possibility of shifting from a simulation of one family to a simulation of family typologies or coherent groups. Finally, the third orientation, of a similar nature, will look at the diversity and multiplicity of environments (habitat, appliance, climatic variation) for a given population.

This scaling-up changes the issue and will cause us to reassess the models and solutions that we have just presented. However this evolution will always take place on the basis of a structural intersection with human activity. We will then need to move ahead with the responses that we give to three epistemological questions underpinning social simulation: 1) What pertinent numerical reduction will account for human experience? 2) What level of proof is required to validate a simulation relating to human experience? And finally, 3) What knowledge (simulation/human experience) is needed to construct energy-related studies?

REFERENCES

Amblard, F., and Dugdale, J. 2011. Simulation sociale orientée agent. *Revue d'Intelligence Artificielle (RIA)* , vol.25, 1, Hermés.

Amblard, F., Ferrand, N., and Hill, D. R. 2001. How a conceptual framework can help to design models following decreasing abstraction. 13th SCS-European Simulation Symposium. Marseille, France.

Bahaj, A., and James, P. 2007. *Future Energy Solutions.* Retrieved 01 29, 2012, from http://www3.hants.gov.uk/fes.pdf

Brax, N., Amblard, F., and Becu, N., et al. 2010. When predictive modelling meet participatory simulation: a feedback on potential and issues of a combined approach. MAPS2 Conference, Teaching of/with Agent in the Social Sciences. Paris. France.

Drogoul, A., and Ferber, J. 1994. Multi-Agent Simulation as a Tool for Modeling Societies: Application to Social Differentiation in Ant Colonies. *Lecture Notes In Computer Science,* Volume 830/1994: 2-23.

Gil-Quijano, J., and Sabouret, N. 2010. Prediction of humans' activity for learning the behaviors of electrical appliances in an intelligent ambient environment. *IEEE International Conference on Web Intelligence and Intelligent Agent Technology.* IEEE, Computer Society, ACM: 283-286

Guibourdenche, J., Vacherand-Revel, J., and Grosjean, M., et al. 2011. Using multiple scores for transcribing the distributed activities of a family. CSCW '11 Conference on Computer supported cooperative work . New York: ACM.

Haué, J.-B. 2004. Intégrer les aspects situés de l'activité dans une ingénierie cognitive centrée sur la situation d'utilisation. Integrating situated aspects of activity in a cognitive engieneering approach centered on situation of use. *@ctivités, 1(2)*: 170-194.

Heath, B., Hill, R., and Ciarallo, F. 2009. A survey of Agent-Based Modeling Practices (January 1998 to July 2008). *Journal of Artifical Societies and Social Simulation* . 12(4)9: http://jasss.soc.surrey.ac.uk/12/4/9.html.

Horcik, Z., and Durand, M. 2011. Une démarche d'ergonomie de la formation: un projet pilote en formation par simulation d'infirmiers anesthésistes. Ergonomics for simulation-based training: a pilot project for anaesthetic nurses training. *@ctivités* , *8 (2):* 173-188.

Hutchins, E. 1995. *Cognition in the wild.* Cambridge: MIT Press.

Kashif, A., Le, X. H., and Dugdale, J., et al. 2011. Agent based framework to simulate inhabitants' behaviour in domestic settings for energy management. ICAART, International Conference on Agents and Artficial Intelligence. Roma, Italy.

Manzo, G. 2007. Variables, mechanisms, and simulations: can the three methods be synthesized? A critical analysis of the literature », . *Revue Française de Sociologie - An annual English Selection - 48, Supplément*: 35-71.

Phan, D., and Amblard, F. 2007. *Multi-Agent Modeling and Simulation in the Social and Human Sciences.* Bardwell Press, GEMAS Studies in Social Analysis.

Phan, D., and Varenne, F. 2010. Agent-Based Models and Simulations in Economics and Social Sciences: From Conceptual Exploration to Distinct Ways of Experimenting. *Journal of Artificial Societies and Social Simulation.* 13 (1) 5. http://jasss.soc.surrey.ac.uk/13/1/5.html.

Poizat, G., Fréjus, M., and Haradji, Y. 2009. Analysis of collective activity in domestic settings for the design of Ubiquitous Technologies. Proceeding European Conference on Cognitive Ergonomics: Designing beyond the Product. Otaniemi, Finland.

Salembier, P., Dugdale, J., and Fréjus, M., et al. 2009. A Descriptive Model of Contextual Activities forthe Design of Domestic Situations. Proceeding European Conference on Cognitive Ergonomics: Designing beyond the Product. Otaniemi, Finland.

Sanders, L. 2007. *Models in Spatial Analysis.* London: ISTE.

Sempé, F., and Gil-Quijano, J. 2010. Incremental and Situated Modeling for Multi-agent Based Simulations. International Conference on Computing & Communication Technologies, Research, Innovation, and Vision for the Future, Hanoï, Vietnam.

Theureau, J. 2003. Course of Action Analysis & Course of Action Centered Design. In E. Hollnagel, *Handbook of Cognitive Task Design* Mahwah. Lawrence Erlbaum Ass: 55-81.

Theureau, J., and Pinsky, L. 1984. Paradoxe de l'ergonomie de conception et logiciel informatique. *Revue des Conditions de Travail, n°9*: 25-31.

Varela, F. J., Thompson, E., and Rosch, E. 1991. *The Embodied Mind: Cognitive Science and Human Experience.* Cambridge: MIT Press.

CHAPTER 43

Effects of the Order of Contents in the Voice Guidance When Operating Machine

Shintaro Tahara*, Masato Sakurai*, Sayuri Fukano**, Masahiko Sakata***, and Sakae Yamamoto*

*Tokyo University of Science, Japan, ** Mitsubishi Electric Corp., Japan, ***Mitsubishi Electric Group, Central Melco Corp., Japan

ABSTRACT

The purpose of this paper is to examine qualitatively the effects of the order of contents in the voice guidance when pressing the button according to the instruction of voice guidance. It is also to investigate the relationship between the order of contents in the voice guidance and the area of visual searching when pressing the button.

Keywords: order of contents, voice guidance, area of visual searching, touch operation, subjective evaluation

1   INTRODUCTION

Recently, many machines are installed the voice guidance for supporting the operation when users use those machines. Particularly, for elderly people, it is considered that the voice guidance is useful for operating the machine properly. The voice guidance mainly plays a role of notifying how to use the machine auditorily.

Some of the previous studies examined the effects of the speech speed, the loudness, and the pitch for the voice guidance during the task [1-3]. Those studies focus on the physical characteristics of the voice in terms of applying to human beings. There is little study for examining the effects of the contents in voice guidance although the contents are one of the important factors for the voice guidance as well as the physical characteristics.

From the survey of the contents in the voice guidance for one of the machines which installed a lot of voice guidance, it consists of three contents. One of them indicates the place where to operate when users use it. The other one represents what to operate. The last one is how to operate as action. In this paper, these three contents are defined as "place content", "what content", and "action content", respectively. There is no suggestion for the proper order of the above three contents in voice guidance when user uses the machine.

2   EXPERIMENT 1

2.1   Task in Experiment 1

The task was to press the button on the touch panel according to the instruction of voice guidance. Fig. 1 shows the screen of the touch panel in this experiment.

Fig. 1 Touch panel screen in Experiment 1.

Voice guidance flowed by pushing the "power" button. The subjects were asked to press the proper button on the touch panel with the instruction of voice guidance. Then, the subjects evaluated subjectively the how easy it was to understand the voice guidance on a scale of 1 (not easy to understand) to 5 (easy to understand). The order of contents (place, what, and action content) in voice guidance was randomly changed for each trial. Moreover, even while the voice guidance is flowing, it tells that button operation can be performed, and the subjects enabled it to perform the button operation up to the subjects' timing.

428

2.2 Voice Guidance

The order of three contents (place, what, and action contents) included in voice guidance was replaced, and voice guidance of six conditions was created as shown in Table 1. As the explanation of the place content, the voice guidance which directs "right", "grill", and "left" was created. As the explanation of the what content, the voce guidance which directs "ON/OFF", "deep-fried dishes", "dried fish", and "timer" was created for corresponding to each button. As the explanation of the action content, the voice guidance which directs to push a button was only created. The voice guidance which directs a total of 36 kinds of operations on each of six buttons except the "power" button of Fig. 1 was created.

Table. 1 Voice guidance of six conditions.

Conditions	The order of contents
Condition1	Place-What-Action content
Condition2	Place-Action-What content
Condition3	What-Place-Action content
Condition4	What-Action-Place content
Condition5	Action-Place-What content
Condition6	Action-What-Place content

2.3 Procedure

The flowchart of the procedure is shown in Fig. 2. The subject performed the calibration so that a touch panel could be correctly operated before the task. After the calibration, the subject performed the task and the subjective evaluation about its voice guidance. Each subject randomly performed the task for 36 times (for six conditions in the voice guidance and for six buttons in Fig. 1).

2.4 Apparatus

The experiment was run on a LENOVO Think Pad X200 Tablet. The program was developed in Visual Basic with Visual Studio 2010. The voice guidance commands were in female voices. The camcorder

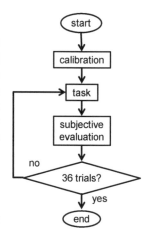

Fig. 2 Procedure.

(SONY Handycam HDR-S12) was used to record the subjects' operation during the task.

2.5   Analysis in Reaction Time

The reaction time in performing the task was measured from the images recorded by the camcorder for each subject. Fig. 3 shows the reaction time in this experiment. The reaction time was defined as the time from the pressing the power button to pressing the proper button instructed by the voice guidance in Fig. 1. The positive value is if the subject presses the proper button after finishing the voice guidance. The negative value is if he/she presses the proper button before finishing the voice guidance.

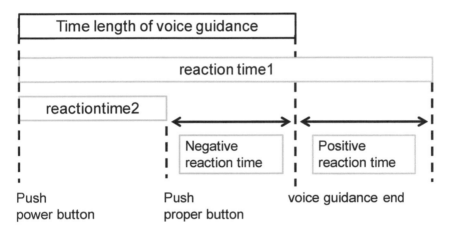

Fig. 3 Reaction time.

2.6   Subjects

Twelve subjects participated in this experiment. The average age was 23.7 years (S.D. 4.0).

3   RESULTS AND DISCUSSIONS IN EXPERIMENT 1

3.1   Subjective Evaluation

Fig. 4 shows the results of the subjective evaluation regarding the ease of understanding the voice guidance over all the subject responses for six conditions in the voice guidance. The horizontal and vertical axes indicate each condition in the voice guidance and subjective evaluation value, respectively. Each bar represents the average results including the standard deviation within the subjects.

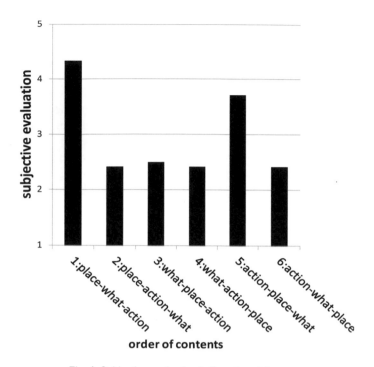

Fig. 4 Subjective evaluation in Experiment 1.

In this figure, the values of condition 1 and 5 are relatively higher than the others, especially the condition 1 is the highest in all. In terms of a feature for the explanation of the voice guidance, the conditions 1 and 5 are composed that "what content" is given to the next of "place content". The voice guidance of "what content" can extract three from six buttons for the selection of the buttons according to the voice guidance. On the other hand, the voice guidance of "place content" can focus on two from six buttons for the button selection in the task. It is considered that it had better give the order of contents for the voice guidance in such a way as to reduce as much as possible the number of buttons which can select.

The difference between the conditions 1 and 5 is a position of "action content". Unlike other contents in the voice guidance, for "action content", there is only one explanation for pressing the button. Therefore, the importance as content is low. It is suggested that the evaluation of the voice guidance becomes high when the minor priority content places behind in terms of the order of voice guidance.

3.2   Reaction Time

Fig. 5 shows the results of the reaction time in this experiment over all the subjects' responses for six conditions in the voice guidance. The horizontal and vertical axes indicate each condition in the voice guidance and the reaction time, respectively. Each bar represents the average results including the standard deviation within the subjects.

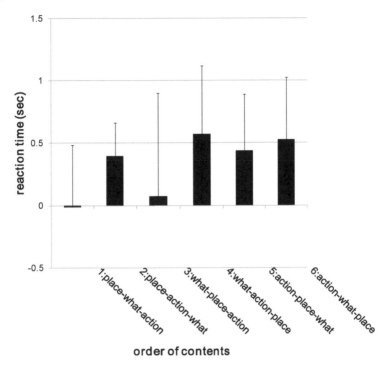

Fig. 5 Reaction time in Experiment 1.

The reaction time of conditions 1 and 3 are relatively shorter than those of other conditions. Since the condition 1 is negative value, as the voice guidance is flowing, it turns out that button operation is performed. From the images recorded by the camcorder, it is confirmed that the subject presses the proper button while the voice guidance is flowing. For both the conditions 1 and 3, the "action content" is given at the end for the voice guidance. It is considered that the subjects press the proper button immediately when they get the information needed to operate it from the voice guidance even though the voice guidance is flowing.

4 VERIFICATION EXPERIMINT (EXPERIMENT 2)

From the results of Experiment 1, it was examined that whether it had better give the order of contents for the voice guidance in such a way as to reduce as much as possible the number of buttons which can select as Experiment 2. Experiment 1, the number of buttons was able to be extracted to two by "place content", and was able to be extracted to three by "what content", respectively. In one side and the experiment 2, the experimental condition was changed so that the number of buttons could be extracted to three by "place content" and could be extracted to two by "what content". The procedure, apparatus, and how to calculate the reaction time were the same as Experiment 1.

4.1   Task and Voice Guidance in Experiment 2

The task was to press the button on the touch panel according to the instruction of voice guidance as well as Experiment 1. Fig. 6 shows the touch panel screen in Experiment 2. The subjects were asked to evaluate the how easy it was to understand the voice guidance on a scale of 1 (not easy to understand) to 5 (easy to understand) as well as the task in Experiment 1.

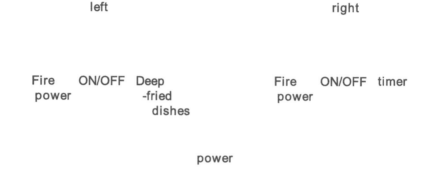

Fig. 6 Touch panel screen in Experiment 2.

The order of three contents (place, what, and action content) included in voice guidance was replaced, and voice guidance of six conditions was created in the same as Table 1 in Experiment 1. As the explanation of the place content, the voice guidance which directs "left" and "right" was created. As the explanation of the what content, the voce guidance which directs "fire power", "ON/OFF", "deep-fried dishes", and "timer" was created for corresponding to each button. As the explanation of the action content, the voice guidance which directs to push a button was only created as well as Experiment 1. The voice guidance which directs a total of 36 kinds of operations on each of six buttons except the "power" button of Fig. 6 was created.

4.2   Subjects

Twelve subjects participated in this experiment. The average age was 23.4 years (S.D. 4.0).

5 RESULTS AND DISCUSSIONS IN EXPERIMENT 2

5.1 Subjective Evaluation

Fig. 7 shows the results of the subjective evaluation regarding the ease of understanding the voice guidance over all the subject responses for six conditions in the voice guidance including that of Experiment 1. The horizontal and vertical axes indicate each condition in the voice guidance and subjective evaluation value, respectively. The black and white bars represent the average results of Experiment 1 and 2 including the standard deviation within the subjects, respectively.

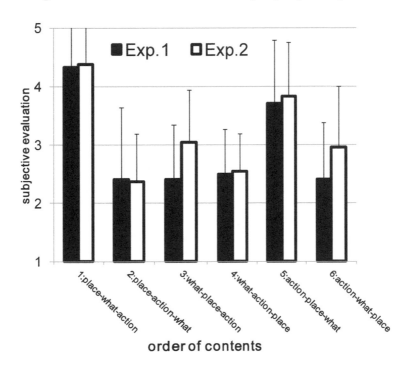

Fig. 7 Comparison of the subjective evaluation in Experiment 1 and 2.

The values of conditions 3 and 6 in Experiment 2 are relatively higher than those of Experiment 1. It means that the subjective evaluation value becomes high when flowing with the order of contents for the voice guidance in such a way as to reduce as much as possible the number of buttons which can select. Still, the value of condition 1 is the highest in all the conditions and similar with the results of Experiment 1. The voice guidance of condition 1 consists in the order of "place content", "what content", and "action content". In the touch panel screens in

Experiment 1 (Fig. 1) and 2 (Fig. 6), the explanation of "place content" can narrow the area of visual searching for the button selection compared with that of "what content" in the voice guidance. Therefore, it is considered that it is important for the voice guidance not only to flow with the order of contents in such a way as to reduce as much as possible the number of buttons which can select but also to show the voice guidance to narrow the area of visual searching for button selection. From the results that the value of condition 1 is higher than that of condition 3, it is important to show the voice guidance to narrow the area of visual searching for button selection rather than to flow with the order of contents in the voice guidance in such a way as to reduce as much as possible the number of buttons which can select.

5.2   Reaction Time

Fig. 8 shows the results of the reaction time over all the subject responses for six conditions in the voice guidance including that of Experiment 1. The horizontal and vertical axes indicate each condition in the voice guidance and the reaction time of sec, respectively. The black and white bars represent the average results of Experiment 1 and 2 including the standard deviation within the subjects, respectively.

In comparison of the results of Experiment 1 and 2, the reaction time of condition 3 and 6 in Experiment 2 is relatively shorter than that of Experiment 1. It is considered that the reaction time become short by flowing the voice guidance in such a way as to reduce as much as possible the number of buttons which can select. Although the reaction time of condition 3 is shortest in all the conditions in Experiment 2 that of condition 1 is relatively short and its value of the subjective evaluation is highest in all the condition from the results of Fig. 7. Therefore, it is suggested that it is important to show the voice guidance to narrow the area of visual searching for button selection rather than flowing with the order of contents in the voice guidance in such a way as to reduce as much as possible the number of buttons which can select.

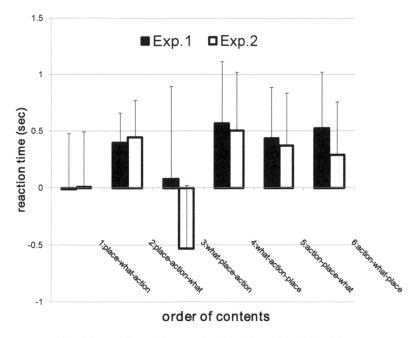

Fig. 8 Comparison of the reaction time in Experiment 1 and 2.

6 CONCLUSION

It is found that the value of subjective evaluation is high when the lower priority content is behind the others as the order of contents in the voice guidance. In addition, the subjects highly evaluate the effectiveness when showing the voice guidance to narrow the area of visual searching for button selection rather than flowing with the order of contents in the voice guidance in such a way as to reduce as much as possible the number of buttons which can select. This is important for designing the voice guidance from the practical point of view.

7 REFERENCES

[1] Kobayashi, M., Shikama, N., Morimoto, M., Sato, H., Sato, H.: Optimum speaking rate and speech level for speech communication in public spaces, *Architectural Institute of Japan*, D-1, pp.269-272, 2008.(in Japanese)
[2] Kawanami, H., Hirose, K.: Analysis and Synthesis of Speech Rate in Dialogue Speech Based on Prosodic Structures, *Information Processing Society of Japan*, Vol.22, No.10, pp.49-54, 1998.(in Japanese)
[3] Uchida, T.: Effects of the speech rate conversion on the impressions of pitch and the images of speaker's personality, *Acoustical Society of Japan*, Vol.56, No.6, pp.396-405, 2000.(in Japanese)

CHAPTER 44

Framing the Socio-cultural Context to Aid Course of Action Generation for Counterinsurgency

Mike Farry, Bob Stark, Sam Mahoney, Eric Carlson, and David Koelle

Charles River Analytics Inc.
Cambridge, MA

ABSTRACT

Course of action (COA) generation is guided by the Military Decision Making Process (MDMP), which is meant to guide warfighters through a linear, quantitative process to complete these missions effectively. However, since the MDMP has its roots in purely kinetic operations, a strict adherence to that process is not necessarily a good fit for the counterinsurgency (COIN) operational environment. Specifically, commanders must recognize the added importance of measuring mission *effectiveness* in addition to mission *performance*. While more objective measures of performance (MOPs) are suitable for kinetic operations, these MOPs are poorly suited to COIN, where information is frequently more qualitative, subjective, and difficult to directly observe. Examples of kinetic MOPs include, "Destroy 90% of enemy buildings" or, "Maintain friendly combat power above 50%." In COIN, similar MOPs are still relevant but do not address the root issues of the insurgency. An analogous example for COIN is, "Decrease frequency of IED attacks from 20 per month to five within our Area of Operations (AO)." This MOP represents a readily observable, quantitative value that can be tracked over time, but alone, it does not contribute to longer-term goals such as insurgent groups' strength and the region's long-term stability. To achieve that MOP, coalition forces (CF) might focus on finding and eliminating IED caches. In response, insurgents rapidly change their tactics, techniques, and procedures (TTPs) to focus on kidnappings and intimidation of the local population rather than direct attacks on CF. While the single MOP is achieved, the insurgent network continues to have a harmful effect on the AO.

Despite the difficulty in determining MOPs and MOEs, the MDMP is somewhat reliant on them since it is inherently a quantitative process. At its core, the MDMP is essentially a procedural cost-benefit analysis that enables commanders to apply resources and manage risk effectively to achieve the desired result in the AO. While the explicit inclusion of socio-cultural factors in the MDMP is difficult, commanders already do so implicitly. As revealed by a Work Domain Analysis (WDA) that we describe, commanders rely largely on subjective judgments and intuitions to bridge this gap. This paper provides an analysis of cognitive functions executed by warfighters who plan and execute COIN missions, a workflow that combines the doctrinal intent of the MDMP with best practices from the field, and a discussion of implications for intelligence analysis.

INTRODUCTION

Course of action (COA) generation is guided by the Military Decision Making Process (MDMP), which is meant to guide warfighters through a linear, quantitative process to complete these missions effectively. However, since the MDMP has its roots in purely kinetic operations, a strict adherence to that process is not necessarily a good fit for the counterinsurgency (COIN) environment. Specifically, commanders must recognize the added importance of measuring mission *effectiveness* in addition to mission *performance*. In this paper, we consider two major differentiating factors between COIN operations and kinetic or Major Combat Operations (MCO): (1) the increased importance of socio-cultural factors in COIN operations; and (2) an increased focus on *tactical* decision making, mission execution, and intelligence collection within COIN operations. As orders flow down the chain of command, they may not include sufficient socio-cultural contextual information relevant to the local Area of Operations (AO). While operational orders (OPORDs) usually contain some *a priori* knowledge of socio-cultural information, this information tends to be at a national and regional level and may not translate well to local missions and situations. Lacking localized knowledge, tactical warfighters may fail to consider potential secondary and tertiary effects of their actions, resulting in high mission performance but low mission effectiveness. Tactical warfighters must carefully execute the MDMP in these missions, since that linear, quantitative process is frequently at odds with the environment. That is, the socio-cultural context is continuously evolving, frequently qualitative, and often complex and difficult to understand. A thorough assessment of relevant socio-cultural factors is a prerequisite for "socially aware" COA generation so that the effect of the COA on each of those factors can be explicitly considered, resulting in true mission effectiveness. Given the complex and unpredictable nature of human behavior, however, it is unrealistic to expect that those effects will be predicted with high levels of accuracy and precision. However, as long as the effects are considered in a thorough, objective, and disciplined way, warfighters will have the information they need to select an effective option among a variety of candidate COAs.

While the socio-cultural context can be difficult to integrate explicitly into the MDMP, tactical warfighters already do so implicitly. That is, tactical warfighters are well-suited to observe the general qualitative, "atmospheric" factors of their AO, detect subtle changes that provide feedback for decision making, and exploit their intimate knowledge of the AO to advance their commander's objectives (Flynn, Pottinger, & Batchelor, 2010). Direct observation of the AO, along with the study of intelligence briefings (and in some cases, raw data such as UAV imagery or patrol reports) combine to form the tactical warfighter's understanding of the socio-cultural context. As shown in Figure 1, tactical warfighters must bridge a critical gap when implementing the MDMP: uniting the mission requirements from OPORDs while appreciating the unique socio-cultural context of their AO.

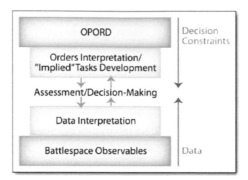

Figure 1: Bridging the gap between orders and the socio-cultural context indicated by battlespace observables

Currently, tactical warfighters rely largely on subjective judgments and intuitions to bridge this gap, using the MDMP to guide them through the doctrinally specified steps of mission analysis, course of action (COA) development, COA analysis, COA comparison, COA approval, orders production, and COA monitoring. For the COA development, analysis, comparison, and approval steps, warfighters must interpret *orders* and information requirements (IRs) through the lens of their knowledge of their AO. Warfighters must also interpret *data* collected in the field, such as reports, to relate their socio-cultural observations and intuitions to the quantitative measures of effectiveness (MOEs) used to assess COAs. In this paper, we provide an analysis of the cognitive complexities and common shortcomings commonly encountered in that process, a sample workflow designed to assist tactical warfighters execute the MDMP and focus on mission effectiveness, and discuss how intelligence collection and analysis processes can be integrated to measure mission effectiveness continuously. We conclude by discussing future work, including example visualization and decision support concepts to support the workflow we discuss, and a discussion of the role of socio-cultural modeling technologies to enable forecasting and prediction.

COIN PLANNING COGNITIVE FUNCTIONS

One of the key goals described in this paper was to discover the true cognitive functions warfighters perform in executing missions in COIN environments, to provide a basis for comparison with the doctrinal MDMP. To achieve that goal, we used Work Domain Analysis (WDA), a technique commonly employed by the Cognitive Systems Engineering (CSE) community to characterize the work environment, decision and action constraints, and the nature of cognitive tasks performed (Rasmussen, 1986; Rasmussen & Vicente, 1989; Roth et al., 2000; Roth & Mumaw, 1995; Vicente & Rasmussen, 1990). As part of this WDA, we conducted more than 400 hours' worth of knowledge elicitation (KE) sessions with Army subject matter experts (SMEs), all of whom have experience conducting the MDMP in COIN environments at varying echelons of command and within different roles (e.g., intelligence, operations). We also conducted a series of observations (totaling approximately 21 days) at the National Training Center (NTC) at Ft. Irwin, CA, to obtain a first-hand perspective on difficulties encountered by practitioners in a controlled training environment. This analysis serves as the foundation for a workflow that warfighters can follow to be compatible with the MDMP while recognizing the realities of their operational environments.

To conduct our analysis effort, we began by scoping the bounds of our analysis appropriately. The need for greater mission effectiveness and awareness of socio-cultural factors permeates COA generation at all echelons of operational command. Higher echelons (e.g., Army divisions or even the COCOM level) face significant difficulty integrating socio-cultural factors into wide ranging campaigns since they lack the insight gained by months and years of direct exposure at the tactical level. These difficulties are distinct from the issues at the tactical level, where warfighters must implement specific tasks to achieve the goals specified by higher echelons. Tactical warfighters must interpret orders from higher echelons in light of the local atmospheric and intuitive knowledge of their AO. No matter the echelon, this intelligence—whether it is contained in vetted reports, raw data, or in warfighters' mental models—must be properly exploited to support effective COA generation and mission execution. While our analysis reflects the reality of the evolving COIN mission by focusing largely on tactical issues, we have also explicitly considered the role of communication and interchange among echelons. The cognitive functions and workflow we describe below are relevant to all echelons of command. It is also critical to note how subjective notions of the operational, socio-cultural context permeate shared documents as well as warfighters' own mental models. While these notions are necessary, they may also lead to biases that detriment the mission planning task. Given their close relationship to the AO and immersion in the culture, tactical warfighters are the most well-suited to observe the general "atmospheric" factors of their AO, but they may lack the experience and knowledge to properly exploit their perspective. Their observations are useful to detect subtle changes that provide feedback for decision making, and they can exploit their intimate knowledge of the AO to advance their commander's objectives. If units are able to piece those atmospherics together into a rich socio-cultural context, commanders will have a solid foundation to perform the MDMP in a more adaptive manner and account for the dynamism inherent in the socio-cultural operating environment. Our analysis sets the tone for a practical, detailed perspective on executing the MDMP in a COIN environment, as well as strategies for observing, interpreting, representing, and projecting that contextual information.

Based on our KE sessions, observations, and analysis, we synthesized the observed cognitive complexities into categories and bundled those categories into a collection of higher-order cognitive work functions that are common across warfighters executing COIN missions. Four main functions that must be supported for all warfighters, whether they are responsible for generating the COA, executing it, or reporting on its status are: (1) situation assessment, (2) action selection, (3) communication, and (4) continual monitoring. These functions extend beyond the scope of the doctrinal MDMP, but each of them must be considered for COIN environments. The functions are not necessarily sequential, but a sequential ordering is logical for this discussion. The first function is **situation assessment**. In this function, warfighters "bootstrap" to construct an initial socio-cultural context. To perform this function, warfighters may rely on existing intelligence (e.g., reports of key leader engagements (KLEs) with local tribal leaders), existing personal or institutional knowledge, their own research, and visualizations of available historical data. Once situation assessment is complete, warfighters have a working socio-cultural context they can use to support the other functions. This working context is subject to constant revision. The second function is **action selection**. To generalize, this step may be considered equivalent to COA generation—that is, given an awareness of mission goals and the working socio-cultural context, warfighters develop a COA. However, we use a more abstract term for this function to accommodate the differences in terminology; while every echelon of command necessarily decomposes and interprets orders, they are not necessarily performing strict "COA generation." The third function is **communication**. For higher echelons communicating downward, this function includes the composition of orders (e.g., an OPORD) and communication of the working socio-cultural context. However, this function also includes intelligence reporting of the chain of command, enabling tactical warfighters to contribute to a refined, shared socio-cultural context. The fourth and final function is **continual monitoring**. In this function, the work performed by warfighters comes full circle. Warfighters must closely monitor the performance and effectiveness of the selected COA over time. Many of the tasks performed under this function are similar to those in the first function of situation assessment, and this function is largely a continual version of situation assessment overall. This continued assessment provides the basis for selecting new actions and communicating new updates.

A WORKFLOW FOR MDMP IN COIN ENVIRONMENTS

Based on the four cognitive work functions described above and a comparison against the doctrinal specification of the MDMP, we describe a workflow that reflects a realistic approach for tactical warfighters to execute the MDMP in COIN environments. Our design process for this workflow included the advice and feedback of SMEs at every step. During iterations, we verified the operational relevance and validity of the steps and their relationships. Based on this analysis, we determined that the linear and discrete (non-continuous) typical interpretation of the MDMP did not effectively support COIN in practice. For example, the COAs that support mission orders are often reconsidered due to changing environmental conditions, which then necessitate re-planning.

This workflow focuses on the natural tension between the expressed orders and the available data as illustrated in Figure 1, and enables warfighters to incrementally contribute to work products in a free-form manner as needed in COIN environments. This approach contrasts with the traditional, MCO-focused execution of the MDMP, which is more linear and focuses on a specific kinetic engagement with a well-defined start and end point. The new workflow respects the complexity in the relationships between COAs, MOEs, and other environmental, socio-cultural, and behavioral factors that may affect MOE values over time. That is, this approach accommodates the generation (through inference or interpretation) of COA elements from OPORDs as circumstances change, rather than trying to apply previous COAs and MOEs to new situations. For instance, to measure stability an OPORD may specify an MOE concerned with the number of active infrastructure projects in an AO since insurgent groups target infrastructure projects to sow uncertainty and doubt in the host nation's (HN) population. However, that MOE may be rendered ineffective if summer heat, monsoon season, and other weather factors force infrastructure projects to halt temporarily. Instead of force-fitting this MOE into the situation, this iterative and free-form workflow enables an analyst to redefine MOEs on the fly. This workflow focuses on the decomposition and interpretation of OPORD elements for the given AO and supports the seven steps of the MDMP:

(1) Receipt of mission
(2) Mission analysis
(3) COA development
(4) COA analysis
(5) COA comparison
(6) COA approval
(7) Orders production

The doctrinal MDMP dictates decomposition of the OPORD into its components: situation, mission, execution, sustainment, and command and control. In this workflow, warfighters are guided to use the information in these components to drive the development of COAs and their associated MOEs. The domain model in Figure 2 shows how this decomposition can be performed.

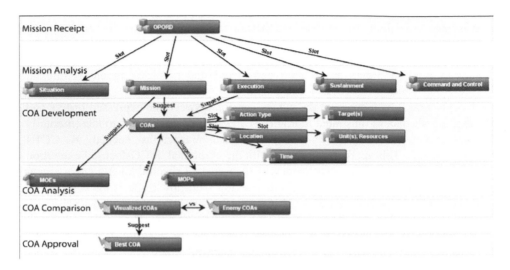

Figure 2: Expanded domain model including COIN-specific elements for the MDMP

Some MOEs and mission tasks must be inferred in addition to those specifically mentioned in the OPORD. These facts suggest that in practice, MDMP decision-making is not only based on the OPORD, nor only on battlespace observables, but rather on both of them simultaneously. Furthermore, during these decision-making processes, we determined that it is possible that a warfighter may not necessarily start with mission task analysis. Instead, the warfighter may begin by determining MOEs from mission goals—when bootstrapping and setting up mission parameters for the first time, they may naturally gravitate toward a specific step of the workflow that is familiar to them to gain a foothold. The user may even begin by considering the capabilities of available resources and what COAs they enable, viewing the MDMP as a cost-benefit analysis and recognizing the constraints put forth. In addition, the continual monitoring (and replanning) cognitive function mentioned above indicates that the mission analysis work products are static, and that the mission plan must evolve to the ever-shifting environment. As a result, our representation of the workflow is circular rather than linear, as illustrated in Figure 3.

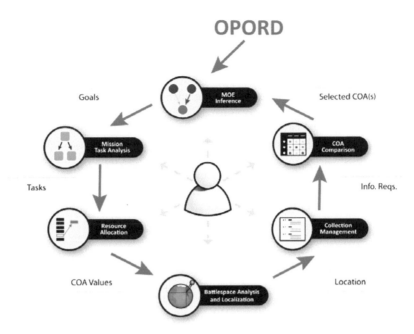

Figure 3: The MDMP/COIN workflow, represented as a series of repeating steps

THE PARALLEL PROCESS: MANAGING COLLECTION

The workflow designed in this paper relies on a dynamic and continuous assessment of MOEs to ensure that warfighters are executing their orders with an appreciation for their AO. This approach has potential benefit for warfighters in COIN missions, but places a burden on them to collect and interpret those MOEs as they evolve over time and link them to their ultimate causes (whether they were intentional elements of the COA or they resulted from environmental or behavioral factors). As a result, our discussion of this workflow is not complete until we consider the implications for intelligence collection and analysis. We consulted with our SMEs on how this process currently occurs in COIN environments, and how our new workflow may organize, enhance, support, and guide intelligence collection and analysis activities. Since the workflow encourages warfighters to spell out MOEs in a quantitative manner along traditional MDMP lines, and the effects they wish to achieve through their COAs, those projected changes provide guidance for intelligence processes. This insight, in turn, drives us to discuss the relationship between MOEs, MOPs, and intelligence analysis terminology such as Information Requirements (IRs).

Table 1 covers the relationships we posit between measures and IRs for this workflow, developed in conjunction and with the approval of our SMEs. We determined that MOEs may be approximately mapped to Priority Intelligence Requirements (PIRs), that MOCs may be mapped to Specific Intelligence

Requirements (SIRs), and that MOPs may be mapped to Specific Orders and Requests (SORs). The rationale for this mapping is covered partly in the table, and considers the scope of the IR or measure and its use within evaluating intelligence hypotheses or for re-planning. At the lowest level, SORs and MOPs both provide information on specific quantitative values that require little interpretation and may or may not be useful for higher-order assessment and planning. At the highest level, PIRs and MOEs are similarly linked: they may be comprised of multiple quantitative and qualitative values, require interpretation by warfighters, and are more directly linked to analytical hypotheses and planning tasks. At the middle level, intelligence analysis doctrine already has a well-defined term that links SORs to PIRs: SIRs. This median level of specification is required to link the very basic information from SORs to higher-order elements of information suitable for briefing to a commander in the form of PIRs. SIRs enable collection managers and intelligence analysts to collaborate on analytical conclusions and hypotheses at that intermediate level. Neither doctrine nor practice contains an existing analog between MOEs and MOPs. In our discussions with SMEs, we have found that proposing the new term "measures of change", or MOCs, resonates and fosters an intuitive understanding. MOPs represent simple observable factors collected over time, while MOEs represent compound variables subject to interpretation and reasoning. In the middle, MOCs assist analysts and planners in assessing how MOPs and other contextual variables change over time with respect to one another, while still maintaining a level of abstraction below MOEs. While this reasoning process and the resulting mapping seems intuitive, future research, design, and validation efforts will determine the benefits of and possible problems with it in practice. Once a COA has been developed, this mapping and linkage between planning and analysis ensures that the limited collection assets available are focused on the highest-value information.

Table 1: Mapping of IRs to measures (MOEs, MOCs, and MOPs)

Question Type	Use for Input	Use for Evaluation
PIRs	Ask "big" questions to motivate missions	MOEs
SIRs	Create PMESII trends to answer PIRs	New PMESII trends to answer MOEs; "measures of change" (MOCs)
SORs	Provide specific events/HVIs supplementing PMESII knowledge	MOPs

FUTURE WORK

In this paper, we have described a workflow for executing the MDMP within COIN missions with an explicit appreciation for the socio-cultural context of the environment, validated and supported by experts. In discussing future work and research in this area, we can consider how this workflow and associated processes may be supported by technical systems and tools, including visualization and planning tools, as well as modeling and simulation capabilities.

A key benefit provided by the workflow described in this paper is more quantitative COAs for COIN missions. To perform effective COA generation for COIN, warfighters need to be able to assess the advantages and disadvantages of a set of candidate COAs with respect to socio-cultural information. The MDMP is a consistent and thorough process for COA generation, but it also provides all units with the flexibility they need to adapt the process to their use cases. Visualizations should mirror this consistency, thoroughness, and flexibility by providing representations of socio-cultural factors grouped by simple parameters: temporal windows, specific regions, demographic lines, and blue and red COAs. To be maximally flexible, analysis tools should not impose many specific requirements about what comprises a "COA" for their representations, and few restrictions on what model is used to calculate values for socio-cultural factors, in cases where modeling and simulation technologies are brought to bear. Additionally, this workflow comprises explicit steps, in which warfighters are able to iteratively contribute new elements to the constantly evolving COA. Since the resulting COA is so detailed, planning tools must effectively support switching among frames of references smoothly, grounding the warfighter and supporting them in maintaining awareness of the higher-order goals of the COA as they manipulate details. Based on guidelines provided by David Woods (Woods, 1984), we considered how planning tools might optimally support visual momentum for users managing task work in multiple views: spatial cognition. Spatial cognition is supported when the interface is based around spatial visualizations, controls, and interactions, which allow the users to use their spatial cortex to process the information. Therefore, to support these dynamic decision-making processes, planning tools should exploit spatial cognition to enable warfighters to move between these steps at their own discretion, consider their work complete at any time, and exit the workflow.

The explicit definition and tracking of quantitative MOEs is of benefit to the MDMP, but can be difficult to establish and maintain. This capability is described as the Collection Management step of the workflow, and Figure 4 illustrates a sample interface to support the warfighter in Collection Management. This interface uses the relationships between measures and IRs, provided in Table 1, to equate the two concepts and uses a hierarchical format to illustrate how collection tasks contribute to the higher-order MOEs that formulate warfighters' decision points. It also enables warfighters to task assets to satisfy the collection tasks, enabling warfighters to construct a plan to optimally use their assets to obtain the most relevant and insightful intelligence.

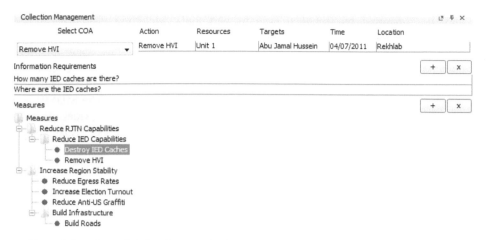

Figure 4: Sample user interface for managing intelligence collection in tactical environments

REFERENCES

Flynn, M. T., Pottinger, M., & Batchelor, P. D. (2010). *Fixing Intel: A Blueprint for Making Intelligence Relevant in Afghanistan.* Washington, D.C.: Center for a new American Security.

Rasmussen, J. (1986). *Information Processing and Human Machine Interaction: An Approach to Cognitive Engineering.* New York: North Holland.

Rasmussen, J. & Vicente, K. J. (1989). Coping With Human Errors Through System Design: Implications for Ecological Interface Design. *International Journal of Man-Machine Studies,*(31), 517-534.

Roth, E., Gualtieri, J., Easter, J., Potter, S., & Elm, W. (2000). Bridging the Gap Between Cognitive Analysis and Cognitive Engineering. In *Proceedings of 5th Naturalistic Decision Making Conference.* Sweden.

Roth, E. M. & Mumaw, R. (1995). Using Cognitive Task Analysis to Define Human Interface Requirements for First-of-a-Kind Systems. In *Proceedings of Proceedings of the Human Factors and Ergonomics Society 39th Annual Meeting,* (pp. 520-544). Santa Monica, CA: Human Factors and Ergonomics Society.

Vicente, K. J. & Rasmussen, J. (1990). The Ecology of Human-Machine Systems: II. Mediating Direct Perception in Complex Work Domains. *Ecological Psychology,* 2207-249.

Woods, D. D. (1984). Visual Momentum: A Concept to Improve the Cognitive Coupling of Person and Computer. *International Journal of Man-Machine Studies,* 21229-244.

CHAPTER 45

The *vmStrat* Domain Specific Language

Jonathan Ozik, Nicholson T. Collier, Michael J. North,
W.R. Rivera, Eric Palomaa and David L. Sallach

University of Chicago
jozik@uchicago.edu

ABSTRACT

Domain specific languages (DSLs) are increasingly used to enable subject matter experts in a variety of fields to take advantage of the power and convenience afforded by advancements in computation. A DSL is a language and associated idioms and concepts developed for a specific domain. A DSL can simplify programming in a given domain by focusing on the concepts and constructs relevant to that domain. DSLs can also introduce paradigm changing research methodologies into areas that have not traditionally relied on computation.

In this article we present the Versatile Multiscale Strategist (*vmStrat*) DSL. The *vmStrat* DSL was developed to allow analysts to represent a range of geopolitical scenarios with relevant actors, situations, and outcomes. *vmStrat* includes a set of well-defined programming constructs that embody carefully chosen abstractions based in social theory. These abstractions bridge the gap between the concreteness of software coding and the fluidity of social science. This bridge allows domain analysts to determine which social theories are relevant for a given problem, to implement working software to represent the chosen theories, and then to execute the software to derive useful conclusions from the theories.

The *vmStrat* ontology supports multiple types of social actors at different scales with different capabilities. The *vmStrat* approach is to represent social actors in general form and then customize only those aspects that are unique. Strategic actors may be nations, institutions, political parties, movements, or factions. All of the actor types draw upon the same core mechanisms, even though their particular choices are distinguished in appropriate ways. Actors of different scales can and do populate common scenarios and participate in shared interactions. *vmStrat* allows analysts to represent each actor's affect and strategic propensities as well as how these factors map into outcomes. Analysts can also construct

448

multiple models of the same scenarios for comparative analyses.

The **vmStrat** DSL leverages the Repast Simphony agent-based modeling platform and its ReLogo DSL. Repast Simphony is an integrated, richly interactive, cross platform Java-based agent-based modeling system that runs on most major computing environments. It supports the development of extremely flexible models of interacting agents for use on workstations and small computing clusters. The ReLogo agent-based modeling DSL is a Logo based dialect within Repast Simphony. It enables intuitive and rapid model development for beginning to advanced agent-based modelers. The **vmStrat** DSL substantially extends ReLogo by adding the nuanced support for modeling social actors described above.

Keywords: domain specific languages, agent-based modeling, modeling toolkits

INTRODUCTION

As a modeling framework and decision tool, **vmStrat** is designed to support the analysis by area specialists and subject matter experts (SMEs), with varying backgrounds and technical proficiencies. The tool is used for representing and exploring scenarios, both historical and hypothetical, including their strategic contexts and consequences.

One aspect of **vmStrat** is a domain specific language (DSL) focusing on the domain of dynamic strategy in geopolitical settings. DSLs are increasingly used to enable subject matter experts in a variety of fields to take advantage of the power and convenience afforded by advancements in computation (see e.g., Fowler et al. 2010, Dearle 2010). A DSL is a language, along with associated idioms and concepts, developed for a particular area of application. A DSL can simplify programming in a given area by focusing on the concepts and constructs relevant to that domain. DSLs can also introduce paradigm-changing research methodologies into areas that have not traditionally relied on computation.

The DSL developed for the **vmStrat** modeling framework builds on *Repast Simphony* (North, Howe, Collier, and Vos 2005) and *ReLogo* (Ozik 2011). Repast Simphony is an integrated, richly interactive, cross-platform Java-based agent-based modeling system that runs on most major computing environments (Microsoft Windows, Apple Mac OS X, Linux). It supports the development of extremely flexible models of interacting agents for use on workstations and small computing clusters. *Repast Simphony* models can be developed in several different forms including the ReLogo dialect of Logo, point-and-click flowcharts translated into Repast Simphony models, Groovy, or Java, all of which can be fluidly interleaved.

The ReLogo agent-based modeling DSL is a Logo-based dialect within Repast Simphony. It enables intuitive and rapid model development for beginning to advanced agent-based modelers. ReLogo was developed using the Groovy and Java programming languages. For more information on ReLogo see Ozik (2011). The **vmStrat** DSL substantially extends ReLogo by adding nuanced support for modeling social actors as described below.

VMSTRAT CONCEPTS AND ABSTRACTIONS

In **vmStrat,** models are designed to represent variegated situations and scenarios in a way that integrates the insights of analysts with diverse specialties, and supports the comparative exploration of their insights and assumptions. The key insights underlying such models are drawn from a wide range of socio-historical theories. The foundations of all strategic scenarios are grounded in relationships between public (official) and private (unofficial) affect and dynamic strategic interaction. Within **vmStrat** models, these elements are represented in a virtual space of two or more dimensions. Given the emergence of political instability (or even stability) within various strategic contexts, as well as the propensities of strategic actors (whether persistent or emergent), to engage in strategies, form alliances and, through their efforts, to shape the scenarios unfolding around them, it is intended that regional specialists will find this tool to be useful in achieving strategic insight.

Scenarios are constructed by selecting pertinent actors and identifying their key attributes in relational terms. These relations currently represent the dynamics of affect, official and unofficial; strength; and strategy.

Strength is characterized as the ability to bring resources to bear in such a way that capitalizes on strategic resources to enable action that either supports the actor and/or allies, or weakens adversaries. Strength as a bridge between resource and action is articulated in large measure through the ability of Actors to make or stake a claim. Claims, rooted in the theory of contentious politics (McAdam, Tarrow & Tilly 2001; Tilly & Tarrow 2006), bring resources—natural, political, social, material, or discursive—into contention where at least one of the Actors is a government and where the claim to this resource would strengthen one Actor at the expense of another.

Affect is characterized as both official and unofficial as a way to account for the difference between publicly expressed and privately held emotive factors that may result in preference falsification—the theory that deals with behaviors that are significantly different from expressed sentiment (Kuran 1997). Affect is based on a scale that ranges from *ally* to *adversary*.

Strategy captures the current relationship between pairs (and clusters) of actors on a scale that ranges from *beneficent* to *aggressive* within a given context.

One of the main strengths of **vmStrat** is its ability to process a diverse number of factors, as mentioned above, and produce easily readable representations and various types of data-driven outputs. Another strength of **vmStrat** is the ease with which area specialists will be able to input data since this is done using standard comma separated value (CSV) files. In other words, from Input to Output, the sophisticated machinery of **vmStrat**, while available to users, can also be largely invisible, therefore creating a user-friendly interface for non-programmers.

Scenario

A **vmStrat** Scenario involves the creation of a simulation model in the **vmStrat** DSL. Building on the modeling idioms of *ReLogo*, the model will have an observer (or controller) defined, which coordinates the progress of the simulation. An observer contains generic behaviors (or *methods*) defining the setup of a model, as well as its evolution. The setup of a model generally involves the use of external data to initialize the simulation (see Model Setup and Data Loading, below). The evolution of a model is specified via scheduled actions (see Scheduling Simulation Actions).

Propensity Spaces

A Propensity (p) space is an n-dimensional space used to track actor propensities. P-spaces can be used for actor action selection, actor state maintenance and visualization, as well as many other purposes. The **vmStrat** framework currently provides two p-space types, *strategy/affect* (SA) and *national instability* (NI).

The NI space is a one-dimensional p-space. It is mainly concerned with tracking a single government Actor's *upheaval* and *government change* propensities. Based on these propensities, connections are made to a hybrid outcome space.

The SA space is a 2D p-space. It keeps track of relations *between* actors, rather than information on a single actor. In addition to maintaining information on actor relations, the SA space is also provided with SA Triggers which allow for it to be used as an action-selection mechanism during an engagement between actors. The SA space can also be used to display relationships between actors, as is done in the 2D and GIS displays within **vmStrat**.

P-spaces are not restricted to the two types described. Custom p-spaces can be developed within *vmStrat* to be used as necessary. For information on propensity theory and propensity spaces, see Popper 1959; 1990; Urbach 1980; Gillies 2000; Albert 2007; Belnap 2007 and Bartelborth 2011.

vmStrat Actors

A **vmStrat** scenario involves the interaction of actors. The basic actor type is the PoliticalActor. In addition to the basic actor type there are two default actor types which *inherit* from this basic actor type, StrategicActor and Government. The basic actor type hierarchy is shown in Figure 1.

Figure 1: Default vmStrat actor type hierarchy.

Strategic Actor

The StrategicActor actor type is the basis for engagement between actors. Each StrategicActor possesses a *strategy/affect* (SA) space, which, as explained above, is a *Propensity Space* of dyadic type. The SA space is used to keep track of strategies employed against, and affect (both official and unofficial) held towards, other actors in the simulation.

The Strategy variable is explicitly relational. Strategy, for our purposes, measures the types of moves Actors will take in support of, or against, other Actors. This is built on the concept of prototype multigames (promulgames) developed by Sallach (2000; 2006; Sallach, North & Tatara 2011). In this view, Actors are engaged in multiple games at the same time and, accordingly, a single move can have multiple ramifications. In this instance, the game space is construed as moving from *beneficent* to *neutral* to *aggressive*, with each move also being interpreted by other Actors in the space. A high strategic value means a high level of positive Strategy moving towards or operating in the *beneficent* range. Conversely, a low strategic value indicates a high level of negative Strategy moving towards or operating in the *aggressive* range. These values are determined by expert analysis based on empirical information and data-based inference.

Official and Unofficial Affect are evaluations of the public record made by content experts. Content experts are called upon to examine public statements and official positions and/or actions taken by the pertinent Actors in a given Scenario to evaluate the Official Affect between two Actors. This is a fairly straightforward process. Evaluating the Unofficial Affect requires a little more judgment. Unofficial Affect is meant to capture the privately held Affect of one Actor towards another. As such, the public record is not always instructive in this regard. However, this area is too important to real-world outcomes to be ignored. Analysts are tasked with "reading between the lines," looking at comments made off the record, leaks from well-placed officials, and general content area specialized knowledge to make these important judgments. (Under some circumstances, if the gap between official and unofficial affect becomes too large, the **vmStrat** DSL allows for reducing the discrepancy.)

For each actor, their SA space will contain two separate two-dimensional points (*strategy*, *officialAffect*) and (*strategy*, *unofficialAffect*) for each other StrategicActor in the simulation. Three example SA spaces are shown in Figure 2, where vertical lines are used to connect the official and unofficial *strategy/affect* points associated with an actor.

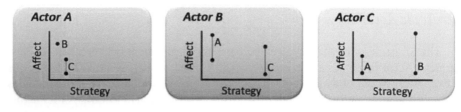

Figure 2: Example Strategy/Affect spaces for the Actors Actor A, Actor B, and Actor C.

Strategy/Affect values are in the range -1.0 to 1.0. Prototypical strategy values are *Aggressive* (-0.66), *Neutral* (0), *Beneficent* (0.66) while prototypical Affect values are *Adversary* (-0.66), *Neutral* (0), *Ally* (0.66).

The **vmStrat** DSL allows the definition of *Strategy/Affect Triggers* within the SA spaces. These Triggers support behaviors that are associated with points within the SA spaces and which enable situated action selection within strategic games.

Government

The government actor type encapsulates the notions of national *upheaval* and *government change*. Each government possesses a *National Stability* (NS) space, which is a *propensity space* of monadic type, in that rather than tracking relationships between actors, it tracks a propensity state of an actor. The NS space is used to keep track of *upheaval* and *government change* for the actor. As such, for each Government actor, their NS space contains one two-dimensional point (*upheaval*, *government change*). Two example NS spaces for two Government actors are shown in Figure 3.

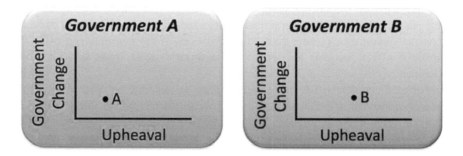

Figure 3: Example *National Stability* spaces for the actors *Government A* and Government B.

The *upheaval/government change* point values are in the range 0 to 1, where 0 indicates status quo and 1 indicates drastic change.

The **vmStrat** DSL defines mappings between *strategy/affect* spaces and *national stability* spaces. For each Government actor, the *upheaval* and *government* *change* is determined by considering all Strategic Actors' *strategy, official affect*, and *unofficial affect* toward the Government actor. For each Strategic Actor, the contribution to the *upheaval* calculation for Government actor a is denoted $U_i(a)$, which is dependent on S_{ia}, the *strategy* employed by actor i toward actor a.

The contribution to the *government change* calculation for Government actor a is $G_i(a)$, which is calculated using the unofficial affect and official affect for each Strategic Actor considered. The upheaval $U(a)$ for actor a is calculated as a weighted $w_i(a)$ average of the contributions $U_i(a)$ from each of the other Strategic Actors.

$$U(a) = \frac{\sum_{i=other\ actors} U_i(a)w_i(a)}{\sum_{i=other\ actors} w_i(a)}$$

Similarly, the government change $G(a)$ for actor a is calculated as a weighted $w_i(a)$ average of the contributions $G_i(a)$ from each of the other Strategic Actors.

$$G(a) = \frac{\sum_{i=other\ actors} G_i(a)w_i(a)}{\sum_{i=other\ actors} w_i(a)}$$

vmStrat Propensity Space Relations

Each **vmStrat** scenario uses a set of interrelated propensity spaces. As detailed above, *strategy/affect* spaces account for each agent's Official and Unofficial views of other agents. *National instability* spaces incorporate all Unofficial views regarding each nation. The NI spaces are individually mapped to Outcome Propensity Bridge (OPB) tables. The bridge module (or, by extension, an alternate outcome space) then determines an outcome based on the OPB being used. Figure 4 shows an integrative overview for a **vmStrat** scenario, including the relationships between the various propensity spaces and the OPB tables.

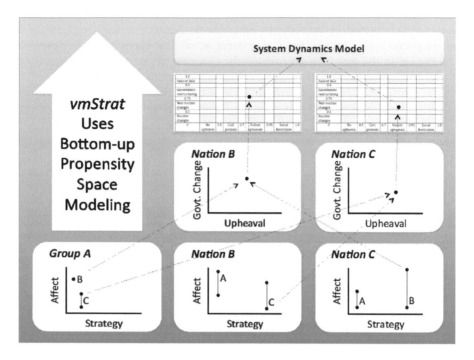

Figure 4: Example *vmStrat* Propensity Space Relationships

Strength

Strength is one of the key variables for modeling Strategic Contexts and political instability. Strength is conceptualized as a link between the resources an Actor has at its disposal and the actions it wants to take. For example, a large reserve of oil or natural gas deposits is a great resource to have but it is considered Strength when an Actor can bring extraction, production, and export of this resource to influence market conditions, destabilize a competitor through price manipulation, maintain the status quo, or contribute to other strategic goals that an Actor may have. Further, social or political instability can sometimes be seen as imposing a substantial diminution of Strength. As governments grow less able to marshal resources towards actions, other Actors, whether domestic resistance forces or outside Actors, may lay claim to those resources, seek to be the legitimate Actors to achieve the actions in question, challenge, or make new claims that further weaken the government in question.

Conceptualizing Strength in this way enables modelers to bridge key concepts and actions. But in order to operationalize these concepts it is necessary to specify the Strength variable. Depending on the nature of the study, there are varying approaches to do this. A robust method involves identifying many indicators of Strength in various areas such as economic, military, energy, diplomatic, and geo-strategic, and populate these with database records in a data repository. These data can then be normalized to a unit range (with values from negative to positive one [-1.0,+1.0], and then weighted and/or adjusted as required for modeling purposes. An alternate method, depending on available SME resources, is to take the relevant categories of geo-strategic, economic, military, energy and diplomatic values, as well as others that may seem pertinent, and use area knowledge to attribute these values to the appropriate Actor. In either case, weighting the values of various data will require expert judgment.

Model Setup and Data Loading

A typical **vmStrat** simulation contains a setup method which sets up the simulation with appropriate initial conditions. The setup method is usually where any external data used to define actor attributes and actor relations are loaded. The **vmStrat** framework uses comma separated value (CSV) files as data input for the simulations. This allows for the use of standard spreadsheet applications to directly interact with **vmStrat** simulations. The following is an overview of the types of data that can be used as input to a **vmStrat** scenario:

Actor Attributes

An *Actor Attributes* CSV file is used to create actors. The file is comprised of columns with headers identifying each attribute type and rows associated with each actor to be created.

Strategy Affect

Three CSV files are used to define the Strategy Affect (SA) spaces of Strategic Actors within a simulation. All three CSV files, *Strategy, Official Affect, Unofficial Affect* are used for initializing the actor SA spaces.

National Stability

A *National Stability* CSV file is used to initialize the National Stability (NS) space of Government actors.

Exogenous Events

An *Exogenous Event* CSV file can be used to specify external drivers for a simulation. External drivers are defined as scripted changes to the simulation which occur at specified times.

Scheduling Simulation Actions

A *vmStrat* simulation is evolved by scheduling actions, typically in the observer. Any method can itself schedule other methods so, for example, a method which is scheduled to occur yearly can schedule within itself a *winter, spring, summer*, or *fall* method.

Multigame Interaction

The interaction between Actors is mediated by the *srategy/affect* (SA) spaces of the Actors. Any Actor can engage in protogames with and/or against each of the others. Alternatively, a single Actor can be chosen to engage all other Actors and a pair of Actors can be chosen to play against each other.

When an actor (*A*) plays against another actor (*B*), *A* and *B* take turns moving against each other. For each *A* protogame turn, *A* inspects its SA space and determines the two points related to *B*.

> (*strategy, officialAffect*)
> (*strategy, unofficialAffect*)

As mentioned above, each actor has SA Triggers defined. These Triggers are used to determine each actor's action selection. The SA Triggers are represented as regions within the SA space and the closest applicable Trigger to either of the two *strategy/affect* points above determines the Trigger, and hence action, that *A* takes relative to *B*.

As a result of the play of the game, properties of both *A* and *B* are subject to modification, resulting in, for example, new *strategy/affect* points. When the next round of multigames is played, the actors will potentially act very differently toward each other. This feedback yields dynamically iterated multigame interaction.

VISUALIZING VMSTRAT OUTPUTS

The following displays of 2D strategy/affect (Figure 5), GIS visualization (Figure 6) and logging (Figure 7) are representative examples of output available to the analyst from **vmStrat**.

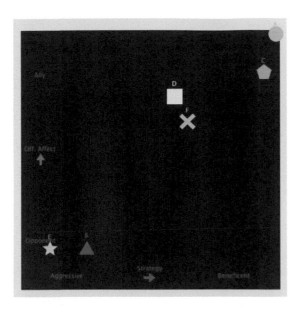

Figure 5: Strategy affect space from the perspective of actor A.

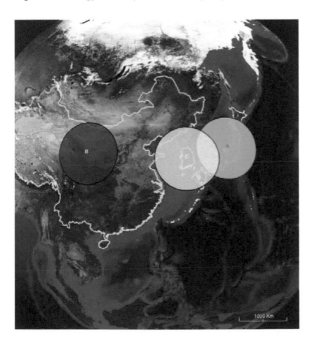

Figure 6: Sample GIS map display.

458

```
LOG - A playing game with B
LOG - A strategy official affect point for B : (-0.5, -0.7)
LOG - A strategy unofficial affect point for B : (-0.5, -0.7)
LOG - ao triggered on A with B as argument
LOG - B playing game with A
LOG - B strategy official affect point for A : (0.1, 0.16477968981440538)
LOG - B strategy unofficial affect point for A : (0.1, 0.1187718757717473)
LOG - nr triggered on B with A as argument
LOG - A playing game with C
LOG - A strategy official affect point for C : (0.9, 0.7)
LOG - A strategy unofficial affect point for C : (0.9, 0.7)
LOG - ba triggered on A with C as argument
LOG - C playing game with A
LOG - C strategy official affect point for A : (0.5, 0.7)
LOG - C strategy unofficial affect point for A : (0.5, 0.7)
LOG - ba triggered on C with A as argument
```

Figure 7: Sample logging output.

CONCLUSIONS

vmStrat is a domain-specific language that combines technological innovation with social depth. On the technical side, it draws upon the strengths of innovative programming paradigms as they become available. This allows the language to build upon effective representations that provide generic capabilities while, at the same time, producing highly situated strategies and consequences. The interactive dynamics supported by **vmStrat** are innovative due to their extensive grounding in social theory.

The result is a tool, a language, and an associated modeling strategy, that together can more effectively address complex socio-historical and policy-oriented issues. What has been summarized here, can best be understood as a down payment on future modeling frameworks that can be constructed on these foundations.

ACKNOWLEDGEMENTS

The authors gratefully acknowledge support for this project from the Office of Naval Research, Award No. N00014-09-1-0766.

REFERENCES

Albert, Max. 2007. The propensity theory: A decision-theoretic restatement. *Synthese* 156:587-603.

Bartelborth, Thomas. 2011. Propensities and transcendental assumptions. *Erkenn* 74:363-381.

Belnap, Nuel. 2007. Propensities and probabilities. *Studies in the History and Philosophy of Modern Physics* 38:593-625.

Dearle, F. 2010. *Groovy for Domain-Specific Languages*, Packt Publishing.

ICG. 2011. Popular Protests in North Africa and the Middle East III. Middle East/North Africa Report No. 105-6 (April). Washington, DC: International Crisis Group.

Fowler, M. & Parsons, R. 2010. *Domain-specific languages*, Pearson Education.

Gillies, Donald. 2000. *Philosophical Theories of Probability*. New York: Routledge.

Kuran, Timur. 1997. *Private Truths, Public Lies: The Social Consequences of Preference Falsification.* Cambridge, MA: Harvard University Press.

McAdam, Doug, Sidney Tarrow & Charles Tilly. 2001. *Dynamics of Contention.* New York: Cambridge University Press.

North, Michael, Thomas Howe, Nicholson Collier, and Jerry Vos. October 13–15, 2005. Repast Simphony runtime system. Proceedings of the Agent 2005 Conference on Generative Social Processes, Models, and Mechanisms, ANL/DIS-06-5, Argonne National Laboratory, The University of Chicago. Chicago, IL.

Ozik, J. 2011. *ReLogo Getting Started Guide.* http://repast.sourceforge.net/docs/ReLogoGettingStarted.pdf.

Popper, Karl. 1990. *A World of Propensities.* Bristol, UK: Thoemmes.

Popper, Karl R. 1959. The propensity interpretation of probability. *British Journal for the Philosophy of Science* 10:25-42.

Sallach, David L. 2000. Games social agents play: A complex form. Presented to the *Joint Conference on Mathematical Sociology in Japan and America.* Honolulu.

—. 2006. Complex multigames: Toward an ecology of information artifacts. Pp. 185-190 in *Proceedings of the Agent 2006 Conference on Social Agents: Results and Prospects,* edited by D. L. Sallach, C. M. Macal, and M. J. North. Chicago: Argonne National Laboratory.

Sallach, David L., Michael J. North & Eric Tatara. 2011. Multigame dynamics: Structures and strategies. Pp. 108-120 in *11th International Workshop on Multi-Agent-Based Simulation,* edited by T. Bosse. Toronto: Springer.

Tilly, Charles & Sidney Tarrow. 2006. *Contentious Politics.* Boulder, CO: Paradigm Publishers.

Urbach, Peter. 1980. Social propensities. *British Journal for the Philosophy of Science* 31:317-328.

CHAPTER 46

Decision Support System for Generating Optimal Sized and Shaped Tool Handles

Gregor Harih, Bojan Dolšak, Jasmin Kaljun

Laboratory for intelligent CAD systems, Faculty of Mechanical Engineering
University of Maribor
Smetanova ulica 17, SI-2000 Maribor, Slovenia
Email address: gregor.harih@uni-mb.si

ABSTRACT

Most authors have provided recommendations for cylindrical handle design in order to increase performance, avoid discomfort and cumulative stress disorders, but none of them considered optimal shape of the handle and no systemization of the knowledge was done at this field. Therefore most small companies, who do not employ an ergonomist, do not have the knowledge for proper implementation of ergonomic principles in tool handle design. As a result, the developed products have poor ergonomics or leave potential for improvement. To overcome these limitations, a decision support system is proposed which can provide competent advice on the handle design process and can generate optimal sized and shaped handle for a target population. Therefore new methodology was developed to generate optimal handle size and shape based on hand anthropometrics and magnetic resonance imaging. The resulting handles consider optimal diameters for each finger which maximizes the possible finger force exertion, comfort and contact area and thereby preventing many task related disorders. An increase of 25% in contact area in comparison to an optimal cylindrical handle was measured, which effectively decreases overall contact pressure. Based on interviews, the subjective comfort rating is also increased. Proposed decision support system decreases the used time and therefore also money while designing a tool handle. The system allows development of optimal tool handles with almost no prior ergonomics knowledge, thereby no iterative design process is needed for the ergonomic analyses of the tool handle.

Keywords: ergonomics, tool handle design, decision support system

1 INTRODUCTION

Human interaction with the physical environment is a natural occurrence, therefore designer has to consider the product-human interaction in order to develop products with high rate of efficiency and comfort. Modern CAE and CAD software allow the designer to evaluate the new product virtually. If a product will be human operated, the system performance is also human dependent, therefore designer has to consider ergonomics in order to achieve the expected system efficiency (Hogberg et al., 2008). Utilization of ergonomic principles in the design process is a very important task. In the field of workplace ergonomics many software solutions exist, although there is still a lack of dedicated ergonomics software in the field of product ergonomics which would make evaluation and analyzing of the proposed design at the virtual stage possible. In a human - product interaction, the designer has three constraints which have to be considered to design an efficient product. Design attributes of the product define the Task and Product constraints, the Cognitive and bio-mechanical constraints are defined with the user. If there is a viable human-product interaction possible, all three constraints must overlap to some extent (Figure 1). Somewhere inside the intersected area is the optimal human-product interaction. To find the optimum, the designer has to set his objective function and perform optimization. Task and product constraints can be altered with different design attributes, therefore designer has to have knowledge about the target population's cognitive and bio-mechanical constraints in order to adapt the product design for the optimal human-product interaction.

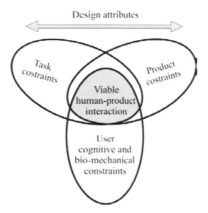

Figure 1 Viable human-product interaction and its constraints.

Since most of the products are designed to be human hand operated, many authors have researched the topic of tool handle design to define the optimal size and shape of a tool handle, since a correctly designed handle can provide safety,

462

comfort and increased performance. Different authors used different criteria to determine optimal cylindrical handle: subjective comfort rating (Yakou et al., 1997, Hall and Bennett, 1956); finger force measurement (Amis, 1987, Chen, 1991); muscle force minimization (Sancho-Bru et al., 2003) and hand anthropometrics (Grant et al., 1992, Oh and Radwin, 1993, Johnson, 1993, Yakou et al., 1997, Blackwell et al., 1999, Garneau and Parkinson, 2010, Seo and Armstrong, 2008). Few studies also used two or more criteria: finger force measurement and muscle activity (Ayoub and Presti, 1971, Grant et al., 1992, Blackwell et al., 1999); subjective comfort rating, finger force measurement and electromyographic efficiency of muscle activity (Kong and Lowe, 2005). Since many different user cognitive and bio-mechanical constraints were considered in mentioned papers, therefore also used methods for the determination of optimal cylindrical handle differ very much. Resulting diameters of the cylindrical handles vary from 25mm to 60mm and above.

Mechanical behavior of the skin and subcutaneous tissue is very important in gasping tasks, since it is in direct contact and the forces and moments are transferred from the tool to the whole hand-arm system. Skin and subcutaneous tissue have nonlinearly visco-elastic properties, where skin is stiffer than subcutaneous tissue (Wu et al., 2007). Both have low stiffness region at small strains followed by great increase of the stiffness by increasing the strain. It has been shown, that higher contact pressures than allowed for specific time can result in discomfort, pain and ischemia. A power grasp can yield in contact pressure of the fingertip of 80kPa, which is excessive loading for skin and subcutaneous tissue (Gurram et al., 1995). Excessive loading can result also in other cumulative trauma disorders such as carpal tunnel syndrome (Eksioglu, 2004). Many of powered hand tools produce vibrations, which are transferred from the handle to the hand. Deformations of skin and subcutaneous tissue while holding the tool and vibration induced by the tool can lead to disorders called hand-arm vibration syndrome which may cause vascular, sensorineural and muscolo-sceletal disorders (Bernard et al., 1998, Youakim, 2009). Objects that follow the shape of the hand result in much lower local contact pressures of the soft tissue, which can prevent discomfort and several disorders (Wu and Dong, 2005).

Most of the authors have focused on cylindrical or elliptical shapes of the handles, but none of them considered the anatomical shape of the hand in the optimal power grasp posture. Handles should vary in size between hand and finger size, since maximum possible exerted finger force is diameter dependent (Kong and Lowe, 2005). Authors suggested that further research of this topic should consider the shape of the hand in the optimal power grasp posture since it could improve the ergonomics of the tool handle (Garneau and Parkinson, 2010).

2   METHODS

While designing for specific target population, the user cognitive and bio-mechanical constraints are even more rigorous, therefore the designer has to carefully determine design attributes in new target population customized product design in order to define the optimal human-product interaction (Merle et al., 2010). Traditional methods such as designing with anthropometric data and simple mathematical models do not allow high consideration rate of constraints and thereby also the optimization of human-product interaction is not possible. Therefore this paper presents new methodology for designing tool handles for target population based on medical imaging as described in following sections which allows high consideration rate of bio-mechanical constraints and tries to reach the optimal human-product interaction for increased performance and comfort and prevention of task related disorders. The findings are integrated into an intelligent decision support system which provides comprehensive design methods for hand tool design process to improve ergonomics with reduced designing time and thereby also reduced costs.

2.1 Determination of optimal sized and shaped tool handle

In this paper most recent studies have been considered to determine the optimal handle diameters (Garneau and Parkinson, 2010, Kong and Lowe, 2005). To cover the wide range of statistical population, ten subjects with least deviation from mean anthropometric measurements according to predefined percentile were used in this study. Anthropometric measurements of selected male subjects with no hand injuries or disorders were performed on the hand. Afterwards optimal diameters for each subject were calculated based on the anthropometric measurements to manufacture a custom optimal cylindrical pre-handle with variable diameters out hard plastics for each subject. To maintain the shape of the hand in its optimal power grasp posture with undeformed skin and subcutaneous tissue, an outer hand mould out of orthotic material Orfilight® (Orfit Industries, Belgium) was manufactured softly holding the optimal cylindrical pre-handle with variable diameters. To obtain the shape of the hand in its optimal power grasp posture with undeformed soft tissue, magnetic resonance imaging (MRI) was performed. The MRI machine was a GE medical systems Signa HDxt 3.0T. Slice thickness was set to 1mm to avoid unnecessary small anatomical structures and surface details. The subject was told to hold the hand in opened position touching the outer hand mould during the scanning to maintain the proper diameters and shape of an optimal power grasp. Scanned images were provided in the DICOM format and were afterwards segmented and 3D reconstructed with a professional medical imaging and editing software Amira® 5.3.3. (Visage Imaging®). Segmentation was done based on thresholding technique since no inner anatomical structures are necessary. The resulting 3D model of the subject's hand does not include any geometric topological relations, therefore no feature based CAD solid modeling techniques and no FEA are possible which requires vector-based modeling environment (Sun et al., 2005). Therefore the STL file obtained in Amira® software was imported into CATIA® V5R20 and the point data form of the STL file was then triangulated to form a freeform NURBS surface model. Since most of the human anatomical parts are organic shapes, feature recognition or reverse engineering on the CAD model is not necessary, therefore a solid model is sufficient. To get the optimal custom handle for power grasp posture, an elliptical cylinder was modeled, since it follows the shape of the hand in its power grasp posture best and it provides greater contact area in the palmar region. The size and the position of the elliptical cylinder were determined so the cylinder fully overlapped the palmar empty volume created by the hand in optimal power grasp posture. To get the shape of the handle, Boolean operation "Remove" was used, which removed the cylinder model volume, which was in overlap with the hand model volume. The overview of the used methods is seen in Figure 2.

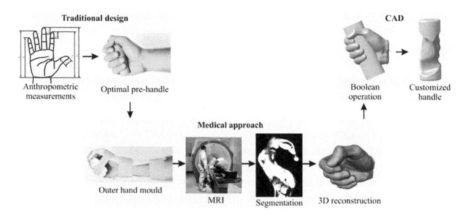

Figure 2 Used methods for obtaining optimal sized and shaped tool handle.

2.2   Mathematical model

For the database which is the base for mathematical model development, it is crucial to have a representative statistical population. To maintain the consistency of the mesh topology, landmarks were placed on the obtained handles. Based on the radial basis function the handle warping and volume parameter extraction was done in commercial numerical software Wolfram Mathematica® 8. Mathematical model is derived from the volume parameterization where algorithm recognition was used (Figure 3). The mathematical model proved to provide sufficient accuracy, since surfaces of the mathematical generated tool handle and tool handle obtained with the medical and CAD approach matched within small deviations.

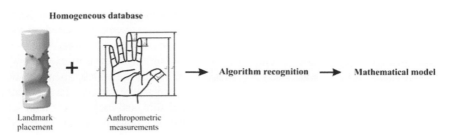

Figure 3 Used methods and derived mathematical model.

2.3 Decision support system

The proposed mathematical model generates optimal sized and shaped tool handles based on target population, but it does not consist of basic ergonomic recommendations which are also important in the design process of the hand tools and especially tool handles. Therefore the mathematical model was integrated into an existing intelligent decision support system OSCAR (Kaljun and Dolšak, 2012). The intelligent decision support system consists of general ergonomic and also aesthetic recommendations for the hand tool design process. Within the system different classes are interconnected with various attributes and their values at the head of the rules in order to describe case specific situations, in which those recommendations for product design should be taken into consideration. The system consist of most important ergonomics design goals and respective design recommendations relating to hand tool design process from which to list just a few: dimensioning and configuration, wrist positioning, tissue compression, excessive forces reduction, etc. The structure and usage of the intelligent decision support system is seen in figure 4.

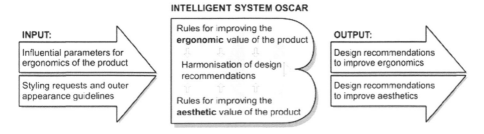

Figure 4 Intelligent decision support system OSCAR.

3 RESULTS

The result is a comprehensive intelligent decision support system which consists of general design recommendations and the mathematical model for generating optimal sized and shaped tool handles. Based on the design requirements of the new hand tool, the designer has to determine the target population which can be very specific or more general according to anthropometric measurements. Based on the target population, the system provides basic design recommendations, such as sizing, configuration, clearance, etc. Afterwards the anthropometric measurements which are needed for optimal sized and shaped tool handle are entered manually or selected from the ANSUR or NHANES anthropometric database. The system suggests a factor of generality based on the hand tool and task. The factor of generality is a smoothing function which is used to smooth the handle generated with the mathematical model based on the anthropometric measurements for the chosen target population. The result is optimal sized and shaped tool handle for specific target population with provided design recommendations which can be directly used in the design process of new hand tool. Schematic usage of the decision support system is seen in figure 5.

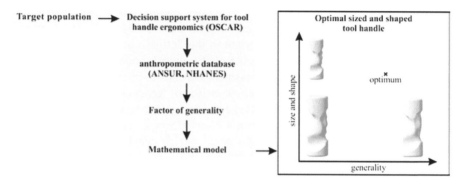

Figure 5 Usage of the decision support system.

4   DISCUSSION

Existing mathematical models for tool handle design determination consider only the diameters for a cylindrical or elliptical handles and therefore do not allow the shape determination which could additionally improve the ergonomics of the handle. Therefore new methods were developed to determine the optimal shape of the handle for a target subject. Firstly, the methods were verified for the accuracy, since the accuracy of the proposed mathematical model is strongly dependent on the accuracy of the used methods. Since the perceived comfort and maximum possible finger force exertion is diameter dependent, the diameters of the handles were virtually checked in CATIA® and compared to the calculated. It has been shown, that the diameters were kept by the outer hand mould during the MRI within small deviations, therefore the maximum possible grip force can be exerted and subjective comfort rating which is diameter dependent is also maximized. Thereby it can be assumed, that the optimal power grasp posture was withheld during the 3D hand shape acquisition with a high rate of accuracy. Optimal cylindrical handles were manufactured using hard plastics to compare the contact area with the optimal shaped and sized tool handles. It has been shown, that the optimal sized and shaped handles provide on average 25% greater contact area than optimal cylindrical handles. The increase of the contact area is due to the fact, that the optimal sized and shaped handles exactly follow the shape of the hand in its optimal power grasp posture and thereby provide the greatest contact area possible for a target subject. While gripping the handle high local and overall contact pressure can be avoided. This provides greater comfort rate and likely prevents cumulative stress disorders compared to a cylindrical handle. Contact area maximization is not possible with cylindrical handles since the greatest contact area can be only achieved with diameters, which are bigger than the optimal diameters for grip force and comfort maximization. Therefore it is necessary also to consider the anatomical shape of the hand in its optimal power grasp posture to maximize the contact area. Target subjects provided also subjective rating on stability and comfort based on an interview which was then compared to the optimal cylindrical handles. Target subjects described the optimal sized and shaped handles as more stable, since stability is greatly increased especially in the axis of the handle where outer torques are transferred with shape of the handle and not mostly with friction as with the cylindrical handle. Therefore a lower normal grip force can be exerted in comparison to the cylindrical handle and the tool can be stable held in the hand. Target subject described the optimal shaped tool handle as more comfortable since the handle suits the hand better with the anatomical shape.

Since the new developed methods for obtaining optimal sized and shaped tool handles allow high consideration rate of user cognitive and bio-mechanical constraints and proved to deliver accurate results, they were utilized to develop the mathematical model for the optimal sized and shaped handle generation based on anthropometric measurements. For the mesh consistency, landmarks were placed on the obtained handles. Proposed volume parameter extraction method proved to deliver consistent mesh topologies, therefore radial basis function was successfully used for 3D model warping. Recent studies provided different statistical methods for the volume parameterization, therefore other parameter extraction and statistical methods will be tested in future to improve the accuracy and robustness and lower the computation time.

Basic handle recommendations and existing mathematical models for diameter determination do not provide comprehensive aid in the design process of tool handles, therefore non-experienced designers are confronted with a challenging task. To aid the designers within the design process, many intelligent decision support systems were developed, although in the field of product ergonomics just few exist. Nevertheless also the decision support systems do not guarantee the achievement of high ergonomic values in tool handle design since the design recommendations are mostly to loose and the end design is mostly dependent on the designer's experience and creativity. This is especially prominent in the shape determination in handle design process. Therefore the proposed mathematical model is integrated into the existing decision support system OSCAR. This way the system provides a comprehensive design aid from selecting the target population, basic recommendations, to generating optimal sized and shaped tool handles for the selected target population. The value of the factor of generality as a smoothing function is based on the product being developed and the associated task.

Based on mathematical model for generating optimal sized and shaped tool handles integrated into an existing intelligent support system it is possible to design hand tools and tool handles without any prior ergonomic knowledge, since the designer is provided with needed design recommendations and optimal sized and shaped tool handle for target population. Therefore the proposed decision support system allows small companies to develop tools with good ergonomics without employing a specialist in ergonomics. With the use of the decision support system and especially the mathematical model the designing process time is reduced which was also shown in several preliminary tests of the system. Thereby companies utilizing proposed decision support system can develop products with better ergonomics and are therefore also more competitive.

Although the system proved to deliver accurate results, the mathematical model could be improved considering more test subject. Future work should also revise the factor of generality in order to provide a more precise amount of smoothing. The smoothing function is applied to whole handle, although future work should also consider the fact that different surface parts of the handle require different quantity of smoothing to achieve the desired generality. Further testing of the decision support system and the mathematical model will be also conducted to evaluate and analyze the impact on the hand tool design process.

5   CONCLUSION

Newly developed methods consider also the anatomical shape of the hand in development of the optimal shaped handle which increases the contact area, gripping force, stability and subjective comfort rating in comparison to traditional cylindrical handles. The proposed intelligent decision support system provides comprehensive aid in ergonomic and aesthetic design process in form of recommendations and in generation of optimal sized and shaped tool handles. Thereby small companies, who do not employ an ergonomist, can develop hand tools considering ergonomic principles for improved performance, comfort and avoidance of cumulative stress disorders.

ACKNOWLEDGMENTS

We would like thanks to prof. Vojko Pogačar from Faculty of Mechanical Engineering University of Maribor, to prof. dr. Andreja Sinkovič and prof. dr. Andrej Čretnik from University Medical Centre Maribor who have helped to establish the cooperation between this two institutions and for selfless help.

REFERENCES

Amis, A. A. 1987. Variation of finger forces in maximal isometric grasp tests on a range of cylinder diameters. Journal of biomedical engineering, 9, 313-20.

Ayoub, M. M. & Presti, P. L. 1971. The Determination of an Optimum Size Cylindrical Handle by Use of Electromyography. Ergonomics, 14, 509-518.

Bernard, B., Nelson, N., Estill, C. F. & Fine, L. 1998. The NIOSH review of hand-arm vibration syndrome: Vigilance is crucial. Journal of Occupational and Environmental Medicine, 40, 780-785.

Blackwell, J. R., Kornatz, K. W. & Heath, E. M. 1999. Effect of grip span on maximal grip force and fatigue of flexor digitorum superficialis. Applied ergonomics, 30, 401-5.

Chen, Y. 1991. An evaluation of hand pressure distribution and forearm flexor muscle contribution for a power grasp on cylindrical handles. Ph.D. Ph.D., University of Nebraska.

Eksioglu, M. 2004. Relative optimum grip span as a function of hand anthropometry. International journal of industrial ergonomics, 34, 1-12.

Garneau, C. J. & Parkinson, M. B. 2010. Optimization of Tool Handle Shape for a Target User Population. Proceedings of the Asme International Design Engineering Technical Conferences and Computers and Information in Engineering Conference, 5, 1029-1036.

Grant, K. A., Habes, D. J. & Steward, L. L. 1992. An analysis of handle designs for reducing manual effort: The influence of grip diameter. International Journal of Industrial Ergonomics, 10, 199-206.

Gurram, R., Rakheja, S. & Gouw, G. J. 1995. A Study of Hand Grip Pressure Distribution and Emg of Finger Flexor Muscles under Dynamic Loads. Ergonomics, 38, 684-699.

Hall, N. B., Jr. & Bennett, E. M. 1956. Empirical assessment of handrail diameters. Journal of Applied Psychology, 40, 381-382.

Hogberg, D., Backstrand, G., Lamkull, D., Hanson, L. & Ortengren, R. 2008. Industrial customisation of digital human modelling tools. International Journal of Services Operations and Informatics, 3, 53-70.

Johnson, S. L. 1993. Ergonomic Hand Tool Design. Hand Clinics, 9, 299-311.

Kaljun, J. & Dolšak, B. 2012. Ergonomic design knowledge built in the intelligent decision support system. International Journal of Industrial Ergonomics, 42, 162-171.

Kong, Y. & Lowe, B. 2005. Optimal cylindrical handle diameter for grip force tasks. International Journal of Industrial Ergonomics, 35, 495-507.

Merle, A., Chandon, J. L., Roux, E. & Alizon, F. 2010. Perceived value of the mass-customized product and mass customization experience for individual consumers. Production and Operations Management, 19, 503-514.

Oh, S. & Radwin, R. G. 1993. Pistol grip power tool handle and trigger size effects on grip exertions and operator preference. Human factors, 35, 551-69.

Sancho-Bru, J. L., Giurintano, D. J., Pérez-González, A. & Vergara, M. 2003. Optimum tool handle diameter for a cylinder grip. Journal of Hand Therapy, 16, 337-342.

Seo, N. J. & Armstrong, T. J. 2008. Investigation of Grip Force, Normal Force, Contact Area, Hand Size, and Handle Size for Cylindrical Handles. Human Factors, 50, 734-744.

Sun, W., Starly, B., Nam, J. & Darling, A. 2005. Bio-CAD modeling and its applications in computer-aided tissue engineering. Computer aided design, 37, 1097-1114.

Wu, J. Z., Cutlip, R. G., Andrew, M. E. & Dong, R. G. 2007. Simultaneous determination of the nonlinear-elastic properties of skin and subcutaneous tissue in unconfined compression tests. Skin Research and Technology, 13, 34-42.

Wu, J. Z. & Dong, R. G. 2005. Analysis of the contact interactions between fingertips and objects with different surface curvatures. Proceedings of the Institution of Mechanical Engineers Part H-Journal of Engineering in Medicine, 219, 89-103.

Yakou, T., Yamamoto, K., Koyama, M. & Hyodo, K. 1997. Sensory Evaluation of Grip Using Cylindrical Objects. JSME International Journal Series C, Mechanical Systems, Machine Elements and Manufacturing, 40, 730-735.

Youakim, S. 2009. Hand-arm vibration syndrome (HAVS). British Columbia Medical Journal, 51, 10.

CHAPTER 47

Ergonomic Work Analysis Applied to Chemical Laboratories on an Oil and Gas Research Center

GUIMARÃES², C.P., CID², G.L., PARANHOS², A.G., PASTURA², F., FRANCA, G.A.N., SANTOS¹, V., ZAMBERLAN², M.C.P., STREIT², P., OLIVEIRA², J.L, CORREA,T.G.V. ²

¹ Pontifícia Universidade Católica do Rio de Janeiro - PUC-Rio
² Instituto Nacional de Tecnologia - INT/ MCT
carla.guimaraes@int.gov.br

ABSTRACT

The aim of this paper is to present an ergonomic study applied to chemical laboratories of an oil and gas research center. The Ergonomic Work Analysis was the main methodological approach. The study was conducted on thirty laboratories of different areas. The Ergonomic Work Analysis methodology involved three stages: 1) reference situation diagnosis and recommendations; 2) ergonomic design concept establishment and 3) evaluation of the new work condition. The reference situation diagnosis is concerned with the analysis of different work activities conducted by laboratories' teams. That work activities were previous selected by the laboratories technicians and managers. For the data collection a questionnaire previous tested by the ergonomics group and also video capture were applied. Data analysis was conducted at the Ergonomics Laboratory of INT. The results showed that even with each chemical lab specificities, some ergonomic problems were common, such as the lack of space between workstations and in the workstation; repetitive and static awkward postures as squatting, trunk and neck forward bending, shoulder flexion and abduction over 90 degrees; manual material handling activities, as lifting, carrying, pushing and pulling heavy loads. Other risk and safety problems were also found, as examples: workstations and trashcans located on escape routes; oil samples handling outside chemical chapels and intermediate oil sample storage in chapels and cabinets. Some environment conditions were also pointed out as common problems in special with reference to heat and noisy. The recommendations based on the diagnosis were presented as two different tools: the Reference Human Activities Database tool applied to the work-related situation diagnosis for storing and organizing the reference situations collected data. These recommendations were discussed by the ergonomic group, designers and the laboratories' teams to establish ergonomic design concepts that were applied on 3D simulation tools. Those tools facilitate project validation process, minimizing design errors and conflicts between users and designers.

Keywords: chemical lab, ergonomic work analysis, oil and gas industry

1 CONTEXT

The aim of this paper is to present an ergonomic study applied to chemical laboratories of an oil and gas research center. The Ergonomic Work Analysis was the main methodological approach. The Ergonomic Work Analysis (EWA) is a methodology which, as the result of studying behaviors in the work situation, provides an understanding of how the operator builds the problem, indicates any obstacles in the path of this activity, and enables the obstacles to be removed through ergonomic action (Wisner, 1995). The central point of this methodology is the analysis of real work activities performed by workers. Based on that, some ergonomic risk interactions have to be included in the study:

- Risk aspects inherent to the worker - involve physical, psychological and non-work-related activities that may present unique risk factors;
- Risk aspects inherent to the job - concern work procedures, equipment, workstation design that may introduce risk factors;
- Risk aspects inherent to the environment - concern physical and psychosocial "climate" that may introduce risk factors.

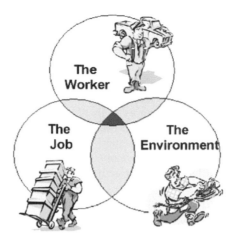

Figure 1 - Ergonomic risk aspects interaction. From Introduction to Ergonomics. How to identify, control, and reduce musculoskeletal disorders in your workplace! OR-OSHA. 101. Revised 01/97 by the Public Education Section Department of Business and Consumer Business, Oregon OSHA, (www.orosha.org/pdf/workshops/201w.pdf) 1997.

The ergonomic design concept is a set of references that should be established in the very first beginning of a new work condition design that defines future work. These ergonomic references give input to work and learning process evolution, work environment design – architectural and layout - and informational and interface devices design, to mention some (Santos and Zamberlan, 1992). These references can be presented as architectural drawings, 3D digital *maquettes* and ergonomic standards that should be adopted by the design team. The ergonomic design helps to define the future work condition concept with respect to human intervention. That guarantees that the design concept match users variability and work performance variability in different use contexts.

In that sense laboratories must be prepared to adapt to changing technologies. Incorporating new technologies can substantially alter demands placed on the laboratory environment. (Mortland, 1997)

Based on our experience in ergonomic studies, we have developed a methodology for work condition ergonomic design that involves three stages: reference situation diagnosis and recommendations; ergonomic design concept establishment and evaluation of the new work condition (Santos and Zamberlan, 2009).

2   METHODOLOGY

The ergonomic study was conducted in thirty laboratories in the oil and gas industry in order to analyze current work conditions, redesign and simulate new work environments proposals.

The Ergonomic Work Analysis methodology involved three stages: 1) reference situation diagnosis and recommendations; 2) ergonomic design concept establishment and 3) evaluation of the new work condition. The reference situation diagnosis is concerned with the analysis of distinct work activities performed by laboratories' teams. Those work activities were previous selected by the laboratories technicians and managers. For the data collection a questionnaire previous tested by the ergonomics group and also video capture were applied. Data analysis was conducted at the Ergonomics Laboratory of INT.

The ergonomic research group visited each laboratory three times a week for one year in a scheduled time. First, the laboratory supervisor was interviewed. After that the research group followed the laboratory technician team to their specific lab where previous selected laboratory test activities were followed and registered by video (figure 2).

The laboratory technician team was also interviewed and the activities were videotaped. The ergonomic group made all activities registration and took measurements of the workplace, tools and equipments (figure 3).

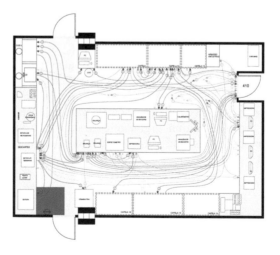

Figure 2 - Laboratory test activity registration

Figure 3 – Example of a circulation study during a sulfuric acid test (H2S)

3   RESULTS

The results showed that even considering the chemical laboratory specificity, some ergonomic problems were common, such as the lack of space between workstations and in the workstation; repetitive and static awkward postures as squatting, trunk and neck forward bending, shoulder flexion and abduction over 90 degrees; manual material handling activities, as lifting, carrying, pushing and pulling heavy loads.

Other risk and safety problems were also found as examples: workstations and trashcans located on escape routes; oil samples handling outside chemical chapels and intermediate oil sample storage in chapels and cabinets. Some environment conditions were also pointed out as common problems in special with reference to heat and noisy.

4   CONCLUSIONS

The recommendations based on the diagnosis were presented in two different tools: a) Reference Human Activities Database tool - applied to work-related situation diagnosis for storing and organizing the reference situations collected data; b) 3D virtual simulation tool

The Reference Human Activities Database tool optimized collected data organization and storage. It enabled workers visualization of real work problems and facilitated company's view of direct costs (errors, re-work, occupational diseases etc) and indirect costs (company image damage, loss of earnings etc). It can also be used as a training tool for new workers (figure 4) (Guimarães, 2012).

The 3D virtual simulation tool (figure 5) helps on design validation process, minimizing design errors and conflicts between users and designers. The use of virtual environments gives the possibility of negotiation, changes and creation of better results due to more graphic and visual interface for non architects and designers professionals. That tool helps on understanding and on discussing of new workspace layout leading to a better design alternative choice.

Figure 4 – Reference data Base

476

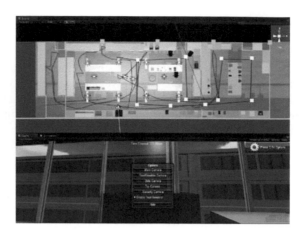

Figure 5 – Example of Lab Simulator

REFERENCES

Guimarães, C.P.; Cid, G.L.; Santos, V.S; Zamberlan, M.C.P; Pastura, F.C.H; Abud, G.M.D; Lessa, C.; Batista, D.S.; Fraga, M. M. (2012) Human Activity Reference Database. 18th World Congress on Ergonomics. IEA 2012, Recife, Brazil.

Guimarães, C.; Cid, G; Paranhos, A.G.; Pastura, F.; Santos, V.; Zamberlan, M.; Streit, P.; Oliveira, J. (2012) Ergonomic Work Analysis applied to Chemical Laboratories. 18th World Congress on Ergonomics. IEA 2012, Recife, Brazil.

Mortland, K.K. (1997) Laboratory Design for Today's Technologies. Med TechNet Presentations.pp.1-14

Santos, V.; Setti E. ; Zamberlan, M. C. P. L. ; Pastura, F. (2004,). Banco De Dados Referencial da Atividade Humana. In: In: XIII Congresso Brasileiro de Ergonomia / II Forum Brasileiro de Ergonomia / I Congresso Brasileiro de Iniciação Científica em Ergonomia - ABERGO Jovem, 2004, Fortaleza. Anais do XIII Congresso Brasileiro de Ergonomia / II Forum Brasileiro de Ergonomia, Fortaleza. In: XIII Congresso Brasileiro de Ergonomia / II Forum Brasileiro de Ergonomia / I Congresso Brasileiro de Iniciação Científica em Ergonomia - ABERGO Jovem, 2004, Fortaleza. Anais do XIII Congresso Brasileiro de Ergonomia / II Forum Brasileiro de Ergonomia.

Santos, V; Guimarães C. P. ; Cid, G. L. (2008). Simulação Virtual e Ergonomia. In: xV Congresso brasileiro de Ergonomia, VI Fórum Brasileiro de Ergonomia, 2008, Porto Seguro. xV Congresso brasileiro de Ergonomia, VI Fórum Brasileiro de Ergonomia, SANTOS, V. ; ZAMBERLAN, M. C. P. L. ; PAVARD, B. . Confiabilidade Humana e Projeto Ergonômico de Centros de Controle de Processos de Alto Risco. Rio de Janeiro: Synergia, 2009. v. 1. 316 p.

Santos, V; Zamberlan, M. C. P. L. (2002). Projeto Ergonomico de Salas de Controle. Sao Paulo: Fundation Mapfre,. v. 1. 200 p

Santos, V. ; Zamberlan, M. C. P. L. (1999). Projeto de Sala de Controle-Estudo de Caso. In: I Encontro Brasil-Africa de Ergonomia / V Congresso Brasileiro de Ergonomia /

III Seminário de Ergonomia da Bahia, 1999, Rio de Janeiro: ABERGO. Anais do Encontro Brasil-Africa de Ergonomia / V Congresso Brasileiro de Ergonomia / III Seminário de Ergonomia da Bahia,

Santos,V., Zamberlan, M.C. Projeto Ergonômico de Salas de Controle. São Paulo: Fundación Mapfre-Sucursal Brasil, 1992.

CHAPTER 48

Evaluation of the Map for the Evacuation in Disaster Using PDA Based on the Model of the Evacuation Behavior

Yoshitaka Asami*, Masato Sakurai*, Daiji Kobayashi**, Kazuo Ishihara***, Naoko Nojima***, and Sakae Yamamoto*

*Tokyo University of Science, Japan, ** Chitose Institute of Science & Technology, Japan, ***Net&Logic Ltd., Japan.

ABSTRACT

The evacuation behavior in the task was measured to reveal which information was useful for the evacuees as addition to the base map when they evacuated using a PDA. As the results, it is found that the grasp of the present location affects planning of the route selection for the subjects in the evacuation task. The combination of the situations of the grasp of the present location and the plans of the route selection can evaluate the quality of evacuation behavior. The evacuation behavior of the elderly subjects is a relatively low quality compared with that of the young subjects. It is considered that they are not familiar with using the PDA.

Keywords: evacuation behavior, elderly, PDA, present location, route selection

1   INTRODUCTION

When a disaster occurs, evacuation routes dependently vary with the fire and other troubles caused by its disaster. It is considered that it is useful to provide the travelable route for the disaster victims when they evacuate. The previous study developed the information system for disaster victims with autonomous wireless network, which is possible to use when a disaster occurs [1]. With using its system, it can provide the information of the evacuation for the disaster using a PDA (personal digital assistant) such as a smartphone (i.e. high-end mobile phone) since many people have had it recently. One of the previous studies has reported that it is not enough to provide the information of only the base map as the evacuation route when they evacuate using a PDA [2-3]. It is necessary to consider which information is useful for the evacuees as addition to the base map when they evacuate.

On the other hand, there is the number of people in age when they evacuate in fact. Particularly in Japan, elderly people increase with year and year, it is important to take into account of the evacuation in elderly people when a disaster occurs. In this study, we focused on examining the behavior of the elderly subjects compared with that of the young subjects during the evacuation task using a PDA in this experiment.

The purpose of this paper is to reveal which information is useful for the evacuees as addition to the base map when they evacuate using a PDA. It is also to study the behavior of the elderly subjects compared with that of the young subjects during the evacuation task using a PDA in this experiment.

2 PRELIMINARY INVESTIGATION

2.1 Target Area

In this paper, we targeted the area (Tansu-machi area, Shinjuku-ku, Tokyo) which is one of the old towns in Tokyo. In one of the reasons why we choose, this area has high proportion of the elderly in living at this area. In addition, this area has many old wooden houses and narrow alleys. Thus, this area is supposed that the evacuation routes are limited by the incidental fire when the disaster occurs. It is necessary to evacuate in a shelter by themselves victims in safety.

2.2 Questionnaire Survey

In order to investigate the awareness for the disaster and the possession rate of the

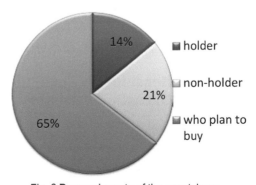

Fig. 1 Do you know where is the shelter?   Fig. 2 Possession rate of the smartphone.

480

smartphone, a questionnaire survey was carried out in this area. The amount of distribution was 3000 pieces, and its collection was 663 pieces (collection rate: 22%). Fig. 1 shows the results of the percentage which people who know where the shelter is. As shown in this figure, the 25% of them is unknown the location of the shelter. Fig. 2 shows the results of the percentage of the possession rate that how many people have smartphones. From this figure, the possession rate of the smartphone is 14% in this area. The total of the possession rate is 35% including the proportion of people who plan to purchase. Therefore, it is considered that it is needed to inform where the shelter is in this area and it is possible to obtain the information of the shelters and the evacuation routes using the PDA such as a smartphone.

3   EXPERIMENT 1

3.1   Methodology

To reveal which information is useful for the evacuees as addition to the base map, it was performed the evacuation task using the PDA. The subject was asked to go to the goal point from the start point on foot using the PDA as the evacuation task. It is supposed that they evacuate to the near shelter using the PDA. The PDA displayed the base map of the area and presented the points of the virtual events such as fire places and closed roads which happen to according to time. During the task, the behavior (including the operation of the PDA) and the utterance of the subjects were recorded by the camcorders and the voice recorder to evaluate the evacuation behavior.

3.2   Apparatus

Two camcorders (SONY, HDR-SR12) and one voice recorder (OLYMPUS, Voice Trek V-61) were used to record the behavior and utterance of the subject during the task. The PDA (DoCoMo, T-01A) was used and the display size was 4.1 inch.

3.3   Subjects

Ten early twenty's subjects participated in this experiment.

4   RESULTS AND DISCUSSIONS IN EXPERIMENT 1

4.1   Relationship between the Travel Distance and the Quality of Evacuation Behavior

From the images recorded by the camcorders in all the subjects, three of ten can evaluate in the shortest route in all the possible routes in the task. They build a plan for the evacuation route with avoiding fires and closures to go to the goal point. It can be said that their behavior is high quality for the evacuation in this experiment and it leads to promptly evacuate for avoiding few dangers such as fires and closure. It is found that the subjects who take a relatively long distance in the task lose their way and misunderstand where they are and the locations for fires, closures, and the goal point on the map displayed by the PDA. And they do not only pass through the locations of fires and closures in the virtual events but also go back along the same route. It can be said that their behavior is not desirable and is low quality for the evacuation in this experiment, and it highly involves the risk by the disaster. It is a strongly relationship between the travel distance and the quality of the evacuation behavior in this experiment.

Table 1 Situations of the grasp of the present location and plans of the route selection.

Class	Situation	Class	Plan
1	Grasp of the present location	A	Decision for the route to the destination concretely
2	Not grasp of the present location	B	Progress to the direction of the destination
3	Misunderstanding of the present location	C	Try to get a clue to grasp the present location
		D	Try to return

4.2 Grasp of the Present Location and Planning of the Route Selection

In order to clarify the factor which causes the evacuation behavior as a not desirable behavior, it was investigated from the utterance of all the subjects in the task. As the results, it is found that the grasp of the present location affects planning of the route selection for the subjects in the evacuation task. According to this, the situation of the grasp of the present location and the plans of the route selection can be classified respectively. Table 1 shows each classification of the situations of the grasp of the present location and of the plans of the route selection. For the situations of the grasp of the present location, there are three classes (1 to 3), and for the plans of the route selection are four ones (A to D), respectively. The combination of the situations and the plans can evaluate the quality

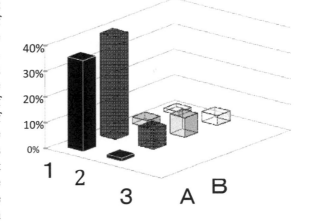

Fig. 3 The combination of the grasp situation of a its present location and the plan of evacuation route selection

of evacuation behavior and the combination of '1' and 'A' in Table 1 is the highest.

Fig. 3 shows the combination results of the situations of the grasp of the present locations and of the plans of the route selection based on the utterances and actions for all the subjects during the task. The x, y, and z axes indicate the situations of the grasp of the present location, the plans of the route selection, and the percentage of the combination of those for the all the subjects' responses. As the results, the combination of '1' and 'A' is the highest quality of evacuation behavior and its percentage accounts about 40%. From the results of the relationship between the travel distance and the quality of evacuation behavior in this experiment, it is important to make the subject perform a high quality evacuation behavior for taking a short distance in the evacuation task. In order to increase the percentage of the high quality evacuation behavior in the combination of 1 and A, it is considered that it is necessary to add the information to easily understand where s/he is for the grasp of the present location on the map displayed by PDA. It is proposed that the information of the name of buildings, passable roads, and the walking distance is added on the map on PDA. In Experiment 2 below, it was used this map in the evacuation task.

5   EXPERIMENT 2

5.1   Methodology in Experiment 2

The map was improved as the above chapter and it experimented by the same method as the experiment 1. Five early twenty's subjects and seven elderly subjects participated in this experiment. The average age of elderly was 68.3 years (S.D. 5.1).

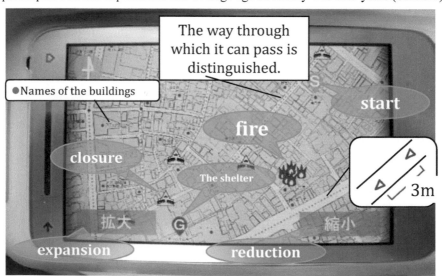

Fig. 4 The screen of the PDA

In addition, we finished the experiment in 30 minutes because the subject is elderly. Fig. 4 shows the screen of the PDA using at this experiment.

5.2 Results: Situation of Grasp of the Present Location and the Plan of Route Selection in Experiment 2

Fig. 5 and 6 show the combination results of the situations of the grasp of the present locations and of the plans of the route selection based on the utterances and actions for the young and the elderly subjects during the task, respectively. The x, y, and z axes indicate the situations of the grasp of the present location, the plans of the route selection, and the percentage of the combination of those for the all the subjects' responses, as well as Fig. 3. As shown in Fig. 5, the percentage of the combination of '1' and 'A' accounts about 80%. In addition, four of five young subjects evacuate in the shortest route. These results indicate the improvement of the map is valid way for the young subjects.

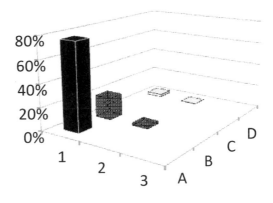

Fig. 5 Combination of the situations of the grasp of the present location and the plans of the route selection (young subjects)

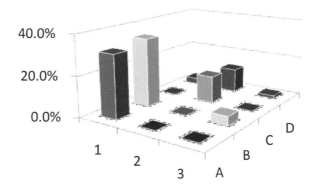

Fig. 6 Combination of the situations of the grasp of the present location and the plans of the route selection (elderly subjects)

On the other hand, as shown in Fig 6, the percentage of the combination of '1' and 'A' accounts about 40% and it is relatively low compared with that of the young subjects. Although the map is improved its evacuation behavior is not so high quality for the elderly subjects.

484

5.3   Discussions in Experiment 2

In terms of the results of relatively low quality evacuation behavior in the elderly subjects, it is considered that it is difficult to grasp the present location since it is not easy to scroll the map of PDA. They are not familiar with using the PDA. In the task, some of them did not reach to the goal point. They did not effectively use the map of PDA to grasp the present location and to plan the route selection for avoiding the disaster events. They did not so scroll the map of PDA and so saw an only small part of the map that they could not find the routes which avoid the disaster events to reach the goal point. In addition, it is mentioned that the size of characters on the map displayed by the PDA is too small to read them easily for the elderly subjects. There was a subject who was bringing it close and looking at the face considerably on the screen of PDA. From these points, the size of characters and of screen of the map in PDA should be large for understanding it easily for elderly people. And it is necessary to be able to operate it easily for the evacuation using the PDA in elderly people.

6   CONCLUSION

The evacuation behavior in the task was measured to reveal which information was useful for the evacuees as addition to the base map when they evacuated using a PDA. As the results, it is found that the grasp of the present location affects planning of the route selection for the subjects in the evacuation task. It is important to make the subjects perform a high quality evacuation behavior for taking a short distance in evacuation. It is considered that it is necessary to add the information to easily understand where s/he is for the grasp of the present location on the map displayed by the PDA. It was, also, to study the behavior of the elderly subjects compared with that of the young subjects during the evacuation task using the improved map of PDA. Although the improved map is effective for the young subjects the evacuation behavior is not so high quality for the elderly subjects. It is suggested that it is necessary to be able to operate it easily for the evacuation using the PDA in elderly people since they are not familiar with using the PDA.

ACKNOWLEDGMENTS

This work was supported by the Strategic Information and Communications R&D Promotion Programme (SCOPE), Ministry of Internal Affairs and Communications, Japan.

REFERENCES

[1] Y. Takahashi, D. Kobayashi and S. Yamamoto, A development of information system for disaster victims with autonomous wireless network, HCI International 2009, Vol.5618/2009, pp.855-864, San Diego, USA, 2009.

[2] S. Yonemura, D. Kobayashi and S. Yamamoto, Study of evacuation strategies for generating support mobile devices, Japan society for disaster information studies 11th convention proceedings, pp.145-148, 2009 (in Japanese).

[3] D. Kobayashi, Y. Asami, S. Yonemura, S. Suzuki, S. Shimada and S.Yamamoto, Study on Evacuation Behavioral Assessment Based on the Concept of Mental Algorithm, The society for Industrial Plant Human Factors of Japan, Vol.16, No.1, pp.52-62, 2011 (in Japanese).

Section IX

New Methods and Modeling in Future Applications

CHAPTER 49

Diagrammatic User Interfaces

Robbie T. Nakatsu

Loyola Marymount University
Los Angeles, CA
Email: rnakatsu@lmu.edu

ABSTRACT

I report on my findings on diagrammatic user interfaces. First, I discuss the essence of diagramming, and present a framework for classifying a great variety of diagramming notations commonplace today. Despite a tremendous variety of usage, in terms of notations and applications, I classify diagrams into six themes: (1) topology, (2) sequence and flow, (3) hierarchy:classification, (4) hierarchy:composition, (5) association, and (6) causality. The framework is largely drawn from my recent book, "Diagrammatic Reasoning in AI" (Wiley, 2010), on which much of this discussion is based.

Second, I present a framework for diagrammatic user interfaces, in which a diagram is intended to be used in a more dynamic way, and hence, can serve as a graphical user interface. I classify such diagrammatic user interfaces along two dimensions: 1) whether the structure is static or dynamic (i.e., can be actively constructed by the end-user) and 2) whether information on the diagram is static or dynamic (i.e., allows for information propagation). While most of the diagrams in use today are static representations, I explore some of these more dynamic and interesting uses, thereby showing how a diagram can transform a rigid black box user interface into one that is transparent, flexible, and easy-to-use.

Keywords: diagrams, diagrammatic user interfaces, system transparency, explanatory power

1   INTRODUCTION

A typical user of computer technologies today is bombarded from all directions with multiple devices and systems. For many users, it has become increasingly difficult to feel comfortable with any of these systems, beyond a rudimentary understanding of how to get basic tasks accomplished. The problem with many of these technologies is that they lack system transparency. By this, I mean a black box system that is difficult to understand because its inner workings are not visible for inspection. These systems are unintelligible to all but a few, difficult to modify, and extremely brittle when something goes wrong.

I argue, therefore, that an appropriate diagrammatic user interface can go a long way toward explaining system actions and providing a deeper understanding of how a system works. A diagrammatic user interface employs a diagram of some kind to aid in system understanding. By diagram, I am referring to a graphical representation of how the objects in a domain of discourse are interrelated to one another. Unlike a linguistic or verbal description, a diagram is a type of information graphic that "preserves explicitly the information about topological and geometric relations among the components of the problem" (Larkin and Simon, 1987). That is to say, a diagram indexes information by location on a 2-D plane—3-D diagrams are also possible, but not yet commonplace. As such, related pieces of information can be grouped together on a plane, and interconnections between related elements can be made explicit through lines and other notational elements. Furnished with such a description, the user is better able to cope with technological complexity

What advantages do diagrams have over verbal descriptions in promoting system understanding? (Nakatsu, 2010, p. 58]. First, massive amounts of information can be presented more efficiently. A diagram can strip down informational complexity to its core—in this sense, it is a minimalist description of a system that can enable users to easily visualize the gist of a problem. Second, a diagram can help users see patterns in information and data that may appear disordered otherwise. For example, a diagram can help users see mechanisms of cause and effect, or can illustrate sequence and flow in a complex system. Third, a diagram can result in a less ambiguous description than a verbal description, because it forces one to come up with a more structured description. By necessity, the notations of the diagramming language, which serve as its vocabulary, circumscribe what is and what is not allowed in the diagram. The advantages of diagrams are well-known in the software engineering discipline, where it is commonplace to draw diagrams to help in the analysis, design, and implementation of information systems.

2 TYPES OF DIAGRAMS

Diagrams are primarily, but now always, made up of geometric shapes interconnected by lines or arrows. (One important exception is the Venn diagram, which shows relationships among sets, not through interconnected lines, but through overlapping circles). The geometric shapes can represent any number of things: people, things, concepts, events, activities, decision points, sets, actions, steps: almost any type of entity can be modeled. The meanings of the shapes are frequently conventions that have been adopted by diagrammers over the years. In general, it is better to stick with standard and conventional notations rather than make up your own.

Lines are used to represent a linkage between two objects on a diagram. Depending on the diagram, the line can take on a different meaning. It could mean any one of the following six linkage types:

1. connected to (topology)
2. precedes in time (sequence and flow)
3. a type of (classification)
4. a part of (composition)
5. related to (association)
6. causes/influences (cause and effect)

Lines can be either directed or non-directed. An arrow is typically used to indicate a directed relationship. Some types of linkages are always directed. Linkages of the type: (2) precedes in time, (3) a type of, (4) a part of, and (6) causes/influences are always directed. For example, the causes/influences linkage is always expressed as a directed relationship, with the cause on the open side, and the effect on the arrowhead side of the line:

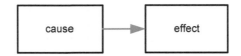

Even with such tremendous diversity in diagramming, in terms of both applications and notations, these six basic types cut across a wide array of diagramming notations, and serve as a framework for understanding and organizing many different types of diagrams. See (Nakatsu, 2010, Chapter 3) as well as my website   http://myweb.lmu.edu/rnakatsu/diagrammaticreasoning/index.html   for examples of each type of diagram.

3   CLASSIFYING DIAGRAMMATIC USER INTERFACES

Sometimes diagrams are constructed informally and created on the fly. At other times, we may wish to commit the diagram to some kind of external representation—most typically we draw the diagram and save it to paper or to a computer file. Because we have stored it to some external media, we preserve it for future use, and don't have to rely on our short term memory. Even better yet, we may create a diagram that serves as a user interface itself—in some cases, intended to be modified and manipulated to help us solve difficult problems or help us better understand a complex system.

	Structure Static	Structure Dynamic
Information Static	**I. Static Representations** Diagram and information cannot be modified.	**IIII. Constructive Diagrams** Diagram is actively constructed but no information propagation.
Information Dynamic	**II. Propagation Structures** Diagram is static but information propagates	**IV. Constructive Diagrams With Information Propagation** Diagram is actively constructed and information propagates.

Figure 1 A Classification of Diagrammatic User Interfaces (Adapted From Nakatsu, 2010)

In Figure 1, I classify diagrammatic user interfaces into four types, based on whether information is static or not, and whether the structure is intended to be modified or not.   The interested reader may wish to consult (Nakatsu, 2010) for more examples that illustrate each of the four types of diagrammatic user interfaces. Here I present two examples taken directly from the book.

The great bulk of diagrams in use today are Static Representations (Quadrant I). Diagrams such as org charts, basic flowcharts, semantic networks and the like are usually not intended to be modified by the end-user. Because of their static nature, these types of diagrams can easily be inserted into the pages of a book, or on a read-only web site. On the other hand, Propagation Structures (Quadrant II) are intended to be modified by an end-user and can serve as a front-end user interface for a system. In (Nakatsu, 2003), I illustrate this type of diagrammatic user interface with an transportation selection application. The system is called TransMode Hierarchy. See Figure 3 for the diagram.

The following problem solving situation is illustrated in the diagram: A transportation broker must make a determination concerning the type of transportation mode (air, trucking, rail, or small package service) to use in order to transport a client's shipment. The determination is made based on the following factors:

- Shipment Weight: What is the weight of the shipment? (in lbs)
- Weight of one item: What is the weight of one item to be shipped? (in oz.)
- Value of one item: What is the dollar value of one item to be shipped?
- Fragility Rating: How fragile is the item? (high, average, low)
- Shipping Distance: How far is the item to be shipped? (distance in miles)
- Transit Time: In how many days must the shipment reach its destination?
- Special Products: Are special products to be shipped? (hazardous-minerals, hazardous-chemicals, agricultural, none of the above)
- Perishable: Are the items to be shipped perishable? (yes/no)
- Door-to-Door Service: Is door-to-door service required? (yes/no)
- Frequency of Service: How frequently is the shipment made? (daily, weekly, monthly, infrequent)

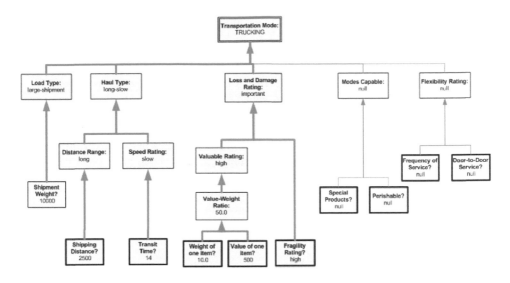

Figure 3 Propagation Structure (Adapted from Nakatsu and Benbasat, 2003)

The root node is the topmost node, which represents the topmost goal of the decision-making problem, in this case, the transportation mode recommend-dation. The root node is represented as a double-boxed rectangle. At the bottom of the hierarchy, the leaf nodes represent the raw, or unprocessed data inputs. They are represented by bold-faced rectangles. For example, Shipment Weight?, Shipping Distance?, and Transit Time?, are all leaf nodes. In between the root node and the leaf nodes are the intermediate nodes. These nodes read in values from the node(s) attached directly below it—their child node(s)—and process this input to produce an output, which, in turn, is passed to the node attached directly above it, the parent node.

A user can visually track a line of reasoning through the hierarchy. The upward-pointing arrows on the hierarchy signify that information moves from bottom to top.

For example, by entering 10,000 as the Shipment Weight?, the Load Type is set to large-shipment. Similarly, when 10.0 is entered as Weight of one item? and 500 as Value of one item?, Value-Weight Ratio is set to 50.0, which will, in turn, set Valuable Rating to high. When the Fragility Rating? is also set to high, the Loss and Damage Rating is set to important. In the diagram above some of the nodes are null, signifying that some information has not been entered (and may not be needed). TransMode Hierarchy will make a recommendation as soon as there is enough information to make a determination: in the case, Trucking is the transportation mode that is finally selected.

A Constructive Diagram (Quadrants III) is dynamic in the sense that the structure itself is meant to be modified. A **Constructive Diagram with Infor-mation Propagation** (Quadrant IV) is dynamic in a dual sense: both the structure itself can be modified, and information can be propagated through the structure. In Nakatsu (2005) I illustrate such a diagrammatic user interface called LogNet. See Figure 4.

At the heart of LogNet is the network model. One way to view and model a logistics environment is as a network of nodes interconnected by transportation links. The problem of specifying the model would be one of specifying the network structure through which manufactured goods flow. To model this environment, three types of nodes are considered: first, the factories, where the products are manufactured; second, the warehouses, which receive the finished products from the factories for storage and possibly for further processing; and third, the customer zones (or markets), which place orders and receive the desired products from the assigned warehouse(s). Product moves through the logistics network via different transportation options (e.g., rail, trucking, shipping, air), which are represented by the connections or links between the nodes. There are two types of transportation links: inbound links move products from factories to warehouses, and outbound links move products from warehouses to customers. In this diagram above, squares represent factories, circles represent warehouses, and triangles represent customer zones.

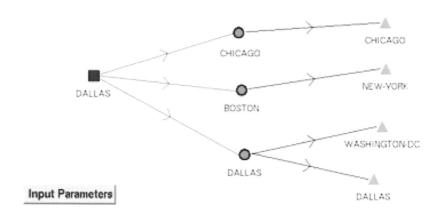

Figure 4 Constructive Diagram (From Nakatsu, 2005)

The network design task involves a tradeoff between consolidation (merging two or more warehouses into one thereby reducing costs) and decentralization (splitting warehouses into two or more separate locations so that they better serve customers but the downside is that the additional warehouse results in increased costs). The diagrams are constructive because they are meant to be actively modified by the user—different network configurations can be drawn (nodes can be added/deleted to the network, and transportation links can be added and deleted to connect and disconnect nodes). By request, the resulting benchmarks (e.g., cost and customer service level) of the network design can be calculated. Furthermore, information propagates in the network, based on how the networks are configured. For example, the inventory carrying cost of a warehouse is updated based on the extent to which inventory is consolidated—the more consolidated the warehouse the less the (per unit) inventory carrying cost, whereas the more decentralized the more the (per unit) inventory carrying cost.

All in all, the creation of diagrammatic user interfaces is really about the use of dynamic techniques, whether through information propagation (Quadrant II and Quadrant IV diagrams) or through constructive diagramming (Quadrant III and Quadrant IV diagrams). These techniques go above and beyond traditional and static uses of diagramming that are more commonplace.

4 CONCLUSIONS

In this paper, I argue that diagrams can be used in more dynamic ways so that they can serve as a graphical user interface of a complex system. Such diagrammatic user interfaces can be used to endow a system with explanatory power, or the ability for a system to explain itself, so that it is no longer a black box. The systems of the future are likely to become more and more complex, so end-users will need more support for understanding them. The use of diagrams can be a powerful tool for furnishing such systems with increased transparency and flexibility. The future of diagrammatic user interfaces looks very promising.

REFERENCES

Larkin, J.H., and Simon, H.A. 1987. Why a Diagram Is (Sometimes) Worth Ten Thousand Words. Cognitive Science. 11, 1 (Jan-Mar 1987), 65-100.

Nakatsu, R.T. 2005. Designing Business Logistics Networks Using Model-Based Reasoning and Heuristic-Based Searching. *Expert Systems with Applications,* 29, 4, 735-745.

Nakatsu, R.T. 2010. Diagrammatic Reasoning in AI. Wiley, Hoboken, NJ.

Nakatsu, R.T. and Benbasat, I. 2003. Improving the Explanatory Power of Knowledge-Based Systems: An Investigation of Content and Interface-Based Enhancements. *IEEE Transactions on Systems, Man, and Cybernetics, Part A,* 33, 3, 344-357.

CHAPTER 50

A Study on Effect of Coloration Balance on Operation Efficiency and Accuracy, based on Visual Attractiveness of Color

Azusa YOKOMIZO, Teruya IKEGAMI and Shin'ichi FUKUZUMI

NEC Corporation
Tokyo, JAPAN
a-yokomizo@ax.jp.nec.com

ABSTRACT

This research aims to study screen design, especially coloration which helps users to find important information on a screen both efficiently and accurately. As a clue to consider of coloration, we focused on visual attractiveness of color. Visual attractiveness is a metric of human's optic attention. We carried out an experiment to verify the effect of visual attractiveness on operation efficiency and accuracy. The result showed that items which are colored with high visual attractiveness color could be found efficiently. However, the effect of visual attractiveness was not significant on operation accuracy. This paper reports the methods, results and consideration of our experiment.

Keywords: usability, visual attractiveness, coloration, efficiency, error

1   INTRODUCTION

In some systems, especially mission critical systems, users' one mistake may cause severe damage (Emerson Network Power, 2011). In these kinds of systems, users are often required to operate not only accurately but also quickly. However, operation efficiency and accuracy is not independent, and some previous researches show that the relationship between operation accuracy and efficiency are often explained as a trade-off (Fitts, 1954). Thus, to make users operate both accurately and quickly, it is important to support users in some way. In this research, we focused on screen design, especially coloration as one of the main elements of screen design (Norman, 1988).

To consider of coloration, we focused on visual attractiveness, which is a metric of human's optic attention (Itti, 1998). It is calculated based on knowledge of human's visual information processing mechanism. Items which have high visual attractiveness are said to pull human's attention strongly. If items which have high visual attractiveness could be found both efficiently and accurately, we can make use of this knowledge to screen design. For example, it would be effective to decide the coloration as important information's visual attractiveness become higher than other information. To confirm this idea, we carried out an experiment which verifies the correlation between operation efficiency, accuracy and visual attractiveness.

2   CALCULATION METHOD OF VISUAL ATTRACTIVENESS

Visual attractiveness is a metric of human's optic attention. A method to calculate visual attractiveness of item's color as a sum of Feature-attractiveness (FA) and Heterogeneity-attractiveness (HA) is proposed (Tanaka, 2000). FA is the item's attractiveness of its own. It is calculated by the item's color attributes, which are hue, lightness and saturation. HA is the item's attractiveness compared with other items on the screen. It is calculated by colors of all items on the screen.

Based on this research, we decided to calculate the visual attractiveness as follows. Calculation of FA based on lightness uses brightness difference from background color (W3C, 2000). Calculation of HA uses colors and sizes of items including background color.

- Feature-attractiveness(FA): A sum of attractiveness based on hue, lightness and saturation

    $FA = |1-H/\pi| + \{|\ L' - L_B'\ |\cdot (2.3 / 255) - 0.65\} + S$

    H: Hue on HLS color space

    $|L' - L_B'|$: Brightness difference from background color

    S: Saturation on HLS color space

- Heterogeneity-attractiveness(HA):

    $HA = |(d-d_m)/std|$

    d: Color difference from average color on CIE LUV color space

    $d = (\Delta L_i^2 + \Delta U_i^2 + \Delta V_i^2)^{0.5}$

    $\Delta L_i = L_i - L_m,\ \Delta U_i = U_i - U_m,\ \Delta V_i = V_i - V_m$

    (L_i, U_i, V_i) : Color of item i on CIE LUV color space

    (L_m, u_m, v_m) : average color on CIE LUV color space

    $L_m = \Sigma(L_i n_i) / \Sigma n_i,\ U_m = \Sigma(U_i n_i) / \Sigma n_i,\ V_m = \Sigma(V_i n_i) / \Sigma n_i$

    n_i : size of item i

    d_m: Average of d

    std: Standard deviation of d

- Visual Attractiveness: It is calculated as a sum of FA and HA.

3 VALIDATION OF CORRELATION BETWEEN COLORATION BALANCE AND OPERATION PERFORMANCE

3.1 Effect on Efficiency and Accuracy

We thought that an item which has high visual attractiveness is easily found quickly and accurately. To verify this idea and apply it to effectual coloration, we made two hypotheses as follows.
- Hypothesis 1: If target items are colored as their visual attractiveness is the highest among all items, users can find the target items quickly and accurately than other items.
- Hypothesis 2: If both target items and other items are colored as they have uniformly high visual attractiveness, target items cannot be found quickly and accurately.

3.2 Experiment

We carried out an experiment of visual search task to verify our hypotheses. In this task, a screen which displays three kinds of items colored with different colors is shown to an examinee. The examinee has to count the number of items colored with target color, which is instructed from an experimenter. The examinee is taught to count the number of item colored with target color, both quickly and accurately.

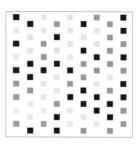

Figure 1 Example of task screen

An example of task screen is shown as Fig. 1. 81 rectangles lined in 9 × 9 are shown on the display. Rectangles are colored with three kinds of colors. The three colors used in the task screen, and the target color among them are previously taught to the examinee. To verify effects of non-target items' visual attractiveness on operation performance, colors used for items in the task screen are fixed as their balance of visual attractiveness becomes like either one of three types shown as Fig. 2.

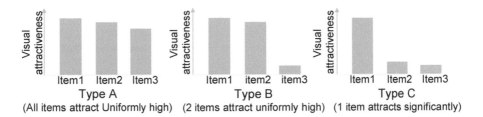

Figure 2 Three types of balance of visual attractiveness

In this experiment, we used a plasma display, whose size is 1104mm × 621mm. It was placed 150cm distant from the examinee. Examinees were six people between 26 to 38 years old, and all of them had no color-vision characteristics. One examinee did 200 tasks. 80 tasks were tasks to verify hypothesis 1, and 120 tasks were tasks to verify hypothesis 2. In each task, we recorded the time to finish the task and the examinee's answer about the number of target items. The correct answer varies from 25 to 29.

3.2.1 Experimental Design to Verify Hypothesis 1

To verify hypothesis 1, we fixed each item's color as their visual attractiveness become like type C, on Fig. 2. We thought that users' optic attention would be attracted to item 1 which is colored with the most attractive color, and item 1 could be easily found than item 2 which is colored with the second most attractive color.

To make variation of task screen on each item's hue and background color, we decided four kinds of color as the most attractive color: red (hue 0), blue (hue 240), magenta (hue 300) and cyan (hue 180). Red and blue are shown on white (255, 255, 255) background. Magenta and cyan are shown on black (0, 0, 0) background. In addition, we decided two kinds of second most attractive color: the same hue as the most attractive color, and 120 degree different hue from the most attractive color. The color of the least attractive color is fixed as gray.

500

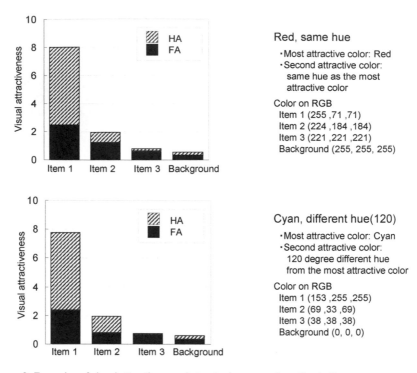

Red, same hue

· Most attractive color: Red
· Second attractive color:
  same hue as the most
  attractive color

Color on RGB
Item 1 (255 ,71 ,71)
Item 2 (224 ,184 ,184)
Item 3 (221 ,221 ,221)
Background (255, 255, 255)

Cyan, different hue(120)

· Most attractive color: Cyan
· Second attractive color:
  120 degree different hue
  from the most attractive color

Color on RGB
Item 1 (153 ,255 ,255)
Item 2 (69 ,33 ,69)
Item 3 (38 ,38 ,38)
Background (0, 0, 0)

Figure 3 Examples of visual attractiveness in two task screens (hypothesis 1)

Since there are four kinds of colors as the most attractive color and two kinds of color as the second most attractive color, task screens have eight variations. The actual colors of items are decided by controlling lightness and saturation as their visual attractiveness become like type C on Fig. 2. As an example, the visual attractiveness of each color on two screens is shown as Fig. 3. Each color on other six screens has almost same visual attractiveness.

On each task screen, examinees are instructed to find either the most attractive color or the second most attractive color.

3.2.2 Experimental Design to Verify Hypothesis 2

To verify hypothesis 2, we fixed each item's color as their visual attractiveness become like either one of type A, B, C on Fig. 2. On screen of type A, user's optic

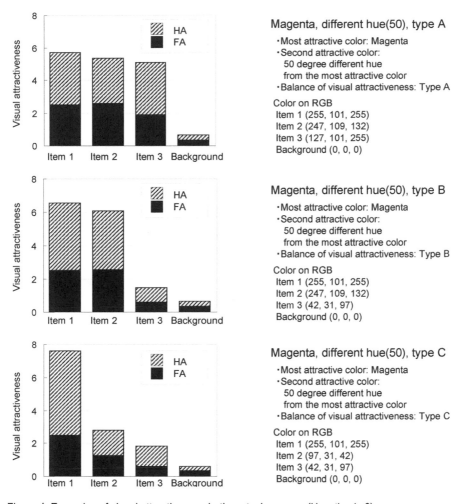

Figure 4 Examples of visual attractiveness in three task screens (Hypothesis 2)

attention would be attracted to item1, but because item 2 and 3 have also high visual attractiveness, user's optic attention would be also attracted to item 2 or 3. On screen of type B, user's optic attention would be attracted to item1, but item 2 would also pull user's attention. Thus, we thought that item1 could be easily found on screen of type C than type B, and item 1 could be easily found on screen of type B than type A.

To make variation of task screen on each item's hue and background color, we decided the most attractive color and the background color as the same as tasks in hypothesis 1. In addition, we decided two kinds of the second most attractive color: 50 degree different hue from the most attractive color, and 120 degree different hue from the most attractive color. The least attractive color's hue is decided automatically from the other two colors' hue. If the most attractive color's hue is x,

502

and the second attractive color's hue is x + a, the least attractive color's hue is x - a. Moreover, there are three types of visual attractiveness balance between colors used in each task screen as Fig. 2.

Since there are four kinds of colors as the most attractive color, two kinds of colors as the second most attractive color and three kinds of visual attractiveness balance, task screens have 24 variations. The actual colors of items are decided by controlling lightness and saturation as their visual attractiveness become like either type shown on Fig. 2. As an example, the visual attractiveness of each color on three screens is shown as Fig. 4.

On each task screen, examinees are instructed to find the most attractive color.

4   RESULT AND CONSIDERATION

We carried out 480 tasks for Hypothesis1, and 720 tasks for hypothesis2. In tasks for hypothesis 1, 26 tasks were mistaken. In tasks for hypothesis 2, 45 tasks were mistaken. Using the result of these tasks, we did analysis of efficiency and accuracy. Efficiency is analyzed from the average time to find one item on each task. Accuracy is analyzed from error rate, which is the percentage of mistaken tasks. For the analysis of efficiency, we used tasks which examinees answered correctly.

4.1   Result of Experiment for Hypothesis 1

4.1.1   Efficiency

The result of efficiency is shown as the left side of Fig. 5. The most attractive color is efficiently found than the second most attractive color. By the Kruskal-Wallis test, the difference of average time to find one item among target item's attractiveness is significant beyond the 0.01 level (χ^2 (1, N = 454) = 28.8, p = 7.97 × 10^{-8} < 0.01).

Next, we focus on hue difference in the task screen. The right side of Fig. 5 shows the result of efficiency, which is divided according to hue difference between the most attractive color and the second most attractive color. In both types of hue difference, the most attractive color is efficiently found than the second most attractive color. This tendency is stronger in screen with same hue than screen with 120 degree different hue. By the Kruskal-Wallis test, the difference of average time to find one item among target item's color is significant beyond the 0.01 level in both type of hue difference in task screen.

 - Same hue: (χ^2 (1, N = 225) = 25.3, p = 4.84 × 10^{-7} < 0.01)
 - 120 degree different hue: (χ^2 (1, N = 229) = 7.26, p = 7.04 × 10^{-3} < 0.01)

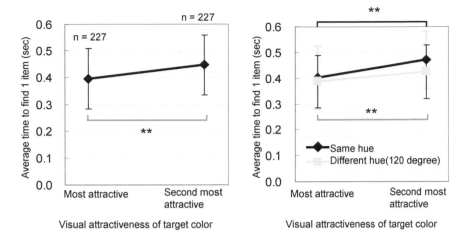

Figure 5 Relationship between visual attractiveness and operation efficiency

4.1.2 Accuracy

Error occurred in 13 tasks of 240 tasks when the target item was the most attractive color, and 13 tasks of 240 tasks when the target item was the second most attractive color. There is no difference in error rate between which colors to find.

4.2 Result of Experiment for Hypothesis 2

4.2.1 Efficiency

The result of efficiency to find the most attractive color in each type of screen is shown as the left side of Fig. 6. Target item in Screen of type C is most efficiently found, and screen of type A is the worst. By the Kruskal-Wallis test, the difference of average time to find one item among types of screen is significant beyond the 0.01 level (χ^2 (2, N = 675) = 42.6, p = 5.70 × 10^{-10} < 0.01). In addition, by the Steel-Dwass test, the difference of average time to find one item is significant beyond the 0.01 level among every pair of screen (type A-B: t = 3.42, p = 1.78 × 10^{-3} < 0.01; type A-C: t = 6.29, p = 9.80 × 10^{-10} < 0.01; type B-C: t = 3.56, p = 1.08 × 10^{-3} < 0.01).

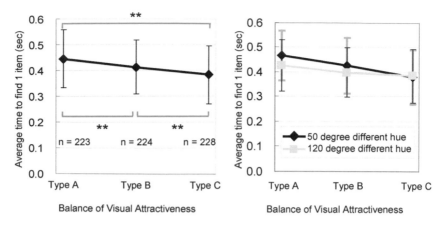

Figure 6 Relationship between balance of visual attractiveness and operation efficiency

Next, we focus on hue difference in the task screen. The right side of Fig. 6 shows the result of efficiency, which is divided according to hue difference between the most attractive color and the second most attractive color. In both type of hue difference, target item in screen of type C is the most efficiently found, and screen of type A is the worst. This tendency is stronger in screen with 50 degree different hue than screen with 120 degree different hue.

By the Kruskal-Wallis test, the difference of average time to find one item among types of screen is significant beyond the 0.01 level in both type of hue difference.

- 50 degree different hue: (χ^2 (2, N = 340) = 35.5, p = 2.00 × 10^{-8} < 0.01)
- 120 degree different hue: (χ^2 (2, N = 335) = 11.5, p = 3.19 × 10^{-3} < 0.01)

In addition, by the Steel-Dwass test, the difference of average time to find one item is significant beyond the 0.01 level among most pairs of screen, except type A-B in 120 degree different hue (n.s.) and type B-C in 120 degree different hue (n.s.).

4.2.2 Accuracy

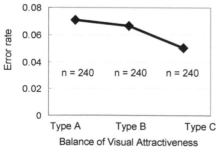

Figure 7 Relationship between balance of visual attractiveness and operation accuracy

The result of error rate to find the most attractive color in each type of screen is shown as Fig.7. Target item in screen of type C is the most accurately found, and screen of type A is the worst. By the 3-sample test for equality of proportions, the difference of error rate among type of screen is not significant (N(0, 1) = 0.996, p = 0.608, n.s.).

4.3 Consideration and Future works

4.3.1 Consideration of the Experiment

[Correlation between visual attractiveness and operation performance]
As the result shown in 4.1.1, the most attractive color is efficiently found than the second most attractive color (Fig. 5). However, as the result shown in 4.1.2, target color's visual attractiveness do not effect on error rate. Thus, hypothesis 1 is supported on efficiency, but it is not supported on accuracy. This means that items which have high visual attractiveness could be found efficiently, without causing error rate increase. If colors on screen are decided as important information's visual attractiveness is higher than other information, users could find important information efficiently without causing more errors.

[Correlation between balance of visual attractiveness and operation performance]
As the result shown in 4.2.1, target item in screen of type C is the most efficiently found, and screen of type A is the worst (Fig. 6). Furthermore, as the result shown in 4.2.2, target item in screen of type C is found in the lowest error rate, and target item in screen of type A is found in the highest error rate, but the tendency of error rate is not significant (Fig. 7). Thus, hypothesis 2 is supported on efficiency, but it is not supported on accuracy. This means that if the visual attractiveness of items on the screen are uniformly high, users cannot find the target item efficiently. It is important to make much difference on visual attractiveness between important items and other items, to make users' operation efficiently.

[Effect on efficiency of hue difference between items on the screen]
As the result shown in the right side of Fig. 5 and the right side of Fig. 6, the hue difference between items on the screen seems to have effect on efficiency. The correlation between visual attractiveness and efficiency is stronger in screen with small difference in hue than screen with big difference in hue. We think this is because, in screen with small difference in hue, the second most attractive color disturbs searching the most attractive color than screen with bigger difference in hue. This result suggests that in visual search tasks, the examinee especially uses the information of hue as an important clue to distinguish the target color among other colors. Thus, it seems effective to consider hue difference between items when deciding coloration.

4.3.2 Future Works

As a future work, we are thinking of revising the calculation method of visual attractiveness. The result showed that even if the balance of visual attractiveness is the same, the operation performance changed by hue difference among items on the screen. Thus, it is expected that if the effect of hue on the calculation method become greater, we could calculate the visual attractiveness more correctly.

Moreover, validations in different kinds of screens are also our remaining question. For example, screen which has many kinds of items colored with different colors such as actual systems' screen, screen with different background color, and so on.

5   CONCLUSIONS

This paper presented the result of experiment to verify the correlation between visual attractiveness of color and operation performance. The result showed that items which have high visual attractiveness could be found efficiently, without causing error rate increase. In addition, to find items which have the most attractive color in the screen, as their visual attractiveness are significantly higher than other items, they could be found efficiently, without causing error rate increase. Therefore, it is effective to consider color design based on the balance of visual attractiveness, for both efficient and accurate operation performance.

As future works, we are expecting to revise the visual attractiveness calculation method, or verify our hypotheses in different situations.

REFERENCES

Emerson Network Power. 2011. Understanding the Cost of Data Center Downtime. Accessed February 29, 2012, http://www.emersonnetworkpower.com/en-US/Brands/Liebert/Pages/liebertwhitepapers.aspx

Fitts. 1954. The Information Capacity of the Human Motor System in Controlling the Amplitude of Movement. *Journal of Experimental Psychology, Vol.47, No.6:* 381-391.

Norman, 1988. "The Psychology of Everyday Things": Basic Books

Itti, Koch and Niebur. 1998. A model of saliency-based visual attention for rapid scene analysis. *IEEE Transactions on Pattern Analysis and Machine Intelligence, 20 (11)*: 1254-1259.

Tanaka, Inokuchi, Iwadate and Nakatsu. 2000. An Attractiveness Evaluation Model Based on the Physical Features of Image Regions. *IEICE A JPN edition, J83-A(5)*: 576-588

World Wide Web Consortium. 2000. Techniques For Accessibility Evaluation And Repair Tools. *W3C Working Draft, Technique 2.2.1*

CHAPTER 51

The Integration of Ethnography and Movement Analysis in Disabled Workplace Development

Andreoni G.[4], Costa F.[1], Frigo C.[2], Gruppioni E.[3], Pavan E.[2], Romero M.[1], Saldutto B.G.[3], Scapini L.[1], Verni G.[3]

[3]Health Care Design, INDACO Department, Politecnico di Milano, Italy
[2]MBMC Lab, Department of Bioengineering, Politecnico di Milano, Italy
[3]INAIL (Istituto Nazionale Assicurazione contro gli Infortuni sul Lavoro), Italy
[4]LyPhe, INDACO Department, Politecnico di Milano, Italy
maximiliano.romero@polimi.it

ABSTRACT

A physical disability should not be an obstacle to participate in the work world, however, some workplaces are not conceived in order to admit disabled workers. The aim of this paper is to present the process conducted by an interdisciplinary team to analyze human-product physical interaction and develop an accessible workplace. We focused our research in a PC-workplace to be used by physically impaired people for their professional reintegration. _In previous experiences [1] simple biomechanical measurements and electromyographic analysis were used to evaluate the physical stress connected to different workplace situations. In the present context we have chosen to apply to occupational ergonomics both a biomechanical and ethnographic approach and then correlate them in an integrated approach. The idea of merging qualitative and quantitative methods has become increasingly appealing in areas of applied research. As Human Machine Interfaces (HMI) and ergonomics are multifaceted issues it is important to approach them from different perspectives and to combine data coming from different methods.

Multicompetence approach integrates different research methods into a research strategy [2] increasing the quality of final results and providing a more comprehensive understanding of the analyzed phenomena.

Keywords: User Centred Design, physically impaired worker, PC workplace, Ethnographic Observations, Biomechanics.

1 INTRODUCTION

Ergonomic evaluation and physical disabled people's workplace development are very difficult issues because standard methods are not applicable: risk analysis is not reliable because tasks are often performed in an unusual way and functional anthropometrical data are difficult to retrieve.

Interesting studies [3] are in progress to develop specific virtual reality approaches for supporting the Design for All approach since occupational ergonomic analysis on virtual mock-up is today not possible due to the fact that the human models within existing applications do not include impaired persons.

Generally disabled workers reintegration is usually faced through ad hoc adaptation of workstation and work environment for each subject [4]. This approach allows obtaining high-customized solutions that are very efficient but involve a great effort and cannot be applied on a large scale. For this reason we decided to focus our research on the development of a standardized adjustable solution as adaptable as possible to different users pathologies and office activities.

Learning from adaptation experiences and virtual reality approaches we based our research on participatory observation [5] and combined it with the development of a proprietary virtual model and its validation through laboratory test.

A participatory phase was planned to directly involve users in product development and evaluation [6].

2 METHODS

The methodological process consists in 3 main steps before a final integration trough product development (Fig.1.).

a. At the beginning we performed ethnographic investigations on subjects affected by spinal cord lesion at different levels to detect their user habits [7] and self made solutions and strategies. Observations accompanied by contextual interview were carried out in the real user environment and involved paraplegic and quadriplegic subjects. Evaluating the collected data we also defined a test setting and a series of motor tasks to be investigated from the biomechanical point of view.

b. The biomechanical analysis based on a model which includes six degrees of freedom (dof) for the upper trunk, three dof for each shoulder, two dof for the elbows, two dof for the wrist was implemented in order to compute the joint moments required to perform the different tasks.

c. The second biomechanical analysis, comprehending real movement acquisition and evaluation, was performed on healthy and spinal cord lesioned subjects to define the strength associated to reaching objects in different positions in the extracorporeal space. A stereophotogrammetric system with eight infrared TV cameras was used to detect the movement of the upper limbs in relation to the trunk, and the movement of the trunk in relation to an absolute reference system fixed within the laboratory. Retro-reflective markers were attached to the head, shoulders (acromions), elbows, wrists, and metacarpal area and dorsal surface of the trunk.

The outputs regarding user behavior and movement strategies were used to define the first design proposals which are the basis for an active involvement of expert users in virtual and real prototypes development. [8].

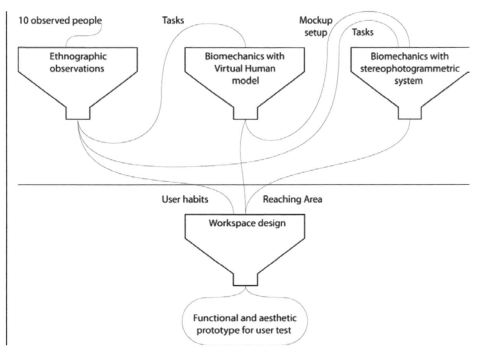

Fig.1. General research process with 3 main steps and integration

The subsequent evolution of the project will include the tests to be performed with the involvement of disabled people and an inquiry about the level of acceptance of the proposed solutions.

Ethnographic investigations

In order to define a specific target group of users, an ethnographic investigation has been performed which was structured in four steps: warm-up, general questions, static analysis of the work environment, observation of the users during their work activity. Sixteen workstations, placed in home and office environment, have been evaluated regarding qualitative issues.

510

We began with 10 ethnographic investigations on subjects affected by spinal cord lesion at different neurological levels to detect their habits and solution strategies during PC-workplace use. Observations together with contextual interview were carried out in the real user environment (6 homes and 4 offices) and involved 7 paraplegic subjects and 3 quadriplegic.

The acquired data have been compared and the real situations have been grouped in 3 different categories related with the level of spinal cord lesion:
— users with high spinal cord lesion, that work with an assistant without moving from their workstation, at home;
— users with middle-high spinal cord lesion, that work without moving from their workstation, at home or in the office;
— users with middle-low spinal cord lesion, that work moving from one workstation to another, in the office, in team with colleagues or with patients.

Fig 2: two different users observed during their work.

In all situations, a great number of object are located on or around the workplaces including obvious and less obvious ones ranging from PC, paper, pencils to mobile phone, pictures and medicaments. A new workstation has to face the problem of organizing all these objects according to user habits.

Size of the work table vary from 100cm to 300cm of width, 70cm to 90cm of depth, 67cm to 85cm of high. Their configurations appear to be affected by working modalities: the desks are predominantly rectangular if interaction with colleagues or patient is needed and "L" shaped in several cases were people work alone and place is available.

The three identified categories were evaluated and we choose the one regarding people with middle/high spinal cord lesion working alone at home or in the office for further development since it is the more statistically frequent situation and it permits to develop solutions that could be also suitable for tele-work.

Deeper analysis of this users group and their needs has been performed through a questionnaire to the users and interviews to experts like occupational ergonomists and disabled worker's reintegration specialists.

On field analysis revealed interesting differences between home and office workstations regarding for example self adaptation solutions, like the placement of the printer under the table at 40cm from the floor, which are more frequent in home workplaces. A careful study of those adaptations could suggest useful solutions to be transferred also in the office environment.

The most important needs detected through on site users analysis regarding the selected category concern:
— avoiding the necessity to shift from wheelchair to operating chair;
— maintaining distances and adjustments in relation with the working area;
— increasing trunk mobility and stretching possibilities;
— increasing trunk balance and facilitating the achievement of an upright posture when it happens to lose it;
— reducing the falling of objects or facilitating their recovery;
— positioning of an easy to reach case for personal items;
— reaching all devices and commands;
— avoiding cable hindrance.

On this basis we defined a test setting and a series of motor tasks to be investigated from the biomechanical point of view on the next step.

Biomechanical analysis with virtual human model

In order to identify the portion of the workplace that can be reached with a certain level of muscular effort, a dynamical model was developed which allowed us to quantify, by simulating several load conditions, the force necessary for completing the task. The model (Figure 3) is composed of a number of rigid bodies corresponding to head, trunk, pelvis and lower limbs, upper arm, forearm, and hand for both sides. The parameters like segments' length and mass, were obtained from anthropometric tables [9]. Location of centers of mass and moments of inertia derived directly from the geometry of the rigid bodies. The focus here was the upper limb movement, and so the following constrains were defined among the segments: three rotational axes at the shoulder representing adduction/abduction, flexion/extension, internal/external rotation, one rotational axis at the elbow, representing flexion/extension, one rotational axis at the wrist, representing pronation/supination of the hand. The trunk was fixed to the backrest of the wheelchair, and the pelvis to the seat. Both inclination of backrest and seat height could be adjusted to test different relative positions between subject and table. The table itself could be raised or lowered and rotated around a horizontal transversal axis to reproduce different slopes. Each point of the extracorporeal space could be reached by changing the angles of the different joints. A limit however was implicit in the total limb length. Additional space could be added by changing the inclination of the trunk. For each position in space of the hand, the corresponding joint angles and joint moments were computed, so that the whole reachable space could be mapped.

In the example presented here, the right hand movement was analyzed, although the procedure could be applied to left-handed subjects as well. The task analyzed was a unimanual task (i.e. without the use of the contralateral arm) and was simulated by leading the hand to reach different points of the desk work plane. The seat height was 0.47 m (forward edge with respect to the ground); the table height was 0.7 m from the ground and the surface was horizontal. The backrest was inclined by 20° on rear, and the relative position between subject and table was such that the lower edge of the trunk (corresponding approximately to the extremity of the rib cage) was at 0.19 m from the edge of the table. A grid of points was defined on the table surface sufficiently close each other as to have a good spatial resolution, within the reaching-area border (namely the limits of the full area where an object can be placed that can be reached by only extending the arm, without moving the trunk).

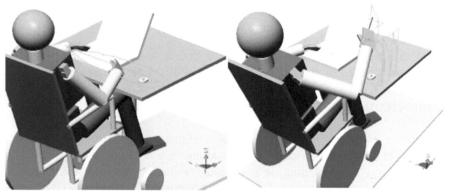

Figure 3-A) The anthropomorphic dynamic model represented in one specific position (see text). The track of the hand centre of mass during systematic analysis of the reaching is reported on the table by a line. B) The same model while scanning the right-hand space on a vertical plane (see the hand track)

Since the same point in the space could be reached in different manners, each representing a diverse combination of rotations about the different axis of the joints, a particular condition was imposed that was a fixed orientation of the hand palm in relation to the horizontal plane. The wrist angle also was kept at a fixed degree. In this way the hand, which originally had six degrees of freedom, is constrained so that only four degrees of freedom are active. These, in our choice, are the three shoulder rotations and the elbow rotation. The goal was thus to associate to each position of the hand, the joint angle and the joint moment obtained from the dynamical simulations, for each of the following movements: shoulder ab-/adduction, flexion/extension, internal/external rotation of arm, elbow flexion/extension.

The results are shown in Figure 4. Here the joint angles and moments associated to each position of the hand in the reachable plane are reported with reference to the shoulder joint. They are represented by three surfaces corresponding respectively to the flexion/extension, abduction/adduction, internal/external rotation degrees of freedom. It appears that the whole positioning of the arm segments has a direct influence on the increase or decrease of any considered moment necessary for reaching a particular point in the space. If a particular joint moment or joint angle cannot be overcome because of limitations in the strength or mobility of the hypothetical subject, different portions of the original space could be identified, which can be reached by applying a moment which is less than the maximum moment the subject can develop. A similar result was obtained for the angular rotations.

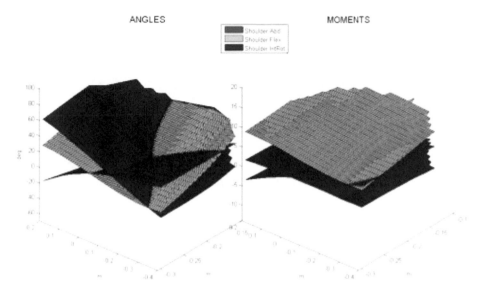

Fig.4) Systematic analysis of the shoulder angles and moments associated to maintaining a given position of the hand on the work plane, supposed horizontal, 5 cm above the table surface. The three intersecting surfaces refer to shoulder flexion (green), abduction (red), internal rotation (blue). The point (-0.4, -0.35) corresponds to the right-rearmost corner of the table surface.

In this way, alterations due to pathology, which imposes limitations in both the range of movement and the moments produced, may be considered in order to identify those parts of the space that could be reached more easily than others and, consequently, in order to consider these limitations during the design process.

This virtual analysis, based on a biomechanical model, was implemented in order to compute the joint moments required to perform the different tasks. The outputs regarding user behaviour and movement simulation were used to define the first design proposals and a functional mock-up of the workstation.

Biomechanical analysis with a motion capture system

A Workspace mockup was then built to perform movement acquisition and evaluation on healthy and spinal cord lesioned subjects in order to quantify the strength required to reach objects with individual motor strategies in different positions in the extracorporeal space. A motion capture system with eight infrared

TVcameras was used to detect the movement of the upper limbs in relation to the trunk, and the movement of the trunk in relation to an absolute reference system fixed within the laboratory. Retro-reflective markers were attached to the head, shoulders (acromions), elbows, wrists, and metacarpal area and dorsal surface of the trunk. (Fig. 5-B)

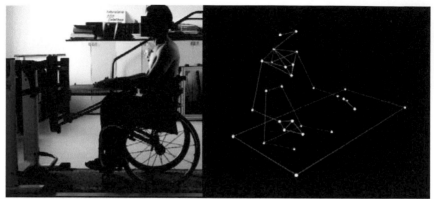

Fig.5- A) Lateral view of a subject placed on the mockup with retroreflective markers. B)

As to the different tests performed, particular attention was paid to the different modes of positioning the trunk. The following conditions were considered:

1) trunk supported by the wheelchair backrest;

2) trunk unsupported, with possibility of supporting the contralateral arm on the table, with slight force in order to help for stability;

3) trunk unsupported, with possibility to apply a relevant support force on the table by the contralateral limb to the purpose of reaching farther positions without falling down;

4) simulation of opening the drawers located laterally under the table surface, as if the subject was picking up objects.

As to the task of reaching positions in the space over the work plane, another test was added consisting in simulating the reaching of objects located on shelves or support planes at a give height above the work plane.

In previous researches [10] we encountered some problems with the analysis of data acquired from natural users because of a large variability in movement behavior that made inter-subject comparison very difficult. On the other hand a too constrained definition of the movements would risk to make them unnatural. In this experience we consequently decided to consider two separate data source: qualitative information from natural movement observed in the first step (Ethnographic Observations) and quantitative information from third step (motion capture), using first step to define which movements to analyze on the third step.

The outputs of the biomechanical tests permitted to fine tune the concept and realize a prototype workstation to be tested regarding usability and acceptability by the end user.

3 RESULTS

Some interesting elements in the reaching strategies have been highlighted by both the observational analysis and the biomechanical analysis with the stereophotogrammetric system. A relevant one was the use of the contralateral limb as a support to help achieving farther objects and balancing the body. In all cases in which the subject adopts this technique, there is a need to provide a strong supporting surface and an area of the table relatively free from objects, in order to allow the limb support. An alternative solution could be to provide special handles directly on the table, where the subject could hang or pull him/herself forward or laterally, consistently with his/her specific needs.

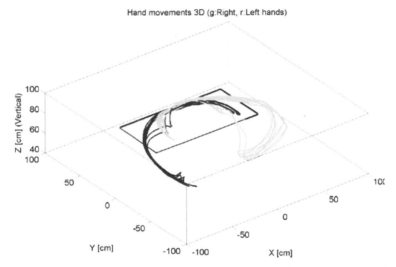

Fig. 6 Positions Tracks of both hands on the space, related with a 70cm height worktop (left hand in darkest line)

In such a way the table surface could be more efficiently used as a work surface. From another point of view the edge of table could provide an accessory support function to improve the trunk stability, for example in the case of an excessive forward bending, when the chest can rest against the table. In this case the border of the table has to be without sharp edges, or even covered by soft material. In some cases a recess tailored on the chest size could help improving his/her medio-lateral stability. The integrated result of this three-steps process was an innovative PC-workplace. Final project has been produced with final aesthetic characteristic and is ready for user tests in a real context of use.

4 DISCUSSION

From methodological point of view, we found a positive compromise to integrate ethnographic qualitative data and physical modeling movement quantitative data in product development defining an proactive approach to ergonomics based on data related to physical interaction between the human and new or existing products. The method was applied to disabled worker PC station development while its extension to other aspects and other concepts is under development. Further developments should also take into account the integration of functional supports needed by persons with spinal cord lesions who sometimes make use of assistive devices and functional electrical stimulation to perform basic functions. [11].

AKNOWLEDGEMENTS:

This work was supported by INAIL (Istituto Nazionale Assicurazione contro gli Infortuni sul Lavoro). The highly professional assistance of occupational therapists of the Spinal Unit of the Ospedale Cà Granda di Niguarda, in particular Mr. Davide Mangiacapra, is warmly acknowledged

REFERENCES:

[1] E.Occhipinti, D. Colombini, C. Frigo, A. Pedotti, A. Grieco, "Sitting posture: analysis of lumbar stresses with upper limbs supported", Ergonomics, vol. 28, n. 9, pp. 1333-1346, 1985

[2] Brannen, J., "Mixing Methods: The Entry of Qualitative and Quantitative Approaches into the Research Process", International Journal of Social Research Methodology, Volume 8, Issue 3, pp. 173–184, July 2005

[3] Aubry M., Julliard F., Gibet S., Interactive ergonomic analysis of a physically disabled person's workplace, European Center for Virtual Reality, Brest, 2007

[4] Andrich R, Bucciarelli P, Liverani G, Occhipinti E, Pigini L, Disabilità e lavoro: un binomio possibile, Fondazione Don Carlo Gnocchi –ONLUS, Milano, 2009

[5] Lifchez, R, and Winslow, B, Design for independent living: The environment and physically disabled people, Architectural Press, Boston, 1979

[6] Helin K., Viitaniemi J., Montonen J., Aromaa S., T., "Digital Human Model Based Participatory Design Method to Improve Work Tasks and Workplaces", Lecture Notes, Computer Science, 2007, Volume 4561/2007, pp.847-855

[7] Amit, V., Constructing the Field: Ethnographic Fieldwork in the Contemporary World, Routledge, 2000

[8] Sanders, E.B., "Design research in 2006", Design Research Quarterly 1, n°1, Design Research Society, September 2006

[9] Clauser C.E., McConville J.T., Young J.W .Weight, Volume, and centre of mass of segments of the human body, AMRL Technical Report (TR-69-70). Wright-Patterson Air Force Base, OH, 1969.

[10] Romero M., M. Mazzola, F. Costa, and G. Andreoni, "An Integrated method for a qualitative and quantitative analysis for an ergonomic evaluation of home appliances", Proceedings of Measuring Behavior 2008

[11] R.Thorsen, M.Ferrarin, R.Spadone, C.Frigo, "Functional control of the hand in tetraplegics based on synergistic EMG activity", Artificial Organs, vol.23 n.5, pp. 470-473, 1999

CHAPTER 52

Multi-Source Community Pulse Dashboard

Timothy P. Darr, Satheesh Ramachandran, Perakath Benjamin, Amelia Winchell, Belita Gopal

Knowledge Based Systems, Inc.
College Station, TX, USA
tdarr@kbsi.com

ABSTRACT

The Multi-Source Community Pulse Dashboard (MSCPD) is capable of: (a) sensing emerging themes, trends and patterns of beliefs, sentiments, rhetoric, etc., from multi-source data covering large social networks; and (b) correlating it with structural and behavioral dynamics within the same or related social networks. MSCPD has successfully demonstrated the ability to sense emerging themes, trends and patterns of beliefs, sentiments, rhetoric, etc. from multi-source data covering large social networks; and correlating them with structural and behavioral dynamics within the same or related social networks. The widget-based MSCPD provides a simple, intuitive and easy-to-use portal that allows analysts to explore the correlations between themes, sentiment and changes in social networks for datasets derived from blogs, chat rooms, and other social networking forums. MSCPD will assist battlefield commanders to achieve optimal course of action planning by enhancing the understanding of the complex dynamics of tightly connected elements spanning demographic, political, military, economic, social, information and infrastructure dimensions.

Keywords: theme detection, sentiment mining, social network analysis, course-of-action planning

1   INTRODUCTION

Modern asymmetric warfare involves operational and strategic decision making under high levels of uncertainty, necessitating the capability for truly effective course of action (COA) planning.   This planning, to be effective, must accommodate and adapt to the state of flux—the enemy's shifting organizational patterns and methods of communication, growth and indoctrination, levels of preparation, policy shifts, advances in capability, new technology acquisition, deployment methods, etc.—that is the hallmark of modern warfare. Most pressing, and most challenging, is the need for accurate situational awareness at many levels, particularly in understanding the complex dynamics of a tightly connected web of elements spanning demographic, political, military, economic, social, information and infrastructure dimensions.

Situational awareness and, consequently, effective COA planning can only be achieved by understanding the mechanics by which these underlying networks operate and change: to, in other words, understand the formal laws and structures that govern the space of the enemy's organizational dynamics. Coming to terms with the state space or "instantiating the dimensions of the state space" will allow planners to not only reason about the current state of the world but also about the world's future state.

The *Multi-Source Community Pulse Dashboard (MSCPD)* is capable of: (a) sensing emerging themes, trends and patterns of beliefs, sentiments, rhetoric, etc. from multi-source data covering large social networks; and (b) correlating it with structural and behavioral dynamics within the same or related social networks. For example,

- Within community X - Are there any new perceptions or beliefs emerging within on-line forums? Are there changes in beliefs and new perceptions within a community as suggested by incidences or intelligence reports?
- Are there any structural or behavioral changes within a social network of interest that is associated with community X?
- What is the relationship between the two changes described above?

An equivalent problem, which is an active topic of interest within the commercial arena, is

- Is there a change in perception or sentiment towards a brand X (Sony or HP, for example) as noticed from an online product review posting or online social networking site? Is there a new aspect regarding a particular product of brand X that is emerging as a topic of discussion (say Sony battery life, HP printer ink costs)?
- Is the change in perception related to a news event associated with the product X? Or to any extraneous event related to company producing brand X? Or to any changes associated with the leadership network for the company producing brand X?

2   MSCPD TECHNICAL APPROACH

Modern asymmetric warfare involves operational and strategic decision making under a high level of uncertainty. A critical need for battlefield commanders is the capability to perform effective course of action (COA) planning that takes into account scenarios that consider variables such as organizational patterns, methods of communication, growth and indoctrination, opponents' level of preparations, policy shifts, advances in capability, new technology acquisition, deployment methods, etc. To achieve the desired COA capability, accurate situational awareness at many levels is needed, particularly in understanding the complex dynamics of tightly connected elements spanning demographic, political, military, economic, social, information and infrastructure dimensions. Understanding the statistical nature and interaction that exist among these dimensions and their relation to future world states is critical to effective asymmetric warfare, like the warfare practiced in SRO type operations in Iraq and Afghanistan.

An overriding fact of asymmetric warfare is that the operational environment is continually in a state of flux, and this flux is reflected in the underlying social and organizational networks that operate in the environment. While dealing with these asymmetric entities, situational awareness and COA planning can only be achieved by understanding the mechanics by which these underlying networks operate and change. Kathleen Carley argues that the basis for understanding organizational dynamics lies in the development of socio-cognitive quantum mechanics, an equivalent of quantum mechanics from the physical world (Carley 1999). In other words, like the universal laws of quantum mechanics that govern the physical world, there are formal laws and structures that govern the space of organization dynamics. Establishing a quantum mechanics' equivalent within the social world requires, first, creating a characterization of the state space within which these entities can be situated, followed by developing computational models that can be used to explain the behavior and evolution of these entities and associated networks. At each point in time, the state of the world can be described by instantiating the dimensions of the state space, and the computational model consists of reasoning about the current and future state of the world.

The concept can be illustrated through the simple example depicted in Figure 1. Here the state of the world is characterized through a single variable: Sentiment towards American Occupation. This variable can be instantiated at any point in time, reflecting the state of the world at that point in time. This figure shows that the sentiment for time T+1 has grown mildly positive from a negative reading at time T. The reason for this shift in sentiment could be that there has been a change in the social network characterizing the leadership of the local resistant militia: the leader C (who was in large part responsible for engineering the local opposition to the SRO operation) was no longer part of the militia decision-making apparatus.

The simple example in Figure 1 illustrates the characterization of the state space within which the state of the world and the entities in this world can be situated, and the rationale for future state behavior. There are some important observations that can be made here. First, the state of the world is represented by a complex and broad array of demographic, behavioral, psycho-social and organizational indicators. Asymmetric threat networks are influenced by a variety of factors ranging from political, social, religious, and economic factors. Accordingly, the motivations, goals, and ideologies of ethnic separatist, anarchist, social revolutionary, religious fundamentalist, and new religious networks differ significantly. Therefore, each network must be analyzed in terms of its demographic, behavioral, cultural, economic, political, and social context in order to better explain and rationalize the motivations of its individual members and leaders, as well as each group's particular ideologies. A second observation is that many of the state space metrics cannot be measured directly, but only through indirect metrics that are observable from disparate structured and unstructured sources.

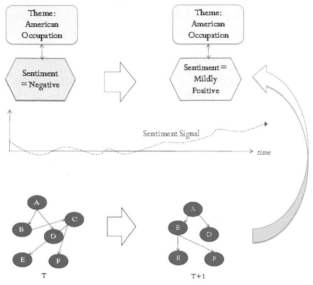

Figure 1 – Sentiment Evolution Example

MSCPD was developed using systematic knowledge engineering to create a suitable state space representation, integrate natural language processing (NLP) and text mining methods for the extraction of themes/topics and sentiments from network data, and establish a formal framework for correlating the changes in network attributes to future state behavior. MSCPD leverages a strong research foundation in the various elements of the proposed approach: text mining and NLP for theme and sentiment extraction (Gopal 2007; Abassi 2008), systems dynamics and Bayesian networks (Cheeseman 1996), knowledge discovery methods (Benjamin and Ramachandran 2002; Benjamin et al 2002), information fusion, optimization modeling, and decision support systems.

3   MSCPD SOLUTION OVERVIEW

The overall conceptual schema supported by MSCPD is shown in Figure 2. The architecture was composed of three sets of modules. The main component of the solution (consisting of the modules on the left side of Figure 2) consists of a theme/topic extraction system that continually monitors unstructured network related data to extract new and existing themes. The themes/topics by themselves do not completely characterize the state of the world – the sentiments and feelings of the community towards the discovered themes offered a more complete characterization of the state of the world. This part of the solution was offered by a sentiment classifier module, which summarizes the community level sentiments toward themes/topics. The topic-sentiment-tuple data provide a state of the world characterization of the community. Another component of the architecture (consisting of the modules on the upper side of Figure 2), provides explanatory metrics and information, such as the metrics from underlying social networks. The social network module together with the module to capture extraneous events and known actions provide the means to capture the explanatory information. Another module provides computational capability to study the relationship between the explanatory variables and the current/future behavior of the world. This computational module provides the means for performing COA evaluation, allowing the planner to evaluate the impact of changes in network attributes on the current and future state of the world, as well as studying the impact of actions (controllable factors).

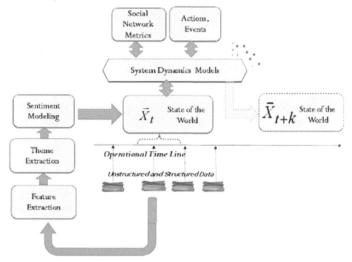

Figure 2 – MSCPD Solution Overview

4   THE MSCPD PORTAL

The MSCPD is a data view widget that is connected with a Bayesian model for COA planning. The design focused on supporting the research and experimentation of relationships between behavioral and socio-structural aspects of communities. The database is extensible enough to allow the inclusion of additional data sources (in the longer term, the system could apply to any analyst/researcher that wants to model and study the phenomena in their respective domains). The widget can communicate in real time with the Bayesian modeler to make predictions based on user-provided values.

The MSCPD portal is shown in Figure 3. The idea is to support an interactive process through a widget-based tool with four inter-related quadrants. Currently, the widget is available on a local server with Internet Explorer and Firefox, but it will be eventually be deployed onto the Ozone widget (Ozone 2011).

The data source is chosen by a drop-down menu at the top of the widget. Most data quadrants contain a date slider that allows the user to indicate a date range over which to view the data. Dates are organized in quarter years, where each quarter is three months long. The check box below the data source selector allows a user to link all of the date sliders so that changes in each are reflected in all graphs.

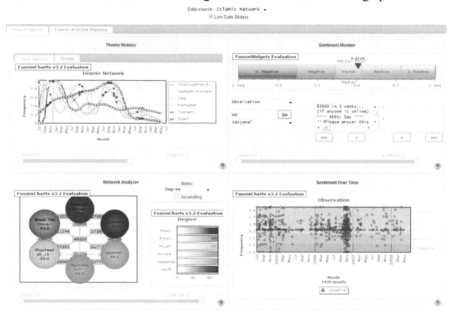

Figure 3 – MSCPD Portal

The upper-left quadrant contains the Theme Monitor, which allows the user to view trends in themes over time. The graph consists of a frequency-versus-time analysis of several hand-picked themes. Vertical lines and red data points indicate that a notable event has happened on that month. Mousing over the data point will display a tooltip with information about the event. By default, all available data sets are displayed. Clicking on a legend entry shows or hides the respective line in the graph. The Data Source tab shows the frequencies of themes which appear in a given data source. The Theme tab shows the frequencies of the theme references made by each data source. Specific themes are selectable by the tab list at the left.

The upper-right quadrant contains the Sentiment Monitor, which displays various sentiment score readings based on particular time frames. The slider bar shows the average sentiment over the selected time period based on a series of given metrics. The first drop-down box selects a metric by which to view the data. The second allows further refinement if available. The left list box contains the possible data items according to the selected metrics. The right contains the data whose sentiment score will be displayed. If the right list box is empty, the average sentiment score is calculated and displayed. The buttons below the list boxes allow the movement of data between the two: ">>>" and "<<<" move all items, while ">" and "<" move only the selected data. The average entity sentiment is available by selecting the "*" option in the drop-down Entity box.

The overall sentiment view takes the average of all sentiment scores in the data source by quarter-year. The entity sentiment view takes the average sentiment score of the observations of a particular entity, as selected in the secondary drop-down box. The list boxes will display the IDs of the observations made by that user in the given time period for which there is data. The group sentiment view takes the average sentiment score of each group of observations. Groups in the given time period are displayed in the list boxes by group title.

The Network Analyzer in the lower-left quadrant visualizes and monitors the social networks with SNA metrics. The drop down menu allows the user to select a particular metric by which to organize the entities of the data source. The top five entities of the given metric are displayed in the graphs. The user may choose to view the entities based on ascending or descending order. The bar graph below the drop-down box shows the entities involved and their respective values based on the given metric. The graph to the left shows each entity and his connectedness to the other involved entities. The number along each edge is the link strength between each entity, or the number of observations which share the same group between two entities. If EntityA and EntityB have a link strength of 1.0, EntityB is a contributor in every group in which EntityA has posted. If the link strength is 0.0, the two entities do not share common groups.

The Sentiment Over Time view is connected with the Sentiment Monitor. While the Monitor shows the average sentiment score over a given period of time, the Sentiment over Time graph shows the specific data points that go into that average.

The Course of Action Planning tab (Figure 5) allows the user to make predictions regarding the social network based on a Bayesian model of the data source. Each slider represents a node in the Bayesian net. Moving sliders on the left (input) half of the screen and pressing the "Plan" button will update the output sliders on the right half accordingly.

The idea behind the Bayesian modeling can be illustrated further using the example below, where a hypothetical relationship is discovered –

"When discussions revolve around social topics (as opposed to political topics) =>

"the network growth gets higher"
"Centrality of the network decreases"

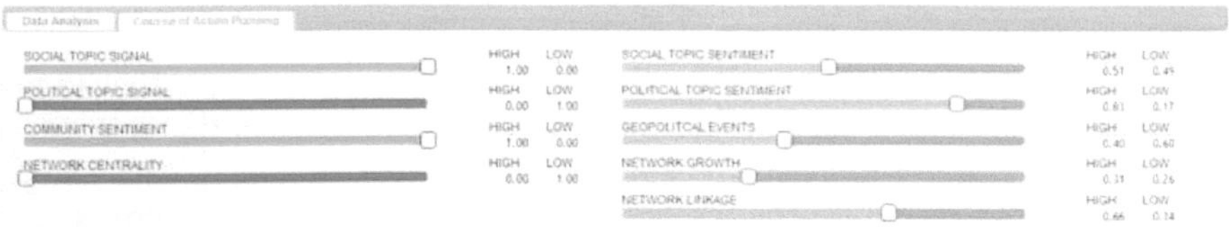

524

Figure 5 – MSCPD Course of Action Planning Dashboard

The relationship between the nature of interactions in the community (social topics as discovered from theme extraction module) and socio-structural property such as centrality and network growth (discovered through SNA modules) is specifically the type of relationship that the MSCPD tool is supposed to discover and parameterize.

5   INITIAL RESULTS (THEME DETECTION)

Themes are tracked using a module that generates a theme metric by using an existing theme detection algorithm to compute a dynamic theme signal that measures the intensity and dynamics of themes over time (Gopal 2007). This theme signal is displayed in the upper right quadrant of the MSCPD portal. This section shows some of the initial MSCPD results for theme extraction using two English language Islamic forum datasets.

To validate the theme extraction and theme tracking components, the following themes were handpicked and studied:

- Jihad justifies killing of innocent people
- Iraq
- Ramadan
- Tsunami
- Islam

Once the theme signal was generated, the analysts studied the signal to determine if there was a correlation between the theme signal and events in the real world.

The trend for the "Ramadan" theme follows a very cyclical pattern since it is an annual event as shown in Figure 6. The same trend is observed for both forums. This shows that the methodology used for monitoring themes over time, once the theme has been properly captured through a bag of concepts, works well for capturing cyclical themes.

The trend for the "Tsunami" theme is shown in Figure 7. As expected for the forum which was created before December 2004, the peak corresponds to the 2004 Indian Ocean tsunami. This again shows that the methodology used for monitoring themes over time, once the themes have been properly captured, works well. On the other hand, for the second forum, which was set up much later, the peaks seem to coincide with postings on threads related to other meanings of "tsunami," such as "Gaza tsunami," "political tsunami in Malaysia," etc.

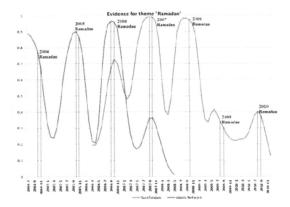

Figure 6 – "Ramadan" Theme Signal

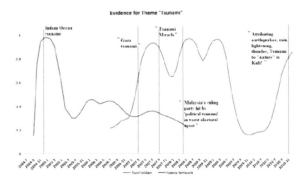

Figure 7 – "Tsunami" Theme Signal

6 CONCLUSIONS

Initial use of MSCPD has demonstrated the ability to sense emerging themes, trends and patterns of beliefs, sentiments, rhetoric, etc. from multi-source data covering large social networks; and correlate them with structural and behavioral dynamics within the same or related social networks. It allows an analyst to explore the correlation between changes in themes, sentiment and social network structure over time. By moving a slider to inspect various date ranges, the user is able to see the "theme signals," representing the changes in themes during the selected time window, for selected themes; the changes in sentiment over the same time interval, and the values of social network metrics for a subset of the social network.

MSCPD has also been used to demonstrate "what if" COA assessment and planning. Based on intuition and the analysis performed on the MSCPD portal, a Bayesian model was created to reflect the changes in social network metrics in response to changes in the themes and sentiments of a community. A user was then able to use the MSCPD COA planning dashboard to observe expected changes in network metrics as controllable theme and sentiment measures were changed.

The widget-based MSCPD portal provides a simple, intuitive and easy-to-use portal that allows analysts to explore the correlations between themes, sentiment and changes in social networks for datasets derived from blogs, chat rooms, and other social networking forums

MSCPD leverages existing NLP technologies to perform the theme extraction and the sentiment mining and analysis. The MSCPD theme extraction demonstrated the capability to correlate changes in themes with events in the world. The MSCPD sentiment extraction demonstrated the capability to correlate changes in sentiment with themes and events in the world.

ACKNOWLEDGMENTS

The research described in this paper was supported by the Army Research Laboratory, Aberdeen Proving Grounds; contract number W911QX-11-C-0061.

REFERENCES

Ahmed Abbasi, Hsinchun Chen, Arab Salem. "Sentiment analysis in multiple languages: Feature selection for opinion classification in Web forums" ACM Transactions on Information Systems, Volume 26, Issue 3 June 2008.

Benjamin, P. and Ramachandran, S. *"Organizational Tools for Knowledge Discovery and Creation"* Proceedings of the World Library Summit - Global Knowledge Renaissance 2002.

Benjamin, P., Erraguntla, E., Ramachandran, S., and Delen, D. Towards a Knowledge Discovery Framework. *Proceedings of the 2002 International Conference on Information and Knowledge Engineering (IKE 2002)*. CSREA Press. Pages 16-22. Las Vegas, June 2002.

Kathleen M. Carley, "On the Evolution of Social and Organizational Networks," in Steven B. Andrews and David Knoke (eds.), vol. 16, special issue of *Research in the Sociology of Organizations* on "Networks In and Around Organizations" (Stamford: JAI Press, 1999): pp. 3-30.

Cheeseman, P. and J. Stutz. 1996. Bayesian Classification (AutoClass): Theory and Results. Advances in Knowledge Discovery and Data Mining, AAAI/MIT Press, Cambridge, MA.

B. Gopal, J. Hamilton, P. Benjamin, and D. Narayanan, "Automated System for the Extraction of Themes from News Articles", *International Conference for Artificial Intelligence*, 8-14, 2007

http://widget.potomacfusion.com/main/home, last accessed Aug 15, 2011.

CHAPTER 53

Development of a 3-D Kinematic Model for Analysis of Ergonomic Risk for Rotator Cuff Injury in Aircraft Mechanics

Edwin Irwin, MSBME, CPE and Kristin Streilein, MSBME, CPE

Mercer Engineering Research Center
Warner Robins, Georgia 31088

ABSTRACT

Ergonomic analysis of the musculoskeletal risk associated with work is a well-established science founded on principles of epidemiological research, with its inherent strengths and limitations. We have found typical ergonomic analysis tools to offer insufficient specificity to explain the high prevalence of rotator cuff injuries experienced by civilian aircraft mechanics employed by a major aircraft maintenance center. A job assessment protocol, utilizing a 3-D kinematic model of the upper extremity is required to address the need for specificity in determining the relative severity of musculoskeletal risk entailed in various job tasks, including the biomechanical characteristics of those risks. The objective of this paper is to describe design parameters developed for a 3-D kinematic model to be used to more fully characterize the ergonomic risks faced by aircraft mechanics. The design parameters and their rationale will be compared and contrasted to previous efforts in this area. The work accomplished to-date on the development of the model will also be presented.

1 IDENTIFICATION OF NEED FOR 3-D MODEL

Mercer Engineering Research Center currently provides ergonomic, biomechanical, and occupational medicine services to a large aircraft maintenance facility as part of a return-to-work program for personnel who experience physical limitations. A recent review of participant data showed that 33% have a history of 1 or more rotator cuff injuries. When broken down by job skill, the highest prevalence (42%) was found in aircraft mechanics. This is 83% higher than the prevalence of rotator cuff injury in males over 50 years of age (Templehof et al, 1999). An effort was begun to identify the ergonomic factors leading to such a dramatically increased prevalence of shoulder injury among aircraft mechanics, and to provide guidance on reducing those risk factors.

The analysis of ergonomic risk in industrial work is a mature science with more than 50 years of epidemiological research to support its application to many different types of industries. It is primarily a science based on population statistics, since it must address the risk relative to a wide anthropometric range and a complex and kinematically indeterminate biomechanical system that easily accommodates sub-optimal and dynamically variable load conditions. Some aspects of ergonomic risk have been well-defined, such as soft-tissue response to vibration and impulse, and the response of the lumbar spine to posture and load. While these analyses cannot provide a definitive dose-response, they can accurately establish the probability of soft tissue or back injury in a cadre of workers performing the types of tasks involving those specific risks.

That is not the case with analysis of shoulder injury. Winkel and Matthieson (1994), among others described the use of surveys, interviews, and observational methods as primary techniques for analyzing the level of ergonomic risk in the workplace. All have inherent limitations in work areas like an aircraft maintenance environment (David, 2005). Symptom surveys can prospectively identify work areas that have underlying ergonomic risks, but have limited task-level specificity due to the range of tasks performed in a particular work area and the aperiodicity and varied duration of task assignments. Event-driven observational posture analysis tools, such as Rapid Upper Limb Analysis (RULA), have good reliability and are applicable to unpredictable task requirements. They have coarse sensitivity to levels of risk, however, and poor specificity of risk to body part. On the other hand, time-driven analysis tools such as the Occupational Risk Analysis (OCRA) Index are very specific and sensitive, but only for highly repetitive tasks (Stanton et al, 2005), which make up a very small percentage of tasks performed by aircraft mechanics.

The lack of specificity of risk assigned to body part not only hampers the quantification of the risk to shoulders involved in different tasks, but also limits the ability to identify appropriate engineering solutions to abate those risks; e.g. working overhead is unavoidable in some areas of large aircraft, requiring more detailed biomechanical data in order to appropriately redesign tooling or procedures and quantify the resulting reduction of risk. Determining the overall risk per shop and per task, and developing ways to reduce the risk is thus currently more heuristic than quantitative, making analysis of return-on-investment for recommended adaptations almost entirely qualitative.

As an example, task analysis for aircraft mechanics in one particular shop (numbering more than 100 mechanics) shows that their work is broken down into 140 non-documentary task categories covering several thousand technical operations, of which only an unpredictable portion are performed on any one aircraft during a maintenance cycle. Seven of the task categories involve overhead work and 8 of them involve some level of hand-arm vibration. Repetitive motion for all tasks is limited to less than 20 cycles without break. Lifting or supporting weight is done below shoulder level with 2 exceptions, both of which involve shared loading of <20 lbf. All work is done with self-regulated breaks. Work planning for aircraft mechanics assumes 4.6 hours of charged time per shift per worker. Thus, while the level of risk for a small proportion of tasks in this shop is elevated, none is highly elevated. Since this risk is amortized across more than 100 mechanics and over the 3 to 6 month time frame during which an aircraft is in the shop it does not appear to correlate with the 80% increase in the prevalence of rotator cuff injury for aircraft mechanics.

Further complicating the quantification of ergonomic risk to aircraft mechanics is the difficulty in differentially assigning levels of risk across the maintenance organization. Rotator cuff injury, particularly sub-acromial impingement syndrome (SAIS) is often associated with repetitive stress over long periods, though the etiology is not clear (van der Windt et al, 2000; Michener et al, 2003). Aircraft mechanics frequently change shops over the course of their employment due to workload changes, promotions, or changes in physical ability. In addition, rotator cuff injury is often not reported as a workman's compensation claim, since the emergent incident is often not work-related, though the underlying cause may be. It is therefore difficult to temporally associate risk at the shop level through the use of cross-section analysis, structured interviews, or anonymous surveys.

An assessment protocol is needed to address the limitations in typical ergonomic surveillance methods for quantifying the risk of shoulder musculoskeletal disease among aircraft mechanics. Such a protocol would incorporate the strengths of existing instruments, but include more detailed analysis of the biomechanical impact of work through use of a 3-dimensional kinematic model of all or parts of the body, to refine the sensitivity of the risk assessment, allowing for more accurate risk stratification, abatement design criteria, and return-on-investment (Winkel and Matthieson, 1994).

2   DEVELOPMENT OF 3-D KINEMATIC MODEL

2.1   Previous Efforts

Kinematic biomechanical models provide the ability to elicit the underlying structural, muscular, and neuromuscular control adaptations that occur in response to external loading requirements. Several such models have been developed for a variety of purposes over the past decade. A representative example includes Morrow et al (2010), Dickerson et al (2007), Holzbauer et al (2005), and Vasavada et al (1998). These models have a number of characteristics in common:

1. segment parameter definitions
   a. coordinate systems based on data from the International Society for Biomechanics (Wu et al, 2005)
   b. scalable anthropometry
   c. customizable inertial characteristics;
2. joint parameter definitions
   a. ball-and-socket characteristics used for the sternoclavicular, acromioclavicular, and glenohumeral joints
   b. scapulothoracic articulation characterized as a 3-dimensional curvilinear sliding joint;
3. muscle parameter definitions
   a. multiple force-generating strings representing the range of fibers for each muscle group using published values for placement, lines of action, and wrapping surfaces
   b. Hill model of activation response using published values for muscle architecture and kinetics;
4. kinematic constraints
   a. fully-defined scapulohumeral rhythm
   b. scapular motion constrained to a geometric representation of the thoracic surface
   c. clavicle motion reduced to 2 spherical degrees of freedom;
5. objective functions, based on published values for each muscle's physiologic cross-sectional area, to resolve force balancing among muscle groups during loading and movement;
6. 3-D body motion data used as input.

The differences among these models reflect their various purposes. Morrow et al and Holzbauer et al focused on the right upper extremity, ignoring the cervical structures and using a limited muscle set around the shoulder girdle to elicit kinetic inputs to the kinematic model in order to investigate the glenohumeral joint forces and the effect of surgical intervention on range of motion. Vasavada et al used a wide array of cervical movers with detailed customization of cervical joint parameters to investigate the contribution of muscles to joint moments and ranges of motion in the neck, but ignored the upper extremities. All of these investigators used the Software for Interactive Musculoskeletal Modeling (SIMM: Musculographics, Inc., Santa Rosa, CA) to formulate and visualize their models.

Dickerson et al explicitly designed their simulation to address the need for providing better analysis tools for ergonomists interested in preventing shoulder injury due to work. The researchers incorporated a set of 23 muscles groups, using 38 individual elements to mobilize 5 segments: torso, clavicle, scapula, humerus, and forearm (radius and ulna combined as one linkage). In addition, Dickerson modified the ball-and-socket joint representing the glenohumeral articulation to incorporate multi-directional glenohumeral joint dislocation forces. This provided optimization constraints that allowed them to investigate glenohumeral motion and joint contact forces. Their software was developed in Matlab in order to interface fluidly with ergonomics software such as JACK or 3DSSPP.

2.2   Design Parameters

While each of these efforts incorporates many useful characteristics, none completely meets the need we have identified within the aircraft maintenance environment. An understanding of the extrinsic biomechanical factors that elicit sub-acromial impingement syndrome (SAIS) in the workplace illuminates some of these requirements. Seitz et al (2011), in a review of the scientific literature, noted several predominant extrinsic mechanisms for mechanical compression of the sub-acromial tendon by other joint structures: faulty posture (i.e. head-forward posture or forward shoulder posture); adverse glenohumeral motion resulting from posterior capsular tightness, shortened pectoralis minor resting length, or altered recruitment of rotator cuff musculature, due to adaptation, fatigue, or weakness; and modified scapular kinematics (i.e. increased anterior tilt, increased scapular protraction, or altered scapulohumeral rhythm) due to altered recruitment patterns resulting from fatigue or weakness. A model that can help identify the presence and estimate the relative impact of these risk factors in a particular task will need to incorporate a significant level of complexity in its representation of muscles, limb segments, joints, and constraints. The parameters in Table 1 below were developed to address these risk factors. In addition, the practicalities of building reliable and validated models for wider use demand the implementation software be mature and widely accepted. For this reason, we chose to develop the model on the SIMM platform, using existing models as a basis.

2.3   Current Progress

The models built by Vasavada et al and Holzbauer et al were combined to form the structural basis for our biomechanical model. This provided a fixed trunk with thoracic spine and rib cage, 7 articulated cervical vertebrae, cranium, right clavicle, right scapula, and a complete and fully articulated right arm from the humerus to the phalanges. The additional structure enabled us to add the extra musculature needed to address the complete shoulder girdle as well as allowing us to investigate the adverse posture issues.

Insight into the impact of modulated scapular kinematics will be supported by the alterations made to some of the linkages, constraints, and kinetic inputs in the assembled models. The kinematics of the clavicle, scapula and shoulder were modified to decompose the shoulder rhythm constraints and add a longitudinal rotation capability to the clavicle. The scapular sliding constraint was removed and an ellipsoidal elastic boundary element added to act as a reaction floor for the scapula. Coraco-clavicular ligaments were added to provide kinematic constraint between the scapula and clavicle, similar to Maksous (1999). The Holzbauer and Vasavada models already incorporated a significant number of muscle groups and elements, some of which do not directly impact the kinematics of the shoulder. Additional muscle groups and elements were put into the model to provide appropriate kinetic input for the decomposed shoulder rhythm and for scapular control. The model now incorporates 60 muscles. Eighteen of these, involving 52 force generating elements, directly motivate the shoulder girdle. Lines of action were adjusted using ellipsoidal or cylindrical wrapping surfaces to provide physiologically realistic action throughout the range of motion. OpenSim software (OpenSim rel. 2.4: Palo Alto, CA) is being used to organize the inverse kinematic solvers, geometric and inertial property modules, muscle parameter functions, and constraint function modules. An image from the resulting model can be seen in Figure 1.

Table 1. Design parameters for a 3-D kinematic model based on known extrinsic biomechanical risk factors.

Biomechanical Risk	Model Requirement
Adverse body posture	As a minimum, skeletal components must include the cranium, the complete spine from C1 to L5, complete rib cage, and one upper extremity (clavicle, scapula, humerus, ulna, radius, carpals, metacarpals, and phalanges).
Posterior capsular constraint or glenohumeral rotation constraint by pectoralis minor	1) The model must accommodate a modified ball-and-socket representation of the glenohumeral joint with modifiable shear force parameters and dynamically variable contact force objective functions. 2) The model must incorporate ligamental structures whose strain response and resting length can be customized.
Adverse rotator cuff muscle recruitment	1) Hill-model muscle representations must have dynamic variability to simulate fatigue response characteristics. 2) Optimization objective functions used to balance stress among the muscles recruited during simulated activity must be adaptable to represent behavioral recruitment patterns.
Modified scapular kinematics	1) Scapulohumeral rhythm must be implemented as an objective function rather than a kinematic constraint. 2) Clavicular dynamics must incorporate 3 degrees of freedom, including long-axis rotation. 3) Sufficient muscle fibers must be incorporated to allow smooth, continuous motion and force.

3.   DISCUSSION

Significant work remains to complete the model. First, the kinematic constraints on scapular movement must be developed. We intend to implement the average scapulohumeral rhythm (Hogfors et al, 1991; de Groot and Brand, 2001) as an objective criterion, rather than a kinematic constraint. An additional criterion will be incorporated to maintain scapular proximity with the bounding surface representing the posterior thorax, in consonance with Maksous (1999).

Another major task required to complete the simulation involves integrating a muscular fatigue and recovery model. Liang et al (2010) described an activation pattern-based model that was developed to evaluate manual handling tasks using digital human modeling programs, such as JACK and 3DSSPP.

Finally, the model must be sufficiently validated to provide reliable information. We intend to compare the kinematic output of the simulation with the scapular kinematics measured by McClure et al (2001), who used bone pins in healthy subjects to directly capture these motions. The fatigue-recovery module will be validated by comparing the output of the model with EMG measures for isolated muscles in experimental subjects. Criterion validity will be developed using the scapular tracking brackets described by Karduna et al (2001) in conjunction with a motion capture system to capture scapular kinematics on experimental subjects performing simulated work tasks matched for tool force, tool motion, and repetition to work performed by aircraft mechanics in the maintenance facility.

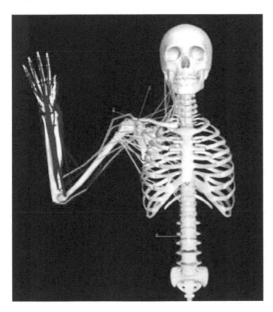

Figure 1. The biomechanical model built in SIMM is composed from models created by Holzbauer et al (2005) and Vasavada et al (1998). Addition musculature was added and kinematics modified to address current needs.

4   CONCLUSIONS

Once completed, the 3-D kinematic model will enable ergonomists to effectively characterize and compare levels of ergonomic risk to the upper extremity in an aircraft maintenance environment. We will use the typical interview and

534

observational techniques to focus the analysis to particular shops and identify candidate tasks for detailed analysis at the biomechanical level using the simulation. This analysis will primarily be done in the biomechanics lab in order to reduce the impact of data collection on productivity. We anticipate this protocol will prove effective in driving down the high rates of shoulder injuries experience by aircraft mechanics.

ACKNOWLEGEMENTS

This work was performed for Warner Robins Air Logistics Center under contract #FA8501-07-C-0054 P00006.

REFERENCES

David GC (2005);"Ergonomic Methods for Assessing Exposure to Risk Factors for Work-Related Musculoskeletal Disorders;" Occupational Medicine, vol. 55: pp. 190-199.
De Groot JH and Brand R (2001); "A Three-Dimensional Model of the Shoulder Rhythm;" Clinical Biomechanics, vol. 16: pp. 735-743.
Dickerson CR, Chaffin DB, Hughes RE (2007); "A Mathematical Musculoskeletal Shoulder Model for Proactive Ergonomic Analysis;" Computer Methods in Biomechanics and Biomedical Engineering, 10(6): pp. 389-400.
Holzbauer KRS, Murray WM, Delp SL (2005); "A Model of Upper Extremity for Simulating Musculoskeletal Surgery and Analyzing Neuromuscular Control;" Annals of Biomedical Engineering, 33(6): pp. 829-840.
Hogfors C, Peterson B, Sigholm G, and Herberts P (1991); "Biomechanical Model of the Human Shoulder—II: the Shoulder Rhythm;" Journal of Biomechanics, vol. 24: pp. 699-709.
Karduna AR, McClure PW, Michener LA, Sennett B (2001); "Dynamic Measurements of the Three-Dimensional Scapular Kinematics: A Validation Study;" Journal of Biomechanical Engineering, 123(2): pp. 184-190.
Liang MA, Chablat D, Bennis F, Zhang W, and Guillaume F (2010); "A New Muscle Fatigue and Recovery Model and Its Ergonomics Application in Human Simulation;" Virtual and Physical Prototyping, 5(3): pp. 123-137.
Maksous M (1999); Improvements, Validation, and Adaptation of a Shoulder Model; Doctoral Dissertation, Chalmers University of Technology, Gothenburg, Sweden.
McClure PW, Michener LA, Sennett BJ, and Karduna AR (2001); "Direct 3-Dimensional Measurement of Scapular Kinematics During Dynamic Movements In Vivo;" Journal of Shoulder and Elbow Surgery, vol. 10: pp. 269-277.
Morrow MMB, Kaufman KR, An KN (2010); "Shoulder Model Validation and Joint Contact Forces During Wheelchair Activities;" Journal of Biomechanics, vol. 43: pp. 2487-2492.
Stanton NA, Hedge A, Brookhuis K, Salas E, and Hendrick HW (2005); Handbook of Human Factors and Ergonomic Methods; CRC Press LLC, Boca Raton, Fl: pp. 8-1, 15-4.
Seitz AL, McClure PW, Finucane S, Boardman ND, Michener LA (2011); "Mechanisms of Rotator Cuff Tendinopathy: Intrinsic, Extrinsic, or Both?" Clinical Biomechanics, vol. 26: pp. 1-12.

Templehof S, Rupp S, and Seil R (1999); "Age-related Prevalence of Rotator Cuff Tears in Asymptomatic Shoulders;" Journal of Shoulder and Elbow Surgery, 8(4): pp. 296-299.

Van der Windt DAWM, Thomas E, Pope DP, de Winter AF, Macfarlane GJ, Bouter LM, and Silman, AJ (2000); "Occupational Risk Factors for Shoulder Pain: A Systematic Review;" Occupational and Environmental Medicine, vol. 57: pp. 433-442.

Vasavada AN, Siping L, Delp SL (1998); "Influence of Muscle Morphometry and Moment Arms on the Moment-Generating Capacity of Human Neck Muscles;" Spine, 23(4): pp. 412-422.

Winkel J and Matthieson SE (1994); "Assessment of Physical Workload in Epidemiologic Studies: Concepts, Issues, and Operational Considerations;" Ergonomics, 37(6): pp. 979-988.

Wu G, van der Helm FCT, Veeger HEJ, Makhsous M, van Roy P, Anglin C (2005); "ISB Recommendation on Definitions of Joint Coordinate Systems of Various Joints for the Reporting of Human Joint Motion—Part II: Shoulder, Elbow, Wrist, and Hand;" Journal of Biomechanics, vol. 38: pp. 981-992.

CHAPTER 54

Analysis of a Procedural System for Automatic Scenario Generation

Glenn A. Martin, Charles E. Hughes and J. Michael Moshell

University of Central Florida
Orlando, FL 32826

ABSTRACT

Within scenario-based training, the creation of said scenarios is a time-consuming and expensive process. Unfortunately, this results in the same scenarios being consistently re-used. While this is likely appropriate for new trainees, it does not provide effective training for continued use. Therefore, the authors have pursued a line of research investigating scenario generation (both assisted and automatic). The goal is to facilitate the creation of qualitatively similar scenarios while still maintaining variety.

The authors have previously documented efforts in reviewing the requirements for a scenario generation system tailored for adaptive training (Martin et al., 2009), in building a conceptual model for scenario representation (Martin et al., 2010) and in creating a procedural system for generating scenarios based around training and learning objectives (Martin et al., 2010). A system now exists to create scenarios based around sets of scenario components. Either the user or the procedural system can select a baseline scenario and add vignettes of varying complexity in order to create a scenario of a goal complexity.

In order to evaluate this system, the authors have prepared a "Turing Test" of sorts. Four scenarios were obtained that were built by human subject-matter experts. The system was then used to create four similar scenarios (i.e., built around the same training objective and at a similar complexity) resulting in four computer-generated scenarios. The authors then presented these four pairs of scenarios to independent subject-matter experts for review. The reviewers were not told the source of each scenario (whether human or computer). In this paper, the authors present the results from this review and provide analysis and conclusions from those results. A brief consideration of the system's shortcomings and potential future enhancements is also included.

Keywords: Scenario-based Training, Scenario Generation, Adaptive Training

INTRODUCTION

The creation of scenarios for training is a time-consuming and expensive process. Unfortunately, this often results in the same scenario library being constantly reused. While this is likely appropriate for new trainees, it does not provide effective training for continued use. Indeed, previous reviews have found five components that lead to effective training scenarios. These are embedded triggers, clearly-defined goals, variety, psychological fidelity and complexity (Martin et al., 2009). Therefore, the authors have pursued a line of research investigating scenario generation (both assisted and automatic). The goal is to facilitate the creation of qualitatively similar scenarios while still maintaining variety.

The authors have created an approach to automatic scenario generation that creates scenarios based around selected training objectives and a goal complexity value. Scenario building blocks are used to create a scenario that satisfies the training objectives and the complexity desired. While the system seems to work well, the purpose of this work is an analysis of the tool by subject matter experts. However, before that is discussed, the system itself is first reviewed.

AUTOMATIC SCENARIO GENERATION

Scenario generation describes the design and development of training events or situations for simulation-based instruction. Traditionally, the creation of a training scenario is performed by one or more subject-matter experts, who manually plan out the scenario (in terms of goals and then locations of entities, actions at certain triggers, etc.). These experts then program their plan into a simulation system. Automatic scenario generation occurs when a computer executes (or assists with the execution of) the scenario design and development process (Martin et al., 2009).

Effective Scenario-based Training

Using scenarios for training alone is not enough. What makes a scenario support effective training? As alluded to earlier, there are five components that lead to effective training scenarios: embedded triggers, clearly-defined goals, variety, psychological fidelity and complexity. Training scenarios should provide training opportunities to practice skills, demonstrate proficiency and receive feedback (Oser, 1999; Issenberg & Scalese, 2007). By designing scenarios that contain these "embedded triggers," the effectiveness of those scenarios is increased.

Scenarios should also contain clear goals, both for the instructor and the trainee (Issenberg & Scalese, 2007). If they are missing, the trainees may not respond in expected manners nor practice the desired knowledge and skills. Related to this need, performance measures should be included.

The need for scenario variety was alluded to earlier. What is meant by variability, however? Variety could be defined as the generation of non-trivially diverse scenarios, meaning that they are not redundant for training purposes (Grois et al., 1998). For example, two scenarios could be defined as not redundant if they fulfill all the requirements of the selected training objectives and differ by at least one significant event. Fundamentally, what is needed is the ability to create scenarios that are qualitatively the same, yet still appear different to the trainee.

Psychological fidelity refers to the "degree to which the trainee perceives the simulation to be a believable surrogate for the trained task" (Beaubien & Baker, 2004). In other words, the scenario has to be believable. This is particularly important as it has been found that scenarios must be believable in order to be effective training scenarios (Cannon-Bowers & Salas, 1998).

Finally, trainees should also be tested on scenarios with varying complexity in order to provide effective learning (Lum et al., 2008). The term "complexity" is used here in order to avoid the subjective term "difficulty." In fact, scenario complexity can be an objective measure of task complexity and structure. Task complexity refers to the number of discrete behaviors that form a task and their integration; task structure refers to the degree of ambiguity within a task (Lum et al., 2008).

Given these goals for the formation of scenarios, before a scenario can be generated by a computer, the question of how to represent one must be answered. This answer includes not only parameters of a scenario, but also a conceptual model of scenarios, an operational definition of scenario complexity, and a framework for linking scenarios to learning objects. Once the concepts are formalized into objective definitions, then the software can generate appropriate scenarios for specific trainees, based upon input such as their past performance and individual characteristics. Previously, a conceptual model using scenario building blocks was created to represent a scenario (Martin et al., 2010).

Conceptual Model

The conceptual model uses notions called baselines and vignettes to represent components of a scenario. Baselines define the environmental setting in which the scenario is taking place. A baseline includes the terrain (which might have been computer-generated, human-generated or based on actual terrain data), time-of-day and weather effects (wind, rain, snow). Since they provide a setting for the training objectives, baselines can support the simplest scenarios. However, a baseline could include a harsh terrain or one in poor illumination or weather; therefore, the training objective requirements will state the need for minimal assets for the scenario itself. Therefore, a scenario with a single baseline will be simplest in assets although the environmental effects may cause it not to be the absolute simplest in nature.

How does a baseline represent different environmental conditions? A notion called augmentations is used. Augmentations "attach" to baselines and add complexity to the scenario and can affect aspects of the baseline itself. Examples include moving the scenario to night or adding rain. Each of these potentially adds complexity to the generated scenario. The adding of augmentations to a baseline begins to add complexity to the training scenario. However, it is important to reiterate that augmentations to baselines simply change the initial environmental situation.

Although a single baseline can support training, these simple scenarios only offer minimally beneficial training experiences. A scenario with only a single baseline supports training the most novice trainees in mostly procedural operations. Baselines lack variability (other than entity location) and additional complexity. In order to expand baselines to better support advanced training, vignettes are used.

Vignettes add content based around learning objectives to the baseline in order to make a larger (and more complex) scenario. Vignettes are essentially pre-packaged alterations and additions to the scenario that are defined as sets of associated triggers and adaptations. Triggers are defined as any kind of check or comparison that returns a Boolean (true or false) value. Adaptations are alterations made to the current situation within the simulation. They can range from those that provide entity manipulations (create, kill, move, fire weapon) to those that provide environmental manipulations (reduce rain, raise sun). Their primary purpose is to adjust the training by causing some event to occur.

Computational Model

The conceptual model was then mapped to a computational model (Martin et al., 2010). Baselines and vignettes are stored as XML files and define scenario parameters (such as specific positions and types of entities) that must be satisfied for each scenario. Complexity is mapped to a concrete notion; currently, scenario components are mapped onto a value between 0 and 100. The value range and specific values selected are assigned in a subjective manner. Thus, there is presently no precise semantic meaning to a value, except that larger values represent greater perceived complexity.

Procedural Generation

With a computational representation in place, a procedural modeling technique known as Functional L-systems (FL-systems) (Marvie et al., 2005) was used to automatically generate scenarios (Martin et al., 2010). FL-systems are well suited for providing such a system and offer two major advantages. First, the extra power provided by having terminal functions (as opposed to terminal symbols) allows higher complexity decision making by the rule system. Second, the terminal functions allow the postponement of resolving requirements. This allows the basic rule system to be written with fewer rules. The parameters of the requirements can be satisfied within a terminal function.

The limitations of FL-systems include the additional work in authoring the rule systems as well as the requirement to write the terminal functions themselves. However, both limitations are only performed once per training domain (for example, only one set is needed for Fire Support Teams).

ANALYSIS OF SCENARIO GENERATION SYSTEM

In order to evaluate the scenario generation approach, a review by subject matter experts was arranged. The goal of the review was to obtain a comparison (by unbiased reviewers who were not familiar with the specific work but were experts within the training domain) between scenario created by expert humans and those created by the automated approach developed.

The Turing Test

Turing asked whether computers could perform well in the imitation game. The imitation game is a game where two hidden players, one male and one female, are interrogated through only written communication by a third player, who must determine which is the male and which is the female (Turing, 1950). In the game, player A intentionally tries to fool the interrogator while player B tries to help. Turing questioned whether a computer could be programmed to play the role of player A.

Over time this "Turing Test" has generalized into an interpretation where one of the two players is a computer and one is a human, and the interrogator must identify them accordingly. Note that the computer must only sufficiently imitate a human; not necessarily actually think. It is this thought experiment that is now commonly called the Turing Test.

The Scenario Turing Test

Given that the goal of an automated scenario generation system is to produce scenarios of sufficient quality, an evaluation was performed to verify its performance using a Turing Test approach. Given a training objective and complexity level desired, a human subject matter expert and the automatic scenario generation system each create a scenario. These two scenarios are then presented to separate, and independent, subject matter experts for review and analysis.

While the goal of this "Scenario Turing Test" is to have the scenarios from the automated system be indistinguishable from those of the human subject matter experts, additional analysis was also performed. The reviewers were asked to evaluate the quality of the scenarios (both human-created and system-created) and to also comment on the conceptual understanding of the automated system itself. In the actual test, four pairs of scenarios were used.

For each scenario pair the reviewers were asked whether one of the pair was easily identifiable as created by a human. If so, the reviewer was then asked to identify it and give a measure of level of confidence in that identification. The reviewer was then asked to give an overall grade for each of the scenarios in the pair.

The reviewers were also asked to identify the strongest and weakest points for each scenario. Any omissions were also to be listed and how each scenario might be improved. Finally, the reviewers were asked if any of the omissions might indicate a weakness in how it was generated.

Expert Review

In the Scenario Turing Test here, two training objectives were identified. Our application domain was that of Fire Support Teams in the U.S. Marines. Therefore, one objective was primarily based upon Indirect Fire (IDF) and one upon Close Air Support (CAS). Specifically, these objectives were:

1. Integrating, coordinating and de-conflicting close air support, indirect fires and maneuver to attack selected targets.
2. Using doctrinal control procedures successfully to coordinate and control attacks from CAS platforms on a visually marked target.

Both "novice" and "expert" complexity level scenarios were produced for each training objective, resulting in four pairs of scenarios being evaluated. For the human-produced scenarios, scenarios added to the new Instructional Support System (ISS) of the U.S. Marines Deployable Virtual Training Environment (DVTE) were used. These scenarios have limited distribution to date and the SMEs had not yet seen them. The scenario generation system then was used to produce corresponding scenarios and these were re-plotted in the same system as the human scenarios to avoid obvious visual differences. The four scenario pairs were randomized and presented to five, independent, subject matter experts as a multi-page questionnaire. Of the five, four returned the document.

The reviewers were asked to give responses to three questions for each pair of scenarios. These questions were:

1. As we informed you at the start, one of each pair of scenarios was generated by a human, and one by our system. In each case, was one of them easily identified as the human-generated scenario? If so, which one? What is your level of confidence in your identification (e. g. 90% sure; 99% sure; where 50% would mean you feel that you're totally guessing).
2. What is your overall assessment of the relative quality of the two scenarios in each pair? E. g. "I would give A1 an A-, and A2 a D+". If you prefer a numeric scale, then use A=4.0, B=3.0, C=2.0, D=1.0 and F=0.
3. For each scenario, what are its strongest and weakest points? Were there any obvious omissions that seemed to you to clearly indicate

some weakness in the process by which it was generated? What were those omissions and how could the scenario be improved?

In addition, the reviewers were asked two questions about the system concept and how well each thought it might satisfy an approach to scenario generation. These questions were:

1. Please discuss your thoughts on the scenario generation model of baselines and vignettes. In what ways does this model correspond with, or differ from, the practice of professionals who generate scenarios? In what ways could this model be improved?

2. Please discuss your thoughts on the vignette representation of sets of triggers and adaptations. Does this representational system accurately and completely capture the corresponding features that are needed for top quality scenario generation? In what ways could our representational system be improved?

Analysis

The four subject matter experts performed a thorough review. Each reviewer was asked to, literally, grade each scenario. The human-generated scenarios resulted in a "grade point average" (GPA) of 2.625 (between a C+ and a B-). This is somewhat surprising as they were based on existing scenarios. However, the result may be more based on the scenario write-ups than the scenario themselves (discussed below). The computer-generated scenarios resulted in a GPA of 2.375 (also between a C+ and B-). While lower than the human-generated scenarios, it is within a half grade. More so, given the reasons offered by the reviewers, this difference can likely be reduced.

In evaluating the automated system, the reviewers most noted that its scenarios often lacked a ground maneuver element as a part of the scenario, particularly within the Indirect Fire scenarios. While the scenarios generated by the automated system do lack this element, it is largely due to the authors' lack of expertise and not the system itself. In fact, it is a straight-forward matter to add creation of this entity to the rules.

Regarding the Close Air Support scenarios, both the human-generated and computer-generated scenarios received low marks (GPA of 1.875 for the human-generation scenarios; 2.0 for the computer-generated). In both cases, this had little to do with the placement of the elements; both sets had a lack of information on sensors and ordnance. The reviewers indicated that both of these elements are critical to proper scenario execution of Close Air Support since both drive what actions the trainee is allowed to perform. For example, an aircraft with laser-guided bombs makes the use of indirect fire unnecessary.

Feedback from the reviewers on the automated system approach focused more on suggestions for how the process could be performed rather than on the system itself. This is likely due to their expertise in training and their vision for the goal of offering the best training possible. Comments fell into two categories. First, the reviewers felt that a more precise identification of the specific training objective (Training & Readiness event in the U.S. Marines) was suggested, including the training goals. For example, is the goal to conduct sequential or simultaneous actions? Once this is understood, the elements of the scenario will flow more naturally to the trainees. The automated system uses this technique conceptually but it may need to be made more explicit.

The second type of feedback on the approach focused on the ordnance and sensor capabilities of aircraft (already alluded to earlier). Given this domain (Fire Support Teams), the ordnance and sensor capability can drive the scenario much more than any sort of geographic concerns (although geography is still a secondary concern). It is interesting that both the human and computer scenarios did not make these elements explicit and may have overlooked it as it does not affect entity placement. However, it is clear from the reviewers that it must be explicit. This leads to a future research item: not only generating the scenario, but also the write-up of the scenario, complete with training objectives and training goals.

Interestingly, the active-duty military reviewer that was lowest in rank thought more highly, qualitatively, of the computer-generated scenarios and felt they provided a good variety of options (as opposed to only allowing a single course of action). The instructors were less satisfied; the retired Lieutenant Colonel was most critical and focused on command-and-control concerns. The lower rank reviewers were also active certified Joint Terminal Attack Controllers (JTACs) as compared to those focused on instruction.

A final analysis of the scenarios concerns the Scenario Turing Test. Recall that each reviewer was questioned as to which scenario within the pair was human-generated. The reviewers correctly chose the human-generated scenario approximately 75% of the time. However, the reason given all but a few times was on the lack of a maneuver element. As discussed, this is correctable with some additional rules in the automated system.

SUMMARY

The Scenario Turing Test provides a compelling review of the scenarios generated by the automated system. While the reviewers were not fooled in many cases about which scenarios were human-generated, the causes are easily addressable. More exciting, the grades assigned to each set of scenarios are competitive, illustrating that an automated approach can indeed generate relevant scenarios. In addition, the reviewers provided very important feedback (some of which also applies to the human-generated scenarios) that will help drive future scenario generation effort.

ACKNOWLEDGEMENTS

This work is supported in part by the Office of Naval Research Grant N00014-08-C-0186, the Next-generation Expeditionary Warfare Intelligent Training (NEW-IT) program, and in part by National Science Foundation Grant IIS1064408. The views and conclusions contained in this document are those of the authors and should not be interpreted as representing the official policies, either expressed or implied, of the Office of Naval Research, the National Science Foundation or the US Government. The US Government is authorized to reproduce and distribute reprints for Government purposes notwithstanding any copyright notation hereon.

REFERENCES

Beaubien, J. M. & Baker, D. P. (2004). The use of simulation for training teamwork skills in health care: how low can you go? *Quality and Safety in Health Care*, 13, i55-i56.

Cannon-Bowers, J. A. & Salas, E. (1998). Team performance and training in complex environments: recent findings from applied research. *Current Dictions in Psychological Science*, 7 (3), 83-87.

Grois, E., Hsu, W. H., Voloshin, M. & Wilkins, D. C. (1998). Bayesian network models for generation of crisis management training scenarios. *Proceedings of Innovative Applications of Artificial Intelligence Conference*, Madison, WI, Association for the Advancement of Artificial Intelligence.

Issenberg, S. B. & Scalese, R. J. (2007). Best evidence on high-fidelity simulation: what clinical teachers need to know. *The Clinical Teacher*, 4 (2), 73-77.

Lum, H. C., Fiore, S. M., Rosen, M. A. & Salas, E. (2008). Complexity in collaboration: developing an understanding of macrocognition in teams through examination of task complexity. *Proceedings of 52nd Annual Meeting of the Human Factors and Ergonomics Society*, New York, NY: Human Factors and Ergonomics Society.

Martin, G., Schatz, S., Bowers, C.A., Hughes, C. E., Fowlkes, J. & Nicholson, D. (2009). Automatic scenario generation through procedural modeling for scenario-based training. *Proceedings of the 53rd Annual Conference of the Human Factors and Ergonomics Society*. Santa Monica, CA: Human Factors and Ergonomics Society.

Martin, G. A., Schatz, S., Hughes, C. E. & Nicholson, D. (2010). What is a scenario? Operationalizing training scenarios for automatic generation. *Proceedings of the Applied Human Factors and Ergonomics Conference*, Miami, FL.

Martin, G. A. & Hughes, C. E. (2010). A scenario generation framework for automating instructional support in scenario-based training. *Military Modeling and Simulation*, Orlando, FL.

Martin, G. A., Hughes, C. E., Schatz, S. & Nicholson, D. (2010). The use of functional l-systems for scenario generation in serious games. *Proceedings of the Foundations of Digital Games (Workshop on Procedural Content Generation)*, Monterrey, CA.

Marvie, J., Perret, J. & Bouatouch, K. (2005). The FL-system: a functional L-system for procedural geometric modeling. *Visual Computer*, 21 (5), 329-339.

Oser, R. L., Gualtieri, J. W., Cannon-Bowers, J. A. & Salas, E. Training team problem solving skills: an event-based approach. *Computers in Human Behavior*, 15 (3-4), 441-462.

Turing, A. M. (1950). Computing machinery and intelligence. *Mind*, 59, 433-460.

CHAPTER 55

Motion Synthesizer Platform for Moving Manikins

Ali Keyvani[1,2], Henrik Johansson[3] and Mikael Ericsson[3]

[1] Innovatum AB, Trollhättan SE-461 29, Sweden
[2] Chalmers University of Technology, Gothenburg SE-412 96, Sweden
[3] University West, Trollhättan SE-461 86, Sweden
Ali.Keyvani@chalmers.se

ABSTRACT

Digital Human Models are getting more applicable in simulation of the human postures/actions during different product development stages. Despite the existence of various modeling techniques, simulated motions are still looking unrealistic. It was identified that one promising way of generating more realistic motions is to use real human motion data as a source. A synthesizer platform is presented in this study which is able to store motions taken from real human in a database and synthesize new motions using already existing motions. The platform is also able to search the database, analyze the data, and feed the data to DHM tools. The platform functionality was tested within two areas: a) by composing new motions by combining the arm motions and walking motions from different subjects and b) by using analysis tools to compare generated motion in a commercial software against captured motions from real human. The obtained results show promising applications of the platform as it can be used to generate new motions and/or analyzing the motions.

Keywords: Digital Human Models, Human Motion Analysis, Motion Synthesizing

1 INTRODUCTION

Competition in global markets and advances in manufacturing technologies have brought new challenges to the realization of new products. Success in product development projects is dependent on competitive prices, high quality and short time-to-market. To reach these goals, it is common to use simulation platforms to predict unknown challenges as early as possible (Rooks, 1998). In order for these platforms to be comprehensive, they need to account for human actions and the presence of humans in the product development process. In such platforms, a human should be addressed both as a customer/user (human-centered design) and as a production/assembly worker (production ergonomics). Within these two contexts, product simulation and digital manufacturing technologies are striving for the integration of human simulation in virtual manufacturing tools (Van der Meulen and Pruett, 2001, Chaffin, 2005).

Human postures/motions are usually simulated with graphical computer manikins called Digital Human Models (DHM). Several commercial DHM products are available on the market to integrate these digital humans into CAD-based product and production environments, such as Santos (Abdel-Malek et al., 2007), Jack (Badler et al., 1993), and RAMSIS (Seidl, 1997). Postures, motions and interactions of the digital manikin with product and/or production lines can then be studied and verified.

Assuming the human as a customer who uses the products, the simulation process involves ergonomics and comfort issues (Woldstad, 2000, Reed et al., 2006). Different user populations with varying anthropometries must be studied and parameters such as reach envelope and line-of-sight need to be checked (Reed et al., 2003). Besides, considering the human as an assembly worker deals with frequent interaction between the worker and working environment, including machines, components, and subassemblies. Several motions are involved in these interactions, and each interaction should be verified carefully. To validate the production processes, different perspectives such as ergonomics, reachability, line-of-sight, and task timing have to be studied (Rider et al., 2004, Zhou et al., 2009, Chang and Wang, 2007). Hence, it is important to have motion modeling techniques that can generate realistic human postures/motions in an accurate, fast and reliable way (Lämkull et al., 2009a).

In spite of the availability of various techniques for the modeling of human motion (Zhang et al., 1998, Park et al., 2008, Wang, 1999, Mavrikios et al., 2007), the functionality of DHM tools is still limited in practice. Generating new postures is mostly done by manually adjusting body joints one by one, which is time consuming, unreliable and inconsistent (Lämkull et al., 2009b). Moreover, even if a simulation is based on realistic postures, the transforming motion between two postures often looks unrealistic when generated by the software. One promising way of improving the functionality of DHM tools is to use real human motion data captured by different motion-capturing techniques (Faraway, 2003).

A Number of methods, such as key-framing, motion warping (Witkin and Popovic, 1995), and motion retargeting (Gleicher, 1998), use real motion data to generate new motions. In addition, it is possible to synthesize a motion scenario using pieces of motions from different sources (Grassia, 2000) and applying them to different parts of the digital model's body if an extensive motion database exists. In many cases, these pieces of real motions can be found in large captured files available on the market (e.g. the film industry).The use of real data is thus promising, but the application of these methods also presents developers with a number of problems:

- The need for an extensive motion database
- Inefficient indexing and poor annotation of motion pieces
- Incompatibility of motion captured file formats
- Lack of detailed analysis on motion characteristics, and
- Variation of skeleton configurations in the captured motions

The platform presented here addresses these problems by providing an integrated data storage schema, data manipulation tools, motion analyzing functionalities, and synthesizing tools. Motions from different sources and with different skeleton configurations can be stored in a unified form and be indexed both by joint values and joint positions. Moreover the motion analyzer toolbox - implemented in Matlab- allows the study of motion characteristics and joint relations. This enables the user to efficiently coordinate and synthesize pieces of motions together without losing the realist of motions.

The platform functionality was tested in two ways: a) by means of new motions of carry-walk developed on the basis of hand motions from external sources and walking motions generated in laboratory and b) by means of analysis tools to compare software generated motions in a commercial product (Jack™) against real motions and to be able to suggest necessary changes in velocity profiles of the motion generator for a more natural look.

2   MOTION SYNTHESIZER

The Motion Synthesize Platform that is discussed in this paper consists of three basic components, called layers, see Figure 1.

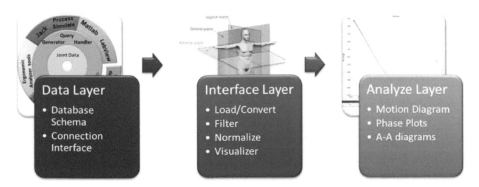

Figure 1. Structural layers of the synthesizer platform

548

2.1 Data layer

The first layer is the foundation of the synthesizer platform and consists of a database schema. The schema can handle the motions data and information about the motions. The database also provides ways to search, retrieve and store motion captured data. This information is then used by the platform to synthesize new motions. Some of the information that the database system can handle are:

- Joint data, e.g. positions and values
- Different configurations of body skeletons
- Customized body zones
- Spatial and temporal annotations

2.2 The Interface Layer

The second layer consists of an interface that can communicate with other programs and software. The interfaces use SQL queries and common database techniques to interact with the database.

The interface layer can load, convert and visualize motion capture data. The load and convert functions in the interface layer makes it possible to import and convert motion captured data from a number of different motion capture systems and file formats. It can thereby synthesize motions from a number of captured motions and from a number of sources. The layer is currently implemented in Matlab and it functions as a toolbox for handling motion captured data. The toolbox contains a number of different functions, such as:

- Load/Convert Motion Data
- Import/Export
- Normalizing Data
- Noise Reduction
- Cut – Add – Stretch – Squeeze of motions
- Split – Insert of body parts
- Visualize

In this paper, the cut, add, stretch and split operations have been used to generate a synthesized motion. The **Cut_Motion** and **Add_Motion** functions are used to extract and paste frames in a motion sequence. The functions **Split_Body** and **Insert_Body** are used to combine actions from different body parts, such as arms, lower body or head.

2.3 Analysis Layer

The third layer is the layer where motion captured data is analyzed. It can be used to perform both single joints analysis and/or relationships between coupled joints. Some of the provided tools in this layer are:

- Joint/Location-Time Diagrams
- Joint-Joint Diagrams
- Joint-Location Diagrams

- Angle-Velocity diagrams (Phase Plots)
- Velocity-Velocity Diagrams

The tests that are conducted in this paper focus on Phase diagrams for single joints. These phase diagrams show the angle-velocity relationship for each joint during a motion.

3 RESULTS

To validate the functionality of the platform, two test scenarios were designed: a) the first test (Motion Synthesizing) focused on the Data and Interface layers and b) the second test (Motion Analysis) examined the analysis layer of the platform.

3.1 Motion Synthesizing

The purpose of this test was to generate new motions by combining parts of existing motions in the database. These parts, called motion pieces, are divided in both temporal, e.g. frame 100 to 200, and spatial, e.g. upper-body sections.

The test scenario is designed as follows: *Find a motion representing the action "Walking Forward" (Motion-1). Then find a motion containing action "Move Box with Both Hands" (Motion-2), with the subject's arm-length as close as to the subject in walking motion. The result of the synthesized motion is a subject carrying a box and walking forward.*

The motion database was fed with a mixture of motions from our laboratory, HUMOSIM motion data (Reed et al., 2006), and other available motions on the web. The database contained in total over 20,000 motions. The data contained motions with different body configurations, different formats/lengths, and different subjects. Information about actions which is performed during each motion was filled in manually and stored in the annotation table of the database. This means that the actions are recognized by human observers and assigned to frame ranges and parts of the subjects' body. Table 1-3 show the specification of chosen candidates for Motion-1 and Motion-2. The search/retrieve of the candidate motions was performed using SQL language and Matlab Database Toolbox.

Table 1. Specifications of candidate motions

	Motion-1	Motion-2
Motion Source	Motion Capture	Motion Capture
Number of Frames	3000	74
Frame Rate	60 Hz	25 Hz
Number of Segments	27	19
Data Format	HTR	LOC
Upper Arm Length	281 mm	279 mm

Table 2. Annotation Table for Motion 1

Frame Range	Tag	Body Zone
1 - 3000	Male	Full Body
1 - 575	Stand up	Lower Body
100 - 490	Arm Swing	Right Arm
580 - 1400	Jump	Full Body
1515 - 1732	Walk Forward	Right Leg + Left Leg + Trunk
1740 - 2314	Sit	Right Leg + Left Leg + Trunk
2320 - 3000	Stand up	Lower Body

Table 3. Annotation Table for Motion 2

Frame Range	Tag	Body Zone
1 – 74	Male	Full Body
1 – 74	Move Object	Right Arm + Left Arm
1 – 3	Grasp	Right Hand + Left Hand
72 – 74	Release	Right Hand + Left Hand

Inspecting the annotation table of source file for Motion-1, it can be seen that frame range 1515-1732 had been annotated as 'Walking Forward'. Therefore, *Cut_Motion* function was used to extract the detected frame range and assign it to Motion-1 (Figure 2-a). Next, Motion-2, in which the subject was moving a box to a lower position, consists of only 74 frames (Figure 2-b & c) . Therefore it is stretched to the same length as Motion-1 using the *Stretch_Motion* function. Next, the arms and head of the subject in Motion-2 (Figure 2-d) and the lower body and trunk of the subject in Motion-1 (Figure 2-e) are split using *Split_Body* function. Finally, two parts are combined and a new motion (Figure 2-f) by using *Insert_Body* function is generated. The resulted motion is a person who is carrying a box and lowers it while he/she is walking. The people who observed the synthesized motion, judged it as a natural looking motion.

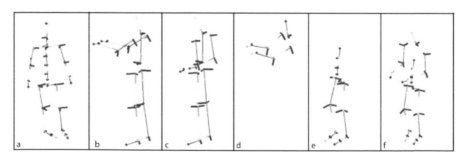

Figure 2. Snapshots from the synthesizing scenario

3.2 Motion Analysis

The same motion (walking normally straight ahead) is generated using two different sources: i) Jack™ software and ii) real human data captured by motion capture technology. Shoulder joint angle in the sagittal plane was selected as the focus of attention during the test. Figure 3 shows screenshots from the test (top) and the plots resulted from each source (bottom). The plots indicate how the shoulder angular velocity changes in relation to shoulder position (shoulder angle). The walking motion in Jack has an unrealistic look. For instance, just in the shoulder joint two differences could be determined: i) The shoulder movement in Jack has symmetric changes around the shoulder resting position (zero degree) and is about ±12 degrees, while this range for a real human is about -13 to +26 degrees, which means that it is asymmetric; ii) The plot corresponding to a real human (Figure 3. bottom-right) shows smooth motions when the arm passes the side of the body and moves to the back. In contrast, there is an abrupt change in the joint velocity in the Jack when the arm passes zero-angle (Figure 3. bottom-left). This will give a motion that looks robot-like. The differences prove that the motion generator algorithms used in the software must be modified in order to produce more realistic motions.

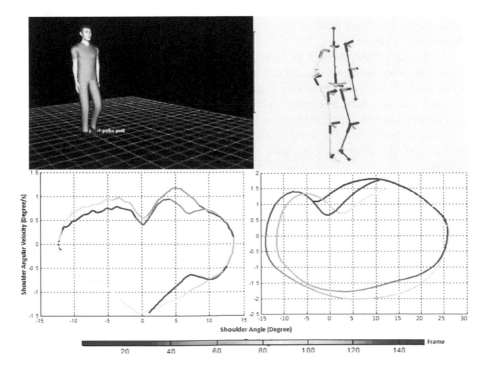

Figure 3. Comparison between shoulder movement in normal walking generated in Jack™ (left) and captured from real human (right). The diagrams plot shoulder angular velocity in relation to shoulder angle. The color bar indicates the time (frame number) progress in the plots.

552

4 CONCLUSION AND DISCUSSION

It is difficult to use existing motion-captured data in DHM tools when these tools are generating new motions. As a result, today's Digital Manikins motions still look unrealistic when you compare it with real human movements. In this study, a new platform has been designed, implemented and tested in order to synthesize new motions from existing human data. The platform consists of a data layer to store motion data and meta-data about the motions, an application layer to establish the connection between data layer and other tools, and an analysis layer to provide tools for evaluating, comparing and synthesizing motions.

To test the functionality of the platform we performed two tests. The first test showed that in certain cases, it is possible to extract some of the joint movements independent of the whole body, such as arm movement. In cases like these, we have shown that motions from different sources can be combined to produce new motions. Several functions such as *Cut, Split, Combine* and *Stretch* were used in this test, see section 3.1. However, the method has some limitations. In many actions, such as running, there are certain dependencies between joints (for example between the shoulder and the hip) which prevent us from applying this method to one part of the body without taking other parts in account. We have addressed this issue by means of a two-dimensional annotation technique. When a new action is annotated in the database, the related body zone is also highlighted. The annotation table can then indicate whether a certain merge can be performed with the current dependency situation or not.

In the second test, the arm movement is compared in a walking scenario which is performed by both a real human and a digital manikin. We have found that DHM tools are still far from generating realistic motions even in simple and straight forward scenarios such as normal walking. The test results show potentials where real human data can improve the realist of the DHM tools.

How people judge if a motion looks natural is not studied in this paper. The verification of the motions is based on visual observations of the synthesized motions.

At the moment, the platform is unable to automatically recognize what actions are performed by each motion piece. Currently, these data are added manually to the annotation table in the database. We suggest continuing this research by implementing functions which can automatically annotate motion pieces.

ACKNOWLEDGEMENTS

The authors gratefully acknowledge the Center for Ergonomics at The University of Michigan for providing the HUMOSIM movement database that was used as one source of human data in this study. This project was funded by CAPE research school at University West, Sweden.

REFERENCES

ABDEL-MALEK, K., YANG, J., KIM, J. H., MARLER, T., BECK, S., SWAN, C., FREY-LAW, L., MATHAI, A., MURPHY, C., RAHMATALLAH, S. & ARORA, J. 2007. Development of the virtual-human Santos™.

BADLER, N. I., PHILLIPS, C. B. & WEBBER, B. L. 1993. *Simulating humans : computer graphics animation and control,* New York, Oxford University Press.

CHAFFIN, D. B. 2005. Improving digital human modelling for proactive ergonomics in design. *Ergonomics,* 48, 478-491.

CHANG, S. W. & WANG, M. J. J. 2007. Digital human modeling and workplace evaluation: Using an automobile assembly task as an example. *Human Factors and Ergonomics In Manufacturing,* 17, 445-455.

FARAWAY, J. J. 2003. *Data-based motion prediction,* New York, NY, ETATS-UNIS, Society of Automotive Engineers.

GLEICHER, M. 1998. Retargetting motion to new characters. *Proceedings of the 25th annual conference on Computer graphics and interactive techniques.* ACM.

GRASSIA, F. S. 2000. *Believable automatically synthesized motion by knowledge-enhanced motion transformation.* Carnegie Mellon University.

LÄMKULL, D., HANSON, L. & ÖRTENGREN, R. 2009a. A comparative study of digital human modelling simulation results and their outcomes in reality: A case study within manual assembly of automobiles. *International Journal of Industrial Ergonomics,* 39, 428-441.

LÄMKULL, D., HANSON, L. & ÖRTENGREN, R. 2009b. Uniformity in manikin posturing: a comparison between posture prediction and manual joint manipulation. *International Journal of Human Factors Modelling and Simulation,* 1, 225-243.

MAVRIKIOS, D., KARABATSOU, V., PAPPAS, M. & CHRYSSOLOURIS, G. 2007. An efficient approach to human motion modeling for the verification of human-centric product design and manufacturing in virtual environments. *Robotics and Computer-Integrated Manufacturing,* 23, 533-543.

PARK, W., CHAFFIN, D. B., MARTIN, B. J. & YOON, J. 2008. Memory-based human motion simulation for computer-aided ergonomic design. *Ieee Transactions on Systems Man and Cybernetics Part a-Systems and Humans,* 38, 513-527.

REED, M. P., FARAWAY, J., CHAFFIN, D. B. & MARTIN, B. J. 2006. The HUMOSIM Ergonomics Framework: A New Approach to Digital Human Simulation for Ergonomic Analysis.

REED, M. P., PARKINSON, M. B. & KLINKENBERGER, A. L. 2003. *Assessing the validity of kinematically generated reach envelopes for simulations of vehicle operators,* New York, NY, ETATS-UNIS, Society of Automotive Engineers.

RIDER, K. A., CHAFFIN, D. B., FOULKE, J. A. & NEBEL, K. J. 2004. *Analysis and redesign of battery handling using Jack and HUMOSIM motions,* New York, NY, ETATS-UNIS, Society of Automotive Engineers.

ROOKS, B. 1998. A shorter product development time with digital mock-up. *Assembly Automation,* 18, 34-+.

SEIDL, A. RAMSIS - a new CAD-tool for ergonomic analysis of vehicles developed for the German automotive industry. 1997. 51-57.

VAN DER MEULEN, P. A. & PRUETT, C. J. Digital human models: What is available and which one to choose. 2001. 875-879.

WANG, X. 1999. A behavior-based inverse kinematics algorithm to predict arm prehension postures for computer-aided ergonomic evaluation. *Journal of Biomechanics,* 32, 453-460.

WITKIN, A. & POPOVIC, Z. 1995. Motion warping. *Proceedings of the 22nd annual conference on Computer graphics and interactive techniques.* ACM.

WOLDSTAD, J. C. 2000. Digital Human Models for Ergonomics. *International Encyclopedia of Ergonomics and Human Factors, ed. Waldemar Karwowski,* Second edition, 3093–3096.

ZHANG, X. D., KUO, A. D. & CHAFFIN, D. B. 1998. Optimization-based differential kinematic modeling exhibits a velocity-control strategy for dynamic posture determination in seated reaching movements. *Journal of Biomechanics,* 31, 1035-1042.

ZHOU, W., ARMSTRONG, T. J., REED, M. P., HOFFMAN, S. G. & WEGNER, D. M. 2009. Simulating Complex Automotive Assembly Tasks using the HUMOSIM Framework.

CHAPTER 56

Human Engineering Modeling and Performance: Capturing Humans and Opportunities

Katrine Stelges, Brad A. Lawrence

Copyright © 2012 by United Space Alliance, LLC
Kennedy Space Center, FL; USA
katrine.s.stelges@usa-spaceops.com; brad.a.lawrence@nasa.gov

ABSTRACT

There have been many advancements and accomplishments over the last few years using human modeling for human factors engineering analysis for design of spacecraft. The key methods used for this are motion capture and computer-generated human models. The focus of this paper is to explain the human modeling currently used at Kennedy Space Center (KSC), and to explain the future plans for human modeling for future spacecraft designs and ground processing tasks.

United Space Alliance's Human Engineering Modeling and Performance program, or HEMAP, uses state-of-the-art motion capture tools to examine the interfaces between people, equipment, and the work environment. In addition to real-time motions and interactions, HEMAP's ability to stream multiple-person tasks and objects simultaneously provides revolutionary real-time feedback of the ergonomic effects on the participants to allow for real-time changes or optimization of the process or task at hand. Use of head-mounted displays, physical mockups, data-tracking gloves, and/or bi-directional collaboration of interactive motion data offer variations to resource and time requirements in meeting customer and engineering needs.

The real-time ergonomic evaluations with ability to infinitely replay and alter the viewing angles of simulations adds increased fidelity to assure the workspace, including human interactions, are validated or adjusted to optimize human-system interfaces to prevent risks to personnel, equipment, designs, and schedules in the long term. This innovative, forward-reaching technology can be a powerful, cost-saving, time-saving tool to customers who want to optimize design prior to build, and to develop processes and procedures with optimal efficiency and safety in mind.

Keywords: human engineering, modeling, simulation, ergonomics, human digital modeling, human factors, virtual reality, collaboration, head-mounted displays

1 HUMAN MODELING AND IMMERSION

Human Factors Evolution to Applied Human Digital Modeling

Using state-of-the-art motion capture tools to examine interfaces between people, equipment, and the work environment can improve efficiency as well as safety in processes, and in design of equipment or facilities. Traditional human factors assessments involve manual measurements, user interfaces, and basic imagery or video recording to assess risks and identify opportunities for improvement. Certain attributes collected can be used to compare against industry standards, such as the National Institute for Occupational Safety and Health (NIOSH) lifting equation, Rapid Upper Limb Assessment (RULA), or other academia or industry standards. Recordings are primarily taken from a stationary angle and playbacks generally consist of showing the same scene repetitively to look for risks. Standard methods of reviewing the collected data do not allow for additional calculations, manipulation of angle viewpoints, or instantaneous ergonomic or human factors readings. Conventionally, Human Factors Engineers manually compiled data in various formats to try to explain safety issues or justification for changes. Contrarily, today's advances in motion capture, combined with ergonomic software applications, can provide real-time risk indicators while also affording the ability to infinitely playback the task scenario from nearly any angle. These innovations in human factors and ergonomics assessments allow Analysts to apply dimensional calculations, collision detection, visibility, and a variety of ergonomic assessments in addition to offering a means for sharing the simulation and ergonomic data with customers or other change-management entities.

Human Modeling and Immersion at Kennedy Space Center

The Human Engineering Modeling and Performance (HEMAP) Lab at Kennedy Space Center (KSC) originated due to the complex and challenging workspace design issues the Space Shuttle posed on technicians performing maintenance, modifications, and repair operations. The HEMAP Lab, initiated in 2006 by United Space Alliance (USA) Industrial and Human Engineering as well as Computer

Sciences experts, includes a motion capture system that captures live biomechanical motions of humans and performs various real-time, detailed ergonomic analyses of potentially high risk or challenging operations. With the previous knowledge, skills and capabilities developed during the Space Shuttle and Constellation programs, the HEMAP Lab has evolved into a one-of-a-kind, state-of-the-art capability at KSC. The system has proved very beneficial in the design of flight and ground hardware plus the related human tasks, which ensure safe, efficient, and effective ground, flight, and non-earth terrestrial habitation and processing of future space systems.

Figure 1: HEMAP Lab: Live Motion Capture/Tracking of Multiple Persons/Objects

Actual Humans and Human Models

HEMAP's motion capture process begins by taking over 2 dozen anthropometric measurements of the human participants in order to develop their corresponding computer model, or avatar. Next, over 50 data markers are attached to their bodies, atop a tight-fitting motion capture suit. As the participants move in a simulated work environment, over 3 dozen cameras track the markers and feed the data instantaneously to HEMAP's motion capture system. The system provides an interactive 3D view of the simulation, including the motions and tasks as they are being performed by one or multiple participants and objects simultaneously. Rather than building the digital humans according to posture libraries within software applications, HEMAP's collects the actual motions of the participants to provide genuine feasibility and risks regarding human motion data.

Compared to tedious key-framing virtual human models, using actual humans to create the simulations allows Human Factors Engineers and Ergonomic Analysts the benefits of identifying and better understanding human motions, interactions, limitations, restrictions, and opportunities for improvement. Merely building or programming individual human motions can be intensely time consuming and may not provide a comprehensive understanding since the animation or simulation results may include the developer's viewpoint on how the task should be performed. This may not take into consideration all restrictions, contact stresses, comfort levels, support mechanisms such as access platforms or tool handling that could be encountered, or natural tendencies of humans including multiple person interactions. When HEMAP asked participants to repeat the same posture several times for validation testing, it was quickly evident that humans, or a group of

558

humans, cannot typically repeat a task the same way more than once. To test repeatability, HEMAP Analysts had to intervene with gosiometers to get desired, repeated back and arm angles of test subjects.

In cases where newly designed hardware has not arrived, being able to use actual Technicians provided HEMAP exceptional insight into tasks that would not have been possible to diagnose with traditional human factors tool suites or with human modeling only.

Props, Objects, Mockups Add Situational Awareness

HEMAP has used partial or true-to-size, wireframe, motion capture friendly mockups to further augment the participants' situational awareness toward the task being assessed. For example, placing the actual human(s) atop a curved capsule floor or between floor structure which will house avionics boxes allows a quick means of integrating the actual and virtual environments. Creating wireframe objects for participants to lift, carry, and transfer among themselves, also add an increased situational awareness aspect to the task assessment. HEMAP has also learned that weighting the carried wireframe objects to match their intended weight is needed to clearly reflect accurate back angles, forces, hand positioning, and footing needed since humans clearly lift and handle lighter weight objects distinctively different than heavier objects. When time does not afford buildup of physical mockups or objects, immersing participants solely in a virtual environment offers improved human factors capability as well.

Figure 2: Motion Capture Grid, Mockup Objects, 3-Person Task Evaluation

Live Motion Capture with Dynamic Ergonomic Monitoring

Motion data of participants being fed into HEMAP's motion capture system is also streamed into ergonomic software, rendering a real-time, interactive 3D virtual view of a customer-specific workplace environment with quantitative human factors results so Analysts can identify risk reductions and efficiencies on the spot. Dynamic ergonomic monitoring and evaluations can include lower back compression forces, RULA, fatigue, static strength, visibility (first person views), and collision detection, to name a few. Depending on the ergonomic software implemented, Analysts may be provided windows showing ergonomic data as it changes, and may also be provided color-coded charts to easily see increasing risks. Ultimately, being able to monitor human factors and ergonomics of multiple participants at a time offers added realism and value.

Changes in the task or workspace environment during motion capture sessions can be made to minimize risks or optimize the process, resulting in dozens of operational scenarios that can be recorded and later played back for further assessment. For example, if back forces are showing increased risks, changes to the task may add another participant to assist, implement a platform to lie upon, break down task steps, or any other method to reduce risks using the models, physical objects, and participants available.

Applications of Real-Time Motion Capture/Ergonomic Monitoring

HEMAP's live motion capture capability and real-time ergonomic evaluation has proven to be an effective tool, with benefits beyond Human Factors' applications. USA has demonstrated its applied human digital modeling capabilities by reviewing training simulations for effective and negative interactions between personnel and their proposed workspaces, assessing ergonomic risk through use of software analysis tools in a safe, controlled environment without exposure to unknown risks and hazards, and evaluating spacecraft flight systems prior to, and as part of, design processes to contain costs and reduce schedule delays. Participating actors, mainly spacecraft Technicians, adapt very easily to the training simulation, also resulting in efficient scheduling and reduced training costs.

Broad examples of HEMAP applications include: Diagnosis of why a particular task has a high injury or fatigue rate; development of safe procedures for handling delicate hardware/equipment, tight spaces, or awkward postures; and development of optimal tools and handling equipment during facility or vehicle design.

To date, HEMAP has evaluated various project-specific applications:

Ground Processing Tasks

Risks, handling large plywood above rocks for Shuttle transporters, were minimized by adding fork lift device to raise plywood, reducing Technician maximum bend distances.

Figure 3: Ground Processing Task Application

Space Shuttle Flight Hardware Task Evaluations

Evaluations for installing/removing wing leading edge panels resulted in a larger, height adjustable work stand to support two personnel task. Proposed, lengthy reach for overhead crew module window protective cover was eliminated due to excessive risks. Stepping over large, raised, upper Shuttle surface aerodynamic doors was eased with HEMAP-validated handhold locations. Awkward postures within the confined aft department were modeled to identify particular task risks and improvements.

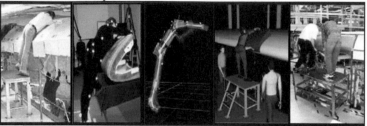

Figure 4: Space Shuttle Flight Hardware Application: Wing Leading Edge Handling

Proposed Designs - Orion Capsule Crew Module Task Evaluations

Installing or removing crew seats, environmental control units, large panels, and avionics boxes within "floor" cavities were evaluated prior to actual hardware being present. Risks were identified and evaluated with HEMAP and recommendations were made to designers, engineers, planners, and operational entities.

During the Orion crew seat HEMAP evaluation, in particular, lower back compression forces were determined to be lower for a 2-person versus a 3-person team, due to limited space in the crew module requiring extended reaches and awkward postures. Based on the simulated training exercises, the HEMAP Team recommended that the crew seat be lifted in halves by a 2-person team and that crew seats integrate fixtures or handholds for improved handling. For avionics boxes, the HEMAP Team learned preferred postures, based on box locations. For nearby objects, Technicians benefited from standing on the curved bulkhead and leaning over to lift/lower boxes. For objects further away from their centers of gravity, Technicians benefited from installed platforms so they could lie prone and easily reach below to lift/lower the components.

Figure 5: Animation Results, Interactive 3D Engineering with Dynamic Ergonomic Data

Human Modeling/Motion Capture Innovations

Motion Capture Variations/Peripherals

Head-Mounted Displays with Full-Body or Simplified Head/Glove

Incorporation of Head-Mounted Displays (HMDs) immerses participants into the virtual environment, sometimes negating the need, time, and costs associated with physical mockups. The participants can solely wear the HMDs to familiarize themselves with the workplace environment. Or, they can also don finger/hand tracking gloves or the full-body motion capture suit to see themselves and others within the virtual environment. Placing the HMD wearer in actual full-size motion-capture friendly mockups further augments HEMAP assessments and provides a superior comprehensive evaluation of human-system integration. Merging the virtual environment with physical objects increases situational awareness as they move and touch mockups while viewing virtual workplaces through the HMDs.

Figure 6: Head-Mounted Displays Immerse Participants, Collision Detection Possible

Alternatives to Standard Motion Capture Suits/Attire

Since tasks are not always performed in normal attire, motion capture and ergonomic evaluations may need to include alternative clothing and equipment options. Innovations to the HEMAP Lab have included markering and motion tracking a participant in a large, bulky, air-supplied protective suit to compare with similar movements performed in traditionally, tight-fitting, normal weight clothing. The technology can be applied to assessments of persons in various protective clothing including astronaut flight suits, helmets, backpacks, boots and gloves.

Figure 7: Alternative Clothing Motion Capture

Collaboration of Motion Data among Motion Capture Facilities

HEMAP has also discovered ground-breaking collaboration that allowed remote motion capture facilities to share the same virtual environment and see corresponding interactions of the moving participants from each location in near real time. This collaborative technology affords concurrent engineering and can reduce travel, other resource costs, and schedule risks. Engineers are provided a realistic view of how humans will interact with their designs, allowing changes before committing resources. With the network collaboration of motion data, HEMAP services can be applied to any location, allowing connection between Human Factors and other entities across vast distances.

Figure 8: Collaboration of Motion Capture Data (2 Locations, Shared Environment)

Iterative Applications of Motion Capture, Peripherals, and Ergonomic Evaluators to Meet Customer Needs

Providing iterative uses of motion capture technologies and ergonomic evaluations can provide varied benefits in meeting vast customer needs. As noted, solely donning HMDs can familiarize participants with CAD environments. Adding head and hand tracking can immerse them and demonstrate visibility, reach, collision detections, and workspace understanding since turning heads or hand motions would match CAD model orientations within users' views. Adding full-body motion captures with use of HMDs can immerse whole human(s) into virtual environments. HEMAP has offered additional value in also placing participants donning HMDs, with or without the motion capture suits, in actual wireframe, motion-capture friendly mockups of the crew module so they can gain comprehensive understandings of merged virtual and physical environments.

Motion capture sessions can be performed at the HEMAP Lab in which customers, engineers, and others may also choose to participate on-site to assist in the motion capture and real-time scenario adjustments and evaluations. Or, collaborating motion data between remote motion capture facilities within a shared virtual environment can show interactions among participants who may never meet physically. Compared to traditional human factors, learning to apply iterative applications of applied digital human modeling, motion capture, and software-supplemented ergonomic applications can provide Human Factors experts and

Ergonomics means to more quantitatively evaluate existing tasks or assess tasks associated with proposed designs or processes.

Future Plans for Human Modeling for Future Spacecraft Designs and Ground Processing Tasks

Augmented Reality Integrated with Motion Capture

Integrating Augmented Reality (AR) into combined motion capture and ergonomic HEMAP assessments will present work instructions and augmentation symbols, specific to the work step or hardware, to the views of the HMD wearers.

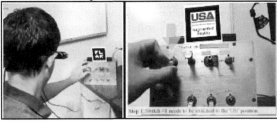

Figure 9: Augmented Reality Showing Work Instructions/Augmentations

Biomedical Monitoring/Streaming Integration with Motion Capture

HEMAP is adding biomedical monitoring to its motion captures to assess the holistic human, combining integral bodily functions such as heart rate and temperature with external or gross motor evaluations.

Alternative Tracking/Peripheral/Evaluation Options

Where customers require assessments near their hardware or mockup builds are not feasible, having camera-less means to track HMDs, head, and hands may be all that is needed. While considering project and facility requirements, having options for camera-less, suit-free, or alternative tracking systems may prove effective.

2 CONCLUSIONS

Use of multiple object motion capture technology and digital human tools in aerospace has demonstrated to be a more cost effective alternative to costs of physical prototypes, provides a more efficient, flexible and responsive environment to changes in the design and training, and provides early human factors considerations concerning the operation of a complex launch vehicle or spacecraft. Using customer interfaces, needs, and relevant computer drawings, exceptional insight into products and processes can be provided to customers before designs are finalized. Using 3D data of existing or proposed designs, integrated with live participants performing simulated tasks or merely immersed in the virtual environment, HEMAP allows customers to see how real people will interact with proposed designs before fabrication or modifications.

Being able to assess real-time motions of multiple persons and objects, with dynamic ergonomic, collision detection, reach, and visibility assessments provide the most quantitative first look at human factors and ergonomics associated with a task. Adding the ability for participants to immerse themselves in the virtual reality using head-mounted displays or to touch real mockups provides the most comprehensive human-system interface assessment.

Looking forward at future designs and means to assess risks in proposed designs of spacecraft or other vehicles/facilities provides avenues for further improvements and advances in applied human digital modeling and motion capture opportunities. Supplementing motion capture with biomedical monitoring, augmented work instructions, and indicators provide the participant and Analysts quick, cutting edge understandings of risk identifications and improvement opportunities. To support engineering and operational entities, KSC's HEMAP Lab has innovated COTS products to create a motion capture/dynamic ergonomic assessment system that captures humans and opportunities to meet existing and proposed designs.

3 ACKNOWLEDGEMENTS

The authors would like to acknowledge the entire Human Engineering Modeling and Performance (HEMAP) Team at United Space Alliance as well as NASA and Lockheed Martin for their continued support.

4 REFERENCES

United Space Alliance Human Engineering Modeling and Performance (HEMAP) Customer Education Capabilities Package; Validation and Test Report. Copyrights © 2010.
http://www.unitedspacealliance.com/stubs/engineeringhumanfactors.cfm
Jeffrey S. Osterlund & Brad A. Lawrence. 61st Virtual Reality: Avatars in Human Spaceflight Training. International Astronautical Congress, Prague, CZ. Copyright ©2010 by the International Astronautical Federation.
Stambolian, Damon. Co-Authors: Donald Tran, Katrine Stelges. KSC Human Factors Lessons Learned. NASA Project Management (PM) Challenge 2012.

Index

Alexander, T., 103, 134
Andreoni, G., 507
Asami, Y., 478
Ashley, K., 191
Avin, K., 327
Baisini, C., 123
Benjamin, P., 517
Benson, E., 374
Bertilsson, E., 235
Bewley, W., 212
Biswas, G., 202
Blackledge, C., 374
Bohlin, R., 265, 275, 285
Bosch, O., 75
Bruzzone, A., 123
Carlson, J., 265, 275, 285
Carlson, E., 436
Chen, W., 337
Cheng, Z., 3
Chuang, C., 147
Cid, G., 471
Cloutier, A., 13
Cole, E., 327
Collier, N., 447
Cook, S., 245, 255
Corner, B., 147, 159, 169
Correa, G., 471
Costa, F., 507
Cowley, M., 317, 374
Darr, T., 517
Delfs, N., 275, 285
Dolšak, B, 460
Durlach, P., 222
Endo, Y., 392
England, S., 374
Enkvist, T., 123
Ericsson, M., 545
Farry, M., 436
Femiani, J., 147
Franca, G., 471
Frey-Law, L., 327
Frigo, C., 507
Fukano, S., 426
Fukuzumi, S., 495
Gopal, B., 517

Gragg, J., 13, 23
Gruppioni, E., 507
Guimarães, C., 471
Gustafsson, S., 275, 285
Hanson, L., 235, 265
Haradji, Y., 416
Harih, G., 460
Hart, J., 113
Harvill, L., 317, 374
Hlaváček, P., 337
Högberg, D., 235, 265
Hudson, J., 159
Hughes, C., 536
Ikegami, T., 495
Inoue, M., 85
Irwin, e., 527
Iseli, M., 212
Ishihara, K., 478
Johansson, H., 545
Johnson, W., 181
Kakara, H., 65
Kaljun, J., 460
Keyvani, A., 235, 545
Kinnebrew, J., 202
Kitamura, K., 85
Kobayashi, D., 478
Koelle, D., 436
Koenig, A., 212
Koshiba, H., 57, 95
Kouchi, M., 392, 401
Lawrence, B., 555
Lee, J., 212
Lei, Z., 33
Li, P., 169
Liang, D., 23, 33, 43
Lin, Y.-C., 357
Long, J., 33
Luximon, A., 295
Lynch, C., 191
Ma, X., 295
Maeno, T., 75
Mahoney, S., 436
Makabe, K., 384
Mårdberg, P., 265, 285
Margerum, S., 317

Marier, T., 327
Marler, T., 364
Marshall, R., 245, 255
Martin, G., 536
Martinez-Carmona, D., 348
Mathai, A., 327
Miyata, N., 392
Miyazaki, Y., 65
Mochimaru, M., 392, 401
Moorcroft, D., 307
Moshell, J., 536
Mosher, S., 3
Motomura, Y., 57, 95
Munro, A., 212
Nakatsu, R., 489
Neuhöfer, J., 134
Nishida, Y., 65, 85
Nishimura, T., 95
Nojima, N., 478
North, M., 447
Nyce, J., 123
Ohba, K., 95
Oliveira, J., 471
Ozik, J., 447
Palomaa, E., 447
Paquette, S., 169
Parakkat, J., 3
Paranhos, A., 471
Park, J., 407
Pastura, F., 471
Pavan, E., 507
Pellettiere, J., 307
Peña-Pitarch, E., 348
Perez, E., 374
Poizat, G., 416
Rajulu, S., 317, 374
Ramachandran, S., 517
Razdan, A., 147
Rhee, J., 407
Rivera, W., 447
Robinette, K., 3
Romero, M., 507
Sagae, A., 181

Sakata, M., 426
Sakurai, M., 384, 426, 478
Saldutto, B., 507
Sallach, D., 447
Santos, V., 471
Scapini, L., 507
Segedy, J., 202
Sempé, F., 416
Shirasaka, S., 75
Sottilare, R., 113
Spain, R., 222
Spurrier, K., 327
Stark, B., 436
Stelges, K., 555
Streilein, K., 527
Streit, P., 471
Sultan, S., 364
Summerskill, S., 245, 255
Surmon, D., 212
Tada, M., 392
Tafuri, S., 275
Tahara, S., 426
Takenaka, T., 57
Ticó-Falguera, N., 348
Tremori, A., 123
Verni, G., 507
Vinyes-Casasayas, A., 348
Winchell, A., 517
Xu, B., 337
Yamamoto, S., 384, 426, 478
Yamanaka, T., 65
Yang, J., 13, 23, 33, 43
Yang, L., 337
Yasui, T., 75
Yokomizo, A., 495
Yoon, S., 65
Zamberlan, M., 471
Zehner, G., 159
Zhang, M., 295
Zhang, W., 337
Zhang, Y., 295
Zhou, J., 337
Zou, Q., 43

T - #0307 - 071024 - C580 - 234/156/26 - PB - 9780367381127 - Gloss Lamination